Future of Fisheries:

Perspectives for Emerging Professionals

Future of Fisheries:
Perspectives for Emerging Professionals

Edited by

William W. Taylor
Center for Systems Integration and Sustainability
Department of Fisheries and Wildlife, Michigan State University
115 Manly Miles Building, East Lansing, Michigan 48823, USA

Abigail J. Lynch
Center for Systems Integration and Sustainability
Department of Fisheries and Wildlife, Michigan State University
115 Manly Miles Building, East Lansing, Michigan 48823, USA

Nancy J. Léonard
Northwest Power and Conservation Council
851 S.W. Sixth Avenue, Suite 1100, Portland, Oregon 97204, USA

American Fisheries Society
Bethesda, Maryland
2014

Suggested citation formats follow.

Entire book

Taylor, W. W., A. J. Lynch, and N. J. Léonard, editors. 2014. Future of fisheries: perspectives for emerging professionals. American Fisheries Society, Bethesda, Maryland.

Chapter in book

Franzin, W. G., P. M. Cooley, and P. A. Nelson. 2014. The scientist's journey: that was then and this is now. Pages 81–89 *in* W. W. Taylor, A. J. Lynch, and N. J. Léonard, editors. Future of fisheries: perspectives for emerging professionals. American Fisheries Society, Bethesda, Maryland.

Cover illustration by Briana N. Bromley, Walden, Colorado

Printed in the United States of America on acid-free paper.

Library of Congress Control Number 2014941912

ISBN 978-1-934874-38-7

American Fisheries Society Web site: www.fisheries.org

American Fisheries Society
5410 Grosvenor Lane, Suite 110
Bethesda, Maryland 20814
USA

Contents

Foreword . . . xi
Preface . . . xiii
Acknowledgments . . . xvii

The Academic Environment

We Are the Apex Predators . . . 3
James L. Anderson

Do Not Believe Your Model Results . . . 9
Steve Cadrin

Mentors Matter: Strategies for Selecting the Right Mentor . . . 15
Steven J. Cooke and Constance M. O'Connor

Have You Prepared Yourself for the Changing Employment Arena? . . . 23
William L. Fisher and Daniel C. Dauwalter

The Intern–Mentor Experience: A Sample of the Real World . . . 31
Molly J. Good and John F. Kocik

Cultivating a Positive and Productive Postgraduate Research Experience: Tools for Students and Supervisors . . . 39
Alistair J. Hobday and Lucy Robinson

Linking Successful Careers to Successful Fisheries . . . 47
Charles F. Rabeni and Shannon K. Brewer

Timing Is Everything . . . 53
Erin J. Walaszczyk, Cory O. Brant, Nicholas S. Johnson, and Weiming Li

Finding that Academic Position . . . 59
David W. Willis and Daniel A. Isermann

Traits to Nurture

Salmon Research in Alaska: My Mentor Sparked My Career . . . 69
Kenneth L. Beal

Give, and You Shall Receive . . . 75
Charles (Chuck) C. Coutant

The Scientist's Journey: That Was Then and This Is Now . . . 81
William G. Franzin, Paul M. Cooley, and Patrick A. Nelson

The Importance of Mentoring and Adaptability among Fisheries Professionals: Learning Survival Skills from Coyotes . . . 91
Wayne A. Hubert, Paula Guenther, and Diana Miller

Mentoring from the Heart 99
Donald C. Jackson

Swift to Hear, Slow to Speak, Slow to Wrath 105
Jud Kratzer

Sustainable Fisheries Is a Worldwide Objective: As a Fisheries Scientist, the World Is Your Oyster 111
Margaret Mary McBride

When Opportunity Knocks, Don't Blink 121
Edward L. Mills and James M. Watkins

So You Say You Love Fish 129
Michael P. Nelson, John A. Vucetich, and Kathleen Dean Moore

Let's Play Two: Optimism Makes All Things Possible 135
Larry A. Nielsen and Gretchen L. Stokes

Curiosity Comes Naturally, but Three Other C's Must Be Learned—the Earlier the Better. 141
Steven G. Pueppke

Facing Reality and Overcoming Adversity 147
Jerry L. Rasmussen

Today's Fisheries Challenges Require Creative, Interdisciplinary Problem Solving. 153
Kyle S. Van Houtan and Eric C. Schwaab

Expand Your Horizons: The Importance of Stepping out of Your Comfort Zone 159
So-Jung Youn, William W. Taylor, C. Paola Ferreri, Nancy Léonard, and Amy Fingerle

Diversity to Appreciate

Succeeding as a Nontraditional Graduate Student: Building the Right Support Network 167
Robin L. DeBruyne and Edward F. Roseman

Understanding Generational Differences in the Workplace. 173
Kelly F. Millenbah and Bjorn H. K. Wolter

Can We Really Have It All?. 179
Jessica Mistak

Reconnecting People to Their Natural Environment 185
Christine M. Moffitt, Zachary L. Penney, and Lubia Cajas-Cano

Changing the Game: Multidimensional Mentoring and Partnerships in the Recruitment of Underrepresented Students in Fisheries 193
Stacy A. C. Nelson, Ernie F. Hain, Brett M. Hartis, and Ashanti Johnson

Openness to the Unexpected: Our Pathways to Careers in a Federal Research Laboratory 201
Kurt R. Newman, David B. Bunnell, and Darryl W. Hondorp

Mentoring Minorities for More Effective Fisheries Management and Conservation 207
Mamie Parker and Dana M. Infante

We Are All in This Together: Capitalizing on Individual Abilities for Collective Benefit 215
William W. Taylor, Nancy Léonard, So-Jung Youn, and C. Paola Ferreri

Skills to Develop

Not Fish, Not Meat: Some Guidance on How to Study Fisheries from an Interdisciplinary Perspective 223
Robert Arlinghaus, Len M. Hunt, John R. Post, and Micheal S. Allen

Make a Science of Communication. 231
Elizabeth L. Beard and Samantha M. Wilson

Creating Professional Networks for Successful Career Enhancement 239
T. Douglas Beard, Jr.

Interviewing Strategies and Tactics for Success. 245
Henry (Rique) Campa, III and Alexandra Locher

Influencing Your Agency's Thinking. 251
Robert F. Carline and David A. Lieb

"If I Know All the Science in the World, I'm Going to Change the World"—The Fisheries Scientist's Fallacy?. 257
Amy Fingerle and William W. Taylor

How to Make a Difference When Fighting for Something You Love 263
Denny Grinold and Marissa Hammond

Resource Management in the Face of Uncertainty. 267
Daniel Hayes, Bryan Burroughs, and Bradley Thompson

Carry a Big Net—Cast It Far and Wide. 273
Robert M. Hughes, Daniel J. McGarvey, and Bianca de Freitas Terra

Evolution of a Fisheries Scientist: From Population Dynamics to Ecosystem Integration 279
Peter C. Jacobson

Casting a Wide Net: Integrating Diverse Disciplines and Skillsets in Fisheries Policy Careers. 285
Kristine D. Lynch and Kelly M. Pennington

Managing Your Career. 291
Jim Martin and Abigail Schroeder

What They Do Not Teach You in Graduate School: How to Be an Effective Listener, Be a Good Participant and Chair at Meetings, and Deal with Opposing Views. 299
Russell Moll and Shauna Oh

A Human Side of Fisheries 303
Shawn J. Riley and Amber D. Goguen

Go Forth and Tell Them What You Have to Say! 311
Kelsey Schlee and Stan Moberly

Learning Is an Ongoing Experience: What Fishers Have Taught Us during Our Careers 317
Vahdet Ünal and Huriye Göncüoğlu

The Odyssey of a Fisheries Scientist in Greece in the 21st Century: When the Journey to Ithaca Is Still Harsh but Probably More Interesting than Ever! 323
Vassiliki Vassilopoulou, Paraskevi K. Karachle, and Andreas Palialexis

Leadership in Practice

Leadership Starts at the Beginning 331
Douglas Austen

Leading for Conservation Success 337
Hannibal Bolton and Cecilia Lewis

Lessons on the Road to Leadership Effectiveness 345
William A. Demmer

Leading with Vision 351
Kenneth Haddad and Jessica McCawley

Considering Habitat in the Interdisciplinary Fisheries Management Profession 357
Terra Lederhouse and Thomas E. Bigford

Lessons in Leadership 365
William F. Porter

Fisheries Decision Making: Advice from the Introduction of Pacific Salmonids into the Great Lakes 371
Howard A. Tanner and Abigail J. Lynch

Saving Fish? Saving Fisherfolk? Reflections on Designing Governance Policies for Fisheries 377
Anastasia Telesetsky and Rebecca Bratspies

Fly-Fishing for the Future: How the Michigan State University Fly Gals Are Mentoring Future Conservation Leaders 383
Kerryann Weaver and Tom Sadler

Emerging Topics

Riding with the Drivers of Change in Fisheries Science: A Holistic Approach for the Future 393
Devin M. Bartley, Nicole Franz, Carlos Fuentevilla, and Koji Yamamoto

How Will Invasive Species Impact the Future of Fisheries? 401
D. Andrew R. Drake and Nicholas E. Mandrak

Three Sides of the Same Coin: Aquatic Animals–One Health–Ecosystem Health............409
Mohamed Faisal

Enforcing Fishery Laws: The Key to Protecting the Commons... 417
Marc Gaden, Jill Wingfield, and Chris Goddard

The Growing Importance of Communication for the Future of Interjurisdictional Fisheries Management, Using Bluefin Tuna as a Case Study 423
Fábio H. V. Hazin and Felipe Carvalho

The Urban Future of Fisheries...429
Sara Hughes

Climate Change and the Future of Freshwater Fisheries 435
Daniel J. Isaak

Making Adaptive Management Work: Lessons from the Past and Opportunities for the Future 443
Michael L. Jones and Gretchen J. A. Hansen

Catfish in the Courtroom: How Forensic Science Catches Seafood Cheats 451
Trey Knott and David D. Stephens

Fisheries as Coupled Human and Natural Systems 459
Abigail J. Lynch and Jianguo Liu

Aquaculture in the 21st Century: Opportunities for the Emerging Professional 467
John R. MacMillan and Eric A. MacMillan

Inland Fisheries in Russia: Tales from the Past and Lessons for the Future......................... 473
Dmitry F. Pavlov and Dmitry D. Pavlov

The Role of Scientists in Public Policy Development and Advocacy: Advice for Scientists and Policy Makers 479
Mark Rey and Adawnice Lucas

Achieving Funding Needs for Fishery Resources Management and Angler Access 483
Gordon C. Robertson, D. Michael Leonard, and Elizabeth A. Yranski

The Changing Landscape in Atlantic Menhaden Assessments: Best Available Science, Uncertainty, and the Tension between Science and Management 489
Douglas S. Vaughan and Amy M. Schueller

Future of Inland Fisheries...495
Robin L. Welcomme

Epilogue 503
Betsy Riley, William W. Taylor, and So-Jung Youn

Foreword

Recently, I was listening to Kenny Rogers' song "If I Knew Then What I Know Now." As I was humming along to the refrain, "And if I knew then what I know now/I'd have found a way/To make things work out somehow," it resonated with me and how much it applies in our life and our profession. My mind quickly started to think of the lessons that I have learned over my lifetime, and that if someone could have guided me earlier, and I had listened to this guidance, it might have made my career, and life, a bit less tumultuous at times.

As I thought about this refrain and the lessons I learned, I realized that I wanted to pass on some of my experiences to my family, friends, students, and colleagues to perhaps make their way through life and profession a bit smoother. Often, we call this mentoring; it is an active, directed interaction that allows both individuals involved to learn from each other and share deeply about their knowledge, experiences, thrills, trials, and tribulations, the hardest to share often being the most meaningful.

Having thought about this in some detail, I started to think about the people who really took the time and interest to truly mentor me, those individuals who allowed me to grow and actuate my potential while being present as a friend and safety net. All of us need a hand at some point in life as we experience the challenges of the unknown that come with having a full and productive life. I have been fortunate to have had many people reach out and take me under their wing, fostering my skills and enthusiasm and, in so doing, helping me to become a better person and professional. These individuals are from all walks of life, including, first and foremost, my spouse Evelyn, my children (the great equalizers in life), my parents, and my siblings. In addition, I was fortunate to have had many caring and talented colleagues throughout my academic and professional career that took time to invest in me. Most notably, in this group were a few key professors at Hartwick College where I completed my undergraduate studies; my mentor for my doctoral program, Dr. Shelby D. Gerking, at Arizona State University; and my students, colleagues, and administrators at Michigan State University and in the fisheries profession, locally, nationally, and globally. I thank them for assisting me and in believing in me when few others did. Without their ongoing interactions, sage advice (not always taken—to my detriment), and unfailing friendship and support, none of my successes would have been possible.

As I reminisced about many of these people in my life and all the mentoring events I have experienced, I realized that we could produce a book of short mentoring vignettes that might provide others an early glimpse of what they might need for support and fulfillment during their career and personal life. The outcomes of these efforts are in this book. I hope the content within these pages provides you with the guidance, confidence, and insight as you navigate through the choppy waters of life. As these vignettes often reiterate, life is a group activity, and joining forces and sharing deeply of yourself will ultimately make your world more exciting and meaningful and your profession better. I wish you much success in navigating through the rapids and hope that this book in some small way allows for great success and happiness in your personal and professional life.

William W. Taylor
May 2014

Preface

On Becoming a Mentor and Why Mentoring Is Important

In this fast-paced world of hectic schedules, impending deadlines, and increasingly long to-do lists, why would anyone want to take the time and effort to be a mentor? Mentoring is a significant time commitment, and it means using the resources and networks that a mentor has accumulated with years of experience to benefit a mentee, hopefully improving his or her life and professional experiences. Mentors are excellent resources to help identify strengths and interests and can offer a clear look into one's chosen field, sharing in both triumphs and failures. A more philosophical view of mentoring can be described as a way to pay it forward and give back to the broader community in an attempt to improve the chances for others to succeed. Mentoring ultimately provides an environment where mentors act as a safety net, encouraging their mentees to take risks and learn from their mistakes, allowing for a better professional and individual. Regardless of how each individual defines mentoring, the product of the relationship is a synergy that empowers people to be or do something they did not recognize on their own.

We all know that a successful career takes work and development of needed skill sets (we are not born into a profession or a position), and to excel, we need someone to make a personal investment in us and provide critical feedback on our performance. Mentors provide that safe space to discover more about who you are and who or what you might be and to help identify how to get to where you want to go. This safe space encourages honesty without incrimination or destructive feedback and provides a confidential and valued source of guidance. Creation of a safe space also promotes an environment where mentors and mentees can forgive each other for mistakes (and perceived mistakes) that will be made as they travel through this unpredictable journey we call life!

Mentors see a unique spark in mentees and are willing to take the time and have the ability to be able to uncover and develop their qualities and potential. In order to do this, mentors are often at a point in their career where they have experienced personal and professional successes and failures and are willing to share those experiences. It is these personal experiences and life lessons that shape and guide their mentoring.

What Makes a Good Mentor?

In the simplest terms, mentors possess the qualities of good role models and have traits worth emulating. In particular, they are trustworthy, honest, approachable, and available; possess a vision for the future; and exhibit a great deal of compassion. Mentoring is a one-on-one process that is based on common experiences, mutual trust, and respect. Mentors generally are further along in their careers than their mentees and may have accrued extensive social and political capital through their substantial personal and professional networks, which they willingly provide to their mentee to engage and learn from. Mentoring is, consequently, an active process that blends a mentor's past experiences and wisdom to influence the development of another individual who is an eager recipient of this feedback.

Effective mentors are able to turn their mentees' missteps into opportunities on which to reflect and learn. Mentors should be willing to speak the truth as they see it because they care and are invested in mentees' personal and professional growth. There are many types of mentors who can serve different roles and bring varying strengths to the development of the same mentee.

With a network of mentors, you can surround yourself with people who can support and stimulate you, as well as provide for alternative solutions and additional milestones to achieve on the pathway to future success and happiness. Your mentors can serve as a sounding board; offer career advice and growth opportunities; provide objective, frank feedback and guidance; facilitate connections; empower; and help craft a road map to success. Mentors have a keen ability to uncover your potential and personal assets and challenge you to pursue actions to push you gently beyond your current abilities to discover new strengths. Mentors often provide a safe haven for you to take risks but do not necessarily prevent the bumps and bruises along the way.

Mentor–Mentee Relationship

Forming a mentor–mentee relationship must be a mutual process, and it takes initiative and time on the parts of both the mentor and the mentee. Both need to be open to share and receive information honestly and with respect, a two-way communication flow. Do not assume a mentor possesses all of the knowledge; mentees also bring their own knowledge and experience. The relationship grows when the mentor begins to see that the mentee is listening and demonstrating that he or she hears and appreciates the conversation and that feedback is given with compassion. This feedback loop reaffirms the mentor–mentee relationship and encourages continued investments. Any mentor can give praise, but it is the critical reflection and feedback that promotes individual and professional growth. Mentoring is not about lecturing the content or materials and taking notes. Instead, a mentee needs to be able to blend the mentor's advice and incorporate it into his or her own style—one cannot learn to swim by just reading a book, but one cannnot swim without learning the techniques of how to swim either. In some cases, mentoring is done through stories or metaphors to symbolize broader aspects of life. These allegories can only be incorporated and internalized if you can understand them and can see the connection to your life situation.

Trust and respect are the cornerstones of a productive mentor–mentee relationship, but both take time and personal interaction to develop. If the mentoring relationship lacks respect, one will not risk telling the truth. Mentees need to feel like they will be accepted no matter what they share or whether they succeed or fail. Mentees should know that their mentor believes in them and will help them when they are in trouble; otherwise, mentees may be reluctant to take risks and expand their abilities. A mentee also must feel heard, as he or she is exploring how to troubleshoot and identify options to solve problems. Even if the solutions do not work as hoped, if a good mentor will believe in the mentee's ability to overcome the problem and together, the mentor and mentee can generate solutions that do work.

If you want to push the frontiers of knowledge and your own abilities, you are going to need to take some risks. And if you are willing to push the envelope and take some risks, you will likely experience some failures along the way. A fatal mistake some mentors fall victim to is their intentional or unintentional eagerness to solve the mentee's problem for him or her or to take responsibility for his or her mentee's failure or mistake. Confident mentors allow mentees to take responsibility for their own mistakes and fix them themselves, providing guidance along the way as requested. During this process, mentors should be inquisitive and ask probing, thoughtful questions to lead the mentee through the decision-making model.

A mentor should be a guide rather than a dictator. A mentor can suggest several possible paths, but the final decision should be made by the mentee. Otherwise, mentees learn how to follow directions but never learn how to be leaders in their own right and are, instead, always dependent on their mentors to make decisions. Mentees should remember to be open to learning from mistakes, clean up what they can, and move forward. Failure is not fatal; failure to keep trying, however, is.

So, what can you do to get the most out of your mentoring experience? It is up to you to take the initiative, challenge the status quo, and develop and nurture your mentor–mentee relationship(s). The vast majority of successful people will credit their mentors with helping them find their way, believing in them, and empowering them to take risks and make a difference. Often the significance of the advice or lessons learned does not resonate until months or years later. Along the way, express gratitude by recognizing and thanking your mentors. A good mentor is like a good dog, always there for you and proud to be so! We encourage everyone, no matter what age and experience, to actively seek future mentoring opportunities. You and the profession will be better for it!

Future of Fisheries: Perspectives for Emerging Professionals

This book serves as a medium for distinguished mentors to share their personal and professional lessons learned, as well as their insight on key concepts that influenced their career trajectory and the profession. More than 70 relatively short, philosophical, mentoring vignettes can be found within the pages of this book. They are broken into six sections:

- The academic environment
- Traits to nurture
- Diversity to appreciate
- Skills to develop
- Leadership in practice
- Emerging topics

In all vignettes, the authors identify challenges they faced and share their guidance on how to remain resilient in the face of similar challenges. In some instances, the vignettes are coauthored by a mentor and his or her mentee(s); in others, the vignette focuses on the mentor's perspective only. We encourage you to read and reread this book throughout your career as some messages will inspire immediate "Aha!" moments while other vignettes may better resonate with you at some point in the future. It is our hope that this book inspires and empowers the emerging professional and also serves as a source of inspiration for the more seasoned of us to recognize that mentoring indeed is important and that our efforts have made a difference in helping others succeed.

We believe that if you are reading this book, you possess an innate desire to move the needle forward, build capacity within yourself, help strengthen others and our profession as a whole, and make a positive impact. Go forth and do great things! Fish On!

Acknowledgments

This preface is the result of extensive discussions on mentoring with many people, including Dr. Fred Poston, dean of the College of Agriculture and Natural Resources at Michigan State University (MSU); Dr. Niles Kevern, professor emeritus and former chairperson of the Department of Fisheries and Wildlife at MSU; Karen Klomparens, dean of Graduate School at

MSU; Lou Anna K. Simon, president of MSU, June Youatt, provost of MSU, and Mr. John Robertson, former chief of fisheries, Division of the Michigan Department of Natural Resources. We thank So-Jung Youn, Betsy Riley, and Molly Good for their thoughtful reviews of this preface and this book.

William W. Taylor
Abigail J. Lynch
Nancy J. Léonard
Jordan Pusateri Burroughs

Acknowledgments

When we first floated the idea of a short volume of mentoring vignettes from seasoned fisheries professionals to those just starting their career, we got overwhelmingly positive feedback. In particular, we thank the American Fisheries Society Past Presidents Council for supporting the project and providing recommendations on the proposal to refine its scope.

We are most grateful to the authors of the vignettes. This book and its format are far afield from a traditional scientific manuscript, and as many authors can attest, the chapters are deceptively difficult to write. Without their expertise, directive, and especially patience, this endeavor could not have come to fruition.

We wholeheartedly thank Marissa Hammond for her coordination of the review process and more than 200 reviewers from countless organizations, agencies, and academic institutions. She may not have known all that she signed up for in the beginning, but she was an excellent team player and completed her role with unquestioningly loyalty and professionalism.

Each vignette in this volume was externally peer-reviewed by at least two individuals, as well as the three editors, for clarity of message; the comments of the reviewers were carefully considered by both the authors and the editors. For the time and consideration they invested as reviewers, we thank Edward Allison, Brenda Alston-Mills, Paul Anderson, Jack Bails, Betsy Baker, Devin Bartley, Bruce Barton, Rick Baydack, Doug Beard, Bonnie Becker, Jim Bence, Mark Bevelhimer, Dean Beyer, John Biagi, Sara Block, Bo Bonnell, Lynne Borden, John Boreman, Stephen Bowen, Mark Breederland, Bradford Brown, Pedro Bueno, Doug Buhler, Scott Burgess, Dale Burkett, Elizabeth Burleson, Bryan Burroughs, Tom Busiahn, Dave Caroffino, Kendra Cheruvelil, Richard Christian, Julie Claussen, Noreen Clough, Elinor Colbourn, Alison Colotelo, Steve Cooke, Tom Coon, Mark Coscarelli, Robert Curry, Andy Deines, Jim Diana, Andrew Dolloff, William Edwards, Karim Erzini, Ron Essig, Sally Fairfax, Frank Fear, Bill Fisher, Daniel Fitts, Jeff Fleming, Gary Fornshell, Ann Forstchen, Ken Frank, Angela Fuller, Marc Gaden, Tracy Galarowicz, Cloe Garnache, Sherman Garnett, Frances Gelwich, Tom Gengerke, Larry Gigliotti, Chris Goddard, Raquel Goñi, Lonnie Gonsalves, Molly Good, Rowan Gould, Peter Gullestad, Ken Haddad, Liz Hamilton, Wade Hamilton, Marissa Hammond, Mike Hansen, Fred Harris, John Harrison, Dan Hayes, Jay Hesse, Tracy Hill, William Hogarth, Wayne Hubert, Dana Infante, Don Jackson, Peter Jacobson, Bill James, Gareth Johnstone, Mike Jones, Philip Kaufmann, Reuben Keller, Ronald Kinnunen, Jim Kitchell, Barb Knuth, John Kocik, Tracy Kolb, Josh Korman, Dan Kramer, John Kurien, Tom Kwak, Mark Laustrup, Joseph Leach, Aaron Lerner, Colin Levings, Cecilia Lewis, Jason Link, Steve Lochmann, Andy Loftus, Ryck Lydecker, Ayman Mabrouk, Eric MacMillan, John Magee, Joe Margraf, David Marmorek, Jim Martin, Bonnie McCay, Bill McConnell, Richard McGarvey, Steve McMullin, Matt Menashes, Andrea Miehls, Kelly Millenbah, Dirk Miller, Ed Mills, Michael Moore, Emilio Moran, Phil Moy, Peter Moyle, Katrina Mueller, Holly Muir, Andrew Muir, Brian Murphy, James Murray, Nancy Nate , Michael Nelson, Genny Nesslage, Kurt Newman, Vivian Nguyen, Sue Nichols, Larry Nielsen, Joe Nohner, Tony Nunez, John Organ, Andrea Ostroff, Eric Palmer, Nick Parker, Frank Parrish, Donna Parrish, Craig Paukert, Andrew Paul, Don Pereira, Jim Perry, Marielle Peschiera, Vickie Pontz, Bill Porter, Seth Powers, Steve Pueppke, Jordan Pusateri Burroughs, Michael Quist, Ron Regan, Henry Regier, Kris Renn, Victor Restrepo, Mark Rey,

Jake Rice, Laura Richards, Betsy Riley, Shawn Riley, John Robertson, George Rose, Joan Rose, Laurel Saito, Gary Sakagawa, Rory Saunders, Don Scavia, Michael Schechter, Kelsey Schlee, Jon Schnute, Carl Schreck, Abigail Schroeder, Amy Schueller, Eric Schwaab, Terry Sharik, Samuel Shephard, Dan Shoup, Joe Smith, Richard Sparks, Richard Stedman, David Stevens, Pat Stewart, Heather Triezenberg, Yin-Phan Tsang, Shawnee Vickery, Ted Vitali, Bruce Vondracek, Kerry Weaver, Robin Welcomme, Randy Westgren, Kevin Whalen, Ross Whaley, Daniel Wieferich, Cindy Williams, Nekesha Williams, Dave Willis, Rolf Willmann, Jack Wingate, Jill Wingfield, Scott Winterstein, Melissa Wuellner, Annie Yau, and So-Jung Youn. We particularly commend the reviewers who generously agreed to review multiple chapters and/or agreed to review when they had also committed themselves to authoring another chapter.

We express our deepest gratitude to Debby Lehman of the American Fisheries Society. Debby trusted us when we came to her with a nebulous concept of a series of short mentoring vignettes and was extremely patient with us when it ballooned into a monstrous project with more than 70 contributions and more than 100 authors. She shepherded this book through many stages of review and revision. We also thank Molly Good for her thorough taxonomic review of each vignette.

Finally, families, friends, and colleagues were particularly important to us in achieving a publication of this scale. While they did not make the decision to pursue this project, they accepted the consequences of the decision. We thank them for providing the needed intellectual and emotional support that is required to accomplish such an undertaking. Specifically, we would like to acknowledge the Taylor Laboratory, the Center for Systems Integration and Sustainability, and the MIRTH laboratory at Michigan State University; Evelyn Taylor and Theodore Roosevelt (Teddy) Taylor, Andy Brackensick, Nasira Léonard-Brackensick, and Horus Léonard-Brackensick; and Eric MacMillan and Ben Lynch-MacMillan.

William W. Taylor
Abigail J. Lynch
Nancy J. Léonard

April 2014

The Academic Environment

We Are the Apex Predators

JAMES L. ANDERSON*
Agriculture and Environmental Services, The World Bank
1818 H Street NW, Washington, D.C. 20433, USA

Key Points

- We are the apex predator. The apex predator is an economic, global, and innovative animal. The question is not how to manage fish, but how to manage the behavior of the apex predator—people.
- Think globally and act globally. Fisheries science and management must be practiced in the context of the dynamic and global socioecological system.
- In addition to solid natural science, learn fundamental economics, policy, and international trade.
- Experience the world outside the lab.

Understanding the Apex Predator

People entering the fisheries science and management profession generally love fish. Some love their intrinsic beauty, some love to eat them, some love to catch them, and some are captivated by complexity of the aquatic systems and fundamental importance of healthy rivers, lakes, and oceans to life itself. I am one of them. When I was a kid, the first big thing I ever bought with my own money was a fishing rod. I had seven fish tanks, I worked in a tropical fish store, and I never missed a Jacques Cousteau documentary. I wanted then, and still do now, to conserve fishery resources and ecosystems. There are many of us with the same motivation. We teach fisheries sciences, work in environmental organizations, and dominate the bureaucracy around fisheries management. However, many who are engaged in fisheries management realize late in their career that they are often not well trained to do their job.

There are many important problems facing fisheries, such as overfishing, habitat loss, postharvest waste, pollution externalities, fraud, illegal fishing, and invasive species. However, the core source of these problems has a simple origin. We have a fisheries problem because we eat fish; we live and work in habitat that is essential for fish to grow; and under open access, we will ultimately overharvest this limited resource, deplete the stocks, and make little net economic return. In order to manage this complex system, we need to understand more than fish biology. We need to understand the dynamic behavior of the organisms in the system, including the apex predators, and we are the apex predator. We dominate any species, and we can modify essentially any habitat. If fisheries managers do not understand the apex predator's behavior, they have little hope of managing the system. The apex preda-

* Corresponding author: janderson8@worldbank.org

tor is an economic, global, and innovative animal. Therefore, the question is not how to manage fish, but how to manage the apex predator—people. All fisheries scientists must understand fish and ecology, but fishery managers need to manage people and their behavior.

To choose not to study and incorporate the behavior of the apex predator in fisheries management is like saying "I want to understand the rabbit population dynamics, but I don't want to consider foxes because I don't like foxes." Well, this is, of course, ridiculous. Being a student of fisheries management and not understanding the importance of economics, trade, and policy is similarly ridiculous. But I have heard students say things like "I hate corporations," "economics is boring," and "it's the government's responsibility to regulate the way scientists advise." This is both naive and dangerous. It is naive because all corporations are not intrinsically bad—most actually want a sustainable business for their children and grandchildren. On the other hand, public sector managers are not intrinsically good; despite their responsibility to the public good, short-sighted, bureaucratic, and sometimes corrupt management agencies have overseen much of the decline in fisheries. Both the public and private sectors need to be held accountable for their action. However, even the most well-intended fisheries managers often come from a traditional science background and they do not really understand the apex predator very well. Because we often do not understand the apex predator or take into consideration the predator's behavior in the natural system, we often end up with a regulatory mess that is composed of a mix of wasteful and ineffective command and control approaches focusing on top-down effort controls and enforcement. This is not good for the fish, the environment, or fishery-dependent communities.

Time and time again, ill-conceived regulations are imposed, and protected areas are defined, with the naive assumption that the governance, enforcement, and community buy-in are all in place. The list of failed fishery management systems and ineffective protected areas is overwhelming, ranging from cod fisheries in the northwest Atlantic, to Bluefin Tuna *Thunnus thynnus*, to marine protected areas in much of the developing world that are designated on the map but ineffectively protected. When you dig—just a little bit—you often find that the managers have a remarkable misunderstanding of the common property problem, human behavior, ocean community culture, and economic incentives as the cause of the failure. Fisheries science and management must be practiced in the context of the dynamic and global socioecological system.

It Is All Interlinked

The first stage is to understand how the planet is interlinked, not just in terms of the biophysical components, but also how the planet is highly interlinked by directed human activity. When we eat shrimp in New York, we influence ecosystems in Thailand, China, Vietnam, Ecuador, or Bangladesh. When we conservatively manage fisheries in the United States, like Swordfish *Xiphias gladius*, it actually tends to increase the harvest of Swordfish outside the United States, where there may be essentially no management at all. When Alaska effectively banned salmon farming in their state with a moratorium in 1989, it encouraged the development of salmon farming in Chile. According to the National Marine Fisheries Service, the United States now imports more than 85% of the fish we eat. This has major implications for fisheries and aquaculture management around the world.

The need to understand and integrate the global socioecological system was recognized as essential (but missing) in developing solutions by a panel of experts, innovators, and leaders from around the world in their report "Indispensable Ocean: Aligning Ocean Health and Human Well-Being" (Blue Ribbon Panel 2013). The panel was exceptionally diverse ranging from environmentalists like Sylvia Earle to Thiraphong Chansiri, president of the world's largest seafood company, Thai Union Frozen Foods, to Jane Lubchenco, former administrator of the National Oceanic and Atmospheric Administration. They recognized that reducing overfishing and biodiversity loss and restoring and protecting habitat is linked to the condition and well-being of the people that depend on ocean resources (all of us). In developing countries, this means fisheries management is tightly linked with food security and sustainable livelihood. You may not fully appreciate this if you have grown up in a rich nation like the United States, where the ocean is rarely a primary source of food and income. However, in the newly developing and less developed parts of the world, this connectedness is obvious. In the developing regions, marine protected areas and fisheries management systems are often not effective, regardless of enforcement, if the people in the surrounding communities are poor and hungry and have little incentive to optimally harvest because governance systems do not deal with the open-access problem. We need to design governance systems and harness economic incentives so that the needs of people and the ecosystem are aligned.

Fisheries Are Global

Understanding a complex apex predator like humans requires understanding more than just the fish harvesters. Human apex predators are also the consumers in New York, Hong Kong, and Paris, as well as all of the participants in the supply chain—retailers, distributors, and processors all over the world. Today's fisheries system is truly global. According to the Food and Agriculture Organization of the United Nations, the value of the international seafood trade is greater than any other animal protein: greater than rice, corn, and wheat; greater than beer and wine; and even greater than coffee. The complexity in the seafood sector is staggering. There are more than 500 commercially traded seafood species. Prices range from a few cents to a single Bluefin Tuna that sold for US$1.76 million at the Japanese Tsukiji wholesale market in 2013. Tons of salmon harvested in Alaska are shipped to China for processing and then sent back to the United States. The point is that the entire global fisheries system is interlinked and that actions by consumers and fishery managers in one country will potentially influence fisheries around the world. For example, highly conservative fisheries management and consumer demand in the United States has contributed to ever-increasing U.S. imports from countries with weak fisheries management, such as Bangladesh, India, China, Vietnam, Indonesia, Malaysia, and Ecuador. The next generation of fisheries managers must think globally and act globally.

The Next Generation

So what should an aspiring fisheries scientist interested in fisheries management do? Of course, I would advise any fisheries professional to take graduate level course work in ecology, fisheries biology, statistics, oceanography, population dynamics, and quantitative modeling. But this is just not enough. I also recommend fundamental course work in microeconomics, natural resource and environmental economics, experimental economics, political economy, game theory, international trade, and policy. Now, I know most institutions may not offer such courses that are taught with fisheries and marine resources in mind. Most institutions are heavily weighted toward economics courses that are theoretical with little context and many prerequisites. Truly multi- or cross-disciplinarians are hard to find. We need more.

Experiential Learning

Take advantage of fellowships, internships, and related programs that offer true experiential learning (learning by doing). This is often missing in programs that emphasize policy, management, and economics. I believe that this is a major weakness in many of these programs. Talk to fishermen and people that depend on fisheries for their survival. They will have knowledge and data that you do not. They will help you generate testable hypotheses.

Related to this is the value in finding a program that attracts high-quality students from across the country and abroad. The culture and intellectual difference between someone from suburban Los Angeles and rural Mississippi can be remarkable. This difference can be almost shocking if they are contrasted with someone from a village in Fiji, Vietnam, or Mozambique. My experience has been that many North Americans and Western Europeans approach fisheries management with an arrogance that undermines their credibility and does not serve the global community and world fisheries well. I was involved with an expert committee where an U.S. ocean expert told a leader from an island nation that their people had no need to eat fish in the 21st century. This culturally insensitive and politically/economically infeasible perspective really does not help solve either the fisheries conservation problem or the challenges facing fisheries-dependent communities.

Final Thoughts

Programs and Advisors

Seek out graduate programs that do not lock you into a major advisor from the beginning. Explore programs that encourage you to meet the faculty, learn about their research, and facilitate the best match in interests and skills. Getting assigned to the wrong major professor is not good for anyone and can be particularly damaging for Ph.D. students. From my experience, one of the most common reasons for a well-prepared student to drop out of a program is often related to a difficult relationship with their major professor. You should seek to work with, not for, your major professor. A good professor will learn from you, as you will learn from him or her. Remember, at the end of your Ph.D. work, you should be a peer, not just a senior research associate.

Question Authority

Learn the difference between advocacy and objective science. Unfortunately, there is too much science that hides behind complex models or glitzy maps of global trouble spots that are not based on solid evidence. Look behind the curtain at the data. Hone the ability to distinguish between solid analysis and pseudoscience that stretches data and manipulates models to yield results that support a personal and/or regulatory agenda.

Summary

The next generation of fishery scientists and managers need to understand not just fisheries biology and ecology, but also the complex behavior of the apex predator–people. Fisheries management is about managing people in an integrated socioecological context. So try to enter the strongest program you can find that offers excellence in fisheries and supporting sciences, strong advising, experiential learning, and a global perspective. We need more fisheries management practitioners with some substantive exposure to natural resource economics, experimental economics, policy analysis, and international trade and markets. The profession is oversupplied with fisheries managers that do not understand human behavior and well-designed incentives. Institutions like the National Marine Fisheries Service, state and country fisheries management agencies, The World Bank, the Food and Agriculture Organization of the United Nations, environmental nongovernment organizations, and the private sector need fisheries professionals who have a global perspective, understand systems, and understand the apex predator.

Biography

James L. Anderson is advisor for oceans, fisheries and aquaculture at The World Bank. Prior to joining The World Bank in 2010, he chaired the Department of Environmental and Natural Resource Economics at the University of Rhode Island. He has published numerous articles related to fisheries and aquaculture management, seafood markets, and international trade, including "The International Seafood Trade" (2003). Recent work has focused on the role of seafood in food security, constraints to aquaculture development, and evaluating how aquaculture and fisheries management reforms are changing the global seafood sector. He was the editor of the international journal Marine Resource Economics from 1999 through 2011 and has served on three National Academy of Sciences National Research Council committees. He earned his B.S. in biology and economics from the College of William and Mary and his Ph.D. in resource economics from the University of California, Davis.

References

Blue Ribbon Panel. 2013. Indispensable oceans: aligning ocean health and human well-being. Blue Ribbon Panel, Washington D.C. Available: www.globalpartnershipforoceans.org/sites/oceans/files/images/Indispensable_Ocean.pdf. (March 2014).

Do Not Believe Your Model Results

STEVE CADRIN*

School for Marine Science & Technology, University of Massachusetts, Dartmouth
200 Mill Road, Suite 325, Fairhaven, Massachusetts 02719 USA

Key Points

- Believe data more than model results.
- Continually scrutinize models to learn from them.
- Do not overcompensate for uncertainty through precautionary science.

Learning from My Mistakes

Early in my career as a fisheries scientist, I recognized that mathematical modeling was a powerful way to study how fish populations behave in response to fishing because fish populations are too remote to directly observe. I took as many population dynamics, stock assessment, and statistical modeling classes as I could, and I became a zealot of stock assessment modeling. It was an exciting time to be learning about population modeling. The field was transitioning from a descriptive exercise to a more analytical challenge. The 1980s and 1990s have been termed the "Golden Age" of fishery stock assessment because statistical modeling was rapidly advancing in the wake of the technological revolution (Quinn 2003).

I was working in New England, contributing to the Northeast Regional Stock Assessment Workshops, and results from our stock assessment models told us that many fish stocks were depleted from overfishing, but we were frustrated that fishery managers were ignoring our advice to substantially cut back on fishing. I developed an extremely technocratic view of how fisheries should be managed. I believed that fisheries would be much better off if the fishery managers would simply follow the recommendations from scientists. I was wrong to think it was so simple. Of course, scientific advice should be considered in resource management decisions, but I believed that our stock assessment models accurately represented the complexity of the ecosystem, and I did not consider the people, their livelihoods, or the policies that attempt to represent a broad set of fishery objectives.

The New England Fishery Management Council has a bad reputation for allowing overfishing. That reputation may have been deserved decades ago, but it is no longer accurate. After the Magnuson-Stevens Fishery Conservation and Management Act was revised in 1996 to require an end to overfishing and legal decisions forced substantial reductions in fishing, some stocks began to rebuild, but many others did not. The reality is that the New England council has followed scientific advice for over a decade, and fisheries have stayed within their regulated limits with few minor exceptions. Despite these efforts, overfishing still appears to be a problem in the region. In my opinion, the cause of continued overfishing in

* Corresponding author: scadrin@umassd.edu

New England was not negligent fishery management or irresponsible fishing. The primary cause of continued overfishing is that the stock assessment model results were misleading.

Despite major investments in fishery monitoring, decades of resource surveys, and extensive peer-reviewed stock assessment modeling, estimates of stock size and projected catches were way off. For example, I coauthored a paper on the recovery of the Georges Bank stock of Yellowtail Flounder *Limanda ferruginea* (Stone et al. 2004) only to learn that the stock had not rebuilt when we updated the stock assessment the next year. Now, nearly a decade later, the updated stock assessment suggests that the stock still has not rebuilt and overfishing continues. If the current stock assessment is correct, each of the previous annual stock assessments was wrong. We are still diagnosing the problems, but in general it appears as though some of the assumed stationarities in the model are invalid (e.g., constant natural mortality when many predator populations are rebuilding and climate is changing).

One attribute that led Darwin to develop his theory of natural selection was his introspection about why he made such fundamental mistakes in classifying species like diverse finches during his voyage on the Beagle. We can all learn from Darwin. We should try to learn from our mistakes, ask ourselves why we made them, and question our science. In retrospect, my technocratic view of fisheries management was wrong because I entirely believed my model results. I should have been scrutinizing the model assumptions and testing them with observations from the field. I should have been listening to my critics, especially those with much more experience on the water in the fisheries I was attempting to model.

My experiences with stock assessment of New England groundfish, particularly Yellowtail Flounder, were literally humiliating to me—in a good way. In retrospect, I now see that I had an audacious belief in our model results, and I defended them dogmatically, rather than recognizing the simplicities and assumptions inherent within the model. Despite decades of research on the species and fishery, reams of monitoring data and advanced statistical treatments, the updated model results refuted the same model's output from the previous year. The only logical conclusion was that the model was wrong, and the peer-review process of producing stock assessments was far from infallible. One or more of the many assumptions in our model was obviously wrong, and we still do not know which one it is. Extensive contortions to the model and the data initially produced more consistent estimates, but those patches no longer work, a retrospective pattern of inconsistency reappeared, and the model no longer fits the data or conforms to fundamental principles of stock assessment. The experience gave me much more humility in the way I interpret results from stock assessment models and changed the way I approach recommendations to fishery managers.

In addition to routine model validation and inspection of how well the model fits the data, we should consider what the data suggest on their own and compare the model results to independent information (i.e., does the time series of fishing mortality reflect the management history?). Inspecting the data with exploratory statistical analysis, visualization of complex data, spatial analysis, and mapping can reveal important patterns to be considered in stock assessment. We should listen to critics of the model and people who do not believe the model results, so that we can scrutinize the model, testing as many of our assumptions as possible, to investigate whether or not the critics might be right. Suppression of dissent is an abuse that should be avoided in science as much as government, so dissenting opinions should be considered in model validation. Even after model validation, we should view the model as an imperfect tool and be open to other perspectives. If a single, optimal model cannot be identified, information from multiple modeling approaches should be considered in scientific conclusions and management advice.

A Common Problem in Modeling Fisheries

My mistake may sound foolish, and the lesson may appear to be simple, but overconfidence in model results is a common pitfall in our profession. Jon Schnute and Laura Richards compared modelers being convinced of their model's reality to the sculptor Pygmalion falling in love with his statue (Schnute and Richards 2001). Kerim Aydin presents a functional relationship between insight gained and belief in the model, in which no insight is achieved from complete belief or disbelief, while maximum insight is gained somewhere in between (see Figure 1, conceived by Aydin).

Although the problem with overreliance on fishery models is well recognized, we continually fall into the same old trap. I often witness scientists who vehemently defend stock assessment model results, and the same scientists defend a revised set of results the next year, apparently forgetting their defense of the previous results. Although defensiveness of model results is intended to maintain scientific credibility, it often has the opposite effect. When scientists defend assumptions that fishermen know are not true from their direct experience, they doubt the entire scientific product. The mainstream of stock assessment is now so focused on advanced statistical parameter estimation that the fundamentals of population dynamics and fisheries are often ignored. For example, management strategy evaluation (the use of simulation to determine performance of alternative assessment methods and policies) has been a fruitful development in our field, but I think we begin to believe that the simulations are actually the truth rather than an artificial construct that attempts to represent the extent of our knowledge about the ecosystem. Most of us have heard it, but we tend to forget the advice that "essentially, all models are wrong, but some are useful" (Box and Draper 1987).

Model results are useful, but they should be taken with a grain of salt and scrutinized. Before we even fit a model, we should ask ourselves what the data say. Are the survey indices greater than or less than they used to be? Are the fishery catch rates increasing or decreasing? Have management actions influenced the catch statistics? Are there informative data sources that are not included in the model? What are the perceived trends in the stock

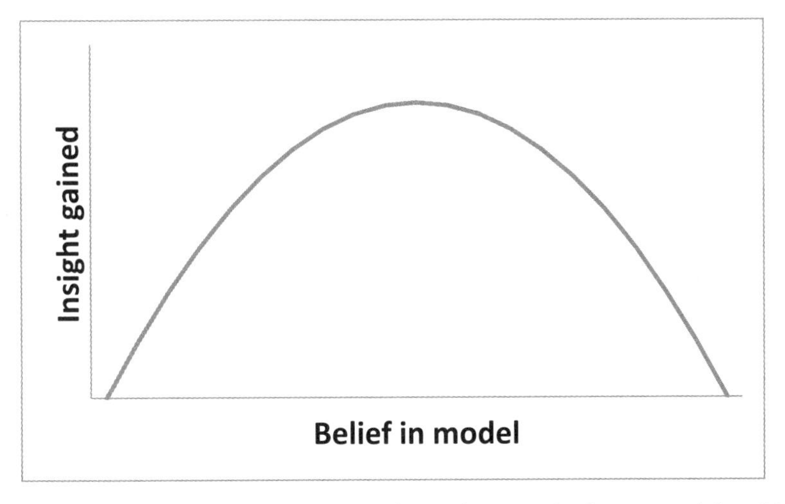

Figure 1. The relationship between how much insight we gain from a model and how much we believe the model and its results, showing that we gain most insight when we do not entirely believe the model.

from fishermen? Do the data sources agree with each other? When developing a model, we should state each assumption and consider whether or not it is a fair representation of the fishery or the fish population. After we fit a model, we should inspect model diagnostics to see if the model fits all of the data and attempt model validation. Finally, we should question if the results make sense. Is the history of fishery development and management reflected in the time series of fishing mortality? Has the population responded to management actions as expected? How do the estimated overfishing limits and rebuilding targets compare with historical estimates of fishing mortality, stock size, and past productivity? Results from a stock assessment model represent just one possible scenario of stock development and exploitation history that fits the data. Usually, the chosen model is the most likely scenario for the data we have, but it is just a story that is consistent with the data. Alternative scenarios or plausible states of nature that have considerably different results may be just as plausible. All of these likely scenarios should be communicated to fishery managers. Many other examples of science-based decision making (e.g., storm tracking) consider perspectives from an ensemble of models and approaches. We should never become fixated on a single model outcome.

Unfortunately, many scientists in our professional community take offense to scrutiny of stock assessment models, either during the peer-review process or after the peer-review process. They feel that critics of the model are making their job more difficult, muddying the waters for making a clear fishery management decision, or even attacking the modelers because they do not like the model results. From my experience, I feel that we can learn from our debates. We can make more informed decisions if we consider uncertainty in stock assessments and results from all valid alternatives rather than just defending one model.

Partially as a response from decades of advocacy for science-based fishery management and nonnegotiable catch limits, many fishery management systems rely on accuracy from stock assessment models that the models cannot deliver. As an alternative, simpler and potentially more robust management procedures are being implemented in several fishery management systems.

Overreliance on Precaution

There are ways to account for uncertainty and communicate it to fishery managers. Unfortunately (and ironically), a common approach to considering uncertainty requires complete belief in the model. The typical way that fishery scientists account for uncertainty in stock assessments is the application of a precautionary approach to fishery management, in which we recognize uncertainty in our decision making. For example, National Standard 1 guidelines for U.S. marine fisheries suggest that annual catch limits should be based on a lower confidence limit of the projected catch associated with overfishing. Unfortunately, such applications of the precautionary approach often only exacerbate the problem of believing stock assessment models. As typically applied, the approach not only assumes that the stock assessment model is correct and the catch projections are accurate, it also assumes that we can evaluate the uncertainty, which we usually cannot. In a precautionary context, fishery managers are led to believe that if they choose a risk tolerance of 25% in their catch limit decisions, then overfishing can be expected to occur on average once every four years. I am concerned that we are doing a disservice to our profession by pretending that our models have such predictive power when our experience shows that they do not. One fruitful approach is to consider model error or among-model variation in the evaluation of uncertainty.

In my opinion, a problem with the precautionary approach is that it can be applied to ridiculous extent. At the extreme, if all uncertainties are considered, we can only avoid overfishing by not fishing at all. Of course, this fails to meet the challenging balance of achieving sustainable fisheries or optimum yield. I submit that we can offer objective and reliable management advice, but we need to understand and communicate the uncertainties.

Perhaps the most damaging problem with the precautionary approach is the tendency for precautionary science rather than precautionary management. Some scientists feel that they should compensate for uncertainty by making biased analytical decisions, so that fishery managers do not make the wrong decision. The problem is worsened when scientists are not transparent about their decisions or justifications. Stock assessment relies on many subjective decisions. Best practices can be used to form reasonable assumptions and make logical decisions, but too often decisions are deliberately made to compensate for uncertainties or suspected biases based on perception rather than logic. For example, stock assessment groups are often reluctant to accept evidence for increasing natural mortality because it would allow for rates of harvest that seem excessive in the context of the existing paradigm. Unfortunately, when fishery managers make precautionary decisions based on precautionary science, the intended magnitude of precaution is artificially amplified. If scientists want to maintain credibility and provide the most informative recommendations to fishery managers, they should base their recommendations on the best estimates and communicate the associated uncertainties, with no hidden assumptions.

Communicating Uncertainty

We can accept and communicate uncertainty without losing credibility. Communicating the strengths of an assessment (e.g., relative trends, relative stock status) and the weaknesses (e.g., absolute estimates of abundance, catch projections) should confer credibility. There is no need to apologize for uncertain assessments. The data available to us are often noisy, and models require simplifying assumptions to represent a complex ecosystem. With this in mind, stock assessment projections are less like predictions from an engineer and more like weather reports. We make informed decisions based on weather reports every day. Some important decisions about livelihoods are based on weather reports, even though we know they are often wrong. If taken to the extreme, a precautionary approach would be to stay home because despite the sunny forecast, there is still a chance of rain.

When we accept that stock assessments are often wrong, we have a greater tendency to scrutinize our models and we are less likely to dismiss information that contradicts our model results, as I did earlier in my career with my technocratic approach to fisheries science and management. If fishermen complain that the model results do not match their observations, we should try to understand why. If there are other conflicting data, we should not dismiss them, but we should consider their implications. If we do not defend conventional stock assessments as though they are gospel, we are more open to applying alternative approaches. For example, critics of ecosystem-based fishery management are concerned that ecosystem science does not have the predictive power needed for fishery management. However, if we accept that we are often failing to realize the expectations from our single-species assessment models, we recognize that the bar is quite low for considering alternatives.

My sense of professional worth in the fisheries science and management system is not diminished by the recognition that stock assessment models are often wrong. On the contrary, I am just as inspired by my work as I was when I was just starting out in my career because this view of stock assessment helps me in my primary objectives: learning how fish populations respond to fishing, and helping to provide information to ensure that we have sustainable fisheries and fishing communities. I do not entirely believe my model results, but models remain an essential component of stock assessment, and through scrutiny of our conventional models, we can usually learn something new and provide useful information for fishery managers.

Acknowledgments

I thank my many colleagues who taught me lessons in stock assessment and who helped me learn from my mistakes. Specifically, my student Dan Goethel has been inspiring in his quest to make stock assessments more realistic. Thanks also to Kerim Aydin and Sarah Gaichas for sharing their insight–belief relationship. Cate O'Keefe, Jon Schnute, and two anonymous reviewers provided valuable feedback on the draft manuscript.

Biography

Steve Cadrin is a professor at the University of Massachusetts School for Marine Science and Technology. He was a stock assessment scientist for 20 years with the National Marine Fisheries Service and state agencies. His accomplishments include the development of precautionary harvest control rules for regional, national, and international fishery resources; advancement of stock assessment methods for a wide range of invertebrate and finfish species; and global leadership in evaluating geographic stock structure and modeling spatially complex populations. His teaching and research agendas focus on population modeling, stock identification, collaborative research with fishermen, and application of advanced technologies for fishery science.

References

Box, G. E. P., and R. N. Draper. 1987. Empirical model building and response surfaces. John Wiley & Sons, New York.

Quinn, T. J., II. 2003. Ruminations on the development and future of population dynamics models in fisheries. Natural Resource Modeling [online serial] 16:341 392.

Schnute, J. T., and L. J. Richards. 2001. Use and abuse of fishery models. Canadian Journal of Fisheries and Aquatic Sciences 58:10–17.

Stone, H. H., S. Gavaris, C. M. Legault, J. D. Neilson, and S. X. Cadrin. 2004 Collapse and recovery of the yellowtail flounder (*Limanda ferruginea*) fishery on Georges Bank. Journal of Sea Research 51:261–270.

Recommended Reading

Dickey-Collas, M., M. R. Payne, V. M. Trenkel, and R. D. M. Nash. 2014. Hazard warning: model misuse ahead. ICES Journal of Marine Science. DOI:10.1093/icesjms/fst215.

Megrey, B. A., and E. Moksness. 2009. Computers in fisheries research, 2nd edition. Springer, New York.

Rose, G. A. 1997. The trouble with fisheries science! Reviews in Fish Biology and Fisheries 7:365–370.

Schnute, J. T. 2005. Mathematics and fisheries: match or mismatch? Electronic Journal of Differential Equations [online serial] 12:143–158.

Schnute, J. T., and L. J. Richards. 2001. Use and abuse of fishery models. Canadian Journal of Fisheries and Aquatic Sciences 58:10–17.

Mentors Matter: Strategies for Selecting the Right Mentor

STEVEN J. COOKE*
Fish Ecology and Conservation Physiology Laboratory
Institute of Environmental Science and Department of Biology, Carleton University
1125 Colonel By Drive, Ottawa, Ontario, K1S 5B6, Canada

CONSTANCE M. O'CONNOR
Aquatic Behavioural Ecology Laboratory
Department of Psychology, Neuroscience and Behaviour, McMaster University
1280 Main Street West, Hamilton, Ontario, L8S 4K1, Canada

Key Points

- Mentors serve important roles, both personally and professionally, particularly for early career professionals, where guidance and counsel from a mentor can encourage success.
- Strong mentors are exceptional listeners and have a willingness to provide advice, without dictating and while still allowing independent ideas.
- A mentee may have many different mentors throughout his or her life. It is the wisdom of an individual, and his or her willingness to share it with others, that makes him or her a mentor, rather than any specific regimented role.

Introduction

Mentors have the potential to play a critical role in the development of fisheries professionals, both personally and professionally. This is particularly the case for early career professionals, where guidance and counsel from a mentor can encourage success in the short and long term (Welch 1997). Failure to find an appropriate mentor may result in a lack of support and encouragement needed to embark on a life-long career in fisheries.

So what are the characteristics of a good mentor? And how or where do you find such a mentor? Here, we tackle those two questions and provide examples from our own positive and negative experiences as mentees. More recently, we find ourselves also holding the role of mentor and are thus able to share experiences from that perspective as well. For context, Cooke has been in the professoriate since 2005 while O'Connor is a postdoctoral fellow. We preface this discussion by noting that we feel that it is entirely appropriate and encourage having multiple mentors to obtain balanced and diverse perspectives. Moreover, in the age of life-long learning, we propose that you are never too old or wise to benefit from a mentor. Although mentoring is recognized as important in fisheries (e.g., Kennedy and Roper

* Corresponding author: steven.cooke@carleton.ca

1990; Kohler and Wetzel 1998; Jackson 2010; Larson 2010; Lang et al. 2010; Boreman 2012; O'Connor 2012), we are unaware of any detailed accounts of what makes a good mentor in the fisheries profession, which makes our discussion of value to early career professionals.

What Makes a Good Mentor?

A report by the National Academy of Sciences et al. (1997) described a mentor as one who "seeks to help a student optimize an educational experience, to assist the student's socialization into a disciplinary culture, and to help the student find suitable employment," a definition that is specific to academic relationships. However, the *Oxford Dictionary of English* defines a mentor (noun) as "an experienced and trusted advisor," which is equally relevant to academic and professional context. In the current discussion, we consider a mentor in this broader sense. Therefore, we also consider that a broad range of individuals can potentially benefit from mentorship. A mentee may be a student but may also be an early career professional or a professional at any stage in his or her career facing a new challenge. Although such definitions are informative in clarifying what we mean by a mentor and a mentee, they do little to identify the traits that are associated with a good mentor. In our opinion, characteristics of good mentors include compassion, fairness, and good listening skills (see below). A mentor at any stage in your career should engage you, learn about your career aspirations, help you identify your strengths and weaknesses, and provide opportunities to help you achieve your goals, while also helping you expand your thinking and horizons. In addition, strong mentors have a willingness to provide advice, without dictating and while still allowing independent ideas. Using our own experiences, we elaborate on some of the characteristics of good mentors: willingness to provide guidance, accessibility, ability to listen, supportiveness, ability to criticize constructively, compassionate, insightful, and attentiveness to psychosocial development. We present these traits as a framework to guide potential mentees seeking an advisor in either an academic or a professional context.

Willingness to Provide Guidance

Mentors must offer guidance to their mentees. Mentors must have some kind of skill or knowledge, and they must be willing to and capable of passing this on to their mentees. While this concept seems a simple prerequisite of a mentor, the ability to teach skills is often harder than the acquisition of skills. Individuals have different learning styles (e.g., logical, aural, and visual) and different approaches to research (e.g., pragmatic, participatory, and hypothesis driven), and mentors may need to work quite hard in order to communicate their skills and knowledge effectively to their mentees. The best mentors are good communicators. That is, they are exceptional listeners, ask the right questions, share perspectives, and make genuine connections. Some universities and academic conferences offer workshops for mentors to improve as teachers, and these workshops can be very useful for both mentors and mentees. For us, some of our most important and memorable interactions with mentors focused on career guidance, especially when faced with multiple options, each with strengths and weaknesses. For example, Cooke received extensive mentoring support when he was looking to join the professoriate. His mentors provided him with the support needed for him to turn down his first academic job offer in favor of holding out for one that was a better fit.

Accessibility

One of the most common problems with academic mentors is that they are busy and may have limited time available for individual mentees. Often the best mentors are also the ones that have the fullest schedules. The best mentors are often the professionals that have been successful in their area, which translates into more commitments, including more mentees. However, a key characteristic of the best mentors is that they are accessible to mentees when the mentees need assistance. Different mentors may use different strategies to connect with their mentees, which can range from scheduling regular face-to-face meetings; staying connected through e-mail, telephone, or video connections; or using a combination of communication methods. With busy mentors, organizing group meetings may be useful. Regardless of the system used, it is important that a mentor respond to mentee concerns. Mentees can help improve this aspect of the mentor–mentee relationship by explicitly asking their mentor which method is best for maintaining communication.

Ability to Listen

It is critical that mentors truly listen to their mentees. Mentees will all have different goals and aspirations, and one mentee may want to learn very different skills than another mentee from the same mentor. For example, one student may be interested in how to engage in science advocacy while another may be interested in learning about experimental design, and another mentee at a different career stage may be more interested in negotiating work–life balance. A good mentor will take the time to discover what a mentee is hoping to gain from the mentor–mentee relationship and tailor his or her guidance accordingly. Along the same lines, a good mentor will take the time to listen to and understand mentee concerns. In this regard, it is also important that mentees clearly communicate their goals or any problems with their current academic or research programs. From the perspective of a mentor, we found that it is frustrating to discover that one of your mentees, especially a student or employee, has been having an ongoing problem or dissatisfaction that you were not told about. It can be challenging for mentees to discuss some issues, but good mentors will recognize the imperativeness of open communication no matter what the topic.

Supportiveness

Not all mentees have the same career goals and aspirations, and a good mentor will be sensitive to these differences. Many professors and professionals have an unfortunate tendency to blindly encourage all students down the same career path. However, the best mentors will attempt to use their knowledge and connections to help individual mentees achieve their individual goals, whatever these goals may be. For example, we have personally interacted with many mentors that have encouraged only an academic path, but not all students trained in fisheries have an interest in becoming professors or even staying in the profession for that matter. We have both interacted with mentees that although interested in fisheries, have also been keen on other career paths such as teaching at the elementary level, physiotherapy, or dentistry. Supportive mentors help mentees identify the path they wish to take and assist them in following said path. In some cases, that may mean supporting a path that takes them out of the fisheries profession or even towards seeking another mentor if that is what is best for the mentee.

Ability to Be Constructively Critical

There is an art to delivering constructive criticism because it is essentially balancing a combination of positive and negative perspectives. The best mentors are masters of this art and have the ability to gauge the cultural context (e.g., receptiveness of the mentee, relationship between mentor and mentee) and adjust their criticism accordingly. Without criticism, there is no growth or development. However, criticism must be expressed in such a way that mentees understand where improvements can be made. Mentors must be able to clearly describe how work can be improved, and not just where there are errors and shortcomings. This may be one of the most important qualities of a mentor, since many mentees feel discouraged by poorly expressed criticism, begin to feel as if there is no way to please their mentor, and ultimately become unmotivated and perform poorly. The best mentors, rather than leaving mentees feeling discouraged, will inspire mentees to improve. Given that being critical is an inherent component of science (e.g., peer review, critical thinking), our mentors have also helped us to do so in ways that are not demeaning to others. For example, when serving as referees for journal articles, both of us are committed to being constructive and respectful, such that we do not deflate but rather inspire. Rather than stating, "My head is pounding, the English is so atrocious that I can't understand the paper," an alternative approach would be to say, "Given the value of clear and concise communication, we strongly encourage the author work with a more experienced writer to address some deficiencies with English." The latter example demonstrates the value of making such improvements and provides direction for doing so without making unnecessary demeaning or flippant comments.

Compassion

For most mentees, especially students, their professional development is only a part of their life. As with the skills related to listening and support, mentors should be aware that priorities differ among mentees and between the mentor and the mentee. This does not mean that a mentor must become involved in a mentee's personal life; while some mentors may offer personal advice, other mentors may find it more appropriate to limit their advice to the professional sphere. In this regard, factors such as gender and age of mentor and mentee come into play. However, all mentors should be compassionate and understanding. Showing concern for a mentee's overall well-being, and making accommodations (e.g., extending a deadline) when necessary, is an important quality of a good mentor.

Insightfulness

This characteristic of a good mentor is less tangible and less easy to define. From a mentor's perspective, this may be the most difficult quality to actively improve on since it incorporates many aforementioned skills, such as listening, guidance, and constructive criticism. Indeed, to some extent it is a composite of other characteristics. However, the best mentors will be able to change the way that students view their research and their professional career through insightful comments. The serendipitous nature of truly insightful comments is one of the main reasons that having multiple mentors can be so helpful. Through quirks of timing and wording, different mentors may be able to offer important insight on different aspects of a mentee's early career, and at different times. By having multiple mentors, a mentee is more likely to benefit from these insights. Multiple mentors mean more advice, but more important are the different perspectives. As noted before, a mentor is not there to make decisions for the mentee, but rather to empower the mentee by providing him or her with information and perspective. As with many decisions, there is rarely a clear right or wrong approach. The answer is always context-specific, with the mentee being a key factor in that context. We both benefited from insight from multiple mentors throughout our early careers. There was certainly no consensus, but the insightful and diverse perspectives enabled us to shape our own perspective. Mentors do not have a crystal ball; they are individuals, and the good ones accept when mentees decide on a different course than the one that they may have suggested.

Attentiveness to Psychosocial Development

Psychosocial development is the development of personality and social attitudes (Kram 1985). The best mentors will encourage not just the development of a skilled professional, but also the development of a colleague who will contribute to the greater fisheries community. This can occur through encouraging outreach and participation in societies and committees. These larger communities both provide opportunities and can act as mentors in themselves, providing new ideas and inspiration for mentees. The best mentors in this case are those who lead by example. Mentees will admire mentors who are committed, passionate, and helpful to others and will be inspired to follow suit. Indeed, both of us have selected mentors along the way that cared about our development and success both outside of fisheries and in it.

Mentorship and Minorities

In addition to career and life advice, mentors also play an important role in increasing ethnic and gender diversity in the natural resources (Lopez and Brown 2011). Finding a mentor with a similar background and life experiences can be the difference between staying in fisheries and picking a new career path, particularly for minorities in fisheries. As a female in fisheries, O'Connor has benefitted enormously from both male and female mentors, but female mentors have been particularly helpful in giving advice on issues such as work–life balance and being a female scientist in a male-dominated discipline.

How to Find a Mentor?

The most common process to find a mentor is to communicate with a broad suite of professionals and look for those that have the traits of a good mentor that we have outlined above. O'Connor's (2012) article in *Fisheries* is aimed towards students seeking a thesis advisor and provides a framework for finding an advisor, including lists of questions to ask potential advisors in order to elucidate whether they have the traits of a good advisor. For example, it is important to speak both to the potential advisor and to current and former students in order to gain a balanced and fair perspective of the potential mentor. However, this process is not always straightforward, and mentors can also be found outside of traditional educational institutions (e.g., resource stakeholders, agency personnel) and outside the fisheries profession (e.g., community elder, athletic coach). It is unfortunately also the case that mentees will find themselves in a situation with a supervisor that they do not get along with or who is not encouraging their professional development. In this case, it is sometimes possible to switch supervisors in order to find one who is a better fit. It is also possible to use a network of mentors to provide the support that may be lacking from the primary supervisor. It is worth mentioning that one is not inherently a mentor just because he or she is a leader or supervisor, and a mentor need not be older or hold a formal supervisory role. Indeed, it is the wisdom of an individual, and his or her willingness to share it with others, that makes him or her a mentor. In some cases, mentors have found us; we have not gone out explicitly looking for mentors, but when we have identified someone with such properties and recognized their potential, we have embraced them as mentors. Identifying a potential mentor is about having respect and admiration for an individual. Quite simply, we cannot think of any mentors that we did not respect and admire first for some reason before they assumed such a mentorship role. We have also had different mentors for different career stages, as well as different mentors for different aspects of our life (e.g., mentor for academic publishing, mentor for development of teaching pedagogy). However, what they have all had in common is that they have been good mentors and exhibited traits that we wished to emulate.

On Being a Mentor

On occasion, we have had mentors that may not have realized that they were serving us in such a capacity. Indeed, when we think of the mentors that we have had and still interact with today, only a small number of them might themselves consider us to be their mentees. Perhaps that is a testament to the notion that some people simply are so natural at mentoring that they do not recognize that they are even doing it. We suggest that fisheries professionals with an active interest in serving as mentors consult the mentoring guidelines for wildlife professionals developed by Suedkamp Wells et al. (2005) or a more general document by Zachary (2002). Indeed, being a mentor enhances personal and professional development, such that both the mentee and mentor benefit enormously. We have found that mentoring is an underappreciated task, so when you have a good mentor, be sure to acknowledge their important role in your career and life.

The Future of Mentoring

Mentoring is changing. For example, e-mentoring, where communication between mentor and mentee is mediated by computer or other electronic technology, is becoming more common (Bierema and Merriam 2002), even in natural resource fields (Kinkel 2011). Neither of us have had experience with such relationships aside from communicating via e-mail, face-to-face meetings (e.g., via Skype), or at least audio communication. However, these relationships may become more common in the future and, particularly with improving video communication tools, have the potential to be valuable and allow mentees to seek multiple mentors without geographic restrictions.

Conclusion

Mentors matter. It has been our experience, after having mentors with a variety of strengths and weaknesses, that it really comes down to individual fit. This is especially the case when your choice of mentor may be only part of a larger decision regarding academic program, institution, research project, and personal or family considerations. A mentor can be the difference between an inspired student who goes on to become a great professional and an unmotivated student who drops out and seeks a different career path. Carefully selecting mentors who will have a positive benefit to your career and professional development is an important task, and given how critical mentors are, it is important to recognize and appreciate a good mentor when you are lucky enough to have one. Finally, those who find themselves in mentorship roles should be open to learning and improving their own skills as advisors and teachings. Mentors are useful throughout one's career, and for different aspects of one's career, and not just for those in the initial stages of their training. In some ways, we are all both mentors and mentees, and embracing the role and making steps to improve the mentorship experience will benefit the fisheries profession as a whole. We have both benefited from long-term relationships with a diverse suite of mentors that have helped shape our careers, character, and personas, and our own mentoring style.

Acknowledgments

We thank the many wonderful mentors (too many to mention) that we have had along the way. Hopefully we can pay forward what you have done for us! SC is supported by the Canada Research Chairs Program. CO is supported by the E.B. Eastburn Postdoctoral Fellowship.

Biographies

Dr. Steven Cooke is the Canada research chair in fish ecology and conservation physiology at Carleton University in Ottawa, Canada. His research is focused on understanding the ecology of stress in wild fish. He also has expertise in recreational fisheries science, biotelemetry, and inland fish conservation. Dr. Cooke is president of the Canadian Aquatic Resources Section of the American Fisheries Society and chair of the Sea Lamprey Research Board of the Great Lakes Fishery Commission. He is also editor in chief of the journal *Conservation Physiology* and an editor for the *Journal of Animal Biotelemetry*.

Dr. Constance O'Connor is currently a postdoctoral fellow at McMaster University in Hamilton, Canada. Her research focuses on addressing theoretical issues related to mechanisms underlying interesting animal behavior and applying these findings to more applied conservation outcomes. Dr. O'Connor is also a regular volunteer with organizations encouraging girls in science and promoting science and outdoor education for youth.

References

Bierema, L. L., and S. B. Merriam. 2002. E-mentoring: using computer mediated communication to enhance the mentoring process. Innovative Higher Education 26:211–227.

Boreman, J. 2012. Developing a legacy through mentoring. Fisheries 37:531.

Jackson, D. C. 2010. Mentoring: a synergistic multi-directional endeavor. Fisheries 35:4:47.

Kennedy, J. J., and B. B. Roper. 1990. The role of mentoring in early career development of forest service fisheries biologists. Fisheries 15(3):9–13.

Kinkel, D. H. 2011. Engaging students in career planning and preparation through ementoring. Journal of Natural Resources and Life Sciences Education 40:150–159.

Kohler, C. C., and J. E. Wetzel. 1998. A report card on mentorship graduate fisheries education: student and faculty perspectives. Fisheries 23(9):10–13.

Kram, K. 1985. Mentoring at work. Scott and Foresman, Boston.

Larson, E. R. 2010. Faculty and graduate student mentoring in the Hutton Junior Fisheries Biology Program. Fisheries 35:350–351.

Lang, T., J. Tiemann, F. Amezcua, and Q. Phelps. 2010. The AFS Young Professional Mentoring Program: our experience as the first cohort of mentees. Fisheries 35:444.

Lopez, R., and C. H. Brown. 2011. Why diversity matters: broadening our reach will sustain natural resources. The Wildlife Professional 5(2):20–27.

National Academy of Sciences, National Academy of Engineering, and Institute of Medicine. 1997. Advisor, teacher, role model, friend: on being a mentor to students in science and engineering. National Academy Press, Washington, D.C.

O'Connor, C. M. 2012. How to find a good graduate advisor and make the most of graduate school. Fisheries 37:126–128.

Suedkamp Wells, K. M., M. R. Ryan, H. Campa, III, and K. A. Smith. 2005. Mentoring guidelines for wildlife professionals. Wildlife Society Bulletin 33:565–573.

Welch, O. M. 1997. An examination of effective mentoring models in academe. Pages 41–62 in H. T. Frierson, editor. Diversity in higher education. JAI Press, Greenwich, Connecticut.

Zachary, L. J. 2002. The mentor's guide: facilitating effective learning relationships. Jossey-Bass, San Francisco.

Have You Prepared Yourself for the Changing Employment Arena?

William L. Fisher*
College of Natural Resources, University of Wisconsin-Stevens Point
800 Reserve Street, Stevens Point, Wisconsin 54481, USA

Daniel C. Dauwalter
Trout Unlimited
910 Main Street, Suite 342, Boise, Idaho 83702, USA

Key Points

- Fisheries education and the fisheries employment arena are changing.
- Fisheries students should seek a core in both the natural and social sciences, and acquire field, laboratory, and communication skills.
- Students need to be persistent in tailoring their degree program and experiences to position themselves for the job they want.

Introduction

Have you properly prepared yourself for employment as a fisheries professional? Fisheries education, employers of fisheries biologists, and issues facing emerging fisheries professionals are changing. Over the past two decades, natural resources education has experienced significant changes not only in the types of students entering undergraduate and graduate degree programs, but also in their academic programs and in the structure of curricula. Likewise, natural resource agencies that traditionally employ fisheries biologists are experiencing changes due to reduced funding and shifting priorities, but new employment opportunities are emerging. This changing landscape presents a challenge for students seeking employment in the fisheries profession. As a student entering or enrolled in a fisheries program, you need to be aware of these changes and be proactive in getting the necessary education and training to help prepare you for the changing employment arena.

Changes in Fisheries Education

Since the mid-1990s, when enrollment in natural resource programs peaked, there has been a general decline in the numbers of students majoring in the natural resource disciplines of fisheries, wildlife, and forestry at many universities. This decline has coincided with an increase in enrollment in environmental science undergraduate and graduate programs, although these trends vary regionally across the United States. Reasons for this change are

* Corresponding author: william.fisher@uwsp.edu

many and were the focus of a 2011 conference on natural resource education and employment sponsored by the Coalition of Natural Resource Societies (CNRS 2012). These reasons include the fact that there are more urban students than rural students entering college, which reflects a general population trend of people moving to cities from rural areas and shifting demographics. Students growing up in urban areas or from certain cultural backgrounds typically do not have traditional outdoor experiences such as fishing and hunting, and their connections with nature are most often through television and media, small urban parks, and zoos. Environmental science programs focused on sustainability and conservation have a broader appeal to urban students than traditional natural resources programs focused on utilization of fish and wildlife and extraction of forest products. Along with these shifting emphases, changes in curricula and declining university budgets are eliminating or restructuring many of the traditional experiential laboratory and field-oriented "-ology" courses, such as ichthyology, limnology, or stream ecology. This often robs students of the hands-on experience of going into the field and results in students who are not sufficiently prepared for jobs as fisheries technicians, biologists, or managers, which is a common complaint expressed by natural resource agencies needing to hire students with fisheries degrees.

Changes in Fisheries Employment

While fisheries education programs are changing, so too are the types of employment opportunities available to fisheries students. The missions and responsibilities of state and provincial natural resource agencies, as well as some federal agencies, continue to focus on managing and conserving fish and wildlife populations and forests. However, most agencies are increasingly focusing on addressing environmental challenges such as climate change, invasive species, and land-use modifications associated with growing urban areas. For example, the program mission areas of the U.S. Geological Survey (an employer of fisheries research scientists) have diversified from traditional disciplines of geology, mapping, water, and biology to climate and land-use change, ecosystems, energy and minerals, environmental health, natural hazards, core science systems (data mapping, analytics, and archiving), and water. At the same time, natural resource agencies are trying to meet the sociocultural and economic needs of special interest groups comprised of various angler specialty groups and other users of aquatic environments (e.g., watersport recreationists). As a result, agencies are increasingly hiring staff with degrees or experience in fisheries-related subdisciplines such as biometrics, conservation biology, environmental science, human dimensions, resource economics, environmental law, and youth education (Table 1). At the same time, agencies are experiencing cutbacks that limit their ability to handle the more diverse workloads associated with expanding agency mandates. This is creating opportunities and challenges for both employers and job seekers.

Although agency mandates have diversified to create nontraditional employment opportunities for fisheries students within those agencies, decreased agency budgets have resulted in a growing number of novel employment opportunities within industry, nongovernmental organizations, and foundations. For example, Trout Unlimited is a conservation organization focused on the protection and restoration of coldwater fishes and their habitats that routinely partners with state and federal agencies on stream restoration, fish passage, and land protection projects. Trout Unlimited has nearly doubled its full-time staff to approximately 170 professionals across the United States over the past five years. This growth has occurred despite a downturn in the U.S. economy and at a time when many public

TABLE 1. Types of full-time permanent jobs in inland fisheries programs within state fish and wildlife agencies, number of states having job types, and average percentage of employees in state fisheries programs by job type. Data from a survey of 41 state fish and wildlife agencies (Gabelhouse 2005).

Job type[a]	Number of state agencies	Average percentage of employees	Range
Fisheries management	41	35.6%	13.5–65.1%
Fish production	41	31.7%	3.5–66.2%
Research	30	6.8%	0.0–32.3%
Environmental science	29	3.9%	0.0–18.5%
Education	36	2.9%	0.0–12.6%
Boat and angler access	28	2.5%	0.0–14.4%
Habitat management	6	2.0%	0.0–43.4%
Computer support	21	0.8%	0.0–4.4%
Information	21	0.7%	0.0–7.9%
Biometrics	11	0.3%	0.0–4.4%
Human dimensions	12	0.3%	0.0–1.9%

[a] Excluding administration, secretarial and clerical, construction and operations, land acquisition, and various other jobs.

agencies experienced budget cutbacks and hiring freezes. While Trout Unlimited employs fisheries biologists, it also employs environmental lawyers, restoration ecologists, engineers, conservation planners, policy makers, and educators, thus requiring its fisheries-educated employees to be able to work in an interdisciplinary environment. Other conservation organizations, such as The Nature Conservancy, also hire fisheries biologists, but these organizations typically have other employees with diverse backgrounds that focus on land protection (e.g., easements), aquatic habitat restoration, policy, and advocacy to conserve aquatic species and ecosystems. Foundations (e.g., National Fish and Wildlife Foundation) and private industry (aquaculture, power authorities, and engineering firms) also employ fisheries biologists to accomplish their missions. Clearly, the natural resource employment arena is changing. However, there is still substantial demand for traditional fisheries students. State fish and wildlife agencies are arguably the largest employers of fisheries students, and fisheries administrators view competency in fisheries and aquatic science as necessary for success in entry-level positions within their agencies (Table 2). In addition, the employers of fisheries students across the different sectors (government, nonprofit, and private) are looking for specific disciplinary expertise, such as fisheries and aquatic sciences, but there is substantial diversity across these sectors. These sectors often want students with additional skillsets and experiences so they can be more proficient at their jobs (Table 3).

Preparing for the Changing Employment Arena

If fisheries education and fisheries employers are changing, how can you prepare yourself for the emerging careers in the profession? Our advice is to be proactive in obtaining the disciplinary skills you need, but also to tailor your individual curricula and experiences outside of your degree program to position yourself for the kind of job you want. First, the fisheries discipline is just that—fisheries. There will always be a fisheries-specific knowledge base that every fisheries professional should have, such as the basic principles of fisheries biology and management, and these will continue to provide the foundation for fisheries

Table 2. Top 15 competencies (courses) for entry-level fisheries management/research biologists and hatchery biologists identified as being important for by state inland fisheries administrators for job success (Gabelhouse 2010).

Rank	Management/research	Hatchery
1	Fisheries management	Fish culture
2	Fisheries field techniques	Fish health[a]
3	Technical writing	Fish nutrition[a]
4	Oral communications[a]	Ichthyology
5	Ichthyology	Fisheries management
6	Basic statistics	Limnology
7	Population and community ecology[a]	Water quality
8	Population dynamics[a]	Oral communications[a]
9	Research methods	Technical writing[b]
10	Experimental/survey design	Genetics[a]
11	Basic ecology	Basic chemistry
12	Limnology	Fisheries field techniques
13	Adaptive management[a]	Basic statistics
14	Geographic information systems[a]	Mathematics/calculus
15	Mathematics/calculus	Basic ecology, invertebrate and vertebrate zoology

[a] Competency is expected to be more important in 10 years.
[b] Recent hires have not been proficient in this competency.

degree program. However, most four-year degree programs also have enough flexibility for students to gain knowledge in specific subdisciplines by taking elective courses. It is here that you can and should tailor your education to the specific job you are seeking. For example, if you want a career in lake and reservoir management, you should take courses in limnology and water quality instead of oceanography and marine biology. If you want a career in fisheries policy, you should take courses in human dimensions and governance

Table 3. Top 10 skills identified within different sectors of conservation (Blickley et al. 2012).

Rank	Government	Nonprofit (nongovernmental organization)	Private
1	Specific/analytical disciplinary[a]	Project management	Specific/analytical disciplinary[a]
2	General disciplinary	Interpersonal	Field experience
3	Project management	Specific/analytical disciplinary[a]	Project management
4	Interpersonal	Networking	General disciplinary
5	Program leadership	Program leadership	Written communication
6	Networking	Written communication	Information technology
7	Field experience	Fundraising, monetary	Interpersonal
8	Written communication	Outreach communication	Personnel leadership
9	Oral communication	Information technology	Program leadership
10	Personnel leadership	Oral communication	Networking

[a] Fisheries competencies (see Table 2) would qualify here for fisheries-type jobs.

rather than courses in cell biology and genetics. The earlier in the education process you identify your desired career path, the more you can tailor your coursework to best position yourself for your dream job.

Much of our advice has been targeted at undergraduate students. However, we both have been actively involved in graduate and postdoctoral education and see the education and employment arenas changing for those of you with or wanting to obtain a graduate degree. Graduate students seeking a career in fisheries often take additional courses to fill their deficiencies. For example, some graduate students that completed their undergraduate degree in a biology program at a small college will likely focus on taking fisheries-specific courses (fisheries management, ichthyology) in graduate school because those courses were not offered during in undergraduate degree program. Graduate students also have greater latitude in tailoring their courses and training to their needs and interests, which is usually centered on their thesis research. A key for graduate students is identifying the universities and programs that fit their interests, the type of job they want, and, most importantly, an advisor with similar interests. There are many career opportunities for graduate students with state and federal agencies and nongovernmental organizations, and Willis and Isermann (2014, this volume) provide excellent advice for graduate students seeking a university faculty position. We do not see the demand changing in the future for fisheries professionals with graduate degrees, but you should expect a more diverse workforce and interdisciplinary workplace. So, begin preparing for the future now by seeking internships (yes, even graduate students should consider being an intern for an agency or organization) and building your professional network through professional societies (e.g., American Fisheries Society, http://fisheries.org) and professional social networks (e.g., LinkedIn, http://linkedin.com).

Tailoring coursework to your subdiscipline of interest may be obvious and easy to do, especially if you are enrolled in a large fisheries program. But what if there is not a course in your subdiscipline of interest, which could likely happen if you are in a small program with only a few core fisheries courses? If this is the case, seek outside opportunities to supplement your degree program and get the experiences you want. If you are interested in youth education, identify an agency or nonprofit organization with an active youth fisheries education program and volunteer time with it. If you are interested in reservoir management, seek internships with the state fish and wildlife agency, federal agency, or private company that manages reservoir fisheries or water quality and quantity. These experiential learning opportunities are critically important in the development of field techniques and experiences seen as desirable skills by natural resources employers. Recognizing the limitations of your degree program is the first step to ensuring that you get the experiences necessary for the job you want. Perhaps most importantly, seek out professors and working professionals who can help mentor you through your degree program and into a career in fisheries.

Gaining fisheries-specific knowledge and technical skills can occur through coursework, volunteering, and internships. However, other skills are required in the diversifying fisheries employment arena. Social, policy, leadership, and project management skills often take a back seat in fisheries degree programs because of limitations in credit hours required for a degree. Although this is may be changing, those skills can be acquired by seeking fishery-related interdisciplinary courses, such as those in sociology and human dimensions, resource economics, law and policy, and communications. These courses will provide information and training for the interdisciplinary issues that natural resource professionals encounter daily, as is evident by their incorporation into the American Fisheries Society's Professional Certification Program designed to set standards and guidelines for recognition of fisheries professionals. To enhance leadership skills, join your university student subunit of the American Fisheries Society (AFS) and become an active participant and leader. Seeking nontraditional volunteer, internship, and seasonal employment opportunities can also provide insights into the increasingly interdisciplinary nature of the fisheries profession. For example, consider working at an administrative headquarters to gain insights into the agency management and policy development. Volunteer at a nonprofit organization to learn new project management and fundraising skills. Different employment sectors hire fisheries students with common skill sets, but they value these skill sets differently (Table 3). You should be aware of what different job sectors are seeking in an employee so that you can tailor your degree program and experiences to get the job you want. The opportunities to gain experiences in fisheries are many, but they require that you be proactive and persistent in seeking them.

So, are you ready for the changing fisheries education and employment arena? Have you imagined your future job in fisheries, thought about the courses you need to prepare for it, mapped out your curriculum, and determined whether you should pursue a graduate degree? Have you identified summer internship and employment opportunities with fisheries agencies or organizations? Are you actively involved in an AFS student subunit, state chapter, committee, or section? Have you started building your network with fisheries professionals? It is never too early to begin these crucial activities that will help you prepare for a career in fisheries. Doing so will not only give you a leg up in being competitive for employment opportunities, but will also develop a sense of personal satisfaction in knowing that you have properly prepared yourself for the ever-changing fisheries profession.

Acknowledgments

Bill would like to thank his mentors, Louis Krumholz, Jim Gammon, and Stuart Neff, the chorus of American Fisheries Society leaders he has worked with, and the many students (including Dan) who helped him understand the parts of the whole and how they all work together. Dan would like to thank Bill Fisher, Frank Rahel, and Jack Williams for helping him understand how to make big-picture issues and landscape-scale science relevant to fisheries management issues on the ground. Both authors would like to thank the editors and the anonymous reviewers for providing constructive comments that greatly improved this vignette.

Biographies

William Fisher is an associate professor in the College of Natural Resources at University of Wisconsin-Stevens Point. He received his B.A. (1976) from University of Louisville, M.A. (1979) from DePauw University, and Ph.D. from University of Louisville (1987). He served as a water quality biologist with the Kansas Department of Health and Environment (1979–1980), stream ecologist with the U.S. Fish and Wildlife Service at Auburn University (1989–1991), assistant leader of the Oklahoma Cooperative Fish and Wildlife Research Unit at Oklahoma State University (1991–2008), and leader of the New York Cooperative Fish and Wildlife Research Unit at Cornell University (2008–2013). He has taught courses in fisheries and aquatic sciences, advised 25 graduate students, authored more than 75 peer-reviewed articles and book chapters, and coedited the AFS book titled *Geographic Information Systems in Fisheries* in 2004. He is an AFS certified fisheries professional, and a past president (2011–2012) of AFS.

Daniel Dauwalter received his B.A. in biology and environmental studies from Gustavus Adolphus College (St. Peter, Minnesota; 1999), M.S. in aquaculture/fisheries from the University of Arkansas–Pine Bluff (2002), and Ph.D. in fisheries and wildlife ecology from Oklahoma State University (2006; Bill was Dan's advisor). From 2006 to 2008, Dan was a postdoctoral research associate at the University of Wyoming where he taught ichthyology and worked with the U.S. Forest Service on long-term monitoring for aquatic species. Since 2008, Dan has been a scientist with Trout Unlimited (TU) working to provide strategic, science-based approaches for TU's varied conservation programs.

References

Blickley, J. L., K. Deiner, K. Garbach, I. Lacher, M. H. Meek, L. M. Porensky, M. L. Wilkerson, E. M. Winford, and M. W. Schwartz. 2012. Graduate student's guide to necessary skills for nonacademic conservation careers. Conservation Biology 27:24–34.

CNRS (Coalition of Natural Resource Societies). 2012. Natural resource education and employment conference: report and recommendations. The Wildlife Society, Bethesda, Maryland.

Gabelhouse, D. W., Jr. 2005. Staffing, spending, and funding of state inland fisheries programs. Fisheries 30(2):10–17.

Gabelhouse, D. W., Jr. 2010. Needs and proficiencies of fisheries hires by state agencies. Fisheries 35:445–448.

Willis, D. W., and D. A. Isermann. 2014. Finding that academic position. Pages 59–66 *in* W. W. Taylor, A. J. Lynch, and N. J. Léonard, editors. Future of fisheries: perspectives for emerging professionals. American Fisheries Society, Bethesda, Maryland.

The Intern–Mentor Experience: A Sample of the Real World

Molly J. Good
Center for Systems Integration and Sustainability, Department of Fisheries and Wildlife, Michigan State University, 115 Manly Miles Building, East Lansing, Michigan 48823, USA

John F. Kocik*
NOAA Fisheries, Northeast Fisheries Science Center Maine Field Station 17 Godfrey Drive, Suite 1, Orono, Maine 04473, USA

Key Points

- Internships provide unique opportunities that contribute to the professional, academic, and personal growth of both intern and mentor.
- As an intern, be open to new ideas, people, and personalities, and make the most of your time by becoming involved in all facets of the workplace.
- As a mentor, be a good listener and offer direction and guidance to your interns, but allow enough freedom for them to learn on their own.

Prologue

The internship experience, a temporary work period lasting three to six months, holds many meanings for undergraduate students. Some view internships as fun times with friends or opportunities to explore new places. Others may dread them because, despite months of time and effort, the monetary benefits are often limited. Mentoring an intern represents opportunities—chances to work with new people and to give back to the profession—and challenges—attempts to balance busy schedules with intern needs while providing both structure and freedom to foster intellectual discovery.

To break down an internship into navigable steps for emerging professionals, we highlight our own story from written application to professional connection. We describe this journey from two different viewpoints: one from Molly Good (MG), a graduate student, and the other from John Kocik (JK), a mid-career fisheries scientist. We hope this combined approach helps both interns and new mentors understand how they can make this journey a smooth and successful one.

* Corresponding author: john.kocik@noaa.gov

Search and Expectations—*Finesse your internship search and look at the internship details. Make expectations clear in your announcements. Find the right match!*

MG: In college, you hear about the benefits of exploring internship opportunities through conversations with professors, graduate students, and peers. To be competitive, I felt pressure to secure an internship as a freshman. From July through October, I began to research fisheries internship opportunities. I searched through agency Web sites; many had an Education section that listed opportunities for students. I consulted with my professors, advisors, and teaching assistants for their personal resources and guidance. Some upperclassmen in my program had already been interns and were willing to share samples of their application packets with me. I also browsed on-campus posting boards for internship advertisements. Early in my sophomore year, I accumulated enough information to prioritize internship possibilities based on my personal capabilities and career interests.

Because employers offer a limited number of positions each year, pursue multiple opportunities for which you believe you are qualified. You will be competing for positions with other highly qualified students who share your interests and academic histories. Do not limit yourself in the number of applications you complete, even though some opportunities might excite you more than others. If you generate questions during your internship search such as, "What daily tasks will I be responsible for?" and "Who will I be working with?" or just wish to learn more about a particular internship position, contact the agency or internship coordinator(s) directly via phone call or e-mail message. In my experience, potential mentors were glad and willing to communicate with me about their available internships. Remember, conduct yourself professionally and demonstrate that you are familiar with the positions being offered—you are making your first impression!

JK: I start to think about internships at the beginning of project and annual planning cycles. What type of help do we need? Will our current project(s) provide professional growth for an intern? Will an intern enhance our group's research capacity? Once those questions are answered, it is important to correctly classify intern roles and expectations. Molly's internship had a mix of field, laboratory, and analysis roles. In internship announcements, I classified the time to be spent in each role with estimated percentages. This allowed candidates to match their interests and enter a position with shared expectations.

Application and Selection—*Stand out in the crowd, make your resume and reference letters pop, and prepare for your interview*

MG: There are many essential pieces to an internship application, and it is crucial that each section is done exceptionally well. To make things easier, I have designed a checklist (see Application Checklist) to help you visualize the application pieces and allow you to budget your time efficiently.

While I was completing internship applications, I focused my time and effort on requesting reference letters and finessing my resume—the two most impactful pieces of the internship application. First, choose your references. People are busy, and you want to be sure to contact them in advance so they have ample time to do an excellent job for you. Choose them wisely because lackluster recommendations could weaken your application. If your references do not know you well, they will be limited in what they can say about your capabilities. So, you will benefit most from choosing a person that is able to confidently comment

Application Checklist

- ☑ Cover letter (not always required)
- ☑ Personal information (name, addresses, family information)
- ☑ Academic information (GPAs, coursework, program information)
- ☑ Official/unofficial transcripts from high school and college
- ☑ Resume or curriculum vitae (include your references)
- ☑ Written statements (likely to be one to three per application, approximately 250–500 words each)
  - ☑ Personal statement—an outline of your diverse background, academic credentials and recognition, and/or other relevant experiences that have shaped your interests
  - ☑ Statement of interest—an outline that describes your area of interest and your career aspirations, and your personal and professional goals or objectives for the future
- ☑ Reference letters (usually one to three)

on many aspects of your academic work and personal attributes. Second, include a current resume. Your resume should be up to date and highlight your past experiences that are relevant to the position. Third, spend time writing, editing, and rewriting statement essays; colleagues, professors, and even your parents can make outstanding editors. Remember, these pieces can be completed ahead of time; adhere to a schedule because the deadlines approach quickly.

JK: I review candidates and internship applications with a team. Each one of us has our own style and intern attributes that we prioritize. I look for eagerness in the cover letter, a concise description of skills, and references that highlight a candidate's experience and passion for fisheries resources. Others are more focused on grades and the comprehensiveness of the resume. We talk through the candidate packages, serving as advocates for our choices, and debate selection criteria. Discussing applicants with a team highlights strengths and weaknesses that not all of us notice. Also, though interns with impressive prior experience are often great assets, I also work to provide candidates without experience an opportunity. This is where the passion of a statement of interest is essential. When MG was selected, we had two other interns with a mix of experience and that added a good balance to our laboratory. A word of caution to interns: deadlines can shift, an application may be incomplete, or an interview window may be very short so monitor e-mail daily.

MG: I knew that my interview was going to be conducted over the phone, so I had time to prepare. My preparation would have been similar for an in-person or video-conference interview, but I would have limited my notes and dressed in appropriate business attire. I made sure to have a comprehensive understanding of the two projects and an idea of the duties expected of me as an intern. I also generated a list of questions about the projects and the people with whom I would be working. These questions show the interviewing panel that you have done your homework and have already started to think about your involvement in their research. Most importantly, I tried to showcase my personality by expressing my sincere excitement about the potential opportunity. I recall that one of the last comments I received from my interviewer was "Your enthusiasm is so refreshing." Make your passions and interests known! Once your interview is complete, the ultimate decision is now out of your hands.

JK: As a mentor, remember it is natural to be nervous, and part of a panel's job is to remove those nerves from the process. Personality types can impact interview dynamics so we try to have extroverts and introverts on our panels. I start with a general question: "Where did you grow up?" This helps the interview become more of a conversation. My goal is to have a structured conversation. I limit questions (six to nine) and use the same order of questions for each applicant. After the interviews, our team ranks and discusses the candidates. Based on the interview, we match the top candidates to specific slots. MG's experience and enthusiasm in the interview led her to a position on the research team that was very interactive with other team members.

Interns should remember that they will have an opportunity to ask questions. Having zero questions leads to an interview ending flat and telegraphs a lack of preparedness. Always be yourself and be honest, and do not overstate experience. It is a delicate balance. An interview is not a time to be humble. So, what is the difference? Overreach: "I have aged fish before and can age any fish without any instruction." Right balance: "I have aged about 300 fish and think I am getting better; I enjoy improving my skills." The second response shows level of experience and the desire to learn more. MG nailed these elements by discussing her stream ecology work. Finally, it is okay to use some fisheries jargon, but there is no need for big words. Relax and tell us about yourself and your experiences; we have all been on the other side of the table before so we really do understand.

The Internship—*Make the most of your time and work, build relationships, and learn about your professional self*

MG: In April of my junior year, I was awarded a Bradford Brown Student Internship position in the little town of Orono, Maine. Three weeks later, I had my driving route set, my car packed to the brim with my belongings, and a smile on my face as my dad and I began the trek from my house in Cincinnati, Ohio to Maine. I was eager for this adventure.

Eleven hundred miles and many Combos snack packs later, I was finally in Orono and ready to begin my internship the next day. By 10 a.m., I was standing outside in the middle of the freezing cold waters of Sedgeunkedunk Stream in waders and raingear while tallying Alewives *Alosa pseudoharnegus* migrating upstream. I was having the time of my life. Sure, the weather conditions could have been better. And yes, I was annoyed that my Smartwool socks were already soaked through. But I was appreciative of this first day. This particular situation taught me, as a new employee, the importance of adaptability in the workplace– it also taught me to keep an extra set of dry clothes nearby! From that point on, whether I was scheduled to tend fyke nets on the Penobscot River Estuary or track salmon movements using ultrasonic telemetry and sonar, I approached each day of my internship with the same earnestness and enthusiasm as I did on that first day in May. You will find that some days are tougher than others, but I wholeheartedly believe that a positive mental attitude is essential for a successful internship experience. You and your fellow interns are all going through the same thing, so rely on each other and talk to each other when the work gets difficult.

My first week in Orono was a busy one with introductions, driving and safety trainings, and daily fieldwork assignments. But once the work started to kick in, it did not take long for me and the other interns to acclimate to a regular schedule. After finally adjusting to my new position, I sat down with JK and his team to discuss the individual research project I would spearhead. I was provided some initial help, but I was also expected to do most of the work on my own. You will realize that you are not your mentor's sole responsibility. To save time on his or her end, try to conduct your research independently, and if questions or complications arise, be prepared to address these on your own before you seek assistance from your mentor. Some days will be hectic, when knowing what to do and sticking to the schedule is nearly impossible. At these times, use your positive mental attitude and help in whatever facet you may be needed. For example, if someone needs you to drop everything and assist in fish surgeries, you do it!

JK: The arrival of interns is both mundane and exciting. First, we have a checklist of must-dos: paperwork and training. Then, it is time for science. I give interns a complete overview of our program. It is a lot of information and can be overwhelming, but experience has shown that understanding the big picture helps them appreciate their vital role. Our interns often form the core of field crews (June through August), enter much of our data, and complete a majority of our scale measurements. So while they learn, they perform essential duties and quickly appreciate that there is no busywork. The first week immerses interns in laboratory culture and hands-on work experience. The next week, I engage them in planning and selecting an independent research project that combines our needs and their interests. The remaining weeks combine assigned tasks and their research projects. For us, MG's project was very successful because of effective communication. I facilitate independence (my travel and workload help, too), but also try to monitor to see if interns are struggling. This is a balancing act—challenging someone compared to overwhelming them is a thin line and highly variable between individuals. Because of my travel schedule, I make a point to stop into the interns' workspace, even it if is only for five minutes each day. Valuable insights for me and lessons for them have happened through this unstructured check-in. My biggest tip for interns is to not be afraid to ask questions–the busier someone is, the more likely you can learn a lot from them. One-on-one time with interns is my favorite part of my job. The energy, enthusiasm, and free-thinking of young professionals gives mentors a boost equivalent to a Starbucks visit and reminds us of why we got into fisheries in the first place. I initiate regular meetings with interns and encourage interns to plan them, too. MG initiated meetings and always had her questions outlined and organized, which resulted in effective outcomes.

We both believe a key to making the most of your internship opportunities is expecting to learn from everyone, not just your mentor, but also biologists, technical professionals, staff from other agencies, and your fellow interns. These people are your resources–use them. An effective intern asks questions of everyone and treats everyone as a valued resource. This leads not only to learning about the science and techniques used, but also to learning about the best ways to deal with people in a workplace.

Beyond the Internship—*Expand your network and maintain your newfound professional relationships*

For many interns, their experience is ephemeral and becomes part of their resume with limited mentor interaction other than a couple of years of references requests. As mentors, we hope all experiences are useful educationally to the intern and provide our team with added short-term capacity. Sometimes, this professional relationship continues and adds to the professional network of both intern and mentor. This has been our experience, and cowriting this personal vignette is possible because of the relationship we have maintained outside of the internship. The mentor–intern relationship can evolve to a more collegial connection over time, but growing this connection involves effort on part of both intern and mentor.

When MG entered graduate school, she worked to build and sustain many professional relationships she made in Maine. She kept JK informed of her latest accomplishments and what was going on in her career and life. In turn, JK kept MG updated on the status of Maine fisheries and passed along notices of professional opportunities. Sustaining these connections is sometimes difficult, but each person gains from an expanded network. From an intern's perspective, when you ask for a reference letter three years after your departure it lets your mentor provide a current perspective. For the mentor, it is great to see the professional growth of the former intern. For both, it may open up new networks in other parts of the country or world.

We both encourage becoming or staying active in professional societies like the American Fisheries Society. At annual meetings, you can catch up with each other and introduce members of your new professional network. You can also volunteer in these groups and sometimes work long distance with each other to further expand your network. The fisheries field is a small world, and before you know it, your network will be a web that truly extends from coast to coast.

Social networking sites now serve as tools for maintaining connections. Although combining your personal and professional worlds can be tricky, it is a current reality. We see a continuum of social networks from the professional LinkedIn to the more unreserved Facebook. Both can have a role in networking, depending on how you use them. LinkedIn is like a living resume, and content there is meant to be profession-related. Facebook is trickier: the type of posts you write and banter you engage in is exposed to both personal and professional acquaintances. Our suggestion is that if you are Facebook friends with your grandparents, then you probably could consider "friending" your mentors and interns.

Concluding Remarks

The process of obtaining an internship seems daunting. We believe that all science-based internships, whether they may fit your interests perfectly or only roughly, are worthwhile for your personal and professional growth. In your internship, remember that a positive attitude in all types of situations, even those unfamiliar and uncomfortable, goes a long way. Be game for anything and everything, and always be willing to pitch in and assist your colleagues and fellow interns. For the mentor, make the internship process a group endeavor. Involve your team in the selection process, encourage them to mentor, and help them become independent mentors. Interns not only add capacity to your team, but also help your team become more cohesive by providing fresh perspectives and new questions. Remember that an internship is usually only a 10-week-sample at one location, but still a meaningful learning experience. We feel that every intern grows and learns about the real world of fisheries and gains an experience of working with folks with diverse personalities and management styles. Discovering that fisheries science offers what you want is important, but discovering what elements are not for you is also valuable. We encourage you to consider an internship that helps you with that realization.

Biographies

Molly J. Good is a doctoral student in the Department of Fisheries and Wildlife and Center for Systems Integration and Sustainability at Michigan State University. She is working with

University Distinguished Professor Bill Taylor and Dr. Ed McGarrell to improve fisheries enforcement strategies on a global scale. Molly earned her M.S. graduate degree in the School of Marine and Environmental Affairs at the University of Washington and her B.S. undergraduate degree in the Department of Fisheries and Wildlife at Michigan State University. After graduate school, she will continue her learning internationally and hopes to one day serve as a mentor to students and emerging professionals in the fisheries community.

John F. Kocik has been a supervisory research fishery biologist at NOAA National Marine Fisheries Service Northeast Fisheries Science Center for more than 20 years. He works on salmon population dynamics and anadromous ecosystems. His current interests include expanding marine animal telemetry networks and data systems to better understand the ecology of anadromous and other fish across seascapes. John earned his biology B.S. at the State University of New York Plattsburgh where his first mentor was Gerhard "Gerry" Gruendling. Also a Spartan, John earned an M.S. and Ph.D. in fisheries science from Michigan State University where his most influential mentor, Bill Taylor, was his advisor.

Cultivating a Positive and Productive Postgraduate Research Experience: Tools for Students and Supervisors

ALISTAIR J. HOBDAY*
CSIRO Marine and Atmospheric Research, Hobart, Tasmania, 7000, Australia

LUCY ROBINSON
CSIRO Marine and Atmospheric Research, Hobart, Tasmania, 7000, Australia
and
The Ecology Centre, The University of Queensland, Queensland 4072, Australia.
and
Institute for Marine and Antarctic Studies, University of Tasmania
Hobart, Tasmania, 7000, Australia

Key Points

- Both students and supervisors can help facilitate a positive and productive postgraduate experience using the planning tools discussed here.
- As a student, making a plan and setting objectives can inform the kinds of skills you want to build that will not only lead to a successful thesis, but will also improve your employment prospects once you finish.
- Awareness of the demands on student and supervisor can build a strong professional relationship during the research training and beyond.

Preparing for the Postgraduate Adventure

Starting a postgraduate research project (master's degree or Ph.D.) in fisheries science is a challenging undertaking and should not be a choice of convenience. Recent years have seen fisheries research careers trend towards specialization with a concomitant need for greater technical proficiency on one hand and interdisciplinarity on the other. How then can both student and supervisor[1] best prepare for the postgraduate research journey, enjoy the experience, and see a student emerge with the in-depth technical abilities and holistic thinking that are critical skills in a modern fisheries science career? Here, we present a number of planning tools that can help you navigate both the challenging and rewarding times as a fisheries research student or supervisor, and suggest some strategies to reduce stress and to manage student–supervisor expectations during a project.

The first step to a great research project is the idea. Students might bring their research idea with them, perhaps from an undergraduate research experience, or it can emerge in

* Corresponding author: alistair.hobday@csiro.au
[1] "Advisor," "mentor," and "supervisor" are all terms used describe a person guiding student research.

discussions with your supervisor team[2]. In many institutions in the United States, where what is often a 4+ year Ph.D. program combines coursework and research, candidates can explore a range of ideas before settling on the final topic after the first one to two years. In other countries, such as England, Australia, Canada, and South Africa, where Ph.D. programs are often 3.5 years or less, students need to hit the ground running. Similarly, in most locations, Master's projects are often predefined given that they last one to two years. Checking that your desired project can also be funded (or supported under existing projects) is important—the budgets for student projects can constrain the scope of the project considerably. Some students will take an existing idea from the supervisory team or spend some months planning the project with the team prior to formal enrollment. We strongly recommend that students understand the research strengths and personalities of the supervisor before formal enrollment. Being able to formulate a solid project requires mutual respect and healthy relationships among your supervisory team. Hence, it is beneficial to invest time in understanding the skills and personalities of potential supervisors, as well as investigating the dynamics between individuals before the team is finalized. Students can learn about the research strengths of potential supervisors by reading their papers and can get a feel for their personalities by organizing at least one meeting (preferably in person or potentially by video chat–seeing faces helps obtain a better sense of the person). Working in a paid or voluntary position with the team before starting may be an option for some potential students and can also give the student and supervisor(s) an opportunity to get to know one another before enrollment. Supervisors can also facilitate contact between prospective students and current or previous students to discuss their experiences, skill development, and dynamics between the members of the potential supervisory team. They may also help you connect with suitable adjunct supervisors from, for example, a government research laboratory, that may help you develop a larger network and increase the relevance of your work to end users.

Students and supervisors need to work at forming (and maintaining) good professional relationships, as during the research you may encounter differences of opinion between you and one of your supervisors (e.g., he or she has different ideas on how your project should be progressing or provide feedback in a highly critical way with little positive feedback and encouragement). This does not necessarily mean you cannot work together, but it will require extra attention and management. Under these circumstances, always try to resolve the issue(s) through discussion, but if it is irresolvable, then try to weigh the extra work (and potential stress) that could manifest from these differences in personality in relation to the skills and expertise provided before you make a decision to change your supervisor team.

Student reflection

In selecting your supervisory team, complementary skills can work well. For example, it is good to have at least one supervisor that can see the big picture and as well as one that can see the "devil in the detail." Additionally, depending on the skills you wish to acquire, you may wish to select at least one supervisor with excellent quantitative skills and another with a history of writing high impact scientific papers (finding both attributes in the one supervisor is even better!).

[2] The supervisory team is also known as a committee, advisory team, or mentor group in different institutions. We mean the group of scientists that oversee a student research project.

Once the supervisory team is formed, we have found that several planning tools can be implemented early in the research design to help ensure that the research idea translates into a successful research project. These approaches are flexible and can be used throughout the degree to focus, reorient, or simply remind both student and supervisor of the pathway that the research will follow.

Refining the Project Idea—Tools for Successful Thesis Planning

Before starting a research degree, some students may have narrowed their research focus to a species or system and may even have the research objectives well defined. However, in the initial stages of a project, we think it is important to establish a problem that interests student and supervisors–students determined to develop and solve a problem too far from supporting interests risk failing to engage members of the supervisory team or may be unable to receive effective guidance and training. With mutual agreement on a project, a good supervisory team should be able to improve the research skill and output of a student at any level (Figure 1).

Once the research problem is defined with the supervisory team, it is important to be flexible in how the problem should be addressed. If the problem is general enough, a switch to new sampling methods, data sources, experimental and analytical methods, study species, or geographical regions will be possible without losing too much time. Such changes should be anticipated in the degree and can occur for a variety of reasons. At the same time, you should ensure that some elements of the research cannot be scooped by others working in the same general area, perhaps by including some data collection that is unlikely to be simultaneously replicated by other researchers. This is less of an issue for fisheries research, as species–geography–fishery combinations tend to be unique, but theoretical issues, such as studies of general life history patterns, may be similar to what others are working on in other laboratories.

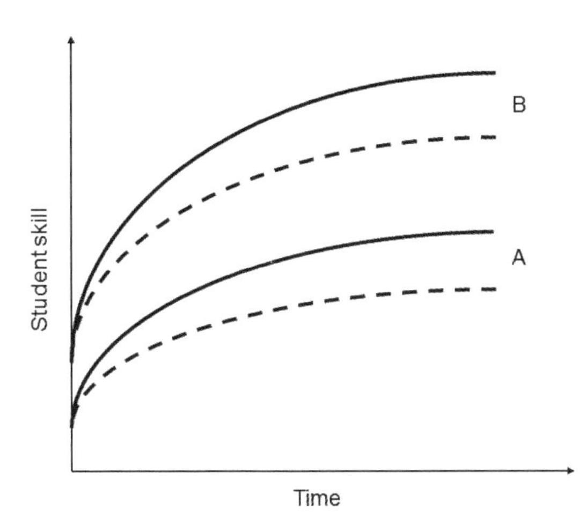

FIGURE 1. Research outputs and student skill development should be enhanced with a good supervisory team that is engaged in the research project (solid line) over self-supervision or poor supervision (dashed line) for both initially average (A) and high-skill (B) students. Of course, initially average students can also become exceptional.

Each research project will differ, but in refining the project, three tools have particularly assisted us: schematic organization, objective setting, and skill mapping. These cannot be done linearly, and we suggest tackling them concurrently, or at least iteratively (Figure 2).

Schematic Organization of the Research Plan

A postgraduate research project should produce a coherent body of work and not a grab bag of interesting findings. Thus, linkages between elements of the research are ultimately desirable, such as experiments that inform a model, or theory that is then tested in the field. Schematic representations can help conceptualize the order to tackle thesis elements and demonstrate a logical flow between the chapters. Devel-

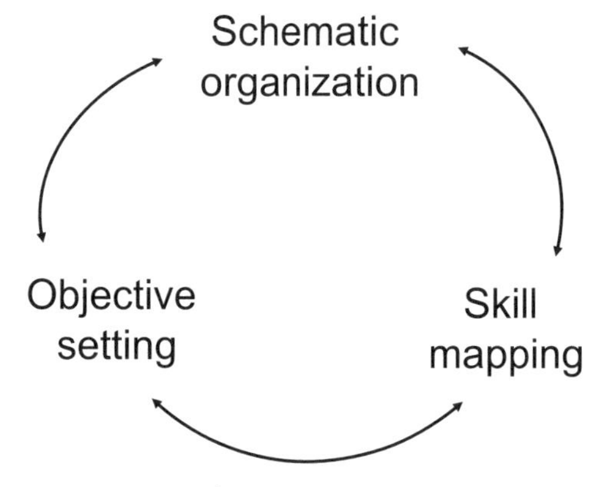

FIGURE 2. Three basic tools to help refine and plan a research thesis: schematic representation, skill mapping, and objective setting. These tools should be developed and refined iteratively during the planning stages and then revisited during the course of the thesis.

oping a schematic thesis plan well ahead of writing your thesis conclusion is advisable! This will provide both the student and supervisors with a succinct summary of the project that is flexible and allows regular updates as the structure changes, as well as making it clear to nonspecialists what the research will achieve (Figure 3). Recognizing that your thesis plan is flexible and can change is important, as it allows discoveries in the first stages of the degree to shape plans for the final years. The approach for the first year might be clearer and more refined than subsequent years–thus, parts of the schematic might be more developed then others, or there may be extras that do not yet fit the plan. Many students now also think about where they might publish elements of the thesis, and potential journals can be listed on such a schematic

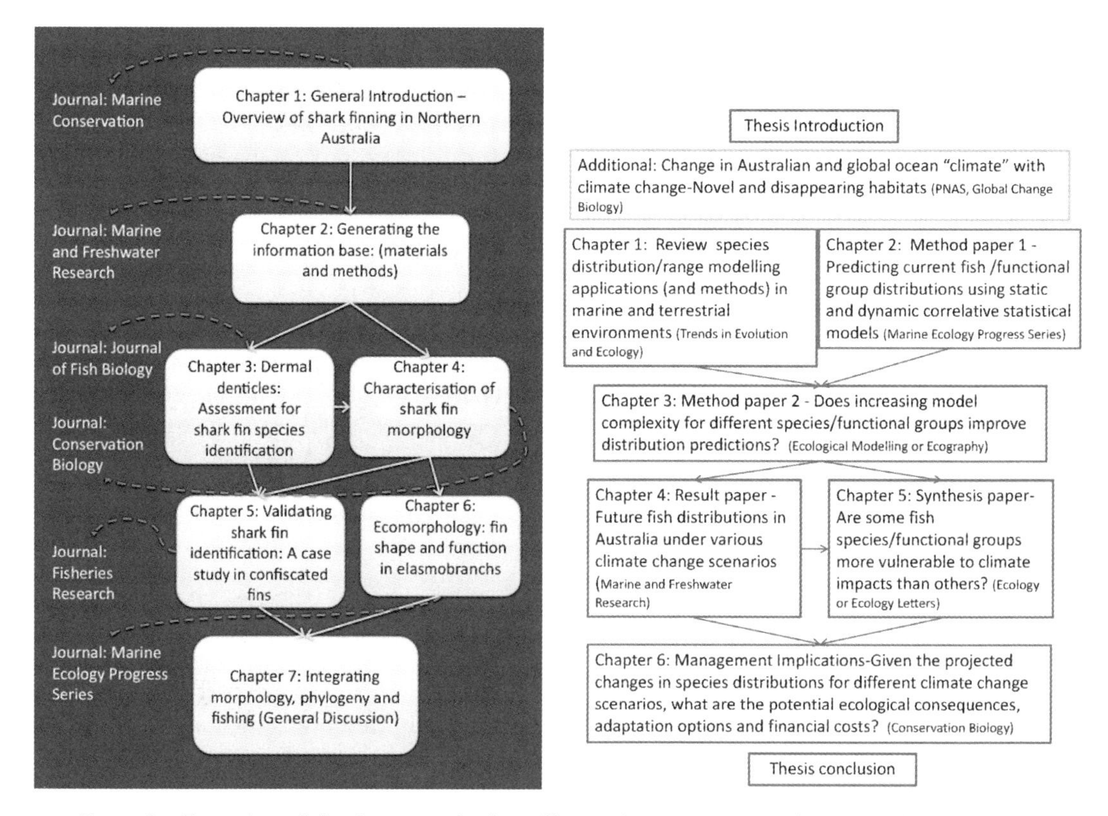

FIGURE 3. Examples of thesis research plans illustrating connectivity between thesis elements. A schematic thesis plan can help to organize student thoughts and be a point of reference for discussions with the supervisory team.

(Figure 3). In time, the status of each element can also be added to the plan (e.g., complete, data collection complete, and 50% completed), and it becomes a valuable summary of progress to date.

Setting Objectives for Research Elements

While it might seem formal, we have found that setting general objectives for each element of the research (e.g., thesis chapters) is also valuable. These objectives provide focus for your thesis—*why* the research matters and *what* you want to achieve (outcomes and outputs)—and can also inform the kinds of skills you would like to improve during your research. These objectives can be mapped onto a thesis plan (Figure 3) to make a one-stop project summary, or might be written in a more discursive manner. Objective setting is also good practice for future writing of grant proposals.

Skill Mapping

A research degree is an apprenticeship of sorts. Students select one or more supervisors to investigate a research problem but are presumably aiming for employment at the end of the degree. If a goal is to work in the applied fisheries science domain, supervisors may be able to provide advice on this directly or refer students to people currently working in fisheries research, management, or industry who can advise on the skills they need for a diversity of careers. Supervisors should be aware that for some students, a research degree represents a chance to gain skills they have always wanted and/or that might be needed for future employment. Thus, a research project should allow for a range of skills to be developed, rather than focus on a single approach.

Finding a balance between skill depth and breadth is recommended as students spending all their time collecting data or learning how to identify 50 species of crustacean larvae will gain great skill depth but may not leave enough time to gain the other skills, such as modeling or statistics, that might also be useful (skill breadth). Alternatively, by reducing depth, the student gains a wider range of skills, and employability. You can continue to develop the skill depth in subsequent research projects. Overall, we suggest finding a skill balance between jack of all trades and master of none (Figure 4).

Mapping the elements of the research project (e.g., chapters) against the skills that will be needed is also a gut check that the whole thesis will be interesting to both student and supervisor and provide the student with a marketable set of skills (Table 1). For example, discovering that a thesis will require years of microscopy may push students to modify the thesis plan. Students may also realize that the problem does not involve much fieldwork and so may look for other ways to obtain or maintain that skill, for example

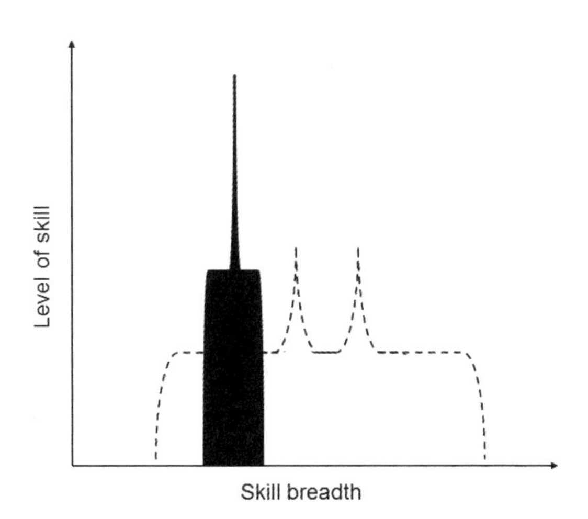

FIGURE 4. A wide breadth of skills with a few specialties (dashed line) will result in more future options than a single highly developed skill (solid area).

TABLE 1. An example of mapping expected skills to be acquired or used in each thesis chapter against a generic list of useful skills in fisheries science. Checks indicate that the skill will be gained in this chapter, and checks in the "Other" column are important to the student, but will have to be included elsewhere in the program of study.

Skill	Chapter 1	Chapter 2	Chapter 3	Chapter 4	Other
Writing	X	X	X	X	
Computer programming		X			
Statistics		X	X		
Experimental design				X	
Fieldwork (e.g., ship-based)					X
Taxonomy					
Databases and data management					
Teamwork and collaboration					
Modelling (many options)			X		
Historical data analysis	X				
Meta-analyses		X		X	
Genetic analyses					
Microchemistry					
Other skills (add as needed)					

by helping other students or attending a class or workshop. A solid grounding in four or five skill areas is realistic for starting in modern fisheries science careers.

The Research Journey

A research degree is a journey that can be stressful at times, and a shared awareness of stressful times can be beneficial to both student and supervisor. In this respect, the final planning tool presented here is more anticipatory and can be used ahead of time to explain how students will, on average, experience stress during the research experience (Figure 5). We have found that a shared awareness of potentially stressful periods leads to more positive responses during these stressful times and can also avoid or reduce miscommunication. Checking in from time to time to see where students feel they are on the psychological stress curve is a valuable exercise for supervisors and can help them to be responsive at appropriate times.

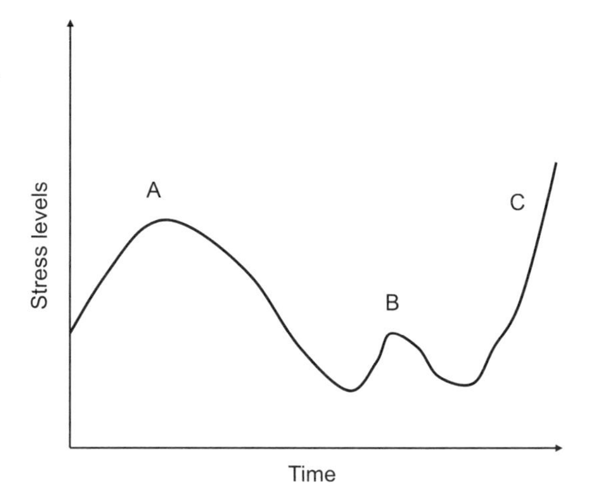

FIGURE 5. Psychological stress typically varies over the course of the research degree. Student stress levels are often high when the thesis begins, as there is uncertainty about the thesis planning (A); decline as the work begins; may be elevated at certain times (e.g., when experiments did not go as well as planned) (B); and almost always rise at the end of the degree (C), as time disappears or students start to worry that the project is not good enough.

Discussing project issues, challenges, or delays (e.g., Figure 5, point B) is not a sign of weakness and can help your supervisors decide when to guide your research rather than encourage you to solve all the problems on your own. Again a good supervisor–student team will mostly have the student solving the problems, with appropriate guidance or direction from the team, but knowing stress levels and how long a student has been independently trying to solve a problem, or struggling with other issues, can help get this balance right.

Students Managing Supervisors

Student supervision is only one responsibility for supervisors, so for students, managing your time with your supervisors is important. Open discussion results in a solution that works for both parties. For example, each supervisor's interest in different thesis elements will vary throughout the project due to his or her own research interests. Discuss the role each supervisor is willing to play as far as the time he or she can invest and which part of the project interests him or her most, perhaps using the thesis schematic (Figure 3). If supervisors are geographically dispersed, this discussion may also guide where the student is based at different times during the thesis. Students can also check in with the skills that they thought they would be getting during the degree and update this during the thesis.

Student reflection

I found it beneficial to spend the first half of my project working with my university supervisors at one location (Brisbane) as I received guidance and instruction on a variety of mathematical and statistical methods that improved my skills and equipped me to solve problems in my project. In the second half of my project, I relocated to the research institution where my other supervisor was based (Hobart) to gain in-depth understanding of the data, species, and fishery with which I was working.

Supervisors often have a number of students and projects. Consequently, their time is divided and stretched. Depending on their level of organization, they may know months and years in advance how much time they will have to contribute to a student project throughout its duration. When developing the schematic research plan, setting objectives, and skill mapping, students can plan with the team to determine when (what months and years) supervisors have time available and how this time can be used most effectively for the project. When scheduling time with supervisors in advance, students should allow for flexibility in the plan and keep supervisors up to date if delays occur.

As a student, it is important to find a balance between respecting how stretched for time your supervisor is and pushing the urgency and importance of your work. Due to the demands of other projects, supervisors can lose track of where your various drafts are and forget their importance. Therefore, it can be helpful to remind them if you need something urgently. Do not be afraid to remind them of this urgency and, if necessary, assert yourself in a respectful and diplomatic way.

Due to the stressful nature of postgraduate research, tension between student and supervisors can occur. Depending on the nature and severity of the problem, it is often wise not to take minor differences of opinion and disputes too personally or to dwell on them. Rather, acknowledge it, and either adapt your plan or confront it. One way of adapting (and taking the pressure off) is working on a facet of your project that is more closely aligned to another supervisor or collaborator. Tensions between supervisors can also occur (perhaps related to other work issues)—these can impact co-supervised students, and efforts should be made to minimize any impacts on student progress.

Concluding Remarks

The conceptual research planning tools described here can be revisited and updated during the thesis; as some research elements will change, new opportunities will present themselves, while some may disappear. Having a clear vision does not prevent deviation from the vision, but if managed well, it will lead to structured deviation, with the support of your supervisory team. A research master's degree or Ph.D. in fisheries science can be a demanding and stressful undertaking, but it is also intensely rewarding, especially when relationships within your team finish at a strong point and students move to the next phase of their career with mutual respect for, and the support of, their supervisors.

Acknowledgments

Lucy Robinson would like to thank her other Ph.D. supervisors, Anthony Richardson and Hugh Possingham, for their guidance and support; her postgraduate coordinator, Elizabeth Aitken; and friends and fellow students in Brisbane and Hobart. We appreciated suggestions for clarity provided by the editors and two anonymous reviewers and the use of Ph.D. schematics provided by former students.

Biographies

Alistair Hobday received a Ph.D. from Scripps Institution of Oceanography in 1998, where he had the support of two supervisors who encouraged independence and his cohort of peers who encouraged collaboration. In the past 10 years, he has supervised honors and Ph.D. students on a diversity of fishery-related projects. He is grateful for what he has learned by working with these committed people. His own research spans a range of topics, including spatial management and migration of large pelagic species and determining the environmental influence on exploited marine species. A focus is investigating the impacts of climate change on marine resources and developing adaptation options to underpin sustainable use into the future.

Lucy Robinson recently completed her Ph.D. at the University of Queensland with support from the Commonwealth Scientific and Industry and Research Organization. She traveled between Brisbane and Hobart to work with supervisors who were based in these two locations, where they provided their guidance and support when it was most needed. Her research projects have predominately focused on developing and applying novel analytical methods to assess the biological, ecological, and economic impacts of human stressors such as climate change, effluent discharge, and recreational activity on marine species and systems. She is also interested in findings solutions to alleviate stress and tension that can result from working in research to maximize the good (and minimize the bad) times.

Linking Successful Careers to Successful Fisheries

CHARLES F. RABENI (RETIRED)
Missouri Cooperative Fish and Wildlife Research Unit, University of Missouri
AB Natural Resources Building, Columbia, Missouri 65211, USA

SHANNON K. BREWER*
U.S. Geological Survey, Oklahoma Cooperative Fish and Wildlife Research Unit
Oklahoma State University, 007 Ag Hall, Stillwater, Oklahom 74078, USA

Key Points

- We define success for future fisheries professions as contributions to the process of sustaining, conserving, or restoring the conditions conducive to healthy fish populations. Sustaining and restoring fish populations require an understanding of holistic problem solving that should be incorporated into a student's graduate education.
- Students need to understand the importance of partnering with people from other disciplines and establishing ecologically relevant objectives, which in turn fosters accountability. Partnering will allow real advances to be made relative to complex and interdisciplinary problems whereas accountability will focus success on biological-based outcomes. If mentoring encourages partnering and accountability, more progress will be made in improving our fisheries.
- Attacking complex environmental problems using this framework will result in leaders whose professional success is synonymous with quality fisheries.

Introduction

The problems afflicting fish populations in the United States (including those of rivers and streams, as emphasized in this essay) are well documented and a major challenge for future fisheries professionals. The situation was personalized for us because in more than 50 combined years of stream research and lots of stream recreation, rarely has either of us visited a stream where overall fish habitat appears to have improved over time, nor have we met a seasoned angler who believes stream fishing is now better than when they were younger. While we recognize that stream degradation in the United States is not a recent phenomenon, and much began in the 1800s as a result of poor land management, it is obvious that considerable recent efforts by extremely competent individuals in our profession have not yielded the desired results, and that for real progress to occur, a different approach might be useful.

We find the environments necessary for healthy fisheries and fish communities to be woefully deficient for meeting the goals expressed more than half a century ago by the

* Corresponding author: shannon.brewer@okstate.edu

federal Clean Water Act to make all United States waterways "fishable and swimmable." Even though this legislation is arguably the most important ever for water quality and fish in the United States, it is not unusual to still have 30–50% of a state's streams and rivers listed on the Environmental Protection Agency-mandated 303 d list of impaired waterways. The conclusion of a recent national study (EPA 2013) suggests that conditions are not improving because now only 21% of the nation's streams are rated at least in "good" biological condition, a 7% decline in "good" streams in just 5 years. Why with all the intelligent people working on this has so little progress been made? Has it something to do with the preparation of these professionals? Does it really need changing? Does it have anything to do with how we measure success? The objective of our vignette is to suggest a new goal for success in the fisheries profession and describe the role of graduate education and mentoring in this model.

What Is Success?

The former leader of the Missouri Cooperative Fishery Research Unit, Dick Anderson, would inevitably greet students and faculty each morning with the question, "Well, what are you going to do for the fishery today?" His salutation was a reminder of the bottom line—that we are in business for the fish. This is the idea we wish to convey in this story, that successful future professionals in fisheries should define their career on results related to fish.

We define success for future leaders in the fisheries profession as contributions to the process of sustaining, conserving, or restoring the conditions conducive to healthy fish populations. Future leaders in our profession will likely have graduate degrees and can provide significant contributions at many levels of a natural resource agency, university, or nongovernment organization. We believe the graduate school experience is a prerequisite to leadership and to success because we have found it is the place where important concepts of career accomplishments develop. If graduate education inculcates broad philosophical ideas that are appropriate to long-term benefits for fish, success in the student's career should eventually follow.

Our current scorecard relative to success leaves much to be desired. The future fisheries professional will continue to face diverse and highly complex problems, among them sustaining lake and reservoir fisheries, reversing declines in most anadromous and marine stocks, rehabilitating stream habitat to support fish diversity, conserving threatened and endangered species, and halting the spread and dominance of nonnative, invasive species. How do we train aspiring leaders to be ultimately successful and not just documenters of deterioration? Our measures of success need to shift from perceived necessary components (e.g., meters of rip rap installed, acres of riparian trees planted, and percentages of landowner participation in land use changes) to improvements in the fitness (e.g., successful reproduction and adequate survival) of fish populations as a result of such activities. If measures of project success were shifted to actual biological outcomes, future professionals would focus on the common goal of sustaining and restoring aquatic ecosystems to a condition where native fish populations thrive.

Are Graduate Programs Doing a Good Job?

Recently, there has been what we view as a positive trend of supplementing traditional fisheries management (with an emphasis on game fish populations and consumption) with bigger-picture, more ecologically oriented education (see Scalet 2007 for a good commentary on this issue). This trend will likely continue; therefore, we must learn how to maintain an emphasis on our main constituency (anglers) while embracing some elements of change that allow fisheries professionals to more actively participate in achieving real successes for aquatic ecosystems, including fishes. Our emphasis over the years for research in Missouri streams has been on sport fish, primarily Smallmouth Bass *Micropterus dolomieu*, as a surrogate for the canary in the coal mine. In other words, what is good for Smallmouth Bass is good for the entire fish community. Therefore, attempts to enhance biological integrity also enhance sport fish populations and vice versa. We feel that elements of conservation biology, stream ecology theory, and landscape ecology, assisted by new technologies and methodologies, are a boon to fisheries management and in no way impinge upon the traditional goal of "making fishing better." The trends we have noticed in some graduate programs are not to the detriment of traditional management, but rather an expansion of the view that populations of interest will thrive in a healthy ecosystem and should be given a high priority. Fortunately, prospective graduate students now can choose from a variety of programs emphasizing fish. We hope that a student searching for an appropriate mentor would give serious consideration to the philosophy of individual faculty and the educational program.

Holistic Science

Sustaining and restoring fish populations involves a focus on the big picture, where spatial and temporal considerations are paramount. One approach to accomplishing this requires a focus on more holistic science (i.e., complex environmental problem solving that is likely never to be improved with individual disciplines operating independently). Rather, the model of success that would allow fish populations to thrive would require true interdisciplinary problem solving.

Holistic science seems an obvious solution, but a nexus of theory, application via technology, and interdisciplinary communication has to exist before the approach can be successful. Consider the basic tenet of interactions important to fish at the watershed and associated nested spatial scales. The theory of this relationship was understood in the mid-1970s (see Hynes 1975), but few advances were made in applying this theory to management problems until we made advances in technology (e.g., geographic information system, modeling packages) and theory (hydrology and stream processes) and improved our communication with terrestrial ecologists. In this example, almost 30 years elapsed before true advances were made. Despite the impressive achievements, our model of success is still missing. A focus on holistic science, combined with true measures of success, would change long-held attitudes and approaches among fisheries managers and researchers. Real progress would result from collaborations between disciplines with two emphases: partnering and accountability. Partnering among research disciplines allows for better understanding of the complexities involved while partnering among agencies and other scientific organizations provides overall leadership and economies of resources. Accountability starts with clearly articulated objectives, such as those emphasized in structured decision making, or if decisions are linked over time, adaptive management. In both cases, monitoring to improve learning is often necessary and hypotheses and models are developed a priori. Ultimately, monitoring is incorporated as a feedback loop to determine if the management actions taken are meeting the desired objectives.

Integrating Holistic Science and Success

Consider the example of a common, and so far intractable, environmental problem in the Midwest and what it would take to make some real headway—something that we have not yet achieved to any significant degree. How do we improve a stream system in an agricultural watershed heavily impacted by sedimentation such that fish populations are degraded? First, biological expertise is needed to examine causes (not correlations) of biotic conditions. This will require examining the linkages between the environment and resulting fitness consequences to fish. In our example, what is it about excess sediment that is ecologically important? Does it affect reproduction, feeding, and overwintering? Is it only important in filling pools, or is interstitial filling key? Is 50% sediment over natural levels impacting fish, or is 100% needed? Once ecological questions addressing fitness are addressed, other expertise is needed to determine the most efficient and cost-effective approach to a solution. What is the source of the sediment? Fluvial geomorphologists can evaluate sediment dynamics of a degraded channel. Plant ecologists and forestry scientists are certainly needed to examine stream-riparian conditions to evaluate mitigation of land-use activities. Integrating the disciplines of soil science and hydraulic modeling and the emerging discipline of ecohydrology could be beneficial to completing thorough watershed assessments. Making results more general by extrapolating to other watersheds (e.g., conservation priority analyses) would require significant geographic information system and hydraulic modeling skills. Accountability and progress could be coordinated and guided by those experts in a structured and interdisciplinary framework, such as provided by the philosophies of adaptive management. This framework also allows important human-dimensions aspects (landowner considerations) and economic realities (cost–benefit analyses) to be considered during the process. Only with this level of effort would we expect important problems to be rationally addressed. It would certainly be expensive and require involvement of numerous partners, with commitments lasting years. We are talking about real interdisciplinary efforts, not only from research and management professionals, but also those dealing with policy. Setting policy on interdisciplinary natural resource issues rarely includes much input from fishery experts, even when large projects obviously have a fisheries component (e.g., early Conservation Reserve legislation, many national EPA monitoring efforts, and many other U.S. Department of Agriculture watershed programs). If one feels it unlikely that the above example could be accomplished, one would also have to feel pessimistic about the future of our aquatic resources because these are the sorts of efforts required to beget success.

In the above example, the fisheries professional is a key component in planning, organizing, and executing the process. The exposure in graduate school to this more holistic approach to tackling problems will hopefully be carried forward by leaders into their career. Remember, top-level administrators, agency heads, and other decision makers were once graduate students. The likelihood of success will increase if biologists and administrators worked together to influence change. Top-level administrators or researchers could influence changes in policies or strategic direction for an agency, influence the focal topics of granting agencies, and conduct exemplary research and improve mentoring the next batch of students. Operational biologists could be very influential in setting biological program objectives and be very proactive in recruiting the interdisciplinary biologists needed to create successful outcomes.

Adjustments to Graduate Education for Students and Faculty

The trend toward more holistic problem solving involves changes for both students and faculty, changes having perhaps both pluses and minuses. First, there needs to be an improved understanding of the big picture. Students need to understand the requirements for a successful project and have a basic understanding of one or more associated disciplines.

Interacting with other disciplines would improve cross-discipline communication early in the student's career. We are pretty certain that students will have to become familiar with more sophisticated technology that may require additional courses or training. If students are required to take additional courses teaching the fundamentals of technological tools or advanced statistical modeling, then their graduate program may take longer to complete or may come at the cost of other conceptual ecological courses. However, proper mentoring could guide fisheries students to an improved understanding of organism–environment relations relative to the overall project objectives. Certainly, a suite of advanced theoretical and statistical models will become important problem-solving tools. Quality mentors will embrace these advanced theoretical and empirical approaches while ensuring the student remains focused on also discovering cause-and-effect mechanisms. Finally, how do we teach students accountability? A major effort in many agencies has shifted toward elements of adaptive management, where projects are periodically evaluated for efficacy and midcourse corrections are applied. Graduate programs that use elements of adaptive management would help students better understand the relevance of their own work relative to the big picture objective. Graduate-level multidiscipline programs in watershed science that incorporate a significant fisheries component would be one type of program where fisheries graduate students could get this training (e.g., University of Minnesota and Utah State University). Similar educational outcomes could also occur in fisheries and wildlife, biology, and zoology programs if there is interdepartmental cooperation with appropriate courses.

Holistic science is also affecting how some university faculty members conduct business of which the prospective graduate student should be aware. Clearly, university pressures on faculty are not aligned with our fishery success model because universities have for many years subjugated valid problem-solving efforts to receiving prestigious grant monies and encouraged excessive publications (see Scalet 2007). There are additional concerns for faculty that will impact a graduate student's experience. More collaboration means more time commitment. Larger collaborative research projects over longer time periods mean more administration will be necessary. These larger projects, combined with the projected state and federal austerity, will likely mean fewer funded projects and increased competition for graduate assistantships. More importantly, larger multidisciplinary research projects will likely have well-established objectives before the student begins their graduate program, leaving less room for student creativity and student input. There will certainly be greater accountability required for the work. However, we hope the result of greater systems understanding and the pride of producing real on-the-ground results that managers can use should override most hesitations, especially for young faculty.

Conclusion

Although the condition of our natural fish heritage has not improved in recent times, we are optimistic that students involved in projects with objective-driven, biologically based outcomes will ultimately benefit fish populations. The process will be slow but rewarding. It starts with the realization that the most important outcome of graduate school is the development of an understanding about what constitutes a successful graduate experience. The mentoring relationship between the student and advisor has always been pivotal to ensuring that the student can see application to conservation and management needs. This set of attitudes and beliefs, developed in graduate school, can then be carried on into professional activities. Future challenges are daunting. Young professionals will soon learn that even the best researched and strategized management plan often meets economic and social obstacles that thwart their best efforts. Nevertheless, we have found a career as fisheries professionals to be both joyful and rewarding. Proper planning before and during graduate school will result in a successful natural resource leader whose career resulted in a legacy of "better fishing and a better quality of life."

Acknowledgments

This is a contribution of the Oklahoma Cooperative Fish and Wildlife Research Unit (U.S. Geological Survey, Oklahoma Department of Wildlife Conservation, Oklahoma State University, and Wildlife Management Institute cooperating) and the Missouri Cooperative Fish and Wildlife Research Unit (U.S. Geological Survey, Missouri Department of Conservation, University of Missouri, and Wildlife Management Institute cooperating). We appreciate suggestions on this chapter from our colleagues Tom Kwak and Craig Paukert.

Biographies

Charles Rabeni received his Ph.D. in aquatic sciences from the University of Maine, researching effects of pulp-mill effluent on benthic invertebrates. After teaching college in Maine, he enjoyed a career at the Missouri Cooperative Fish and Wildlife Research Unit, primarily researching on Ozark and prairie streams, where his many great students did the heavy lifting while he took some of the credit.

Shannon Brewer received her Ph.D. in fisheries and wildlife sciences from the University of Missouri, where her research focused on landscape influences on Smallmouth Bass populations. She then obtained applied experience as a fish biologist with the U.S. Fish and Wildlife Service, working to restore fish populations in the Central Valley of California. Currently, she is the assistant unit leader at the Oklahoma Cooperative Fish and Wildlife Research Unit at Oklahoma State University. Her primary responsibility as a mentor is to promote and support independent thinking and an enthusiasm for science in the next generation of fisheries professionals.

References

EPA (Environmental Protection Agency). 2013. National river and streams assessment 2008–2009. A collaborative survey of the nation's streams, draft. EPA, EPA/841/D-13/001. Available: http://water.epa.gov/type/rsl/monitoring/riverssurvey/upload/NRSA0809_Report_Final_508Compliant_130228.pdf. (July 2013).

Hynes, H. B. N. 1975. The stream and its valley. Proceedings of the International Association of Theoretical and Applied Limnology 19:1–15.

Scalet, C. G. 2007. Dinosaur ramblings. The Journal of Wildlife Management 71(6):1749–1752.

Timing Is Everything

Erin J. Walaszczyk* and Cory O. Brant,
Department of Fisheries and Wildlife, Michigan State University
Room 13 Natural Resources Building, East Lansing, Michigan 48824, USA

Nicholas S. Johnson
U.S. Geological Survey, Great Lakes Science Center
Hammond Bay Biological Station, 11188 Ray Road, Millersburg, Michigan 49759, USA

Weiming Li
Department of Fisheries and Wildlife, Michigan State University
Room 13 Natural Resources Building, East Lansing, Michigan 48824, USA

Key Points

- Time management is key during every stage of your education and career, but how you should prioritize your time can vary considerably across each stage.
- Several tools and strategies are available to assist you in maximizing the potential use of your time and lead you down a successful career path.

Proper time management is unarguably a foundation underlying the success of undergraduate students, graduate students, managers, researchers, and professors alike. Having poor time management can increase the anxiety that comes from the stress of class expectations, meeting schedules, large research goals, publication deadlines, or lingering projects. Graduate school is one critical time that management skills play a vital role in achieving success. The pace of study, schedule, and priorities change considerably from undergraduate to graduate level studies. As you enter graduate school, there are several factors that will allow you to effectively manage and organize your time. Here, personal lessons and experiences will be shared related to the values of time management through several career avenues. It is our hope that the tools and strategies presented here will prove useful for avoiding some of the consequences that inevitably come from poor time management.

Learning how to effectively prioritize tasks is one of the most important tips for achieving successful time management. As a student, you need to be able to identify which tasks to attack first. Undergraduate students may find the solution to prioritizing tasks relatively simple, as they are guided by class schedules and course syllabuses that have clear expectations and deadlines for each class. Undergraduates have the opportunity to constantly compare progress with one another, aiding in the upkeep of studies and the completion of projects. When entering graduate school or a higher career, however, things can change. This transition can be challenging, as you may suddenly have to adapt to a seemingly endless

* Corresponding author: walaszcz@msu.edu

list of tasks in addition to a typical class load. A large majority of tasks may not even have a specific deadline for completion. You must learn to understand that there is no benchmark as to how much you should be working. You may, at certain times, need to work more than it appears your peers are in order to keep up with your project. Only you can determine how much time you must put into a certain semester to do a good job with your course load and individual research project. It can be easy to fall into bad habits of always putting classes first and not saving enough time for research priorities. While classes provide background knowledge and are still a priority as a graduate student, progress in research, hands-on experiences, presentations, and networking also contribute to your success. We urge you to make time for these activities, even if you must settle for spending less time on classes than you are used to. As you advance even further into your career, you will need to focus more on what your job responsibilities are and keep in mind why your level of education was required for the position and the level of effort you should put into your tasks. You may be answering to a larger network of individuals than when you were in graduate school with a single advisor. Smaller tasks, like responding to emails, entering data into a spreadsheet, or building a work bench, may not cause your supervisors to praise you, but these smaller tasks may lead to larger, overarching goals, such as securing research funds that will help you to get recognized for your good work. Accordingly, often your focus should be on larger, multiyear goals that will help to define your career.

It is important that when thinking about priorities you focus your attention on short-term and long-term goals. Creating a specific task list with proposed deadlines can help you to develop a manageable plan to make progress. You should try to organize tasks in order of most important to least important to the best of your ability. Seeking advice from an advisor, mentor, or boss may help you to design this prioritized list. Other time management resources, such as books written by time management experts or workshops on time management that may be offered at your university, can help to provide additional guidance. Many times you may feel overwhelmed by larger tasks or goals, but breaking down bigger projects into doable tasks can help make it easier to make progress. Adding a timeline to goals by creating a schedule allows for better execution of tasks, instead of just trying to wing it. Scheduling can help to avoid procrastination because it gives a certain time to accomplish each task, it helps keep you up to date, and it helps you to avoid rushing work at the last minute in order to meet deadlines. We suggest weekly or biweekly meetings with an advisor or a supervisor, which can help track progress and readjust schedules or priorities.

Of course a schedule is of no value to you if you cannot stay on track. When first starting graduate school or a higher level career, you may find it especially difficult to stay focused. Unfortunately, distractions from the Internet, such as Facebook, endless emails, and YouTube videos, constantly try to steal your attention away from task completion. Distractions like these can be amplified when working in close quarters with several individuals that become your friends and are always eager to share that funny video they have found. Pinpointing major time wasters is critical, but the true challenge is overcoming the temptation once you have identified the source. We suggest several strategies to help battle these temptations:

- When reading articles or reports, turn off the wireless connection to the alluring Internet or print a paper copy and stay away from the computer altogether.
- Turn your phone off or to silent.
- Wear noise-cancelling headphones (with or without music) to block out background noise and alert others to the fact that you are busy.
- Create a to-do list for the day.

- Write things down and try journaling how you are spending your time to see how much free time you really are taking.

We are not machines and unfortunately cannot usually remember every task or meeting time for each day; writing down a schedule is a useful tool. One major mistake is to think to yourself, "How many hours have I worked today? I have put in enough time to be done." How hard you are working cannot be measured by the pure amount of time you spend at work. Even if you sit in your office for 12 hours a day, if you spend your time chatting or wasting time on the Internet, you have not accomplished anything! If you feel your productivity slipping, take a break. We suggest you get up from your desk, stretch, take a short walk, drink some water, or have a snack at least once every couple of hours. We promise that when you sit back down, it will be easier to focus again and to be productive.

Your environment or surroundings can be key to successful time management. As a firsthand example, when we started graduate school, one of the most exciting things for us was that we often had the opportunity to work from several different locations. As coffee lovers, we were immediately drawn to coffee shops. We felt so productive when we would go, sit down with our giant cups of coffee, and pull out our laptops to start working. What we soon discovered, however, was that to truly be productive in this situation requires that an individual not be easily distracted by all of the noises and people. We spent more time people watching than actually working on anything! This type of environment, compared to a library or quiet office, may not allow you to be as productive for certain tasks (e.g., reading that requires a lot of focus, writing a report, and conducting statistics). In contrast, other tasks may be very compatible with a noisy coffee shop (e.g., making figures for publication, writing a presentation, or other more artistic endeavors). In addition, some people actually thrive better in noisier environments. You must adapt and learn what work environments are most suitable for you and your tasks at hand. Are people noisy and distracting or helpful for you? Do you need absolute silence or will music help to keep you driven? Do you find yourself spending more time chatting with coworkers than actually working? Maybe you gain inspiration (leading to higher productivity) from a forest hike or a paddle down a river. These are questions you must answer for yourself. Of course, you must also keep in mind the expectations of your advisor, supervisors, grants, or collaborators for how many hours you are expected to be in the lab, field, studio, or office and adjust accordingly.

We have mentioned the word "adapt" a few times thus far. Specifically, learning to work with and adapt your schedule or work environment to complete tasks. In addition, it is important to learn to adapt your routine to your natural body rhythms. Perhaps there are certain times that work better for you for different tasks. For example, if you are a morning person, you can set time aside for writing early in the day. If you are not, it is probably best not to set tedious or frustrating work at this time. Plan your high energy time for the important tasks that require greater effort and attention while saving routine work for times of low energy. You may want to set aside time in the morning devoted to reading or writing, or set aside certain days of the week devoted to research. If you block off established times for tasks, you are less likely to get fixated on one activity or double-book your time. Developing a routine can make you more focused and allow you to spread your time across the priorities you have. Knowing yourself and monitoring the times you are most productive can help you manage your time appropriately.

There is perhaps no better real life example of the need to develop an adaptive schedule than an extensive and time-limited research field season, a topic that we have much firsthand experience with. Our research focuses on Sea Lamprey *Petromyzon marinus*, which only have one mating season that is followed by natural death. Sea Lamprey live as parasites of fishes in the Laurentian Great Lakes during most of their adult lives, making subjects only available for research during a few months in the spring and early summer during the time that they move up into tributaries to spawn. The limited window of opportunity to study this species requires an excessive amount of planning before migration happens in early April. This involves extensive preparations, such as planning experiments, ordering supplies, interviewing and hiring field technicians, securing land access to field sites, setting up the field sites, acquiring animal use permits, making sure all equipment is working, procuring Sea Lamprey for you and your lab, and so forth. To manage this, writing out detailed timelines with completion dates for each task is very helpful to avoid anxiety. You can watch and manage your progress this way, and if problems arise, you can swap tasks or dates for others.

While an adaptable timeline is useful leading up to field season, this timeline becomes obsolete once fieldwork begins. Several uncertainties found in field work, such as weather events that cause flooding, droughts, or inaccurate responses of animals, require you to adapt to using a form of quick-thinking common sense that is rooted by patience. To assist with this type of field adaptation, we recommend the following:

- Understand the fact that Mother Nature ultimately regulates your progress in the field.
- Familiarize yourself with equipment maintenance and have backups when possible.
- Have a backup experimental plan for curve balls that nature may throw (e.g., another stream to use in case yours becomes bone dry).
- Do not put all your test subjects in one basket (split them up into several tanks in case one fails).
- Include all of this preparation in your timeline so you are ready for disasters before the field season.
- Make preparations early. Making an effort to make a plan and leave enough wiggle room in your field season schedule to manage all of the inevitable uncertainties is a vital part of fieldwork that can help you achieve your field season goals!

One of the greatest challenges, albeit an important one to overcome, is realizing when to say yes and when to say no. As an eager individual starting off on a new career path, you may find yourself wanting and willing to be involved in every task, exciting new research avenue, project, or adventure that arises. This is an easy way to burn yourself out, and you must realize that people can only spread themselves so thin. Too many activities and not enough time focused on the highest priorities may lead to a downward spiral, with lots of anxiety waiting to greet you at the bottom. You must learn to evaluate each request or project and decide which to agree to and which to respectfully decline. Do you have time to participate in something new? How much time will it take? Will this activity get you closer to accomplishing one of your goals? You, of course, do not want to say no to everything, as spending time on things, such as collaborating with other professionals, can lead to an integration of your specialty and ideas with those of others as well as assist you with broadening your interests. Simply allow yourself time to learn to manage your tasks so that you do not bite off more than you can chew. By not overcommitting yourself to others or tasks that are of little importance to you and your priorities, you can spend your time on activities that have more meaning to you.

We believe that if you spread yourself too thin, you leave no time for other needs, such as critical reflective thinking that leads to new concepts and interests. Critical thinking skills can and should be developed during graduate school and into a higher level career. Critical thinking and reflection can allow you to see the larger picture of the work you doing, evaluate all reasonable inferences from data, and remain open to alternative interpretations. During this time, you can focus on placing your work in a larger context, identifying gaps in knowledge, and formulating new questions. Many tasks, such as developing research ideas, often involve a certain level of creativity, whether it is directed towards developing an experimental plan or building something to conduct an experiment. We can all relate that some of the best idea moments have come when reflecting on work already done. Allowing yourself to think critically and creatively, as well as being open to several possibilities of answers, will help drive your work and allow you to carefully evaluate facts leading to your conclusions. Research results, end products, final business presentations, or that final conclusion from your architectural design are not always black and white. For example, unexpected or unclear scientific results have led to several exciting avenues of work and to more than one "aha!" moment. Thinking critically is not limited to our work, but is also important for reflecting upon your life, including personal goals and achievements. It is time well spent that can help you decide what you wish to accomplish in life, what your future career could have in store, and how you plan to get there.

Much like critical reflection is important, taking care of your health, family, social, and spiritual needs are all vital parts of a balanced life. Neglecting personal needs outside of work could have devastating consequences on time management and productivity. As a personal example, there were points in our careers where we pulled all-nighters with the impression we could accomplish more work. This, however, led to reduced stamina and poorer quality work. Getting a proper amount of sleep should never be compromised and is worth so much more than an extra six to eight hours of poor or low-quality work that will inevitably need to be redone. Getting exercise, spending time with your family and friends, having relaxing alone time, and anything else that can contribute to an overall balanced life should be included when planning your work days, weeks, and months in order to stimulate productivity. If you have a major life event or opportunity coming, make sure to plan wisely to have your tasks completed before any deadlines that interfere, so that you do not miss out! There is more to life than work, and your time schedule should reflect this.

Graduate school is a vital time period that you can nurture and develop valuable time management skills. Time management is not necessarily working harder, but rather smarter by utilizing your time for its maximum potential. Using the tips and strategies that are most pertinent to you, make sure you put forth effort to allow your time management skills to develop. In the long run, these valuable skills will help you achieve success in graduate school and in all of your future work endeavors, no matter what profession you choose!

Acknowledgments

This article is Contribution 1832 of the U.S. Geological Survey Great Lakes Science Center.

Biographies

Erin Walaszczyk graduated from Michigan State University in 2007 with a B.S. degree in zoology and in 2008 with a B.S. degree in fisheries and wildlife. She remained at Michigan State and is currently a Ph.D. student researching circadian rhythms and the effects of environmental cues in Sea Lamprey in laboratory and field settings under the mentorship of Dr. Weiming Li. Her research focuses on the effects of pheromones on rhythmicity and behavior as well as the underlying clock genetics and physiology in the Sea Lamprey.

Cory Brant graduated from University of Wisconsin–Stevens Point in 2008 with B.S. degrees in fisheries and water resources, biology, and aquaculture. His current Ph.D. graduate research is under the mentorship of Dr. Weiming Li at Michigan State University. He is conducting stream bioassays to identify functions of new, behaviorally active components of the Sea Lamprey pheromone. The Sea Lamprey, a jawless parasitic fish that represents the most basal lineage of vertebrates, has allowed him an opportunity to view biological systems (molecular to organismal/behavioral scale) from which all other vertebrate systems evolved.

Nicholas Johnson earned his B.S. degree from the University of Wisconsin-Stevens Point, and M.S. and Ph.D. degrees from Michigan State University. He became a research ecologist with the U.S. Geological Survey, Great Lakes Science Center, Hammond Bay Biological Station in 2009 and also functions as adjunct assistant professor in the Fisheries and Wildlife Department of Michigan State University where he mentors graduate students. His research is mostly in collaboration with the Great Lakes Fishery Commission where he is a member of the Sea Lamprey Trapping Task Force and Barrier Task Force.

Weiming Li earned his B.S. and M.S. degrees from Shanghai Ocean University, China, and his Ph.D. in fisheries from University of Minnesota, Twin Cities. He is a faculty member in the Department of Fisheries and Wildlife in Michigan State University and a research scientist supported by the Great Lakes Fishery Commission, Ann Arbor, Michigan. Research in his laboratory is focused on physiology, chemical ecology, and genomics of the Sea Lamprey. Members of his research team have very diverse academic and ethnic backgrounds.

Finding that Academic Position

David W. Willis
Department of Natural Resource Management, South Dakota State University
SNP 138, Box 2140B, Brookings, South Dakota 57007, USA

Daniel A. Isermann*
U.S. Geological Survey, Wisconsin Cooperative Fishery Research Unit
800 Reserve Street, College of Natural Resources
University of Wisconsin-Stevens Point, Stevens Point, Wisconsin 54481, USA

Key Points

- Obtaining an academic position in fisheries is a challenge because competition for these positions is intense.
- To be competitive, students must be persistent and actively seek opportunities that provide experience in teaching, research, and service.
- We provide advice on gaining this experience and also discuss the value of collegiality in the academic setting.

Introduction

Yes, competition can be intense for academic positions in fisheries and throughout the natural resource disciplines. Yes, there is a trend toward fewer tenure-track positions in higher education. Despite these issues, effective, dedicated fisheries professionals with a collegial spirit who enjoy mentoring future professionals are still needed, and these positions are truly rewarding. Interacting with students while working "for the resource" has been a fulfilling activity for both of us. In an academic position you will directly influence future generations of fisheries professionals, and there is no better way to continue the American Fisheries Society (AFS) legacy of mentoring.

Our objective is to provide advice to individuals seeking an academic position at a college or university. We will first discuss the three primary responsibilities of most faculty members, and then offer advice on gaining appropriate experience, the need for collegiality, and the importance of persistence. Our goal is to provide sound advice that should be useful during your progression from graduate student to your first on-campus interview.

A Three-Legged Stool

A useful thought process for dissecting the skill set needed for a fisheries faculty member is the traditional three-legged stool of teaching, research, and service responsibilities. Regardless of whether the academic position is at a large land-grant university, a liberal arts college or uni-

* Corresponding author: dan.isermann@uwsp.edu

versity, or a regional university, some aspects of those three legs will be considered in hiring and promotion decisions. What will differ among institution types is the proportion of time allocated to each leg (Figure 1). Frequently, the proportion of time a professor will allocate to teaching, research, and service is provided in the job announcement, which allows prospective candidates to ascertain the relative importance of each leg of the stool and prepare accordingly. These time allocations are often relatively strict for newly hired professors but can change with accrued years of service. Contacting current professors regarding the realities of time allocation and other matters is always a prudent step in preparing for the interview process, which is an example of an informational interview. The informational interview typically involves a conversation with an individual within your discipline about that person's work. Such an interview is outside the context of a job interview and typically is most effective when you make it clear from the start that you are not asking the contact person to assist with job prospects.

During the job interview process, expect and prepare for visits with various administrators who will have one or more of these three legs as primary responsibilities. The campus interview often transpires over multiple days, and while prospective faculty members might expect the requested research and/or teaching seminars, they may be surprised by the number of visits with administrators, other faculty members, the search committee, a group of graduate students, and even people outside the university, such as a representative from a state natural resource agency. Do not be afraid of repeating yourself to answer questions over the course of these interviews; in fact, such repetition will be a necessity.

Teaching

Teaching should seem like an obvious role for an educator. However, we have seen more interviewees for academic positions fail at this point than any other. Too often, an energetic

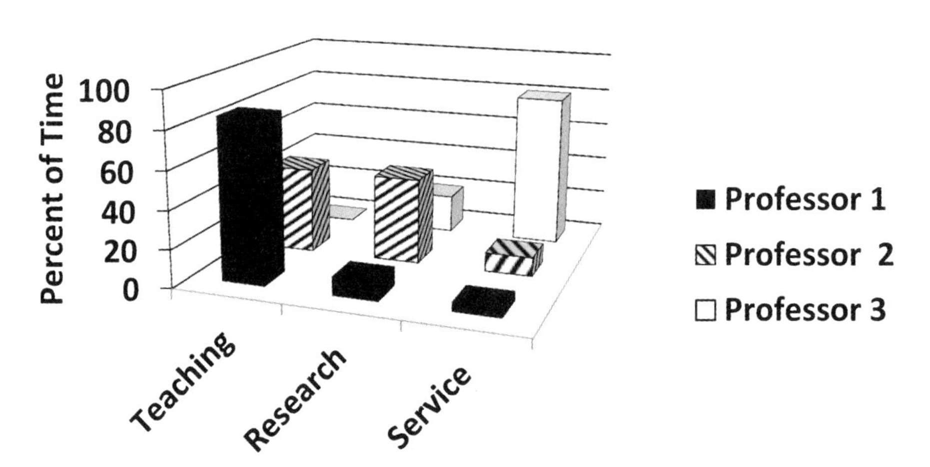

FIGURE 1. Comparison of assigned workloads for three hypothetical faculty members. Professor 1 has a high teaching load and limited time allocated for research and service. Such positions might be common at a liberal arts college or smaller state university where faculty members are primarily charged with teaching undergraduate students. Professor 2 has substantial and equal time allocated to teaching and research and might be employed by a large, land-grant university. Professor 3 has a large service workload, and may be an extension service faculty member working on aquaculture or pond management at a large, land-grant university. These three examples certainly are not the only workloads assigned to faculty members at various colleges and universities but represent realistic scenarios for many faculty positions.

applicant will spend an entire hour with the vice president for academic affairs (sometimes known as a provost) or an associate dean for academic programs and talk enthusiastically and endlessly about her or his research interests and capabilities. One applicant particularly comes to mind because that candidate was clearly the front-runner for the position until focusing almost entirely on research during an hour-long visit with our associate dean for academic affairs. Remember that these administrators deal more with the teaching side of faculty positions, and a substantial part of that interview time should involve discussions of teaching experience and philosophy. Convince such administrators that teaching is as important a component of modern education as research; that is likely to match the philosophy held by administrators who primarily deal with teaching. Research and write a teaching philosophy statement prior to your interview. Such a statement should reflect your beliefs about teaching and learning. In addition to general comments, be sure to provide tangible examples of what you would do in the classroom. Some position announcements may request that you submit such a statement. If not, prepare one anyway, just to organize your thoughts and beliefs on teaching.

Gain as much teaching experience as you can during your graduate-level education. Try to optimize your time, as we all recognize the many demands on graduate students, and seek progressive increases in the quality of your teaching experience. Many students have the opportunity to serve as a graduate teaching assistant, often in a laboratory setting, and such experience will be viewed positively by a search committee. If possible, try to find an opportunity to teach a course where you are instructor of record and students are required to formally evaluate your teaching performance. Such teaching experience is sufficiently important that we recommend that those opportunities be a topic of discussion with potential graduate advisors when you are seeking your graduate positions (i.e., before you make a final decision on which program you will enter). Determine what formal classes in teaching principles and strategies are available to you as you decide your plan of study for your graduate degrees. While the senior author never had a formal course related to teaching and had to learn by the school of hard knocks, the junior author did benefit from coursework on the principles of college teaching. The prevalence of this type of coursework on applicant transcripts has increased, and these courses can improve your personal teaching efforts while also demonstrating dedication to this leg of the stool.

The traditional style of teaching has been a lecture-and-test format; however, most instructors in higher education are moving to more active learning techniques. Active learning is a process in which students engage in activities such as discussion or problem-solving exercises that promote analysis and synthesis of course content. Learning often occurs within teams, which directly applies to the workplace as many natural resource agencies favor a team approach to problem-solving. Active learning improves student retention of information and can be very effective in developing higher cognitive skills such as problem solving and critical thinking. One active-learning strategy favored by both authors is the case-study method. Consider yourself in a classroom with these alternative experiences. First, an experienced instructor provides you with a case history of how one state natural resource agency resolved a question on whether to stock high-elevation mountain lakes that previously were devoid of fish. Given your background in fisheries, you likely are interested in this topic, will pay attention to the information provided, and may even ask a question of the instructor. Now, consider a classroom in which that experienced instructor breaks the class into teams. Each team is given a role to play, such as an animal-rights organization, an anti-fishing organization, consumptive wildlife and fisheries users, a fly-fishing group, an amphibian conservation society, a backpackers association, the state Chamber of Commerce, and an off-road vehicle federation. Each group then provides an overview of its preferred management strategy for these lakes, and discussion ensues. The faculty member may even finish by discussing what this state agency selected as their strategy, which should then lead to another discussion of advantages and disadvantages for that selection. Are you more likely to retain the course material under the first or second teaching strategy? We suspect that after teaching with the case-study method, you will be much more enthusiastic in your discussion with that provost!

Be prepared to provide student evaluations of your teaching performance during an interview process, even if the evaluations are not required materials. Also, try to gain teaching experience at multiple levels, from introductory service courses for nonmajors, to upper-division undergraduate courses within the major, to graduate courses. Not everyone can gain all of these experiences, but ask for such opportunities from your major advisor or department head/chair. Do as much as you can, regardless of whether additional pay can be provided. Yes, we know you are busy; however, you have chosen a highly competitive career path and will need to distinguish yourself from your competitors.

Research

Research is an extremely important component of any academic position. Scholarly activities are expected of faculty members at all colleges and universities, although the time allocated to research will vary by position (Figure 1). Try to establish an expertise that especially interests you and can be a theme that weaves throughout many of your research activities. However, do not forget to add some breadth to your background. Even if your specialty is fish genetics or population dynamics, add some component of general fish ecology to your research portfolio. Standard advice on the importance of peer-reviewed publications and professional meeting presentations apply here; your competition will have stellar curricula vitae (CV) and so must you. While we both cringe at the concept of "working on my resume" or completing a task "because it looks good on my CV," your accomplishments and skills will be discerned from your CV during the selection process for interviewees.

Research activities also help you remain current in your discipline and will enhance the quality of your teaching. In addition, never forget that mentoring others through research efforts is a highly effective form of one-on-one education. While you are a graduate student, mentor your undergraduate students or technicians through their own research efforts. What experience could better prepare you for mentoring of graduate students? Did you note how quickly we slipped back into the teaching component? Despite the convenient categorization system, the three legs of the stool are highly integrated. Finally, learn by collaborating with other students, faculty members, and agency fisheries professionals during your graduate-school years. Collaboration and interdisciplinary research efforts will undoubtedly be a big part of your future academic career. Working with other researchers is a good means by which people early in their careers can meet additional mentors.

An overlooked research opportunity for a fisheries educator is the scholarship of teaching and learning (SOTL), which is scholarly inquiry into student learning that advances the practice of teaching. Most academic programs will recognize scholarly articles on SOTL as peer-reviewed publications when it comes time for your promotion and tenure evaluation, so turn this into one of your strengths. Let it be part of your scholarly activities and it will also improve your teaching. The current generation of students, perhaps more than ever before, responds well to active teaching methods. Plan and use case studies and group exercises. Then, evaluate the outcomes of such teaching methods and publish this information. No need to acquire a large competitive grant to complete such research–it can be part of your day-to-day activities.

Service

The third component of the academic workload is service. Depending on the position you seek, service may involve formal Extension Service programs in traditional land-grant programs, such as an aquaculture specialist or a pond management specialist, where most of your time is spent on such activities. Other faculty members may not have an extension component for their positions but are still expected to provide outreach to the community through volunteer activities, outreach to the profession through volunteer positions or elective offices in professional societies or continuing education efforts requested by a state or federal natural resource agency, and in-house service to the department, college, and university. If you are seeking a position at a liberal arts program, there will be no formal Extension appointment, but there certainly will be expectations for professional service and service at the department, college, or university levels. Again, make this a positive component of your work duties, and discussions with mentors can help a new faculty member decide which service opportunities might provide the best investment of her or his time. Most of us dearly appreciate all that AFS has done for us and our careers, so make time for desirable AFS activities while getting credit for your work duties on your annual evaluation. Those of you who have served as an officer for your student subunit of AFS may wonder if employers will take notice. They certainly will, especially at traditional fisheries programs. Even in the more general ecological or environmental science programs, the value of professional society activities will be appreciated. Take all opportunities during your graduate program years to serve on search committees for open faculty positions; not only will you develop a useful skill set while getting credit for your service activities, but you can also observe prospective candidates and the variety of strategies they employ. The same is true for any other service activities, including departmental committees, college committees, and graduate student councils. The amount of time that you spend on service likely will be influenced by your assigned workload (i.e., Figure 1), which will determine how much weight is given to service activities at the time of your promotion and tenure evaluation.

Postdoctoral Experience

The number of students who step directly from a Ph.D. program to an academic position is relatively low. Thus, fully consider the need for and value of postdoctoral experience. As academic positions become increasingly competitive, an applicant's need for postdoctoral experience increases. Remember that experience in teaching, research, and service will all make you more competitive for a job interview and improve your interview performance because of the practical experience you have gained. We have two primary pieces of advice. First, recognize that not all postdoctoral positions are created equal! When searching for postdoctoral opportunities, try to find those with a breadth of job duties, rather than just writing research publications. Seek experience in writing grants, developing and managing budgets, and personnel management. Some postdoctoral positions might even allow teaching opportunities. Look for breadth in this experience, as well as the depth provided by highly focused research activities. Second, consider state and federal agencies as a source of postdoctoral experience. Before working in academia, we both worked for state natural resource agencies, which provided us with experience in Federal Aid in Sport Fish Restoration project proposals and reporting, interactions with user groups (e.g., anglers), and supervision of permanent or temporary employees. Fisheries faculty members often work closely with state and federal agencies in their research endeavors, and within-agency experience can provide valuable common ground that often promotes continued collaboration. When added to professional skills gained from our fisheries management and fisheries research duties, the overall experience of working for a natural resource agency was highly applicable in our future academic careers.

Collegiality

No discussion of finding an academic position and succeeding in promotion and tenure would be complete without considering collegiality. We often hear from our Ph.D. students that today's faculty members exhibit a high level of stress. We believe that this increased level of stress is common to many fisheries professionals these days, but in the case of faculty members, it likely is compounded by the multi-tasking required to be effective. No longer can faculty members concentrate on one research project and one peer-reviewed publication at a time. They have job duties that are highly varied and constantly overlapping and require the ability to multi-task. Not all people handle this multi-tasking effectively, and stressed individuals may not always interact well with other faculty and students. Most successful fisheries programs that we observe are those where faculty members make a conscious effort to be collegial. Recent faculty searches at both of our universities had interview assessment rubrics that included collegiality as a primary factor in the hiring decision. You will want to demonstrate that you fully recognize the necessity for collegiality during any interview for a faculty position. Academic freedom to express opinions and question established viewpoints is still important but can be accomplished in a positive rather than confrontational manner.

Persistence

Both authors can attest to the need for persistence. We both started our academic careers without the goal of attaining an academic position, but both came to the conclusion that such a career should indeed be our objective. As previously mentioned, both of us accepted positions with a state natural resource agency after completion of our formal education, which did two things for us. First, it continued our informal education, which we subsequently learned would continue throughout our careers. Second, those experiences provided us with postdoctoral experience that likely helped us successfully attain our academic positions. After all, practical experiences undoubtedly make us better instructors in an applied discipline such as fisheries. So, the old axiom applied: persistence pays.

In conclusion, we reiterate that academic careers have been extremely fulfilling for both of us. Do not let a tight job market determine whether you pursue this career option. Instead, decide whether or not this truly is the path for you, and then be sure you plan an appropriate educational pathway to arrive at your goal. Realize that most academic positions will have teaching, research, and service responsibilities and build skill sets related to all three of these duties. Complete coursework in educational principles and strategies, and gain as much practical teaching experience as possible. Ensure that you have the skill sets for research–to be competitive for grant funding, you likely will need both highly specific skills in a particular fisheries discipline, as well as general ecological skills. Do not forget that the scholarship of teaching and learning can lead to peer-reviewed publications, complimenting your traditional fisheries research. Some level of service, whether formal Extension programming or informal outreach efforts, will be expected of all academic professionals. Embrace your professional society, as the real benefits come to those members who are active and involved. Remember collegiality–it can trip up those who are not aware of its importance. Finally, be persistent. Our profession needs you, and we wish you well!

Acknowledgments

David Willis acknowledges mentoring from Steve Flickinger, Donald Gabelhouse, Jr., Chuck Scalet, and Richard Anderson. Dan Isermann has had the opportunity to work with many wonderful mentors, including Roy Heidinger, Phil Bettoli, Dave Willis, Don Pereira, Michael Hansen, and Brian Sloss. They thank Bill Fisher and Mark Pegg for helpful reviews during the U.S. Geological Survey review process and two anonymous reviewers for thoughtful comments during the peer-review process.

Dedication

Dr. David Willis passed away on January 13, 2014. Dave was one of the most respected fisheries scientists of his time and was an exceptional mentor to many students that now work as fisheries professionals. Dave's dedication to training the next generation of fisheries professionals made him an obvious selection as an author for this book, and many of the other authors and the editors benefitted from their interactions with this great man. Dr. Willis left an indelible mark on the fisheries profession, and his many students, friends, and colleagues will ensure that his legacy of excellence continues for generations to come.

Biographies

Dave Willis was distinguished professor and head of the Department of Natural Resource Management at South Dakota State University (SDSU). His B.S. and M.S. were from the University of North Dakota, with a Ph.D. at Colorado State University. Prior to his 26 years of service at SDSU, he worked for the Kansas Fish and Game Commission as both a fisheries management and then fisheries research biologist.

Dan Isermann serves as assistant unit leader of the Wisconsin Cooperative Fishery Research Unit (U.S. Geological Survey [USGS]). He earned degrees from Southern Illinois University (B.S.), Tennessee Tech University (M.S.), and South Dakota State University (Ph.D.). Before his employment with USGS, he was an assistant professor of fisheries at the University of Wisconsin-Stevens Point and previously worked as a fisheries biologist and research scientist with the Ohio Division of Wildlife and the Minnesota Department of Natural Resources.

Recommended Readings

International Society for the Scholarship of Teaching and Learning. Available: www.issotl.org/ (June 2013).

Journal of the Scholarship of Teaching and Learning. Available: http://josotl.indiana.edu/index (June 2013).

Murphy, B. R. 2010. Introduction. Pages 1–15 *in* B. R. Murphy, D. W. Willis, M. D. Klopfer, and B. D. S. Graeb, editors. Instructor's guide to case studies in fisheries conservation and management: applied critical thinking and problem solving. American Fisheries Society, Bethesda, Maryland.

National Center for Case Study Teaching in Science. Available: http://sciencecases.lib.buffalo.edu/cs/ (June 2013).

Vick, J. M., and J. S. Furlong. 2009. How to do an informational interview. The Chronicle of Higher Education. Available: http://chronicle.com/article/How-to-Do-an-Informational/44793 (August 2013).

Wuellner, M. R. 2013. Student learning and instructor investment in online and face-to-face natural resources courses. Natural Sciences Education 42:14–23.

Traits to Nurture

Salmon Research in Alaska: My Mentor Sparked My Career

KENNETH L. BEAL*
549 Washington Street, Gloucester, Massachusetts 01930, USA

Key Points

- Be prepared.
- Improvise when challenges arise.
- Praise work done well.

Field research in fisheries is challenging and rewarding, often demanding innovation in dealing with problems. Summer field work during college is a great place for mentoring and learning professional ethics. Mentors can shape the character of young professionals by their example, and mentees can use this experience to launch their careers. The following experiences are taken from my first professional fisheries job, illustrating the importance of being prepared and innovative. My mentor laid the groundwork for not only that summer's field work, but for my career as well.

In 1962, after finishing my sophomore year in wildlife management at the University of Maine, I headed north to Alaska for a temporary summer job as a fishery aide with the U.S. Bureau of Commercial Fisheries (BCF), the predecessor to the National Marine Fisheries Service. The supervisor of field research, Dale Evans, took considerable time to describe the office's ongoing research, and he explained how the work I would be doing with my two partners (Jack and Dick) fit into the big picture. He said we would be collecting 50 adult Sockeye Salmon *Oncorhynchus nerka* from each of the major tributaries of the Copper River, recording a broad array of meristic counts (such as the number of rays in each fin or scales on the lateral line) and morphometric measurements (such as length and weight). Scales would be collected, and we would take blood samples, which would be packed in ice and sent by air to the BCF laboratory in Seattle, Washington for electrophoretic analyses. The data would be used to test the hypothesis that each tributary in the river held its own race of Sockeye Salmon and that there should be some way to differentiate among them. This was a real-life problem, and I was delighted to be involved. Dale gave me a field logbook and encouraged me to record observations, especially anything unusual.

My field partners and I gathered equipment and checked each item to be sure it worked, and secured spare parts or duplicates in case of breakdowns. We then loaded a pickup truck with beach seines, gill nets, fishing rods, rifles, sampling jars for juvenile fish, camping gear, a couple of outboard motors, and a sectional boat that could be nested into itself and stowed in the Cessna 180 floatplane that we would be using for access to isolated areas at the heads

* Corresponding author: barney@kenbeal.com

of several tributaries. Then, with a small house trailer in tow, well stocked with food, we headed for the bush. It seemed like a dream come true: camping, fishing for salmon, boating, hunting, enjoying the beauty of Alaska, and getting paid for it, too! Did it get any better than this?

In fisheries field work, you sample when fish are available. The Copper River spawning run was just beginning, and it was critical to collect the blood samples from live fish before they spawned because all Pacific salmon die after spawning. As the spawning run got into full swing, our sampling was nearly nonstop. We worked for three weeks without a day off (although we took the afternoon of Independence Day to fish for Arctic Grayling *Thymallus arcticus* and Dolly Varden *Salvelinus malma*). One night, we even sampled until 3:30 a.m., using the light of the midnight sun. A few days after our mini-holiday, Dick flew to Fort Yukon to help with another project while Jack and I continued with the Copper River work on our own. We used the gill net to capture most of our fish, but some were taken using the beach seine, and a few fish we caught by hand in shallow water. As work progressed, we realized there were differences in the physical appearance of Sockeye Salmon from different locations and stages of the run, and I included this in the logbook as anecdotal observations.

Although one usually would not think a rifle would be part of the equipment for fisheries field work, in much of Alaska a rifle is essential to being prepared. We carried the rifle, just in case, whenever we were sampling. We saw bears often, some very close, but our only serious problem was back at our little trailer home on Tolsona Lake. A black bear tried to enter through the window over the sink, not once but three times. The first time, it ripped the screen and broke the pane of glass but left without entering; the next day it broke the repaired window again while we were visiting a neighbor. Later that day, Bob, our pilot, arrived and decided to spend the night, sleeping on an air mattress in the narrow corridor in front of the sink as the trailer only had two beds, one at each end of the trailer, which Jack and I had claimed. This bear had been a camp robber along the lakeshore and a regular visitor to the area where the local lodge disposed of its refuse.

Bob concluded that it was necessary to terminate it before something worse happened—to us or to others. Around 2:00 a.m., we three awoke together when the bear gently shook the trailer as it put its forepaws on each side of the window and looked in. Bob slowly lifted his head above the edge of the sink and locked eyes with the bear just four feet away. After a few seconds of staring, the bear dropped to the ground and shuffled off about 50 feet. Bob had the rifle close at hand and when the bear turned to look back, he shot it. While we were prepared for a bear threat, it was still an adrenaline rush when it happened.

The next morning I skinned the bear and cut some of the meat into portions. I salted the pelt and arranged to have it tanned, and still have this as a memento of our little adventure. The meat was quite good, especially with onions and wild mushrooms. But ultimately, we gave much of the meat to our neighbors as they had a screened-in meat-shed.

When our Sockeye Salmon collecting work on the Copper River was finished, Dale explained that my job would now shift to the tag-and-recapture program, the data from which would be used to determine salmon migration patterns and to help statisticians develop rough population estimates to add to the knowledge of their life histories. Dale and I took a johnboat up the headwaters of the Chena River where we spotted many salmon on the spawning grounds sporting colorful plastic discs, known as Peterson tags. We recovered tags from carcasses on shore and we even found some tags in bear feces!

As we motored along, we would occasionally frighten small flocks of red-breasted mergansers *Mergus serrator*, but instead of flying away, they ran at incredible speed across the water's surface. They were flightless until their molted feathers grew back. The wild beauty of the Chena was wonderful and I saw my first wolf track on the south fork.

Besides being prepared for the unexpected, it is essential to be able to improvise when disaster strikes. During the Chena River trip, as I was ascending a very steep-gradient and high-velocity channel, the steering arm of the outboard motor snapped off in my hand, leaving us rudderless and with no speed control. Quickly wrapping my arms around the motor, I was able to steer by wrestling it left and right as I maneuvered toward a sandbar. I pulled the choke to kill the engine just as we slid up the beach. The outboard was not repairable in the field, but fortunately, we had prepared for such a loss by bringing a spare motor, so we were able to continue our trip.

One evening after supper, as we sat watching the cook fire gradually fade to embers, I asked Dale whether many people had applied for the summer job I held and why he had selected me since my home in Maine was more than 4,000 miles away. Dale said that about 40 people had applied for the job and he had selected me because I was an Eagle Scout. He said that if he had to leave me alone in the bush for some reason, he would not worry about me being able to survive on my own. Scouting teaches young men how to live outdoors, offers them a variety of skills to be prepared, and builds self-confidence and sound ethics. I had no idea when I mentioned this Boy Scout accomplishment on my job application that it would turn out to be the critical factor in being selected.

Following the Chena River introduction to the tag-recovery program, I flew to the little village of Rampart on the Yukon River to help Ernie, another fishery biologist, recover tag data from salmon caught in the three fish wheels he was operating downriver from the village. A fish wheel is essentially a paddle-wheel on a log raft tied to the shore but held out in deeper water by long logs called stiff-legs. Each of the four arms of the paddlewheel is formed into a large basket with the mouth facing downstream when under water. As the water current forces the wheel to rotate, fish swimming upstream enter the basket's mouth and are caught as it rises. When each basket reaches its highest point, the fish are guided down a chute into a semisubmerged cage where they are kept alive until removed by dip net. Most of the catch was Chum Salmon *Oncorhynchus keta*, but we measured and recorded all the fish taken. If the salmon were tagged, we recorded their numbers and then released them to continue their journey to their spawning grounds, some of which were in Yukon Territory, 1,500 miles up the Yukon River from its mouth.

Jack, my former partner, and other biologists were now working on this project too, about 20 miles downriver, where they were using a fish wheel like ours to collect salmon, which they tagged and released after recording key data. One morning, Jack and two others arrived at our tent site by boat with three dead 40-horsepower outboards, all of which had broken down on the prior day. They borrowed one of our spares and then went to the village to ship the others to Fairbanks for repair. When they came back downriver, they took one of our fish wheels as theirs was badly damaged from drifting trees during high water. Bad times are sometimes worse than imagined, even when you think you are prepared.

When my summer job ended, I flew to Fairbanks for a final debriefing with Dale. He complimented my field work and the observations I had recorded in my logbook. He even said he hoped to have a job for me whenever I could return. That winter, he wrote me to report that the data we collected from the Copper River Sockeye Salmon had been analyzed and the hypothesis was valid. It would now be possible to sample Sockeye Salmon in the estuary and get a fairly accurate estimate of the portion of the run going to each tributary. He thanked me for my part in the success of the project and mentioned again the importance of our logbook information. Even though I felt my part in this effort was a minor one, I was delighted to receive his letter of appreciation and the information on project results.

I returned to Alaska for the summer of 1965 and once again worked under Dale's supervision, though I spent the majority of that time as captain of a 35-foot research vessel out of Kodiak. I had worked for six summers as a commercial fisherman on the Maine coast during high school and college, and this experience allowed me to pass the test for a U.S. Coast Guard captain's license, a prerequisite for Dale's Kodiak Island vacancy. Our field work was on the north shore of the island, evaluating habitat and salmon runs in the Terror and Kizhuyak rivers.

Ten years later, Dale once again became my supervisor, this time at the Washington, D.C. headquarters of the National Marine Fisheries Service where he served as Environmental Assessment Division Chief. Dale was an outstanding mentor in all of these jobs; his guidance, wisdom, professionalism, diplomacy, sense of humor, and his ability to keep cool under pressure helped to shape my leadership and management style, and I really appreciate the start that he gave my career. He valued hard work and exceptional performance, and his commendations and pats on the back were gratefully appreciated by everyone. I am pleased to say that we still keep in touch, even though it has been more than 50 years since he hired me for my first professional fisheries job.

Each person prepares for his or her career independently, but sometimes one's career path does not proceed in the normal and predictable direction. When unexpected opportunities arise and you are prepared to accept the risk of trying something different, your career may be changed dramatically. My training, education, experience, and job choices may have been uniquely my own, but I believe being prepared and having an innate curiosity about the natural world were key factors. Robert Frost addresses these choices at the close of his poem "The Road Not Taken":

> I shall be telling this with a sigh, somewhere ages and ages hence:
> Two roads diverged in a wood, and I, I took the road less traveled by,
> And that has made all the difference.

My message to young biologists is to seek every opportunity to gain a wide variety of skills and experience wherever you can, even by taking volunteer jobs if necessary, and learn all you can from your mentors. My message to mentoring fishery biologists is that you have the opportunity to shape the careers of young biologists who come to work for you, and both you and your mentee will benefit from this association. I thank you for your leadership.

Acknowledgments

The author gratefully thanks the American Fisheries Society for permission to use portions of his January 2002 President's Hook column titled "Mentoring in the Field" (*Fisheries* 27(1):4).

Biography

Kenneth Beal retired from the National Marine Fisheries Service where he held supervisory positions in environmental assessment, fisheries management, fisheries development, and cooperative research. He has a B.S. in wildlife management from the University of Maine and an M.A. in marine science from the College of William & Mary/Virginia Institute of

Marine Science. Ken has held many volunteer and elected positions with the American Fisheries Society (AFS), including Fisheries History Section president, Marine Fisheries Section president, Northeastern Division president, and AFS president. He also served on the society's governing board for 11 years.

Give, and You Shall Receive

CHARLES (CHUCK) C. COUTANT*
Oak Ridge National Laboratory (Retired)
120 Miramar Circle, Oak Ridge, Tennessee 37830, USA

Key Points

- Showing enthusiasm and interest attracts good mentors.
- We are both recipient and donor of mentoring throughout our careers.
- Mentoring goes beyond hard science.

Introduction

In the winter of 1958–1959, I was a junior-class student stuck in routine undergraduate biology courses when what I really liked was being outdoors poking around for critters in streams, lakes, and ponds. That was something I did in most of my spare time as a kid. A college major in biology would supposedly let me do similar things for some sort of career. Most of my classmates were pre-med, and I sometimes went along with that to justify my biology major. But it was not what I really wanted. When I learned that my genetics professor was doing consulting work in lake and pond field studies, I mustered my naïve enthusiasm for a talk with him about field biology. A career was launched!

Also launched was a life-long appreciation that an enthusiastic and interested student is a magnet for good mentors. We are truly students all our lives, not just as academic students. When you have and show a genuine interest in what others are doing, those expressions will be reciprocated by an interest in you. We have all heard the well-known adage "smile and the world smiles with you; frown and you frown alone." Those who reach out to others, showing interest and enthusiasm, however naïve, will attract helpful mentors and collaborators, whereas those who do not will likely go it alone and miss out on many career opportunities. Give enthusiasm and interest and you are apt to receive it in return.

My Enthusiastic Start

Whether a result of my somewhat naïve expressions of interest that winter, my subsequent willingness to volunteer in field and laboratory with my colleagues, or just plain good luck (e.g., being in the right place at the right time), my career in aquatic ecology was launched and well tended by numerous mentors with whom I worked side by side. Having ploughed through the required courses for a B.A. degree in biology at Lehigh University into my junior year, I was ready for something more focused, more directly related to my real interest in aquatic biology. The "Aha!" moment was the realization that several professors, particularly Dr. Francis Trembley, professor of ecology (a rare title in the 1950s), were doing consulting

* Corresponding author: ccoutant3@comcast.net

work in lake and pond field studies as well as teaching. I asked Dr. Trembley if he had any summer work that I might do with him. At that point, I really did not know much about what he was doing or what that might entail for me: could I volunteer for some fieldwork, be hired for a summer job, or just talk with him? In hindsight, I was really naïve, but I expressed interest and enthusiasm to a person who seemed to be involved with the sort of biology that was my passion. Events flowed from there. Because I had approached him informally, Professor Trembley and I conversed differently than we had as student and professor for a class. I was a colleague. We shared our enthusiasm for "natural science" and out-of-doors biology. As a result, I was introduced to the naturalist tendencies of other faculty, such as Dr. Bradford Owen, who taught comparative anatomy, histology, and embryology (not exactly ecology-oriented!). To my pleasant surprise, he was also a periphyton specialist working on diatoms from the Delaware River. I would later (early 1960s) serve as his teaching assistant, and he would become my Ph.D. dissertation advisor.

As a result of these collegial interactions with many of the biology faculty, I was selected to be one of two participants in a new summer National Science Foundation (NSF) Undergraduate Research Program in ecology between my junior and senior years, for which the university had applied and received funding. At that point, I did not know what NSF was, or anything about applying for grants (I later learned that we were firsts, both for the university and for NSF). It was a life-changing event for me. During my summer of undergraduate research (1959), I was introduced to and mentored by a number of graduate students, particularly Joe Mihursky (now Dr. Mihursky), who had been around the department but out of sight from us undergrads. Joe was a student of Dr. Trembley and soon became my primary mentor for field techniques. I learned to take Surber samples of invertebrates, run gill and trap nets, do rotenone sampling, pick bugs, and identify fish—only part of which was for my independent research on Delaware River benthic invertebrates affected by a thermal discharge. We students, graduate and undergraduate, worked together on several projects, helping each other and learning in the process.

Looking back, I have to say that a new world of opportunity had opened up that summer, all because I spoke up and expressed interest and enthusiasm the preceding winter and spring. I then gave that enthusiasm substance with a summer of hard work, which also involved lots of fun in a community of faculty and graduate student mentors.

Not every student will be as fortunate as I was in finding such compatible mentors so readily. It takes knowing what really interests you (not so much as a career path, but as a personal interest), looking for faculty or researchers with common interests, and expressing those interests with enthusiasm. Anyone can do it: be brave, approach people, search for common interests, and be adaptive (if wrong, try again; we cannot predict what career doors will open). Follow that with a good work ethic. Finding potential mentors of like mind will likely require digging beyond their course offerings and into their research and consulting activities. This should not be difficult for any student to do.

Research for National Needs

Close and congenial mentors guided me into topics with national needs, which I embraced enthusiastically and which were perhaps critical for my long-term career success. Faculty, graduate students, and I shared common interests in environmental conservation beyond strict academic disciplines. In my case, it was the emerging environmental issues of thermal pollution and effects of dams on river systems. My NSF undergraduate research project was sampling macroinvertebrates at a power station thermal discharge as part of a broader team effort involving faculty and graduate student research that included fish and periphyton. I published the macroinvertebrate work as my master's thesis. During my master's course work, I volunteered to help faculty and other graduate students with fisheries and water quality studies on Green Lane Reservoir in southeastern Pennsylvania. That led me to apply for and obtain a Public Health Service (precursor of the Environmental Protection Agency) fellowship to pursue my doctoral dissertation comparing stream plankton upstream and downstream of the reservoir. My science was applied, and it generated valuable interactions with (and mentoring by) people outside the university, ranging from environmental activists to regulators and company staffs. Because of close and congenial mentoring by faculty and graduate students, I stayed at Lehigh for all three degrees.

From Mentored to Mentor

With time, we find ourselves shifting from mostly being mentored to being the mentor. It is predictably natural for a well-established, mid-career professional to want to support and nurture the genuine interests and enthusiasms of younger individuals wanting to enter the profession. Mentors *want* to open doors for younger protégés who share their interests. It is not an imposition to try to hook up with them.

As I grew from mentored to mentor, it has been my good fortune and pleasure to host numerous interns and younger colleagues for participation in my research. Those who showed interest and enthusiasm garnered my attention and commitment. For many years, particularly in the 1970s and 1980s, I hosted undergraduates in my research program, inviting them to participate for semester-long research projects through arrangements between my employer, Oak Ridge National Laboratory (ORNL), and several colleges and universities. Students applied for these "science semester" positions at their schools (meeting the qualifications of both the school and ORNL), and I selected among those who expressed interest in aquatic ecology. They came for a semester (or summer) of research for college credit (for details about these programs, you can contact Oak Ridge Associated Universities, www.orau.org). Once on site, we sat down together and discussed the broad scope of our research, and the students chose specific projects of interest to them that would contribute to the overall effort. I would mentor the students through the study (sometimes in the laboratory, sometimes in the field, often both) and guide them through writing a report and giving a seminar. In numerous cases, the research was of high enough quality to publish, and we worked together to create a publishable draft from their research reports (often aggregating several related student studies) for submission to a journal. If you are starting out in your career, you should know that such opportunities are available and understand the steps one goes through to take advantage of them.

I find it interesting to see which students took well to the science semester process and which did not and to reflect on my responses as their mentor. Most students came with the enthusiasm one would expect from anyone who had taken the initiative to apply for a research semester away from their college. They would "jump in with both feet," so to speak, and I would readily help them. They were there every day doing really professional work. There were, fortunately, only a few at the other end of the scale. From the moment these few students came, it was obvious that this type of research was either not to their liking (in which case I helped them find another host) or they had an arrogant attitude. One student in the latter category made it clear from the outset that I was there to serve him, give him a good grade, and advance his career. He came with that attitude, shrugged off my attempted mentoring (admittedly unenthusiastic given his attitude), and left with the opposite of what he wanted. Clearly, this is the opposite of interest and enthusiasm, and it was not a productive experience for either of us.

Lifelong Exchanges of Mentoring

We all need to recognize that we are mentored throughout our lives. The give-and-take mentoring by enthusiastic professional colleagues at all levels, as well as by students, has been invaluable for me. We learn from each other. As in our student years, showing interest and enthusiasm with colleagues leads to collaboration and mutual mentoring. For example, I know I have learned from and (hopefully) helped my most recent work partner, Dr. Mark Bevelhimer, who now is the senior aquatic/fisheries person in my former laboratory. We still stay in touch after I retired.

We never grow out of being mentored, and it is not only in science. My career has often had as many interpersonal and policy interactions as scientific ones. Few of us will be doing only science in the realm of fisheries. We interact with others in both science and its application to the management and utilization of aquatic resources. We develop, implement, or critique resource policies. For example, I have served on a number of advisory committees that dealt with applying science to resource policy issues. Nearly every committee had one or a few members who gently mentored the rest of us in the best way to diplomatically achieve our objectives. Most notable was my 16-year service (1989–2005) on the Independent Scientific Advisory Board (and its predecessor boards) for salmon restoration in the Columbia River basin as mitigation for the effects of the basin's many federal dams. Salmon management in the Pacific Northwest is fraught with contentious policy issues. Over my terms, I developed ever-increasing admiration for a mentor on these boards, Dr. Dick Whitney, a retired professor from the University of Washington, who would remind us in the midst of fisheries policy debates that "Fisheries is a social science." And, truly it is, especially because success requires learning all sides of contentious policy issues and coming to reasonable conclusions. An enthusiastic interest in the views of others will encourage them to inform you, not just debate you.

Summary

Expression of interest and enthusiasm is extremely important for receiving career opportunities. My examples encompass and demonstrate both being mentored as a student and mentoring others as a committed professional, for the message is the same. As a student, having an idea of the type of career work you find interesting, finding a faculty member (or other potential mentor) with similar interests, and making your interests known in an enthusiastic manner can lead to unexpected career opportunities. It helps to seek mutual mentoring from other students by volunteering to help them with their projects (and they with yours). Although we tend to think of mentoring as a faculty-to-student activity, we will be mentored and do mentoring throughout our careers. Some of it will be in science (e.g., research, teaching) while other opportunities will likely arise in applying science to social needs (e.g., resource policies). Through it all, the interest and enthusiasm that one gives to others is like a gateway to receiving opportunities and career success: give, and you shall receive!

Biography

Chuck Coutant retired in 2005 from the Department of Energy's Oak Ridge National Laboratory in Tennessee as a distinguished research ecologist. His research specialties were the aquatic environmental effects of power generation, including thermal and hydropower. He focused especially on Pacific salmon and Striped Bass *Morone saxatilis.* He was president of the American Fisheries Society in 1996–1997. His career continues through part-time consulting.

The Scientist's Journey: That Was Then and This Is Now

WILLIAM G. FRANZIN*
Laughing Water Arts & Science, Inc.
1006 Kilkenny Drive, Winnipeg, Manitoba R3T 5A5, Canada

PAUL M. COOLEY AND PATRICK A. NELSON
North/South Consultants, Inc.
83 Scurfield Blvd, Winnipeg, Manitoba R3Y 1G4, Canada

Key Points

- Scientists evolve as they complete formal education and move on to positions in their professions.
- Recognizing promising opportunities and taking advantage of them is a hallmark of successful scientists.
- Be true to yourself, find a good mentor, be an attentive mentee, nurture and improve your strengths, and acknowledge and correct your weaknesses.

Introduction

Successful scientists pursue ideas, both good and bad. They critically review associated historic data and theory and integrate them to formulate testable hypotheses. Becoming a scientist requires intensive study leading to a graduate degree, a unique and existential process both within formal educational institutions and, by virtue of graduate work experience, in the real world. It is a process through which we strive for a high degree of ethical behavior, scientific rigor, and objectivity with, as its reward, the highest principle in education, academic freedom. A flourishing mentoring environment is a critical component of a graduate education and crucial for both mentors and students. Mentoring enriches both students and mentors, as together they realize opportunity.

Life is about opportunity. They often occur in unlikely places or circumstances. To be successful in life, one must learn to recognize and take advantage of promising opportunities whenever possible. A successful mentor–student relationship is the mutual realization of opportunities through the process of self-examination and growth. Perhaps the single largest contribution that mentors make to students is the understanding that there are many gray areas in science, and in that respect, science is no different from life. The process of mentoring in science involves providing the formal evaluation structure, instilling the importance of a larger scientific community, and establishing the importance of historical context. These elements allow the student to not only recognize an opportunity, but also communicate intent and ultimately understand the perspective and scope of issues at hand.

* Corresponding author: wgfranzin@gmail.com

In essence, science provides a structured framework within which we are supposed to learn to be open-minded and objective about problem solving. The headings below represent selected scientific tenets that are not exhaustive but provide a basic flow of science from perception to application as we have come to understand it.

Some Scientific Ideals We Consider Focal

Life provides many opportunities, but not all of them will be obvious. Some of these opportunities may entice you to increase the scope of your work or even change your field of study. These are not trivial decisions as they will influence not only what you do, but may influence how you undertake your work, including how effective you may be. In the following sections, we provide a few key scientific tenets that are common to many scientific approaches. These scientific pillars of understanding should help you learn to recognize and seize opportunities, whether they are long-term research avenues or a moment in a discussion, and yet maintain a balance with the rest of your career.

Perception and Scientific Scope

The scope of scientific interest for many of us is broader than a single researcher can reasonably entertain. The scientist must balance the desire to study everything against the constraints that reality imposes on what can be attained. When new opportunities arise, the scientist must consider the work needed to learn new skills required to take full advantage of the opportunity. When the skills required are near the margins of his or her analytical skill set, he or she must be prepared either to expand or reach out to other scientists. Forming a research group or networking internationally allows the scientist to team up with other like-minded scientists with complementary skill sets. Another fundamental aspect of perception and scientific scope deals with a continuum of ideas that ranges from applied to theoretical. One's perception of scientific inquiry is directly influenced by career path, and the usefulness and appropriateness of the plethora of methods and techniques will vary depending on whether one is focused on management or strict academic research. Both lines of work are required in the fisheries profession and act to keep each other in check insofar as what is considered practical, in terms of management, and what is possible, in terms of research. Typically, management might ask questions and provide funding while academic research would seek answers. Regardless of where you find yourself, the logic of scientific process permeates though all branches and inevitably into our day to day life.

Science as a Process and Way of Life

Science is not for everyone, and the logic of scientific inquiry provides a framework within which scientists ask and answer questions. Scientists are, for the most part, inquisitive people, and the scientific process provides a way of structuring their thoughts on both simple and complex questions. In this regard, the scientific process provides the framework for organizing thoughts but, more specifically, provides a way to ask key questions and determine why certain answers make or do not make sense. This process directs the everyday life of the scientist and offers them a way to deal with life and its challenges. The scientific community is composed of broad and diverse schools of thought, and many of the relationships you will develop will be based on common interests. Some collaborations, however, will be with scientists of different skill sets and different branches of science; these larger collaborations are where the meaning of teamwork is defined.

Science Is Teamwork

When a scientist finds that his or her path leads to interesting opportunities with others who are mutually supporting and become an integral part of his or her skill set, the individual may become inter- or multidisciplinary. Alternatively, one may be the type of researcher that enjoys more focus and depth of study. Both types of scientist are surrounded by the scientific community, which serves as their professional resource, a fabric of knowledge upon which to draw. Their perception and scientific scope may influence where they fit into the scientific fabric and also the type of scientists that are academically adjacent to them. Scientists generally are dependent on communication with colleagues, some more so than others. Communication is of paramount importance in science, but not all scientists communicate well in all mediums. It is essential for scientists to learn to write well because publication in the peer-reviewed literature is the measure of scientists' recognition. It also is important to be able to speak capably, debate with your peers, and prepare good proposals.

Think carefully about the kind of place in which you would like to pursue your career. This choice may determine your research scope or mandate, and with whom you may engage at local, national, or international scales. If the first workplace opportunity does not prove to be for you, you must be prepared to move on to a more appropriate place. Talk with peers and mentors about opportunities that come your way. The decision of your career path or direction will require you to critically assess opportunities as they emerge.

Critical Thinking

Critical thinking characterizes the ability of a scientist to maintain a balanced and objective point of view. Indeed critical thinking is of vital importance to the scientist. Perhaps most importantly, the ability to maintain balance between pursuit of subjective areas of interest and the more pragmatic methodical approach that constitutes research allows a scientist to be fully self-aware and cognizant of the array of choices presented by scientific quest. Somewhere between scientific wishful thinking and real world problems the scientific method is applied to find the truth. Because of this broad spectrum in science, scientists from all the various schools of thought do not always interact smoothly. Critical thinking necessitates that scientists gauge their dependence on colleagues, know where they fit in the larger scheme of science, and realize they may fall short of their objectives. It should not be surprising that if you push the frontier of science forward that it seems to push back. Science is conservative and epitomizes inertia. It requires rigorous, convincing proof to move the needle and change a current theory. This makes communication an essential tool.

Communication

Communication in science includes sharing knowledge and sharing the scientists` points of view. Communication can take many modes: meetings, conferences, scientific publishing, the ultimate completion of a particular piece of research, communication among peers both locally and globally via e-mail or voice, and finally communication to clients, stakeholders, regulators, and the general public. Without the support of the latter groups, very little science will be accomplished as they ultimately pay the bills. Communication is a vital part of the scientific method and the advancement of knowledge. Information once took months or even years to reach all of the intended audiences. Nowadays, one may be in contact with the global scientific community within minutes by posting a communication on popular Internet media such as LinkedIn, Twitter, or Facebook. Even primary scientific publications are submitted, reviewed, revised, and published in the paperless electronic media in a matter of months or less. Communication literally comes down to language and words, and with the diversity of people and the ever growing need for integrated approaches, words can create uncertainty.

Uncertainty

Uncertainty is inherent in all data, but it does not arise only from sampling error or environmental variation. Most often, uncertainty is considered from the perspective of natural variation, data completeness, or model accuracy and precision. The most common sources of uncertainty actually may arise long before sampling is considered; it may be due to subjective judgment, miscommunication, language barriers among disciplines, vague definitions, ambiguous words, or use of words that depend on context. Uncertainty will always be present whether in a discussion or numerical analysis. Try to become accustomed to this; recognize uncertainty when it arises, and manage it (so it does not manage you) by identifying sources of uncertainty explicitly in research plans.

Understanding uncertainty is vital to critical thinking and in communicating ideas and results to your peers. Uncertainty can obscure objective thought if it is left unchecked. As such, it is both an adversary and a close friend. It may create ambiguity that makes progress difficult, although during the course of a career, this challenge builds perceptive skills, and over time, a more objective and forward thinking scientist evolves. Without disclosure of the apparent sources of uncertainty in your science, it is certain that your conclusions and those of other scientists may not be appreciated as well as they otherwise might be. This may mean you have to spend more time in conversation to ensure that an idea is conveyed well, or in a presentation to clearly list the numerical uncertainties as well as the things that are not feasible to sample. Most seasoned scientists will recall specific key moments in time when uncertainty devalued their efforts. This can occur during development of a hypothesis, or an incomplete presentation of a well thought out study, or both. Recognize that uncertainty is an ever-present challenge and that management and discussion of it is a part of the scientific process.

Summary

Your perception of science and the scientific process is likely to change over time as life experiences reshape your understanding. Table 1 shows how our perceptions have changed for a few key scientific tenets at the beginning of our graduate studies compared with our thoughts about 10 years after completing our Ph.D. degrees. This table is neither complete nor exhaustive and does not do justice to the real impact that the mentor–student relationship had. It does show how our perspective changed as our scientific experiences grew, and it highlights our different backgrounds and where our ideas as evolving scientists converged.

Three Scientists' Journeys

What was our vision of the path ahead as we entered graduate school? Discussions with our professors, fellow graduate students, and families probably did not really prepare us for what lay ahead. Serious soul searching is in order as one prepares for a career in science.

TABLE 1. That was then and this is now: the changing perspectives of two scientists over time.

Scientific tenet	Paul Cooley (circa 1994)	Patrick Nelson (circa 1995)
The scientist's journey	Then: The scientist's journey follows suit in an organized world where building blocks of science help society understand complex issues. Now: Science is relatively simple. It is the people that are complicated and slow things down.	Then: The scientist's journey is something between Indiana Jones and David Suzuki. Now: Science is relatively simple. The social context and communication are complicated. People and their motivations are the most complex issues to handle on this journey.
Perception and scientific scope	Then: I am interested most in biology but geography seems to explain many of my ecological questions. Now: I feel the same, but I need colleagues more than ever because my interests are multidisciplinary.	Then: I am a parasitologist who is interested in fish and philosophy. Now: I am a fish biologist and philosopher who is interested in parasites.
Science as a process and way of life	Then: Science is very important, poorly understood, and well respected. I would like to be a scientist because I love water and everything about it and in it. Now: It is important to understand science as a process for your personal journey and your contribution to society as a learned professional.	Then: Science has the answers to the questions (yes, I truly believed this). Now: Science provides answers to questions. Not everyone likes the answers.
Science is teamwork	Then: I can understand most things and am not lazy so I can do pretty much everything myself. When needed I may have to occasionally recruit help. Now: I need and want to interact with peers. Most of them are nice, too.	Then: Peers were friends and classmates. Now: Our successes depend heavily on peers. I consider many of them friends.
Critical thinking	Then: Critical thinking is everything that is analytical. Laboratory and analysis stuff. Now: I did not know that I should be self-evaluating much more, but few opportunities passed me by.	Then: Critical thinking involved a keen mathematical mind, structure, and broad knowledge base. Now: I feel the same.
Communication	Then: I will publish and thereby help others. Now: What was I thinking?	Then: Communication was paramount, particularly the use of the appropriate words. Now: I feel the same.

Table 1. Continued.

Scientific tenet	Paul Cooley (circa 1994)	Patrick Nelson (circa 1995)
Uncertainty	Then: The result of environmental variation and sampling incompleteness. Now: Use of different words in different disciplines for the same thing makes it hard to communicate. This is harder to solve than numerical uncertainty and keeps coming back.	Then: Uncertainty was a mathematical concept that could be demonstrated and overcome. Now: Uncertainty is everywhere and most often has to do with language and use of words. Mathematical uncertainty is rather simple in comparison.

A Mentor's Path of Opportunity: Bill Franzin

> [C]arpe diem, seize the day boys, make your lives extraordinary.
> —*Dead Poets Society*, 1989

An adjunct professorship in the Zoology Department at the University of Manitoba provided my opportunity to engage with graduate students. In turn, the students provided inexpensive, committed, and enthusiastic research assistance. All they required were their stipends and a way to integrate their degree requirements into my research, and then they were let loose. I had the equipment and supplies, and they had the labor and motivation. Graduate students took a little more guidance than trained technicians or experienced biologists, but their dedication to accomplishing their own research and career goals more than made up for the time I spent with them. In addition, through their studies, I learned about the latest approaches and techniques from keen young researchers. Debates about the course of our joint work and ideas for new ventures were a constant feature of our relationships. Graduate students never were concerned about overtime pay, working on weekends, or making tedious journeys by vehicle, boat, or aircraft to remote places.

I did not always directly supervise the graduate students with whom I worked; sometimes I provided the ways and means to do the research with students supervised by other professors in other departments. This article is about the joint opportunity I had in the research careers of two former graduate students, both now colleagues gainfully employed by the same consulting company where they work together on projects related to the research we did together. My modus operandi was and remains "carpe diem," and the following account details this influence from their joint perspectives.

I never thought of myself as a mentor but rather a facilitator, facilitating students in exchange for the assistance I needed and did not have. However, in retrospect, I realize that I accorded the students who worked with me the same considerations I had received from my own mentors a generation before. I did not appreciate my graduate supervisors as mentors at the time, but indeed they were just that and good ones too.

Growth of the Student Mind

> The mind, once expanded to the dimensions of larger ideas, never returns to its original size.
> —Oliver Wendell Holmes

Paul Cooley: I considered my scientific interests most often during my undergrad and the early part of my graduate studies. During this time, I was exposed to new aspects of science and its many opportunities. Many of these occurred outside of university. I was attracted to Bill Franzin's laboratory at the Freshwater Institute as I thought he was a broad thinker. This was an opportunity to work directly with a scientist in the Canadian government because Bill's support biologist position remained vacant due to staff reductions. I was fairly independent and knew that he would not hover over me or pull on the reins too hard. Bill and I work well together and his laboratory would be a place where I could develop and literally follow my nose while supporting Bill's federal government research mandate. At this time, my perception of science was rooted in biology but was trending to interdisciplinary interests. Bill provided two major turning points in my direction; one was a simple habitat-mapping exercise, which, although I did not know it, was the start of my push into geography. The second was when I was told that a habitat study I had designed and partially completed for him was worthy of a master's degree. Neither of these opportunities seemed particularly noteworthy when they first surfaced.

Teamwork, critical thinking, communication, and uncertainty were most important to my journey when I joined an international project in Africa for my doctorate and continue to be so today. The project leaders were of international caliber and far more experienced than I was so it was important to make communications count when I actually had their attention. Near the end of my Ph.D. research, I learned that interdisciplinary study, although rewarding, had spread me thin and was a lot of work. Uncertainty hit hardest at both the beginning and at the end of my Ph.D. I did not see the invitation to join the international project coming (but one of my mentors did). I wrestled with the decision for weeks not only because of the commitment, but because my first trip overseas would be while I was in the middle of my master's program. Later, I accepted a term opportunity to continue the thrust of that project, but funding allocated to the program was later pulled. I felt a sudden awakening and a taste of the uncertainty of international work during a time when I was starting a family. I decided to continue to work in the temperate areas with which I was familiar and where the networks of people were best known to me. Also, my family was becoming paramount. The stakes were too high to continue to independently travel the globe at will. It is advisable to complete your graduate studies as quickly as possible.

Patrick Nelson: I became focused on zoology after seeing my physics laboratory partner studying frantically for a fourth-year biology of fish laboratory exam. I realized that there were whole courses devoted to areas of biology that I had thought were cool like parasites, fish, evolution, and even microscopes (shout-out to Erwin Huebner).The first couple of years of university involved stuffing as much information into my head as possible. My mind wandered quite a bit until I took an introductory parasitology course with Dr. Terry Dick—*ecce voila*! I thought I had found my calling. I worked hard for the next two summers doing parasite necropsies and culturing Lake Sturgeon *Acipenser fulvescens* for Terry`s research. Some uncertainty in funding led me to Bill Franzin's laboratory at the Department of Fisheries and Oceans Freshwater Institute as he had an opening for a master's project on fish. Bill and I got along well from the start. At about the same time, I met Paul Cooley who was working on habitat mapping in lakes. Seeing the spatial pattern of habitats suddenly made sense of all the patterns in fish parasite communities I had noticed. I knew at that moment that Bill's laboratory was where I needed to be. Bill provided me the opportunity to follow my nose and the opportunity to continue my interests in parasitology. He allowed me the mental freedom that I needed to really begin doing science. There were students everywhere at the

Freshwater Institute, and I immediately recognized the importance of teamwork and communication.

My realization of the level of uncertainty in science came when the federal government began downsizing in 1996. There were significant cuts to programs and funding, and it was far more widespread than what had led to me to Bill's laboratory in the first place. I mostly considered uncertainty as simply a statistical thing that needed to be defined and, when possible, accommodated; the adjustment for socioeconomic uncertainty was far more complex. However, this uncertainty also provided opportunities to refocus some of the research to strictly mandate-based work during the summers while working on my thesis on the side. Although this caused inevitable delays in progress on my thesis, I had the opportunity to transition from a master's into a doctoral program. The exposure to larger federal initiatives and the national and international research communities provided a peek into the real world of science and collaboration.

Over the next few years, my research focused on habitat and resource partitioning and their link to biogeography. Of course, parasites were an ever-present part of my research, but they were not focal to my thesis. Life added some complexity with the arrival of children and the disappearance of funding. I immediately dropped off a resume to my present company. They happened to have a unique opening that required my exact qualifications and experience. To my surprise, during introductions to the staff, there was Paul Cooley. The two of us put all of our "how would we do this differently and better" ideas to work and we are still doing it.

Conclusions

We have learned that university training helped all of us to understand science, but that the mentor–student relationship is a key formative period where our mentors showed us science as a process, which is not easily learned from books. Since then, and as practitioners ourselves, we have observed that science in today's world is not a numbers factory or a paper mill. It certainly provides answers, but those answers inevitably carry uncertainty. At the beginning of our undergraduate days, we could not have anticipated how challenging and yet important the larger community of scientists is in moving science forward. People are needed at every step of the way to help develop objectives, provide critical review, and develop and guide policy, as well as design, implement, and evaluate projects in the real world. To be successful as a scientist, one must have a strong foundation developed from a core set of skills that are loosely referred to in the tenets described above. It is important for a scientist to refine these skills so that the opportunities that are unique to their scientific journey can be realized and turned into shared knowledge. It is important to be a mentor on the scientific journey, not only because it creates opportunity, but because mentors help to shape the next generation of scientists. One of the most interesting aspects of mentoring or teaching is the effects mentees can have on mentors; it is a mutual growth process based on feedback from peers and students.

Biographies

William G. Franzin received his Ph.D. degree from the University of Manitoba in 1974 and worked as a research scientist and manager with the Department of Fisheries and Oceans Canada in Winnipeg for 33 years. For about 20 of those years, he also was an adjunct professor in the Zoology Department at the University of Manitoba. Association with graduate students opened unexpected opportunities in research programs, particularly during government downsizing exercises when research assistant positions happened to be vacant. He retired from the federal government in 2008 and presently operates a small consulting firm. He has been active in the American Fisheries Society for more than 30 years and served as the society's president in 2009.

Paul M. Cooley received his Ph.D. degree from the University of Manitoba in 2004. He studied the relationships between habitat and cichlid species distributions and patterns of species richness in rock, sand, and depositional habitats of Lake Malawi, Africa. Before that, he completed his master's degree on nearshore habitat and hydrodynamic models. He also worked with Bill on his undergraduate thesis, which studied the effect of lake size on growth of Walleye *Sander vitreus* at a single latitude. Since 2001, He has worked in the field of environmental assessment where he does research on water level and flow impacts on fish habitat quantity and quality, predictive models for habitat and species, and works closely with hydroelectric utilities and species at risk regulators to understand the suitability of spawning habitat for Lake Sturgeon in the tailrace of hydroelectric stations. He is an adjunct professor, participates as a mentor in the development of student thesis projects, and participates as an advisor on student committees.

Patrick Nelson received his Ph.D. degree from the University of Manitoba in 2005. He had the support of a four-person committee who encouraged him to follow his nose. He began fulltime employment in 2004 and over the past 9 years he has focused on environmental assessment work for major hydroelectric developments. His foci are biogeography, instream flow needs, fish–habitat associations, production models, population genetics, and population models. His research is entirely applied science and has involved species-at-risk issues on Lake Sturgeon, fish-stranding studies, and development of sampling methods for large rivers. The goal of his work is to provide real world answers to applied problems and provide end users the tools they need to make sound management and project decisions.

The Importance of Mentoring and Adaptability among Fisheries Professionals: Learning Survival Skills from Coyotes

WAYNE A. HUBERT*
Department of Zoology and Physiology, University of Wyoming
1000 East University Avenue, Laramie, Wyoming 82071, USA

PAULA GUENTHER
Parks Ranger District, Medicine Bow-Routt National Forests
100 Main Street, Post Office Box 158, Walden, Colorado 80480, USA

DIANA MILLER
Wyoming Game and Fish Department
420 North Cache, Jackson, Wyoming 83001, USA

Key Points

Fisheries professionals must be able to adapt in our current climate of constant and accelerating change. The survival skills needed are not so different from those used by coyotes.

- Critical thinking, cooperation, and communication skills are necessary to adapt and succeed.
- Students and young professionals need to seek mentors who recognize the value of diversity and the need for a wide array of skills in order to adapt to changes into the future.

Introduction

We live in the West where we routinely see and hear coyotes. Without a doubt, their ability to adapt to changes in land use, human encroachment, and climate has allowed them to become one of the most successful species in North America (see Box 1). We have to admire them. Coyotes begin their lives as part of a family group, often called a pack. Family groups are often made up of parents, the new litter of pups, and older siblings. Through this family group, pups learn survival skills. Probably the most important survival skill is the ability to hunt for prey. This is learned by observation of adults and older siblings. Coyotes generally hunt alone for whatever small prey may be available. However, pups are also taught how to cooperate as a group to hunt for larger prey. From cooperative group behavior, the whole pack benefits.

* Corresponding author: whubert@uwyo.edu

Box 1.

Coyote Teaches Salmon to Sing

One beautiful summer day, Coyote trotted along the river, feeling warm sun on her fur and enjoying River's melody among its pebbles. She was also keeping an eye out for food, since she had four hungry pups sleeping nearby in a den beneath an alder tree.

Coyote was putting red raspberries in her pocket for the pups when suddenly she heard a great splashing, and looked out to see Salmon thrashing in the shallows.

"Brother Salmon, what are you doing, working so hard today?" Coyote asked.

Salmon hardly looked up from his efforts, flapping his tail and fanning his fins in the clear water. "I've got to get home. Home is upstream, and I have to swim home," he muttered. Coyote knew Salmon and his family made this journey every year, determined to raise their young where salmon had always hatched.

"Ah-woo, so true," agreed Coyote, and she hopped across River's rocks, singing as she went. "What a beautiful day, ah-woo," Coyote sang, snapping up bright green grasshoppers and putting them in her pocket. When her pocket was full, Coyote turned back toward the river. To her surprise and sorrow, Salmon was very near where she had left him, in even shallower water, his fine colorful skin turning dry and dull. "Brother Salmon, what are you doing, nearly out of River?" Coyote asked.

Salmon hardly looked up in his exhaustion, weakly flapping his tail and fanning his fins in the summer air. "I've got to get home. Home is upstream, and I have to swim home," he muttered again.

"Ah-woo, this isn't working for you, Brother Salmon!" Coyote scolded. "You must try something new!"

Box continues

Box 1. Continued

"This is what we salmon do, swim upstream, and we have never known anything else! But you are clever, Sister Coyote. Will you help me get home?" begged Salmon.

"When I want to go home, I just sing my Home Song," laughed Coyote. "Surely you must have one, too." And she sang, "Ah-woo, it's time to go home, ah-woo."

Salmon thought, "Salmon do not know how to sing, we have never needed to know how to sing!"

But Coyote kept singing, "Ah-woo, we'll sing Salmon home, ah-woo. This song will get us through, ah-woo. Now you sing it, too, ah-woo."

Salmon was hesitant, but then he heard River singing with Coyote, and he tried to hum along. As the melody flowed from him, first softly but then with more confidence, he found he could quiet the drive to swim only upstream, and was able to follow Coyote and her song, flapping and finning back to deeper water in the main channel. The cool water rinsed the dried salt from his skin and the flow smelled strongly of Salmon's family home, so he sang even louder, adding some of his own words, "Ah-woo, this run is so wet, ah-woo. Ah-woo, I'm so glad we met, ah-woo. Ah-woo, I'll never forget, ah-woo."

With a silver flash of scales, Salmon swam away upstream, and as he went, Coyote could hear him still singing, "Ah-woo, I tried something new, ah-woo. Ah-woo, I learned it from you, ah-woo. Ah-woo, and I'll share it, too, ah-woo...."

Coyote shook her fur and grinned a coyote grin, happy that her song had helped Salmon on his way. She turned toward the alder tree to bring the berries and grasshoppers and story to her pups, singing, "Ah-woo, it's time to go home, ah-woo."

~ The End ~

A professional parable written in the style of Native American coyote stories by Paula Guenther, 12/27/2012; inspired by the sculpture "Coyote Leads the Salmon Up the River" by the late Richard Beyer (1925–2012; www.richbeyersculpture.com). Illustration by Briana N. Bromley, Walden, Colorado.

As fisheries professionals, we can learn from animals such as the coyote when contemplating what is needed to survive the constant and accelerating changes occurring in the world and particularly in our profession. We are taught the skills we need from those who have experience. Within the professional world, we call the people who teach us survival skills our mentors. Mentors are akin to the coyote's family group or our pack; we count on them for our survival. Mentoring is critical to the survival of both coyotes and fisheries professionals.

Fisheries professionals cannot expect to spend their careers in positions requiring only the skills acquired during their university experiences. They need to continually learn and adapt. For example, Paula was trained in freshwater fisheries and water resource management and began her career as a fisheries and wildlife biologist in a field office of a federal land management agency. However, as time went by, the desire to assert leadership skills, enhance income, and address new challenges presented her with the need to move into different fields, accept more supervisory duties, add administrative responsibilities, use more multidisciplinary approaches to problem solving, and accomplish higher levels of communication with stakeholders and the public as a recreation manager.

Late-career fisheries professionals who reflect on their careers will readily acknowledge that things change, a lot. Given future changes in technology, knowledge, and the issues facing mankind, the rate of change will accelerate to favor fisheries professionals with highly developed adaptive skills. As an example, Wayne reflects on the fisheries management objectives that drove state agencies in the Midwest and Southeast when he began his career four decades ago. At that time, agencies were driven to create sport fisheries in recently constructed reservoirs. Fisheries managers largely focused on introductions of sport and prey species, supplemental stocking of desired species, management of harvest, and enhancement of habitat. There was little concern for declining native species, invasive species, connectivity of river systems, alterations of riverine habitat within regulated rivers, and other issues that have risen in importance among current fisheries managers. Today, fisheries management objectives in that region are vastly different and require fisheries professionals to apply a different array of skills.

As a young fisheries professional, Diana wonders what survival skills will be most important in her career. What can a young professional do to ensure that she maintains or gains necessary skills? Thinking of the coyote analogy, what can Wayne, as a parent pack member, or Paula, as an older pack sibling, do to help Diana, a young pup, to survive and contribute to the future of the pack and the population as a whole? We have combined our experiences to elaborate on the need for mentoring and adaptability among fisheries professionals.

Survival Skills

Critical Thinking

In order to survive, young coyotes must quickly apply survival skills they learn from other members of the pack and adapt to an ever-changing environment. The process is similar among fisheries professionals who must learn from those with experience and adapt to changes as soon as they enter the workforce. There is an immediate need to adapt their formal educations to address real-world challenges, changing conditions, and organizational protocols. The ability to think critically may be foremost among necessary survival skills. Critical thinking skills allow professionals to assess situations, sort truth from myth, identify assumptions, evaluate available information, recognize information gaps, and identify risks, all essential to survival. Critical thinking skills enable fisheries professionals to apply appropriate tools in differing situations. For example, young professionals may be taught traditional sport and commercial fish population assessment and modeling techniques, such as how to estimate growth and mortality rates, but these skills can be adapted to provide information applicable to the management of native nongame species.

Cooperation

Although coyotes often hunt alone, they also hunt together in a coordinated effort for larger prey, particularly in the winter when survival is most difficult. In today's professional landscape, cooperation is fundamental to the success of fisheries professionals. Administrative specialists, land managers, decision makers, multidisciplinary groups, and the public all have a significant impact on the success of fisheries professionals. Fisheries professionals must learn collaboration skills and adapt them to different individuals or groups. For example, a meeting among fisheries biologists from federal and state or provincial agencies may require skills that promote teamwork and problem solving, whereas a meeting among displeased and disagreeing stakeholders from different public sectors will require use of facilitation and conflict resolution skills. Similar to coyotes, when times are tough, such as when facing tight budgets or complex issues, cooperation is likely to be the key to survival and success.

Communication Skills

Young coyotes enhance their survival by learning how to communicate with other coyotes using body language, vocalizations, scent marking, and other mechanisms depending on the situation. Similarly, young fisheries professionals must become skilled in using a variety of communication techniques and to adapt those techniques to the audience being targeted. Communication among coworkers, peers, or the public will differ dramatically in tone, jargon, and detail. For example, when conducting a public meeting regarding management changes in a particular area, it is important to provide information that is meaningful to the audience and is easy to understand. Often times, this means leaving out the nitty-gritty details. In contrast, in a meeting among natural resource professionals, the details are often critical to the discussion and the outcome of the meeting. In both cases, the information must be presented in an open, honest manner that promotes feedback.

Response to Critical Feedback

There comes a time for every coyote pup when it is disciplined or "put in its place" by a litter mate or adult coyote. How a pup learns from the experience and adapts to fit into the fabric of the pack can determine whether it survives. Similarly, work in the fisheries profession is rarely free from conflict, but effective and constructive use of conflict often determines how successful a professional will be in continuing to influence decisions involving fisheries resources. A common complaint lodged against natural resource management agencies or organizations is that they cannot admit when they are wrong. By refusing to accept criticism or responsibility, fisheries professionals may hinder their own growth, advancement, and survival. Willingness to accept critical feedback and adapt to a situation will lead to more frequent success than repeated growling or showing fangs. Unlike a rigid wolf pack hierarchy where individuals may be driven out or killed for misjudgment, coyotes generally learn, modify behaviors, and survive within the pack. Similarly, young fisheries professionals must be flexible in their interactions with their leaders, mentors, and peers and learn from experiences to improve outcomes and their success.

Adaptation of Specialized Skills

The ability of coyotes to adapt is remarkable, with packs occurring in extreme environments from the southwestern deserts to Wrigleyville, the urban neighborhood home of the Chicago Cubs. When biologists explored genetic variability in animal populations, they recognized the associated array of morphological, physiological, and behavioral characteristics as the material for adaptation. In a relatively stable environment, reduction in variability (specialization) can create advantages to species that enable them to survive and reproduce more efficiently than other competing species in the community. However, biologists have seen what can happen to highly specialized species when the environment changes. Specialists may be ill-equipped to adapt to change and may decline in abundance or even be extirpated through selection processes. Similar selection occurs among fisheries professionals. Current and future changes associated with climate, invasive species, human population growth, and technology will undoubtedly alter the currently most relevant specializations within the fisheries profession. Fisheries professionals unwilling or unable to adapt will have reduced chances of survival, while those who strive to adapt their specializations to new conditions and advances in knowledge and technology will likely thrive.

Learning Survival Skills

Coyote pups learn from older members of the pack, especially those with a wide array of experiences and survival skills. These more experienced pack members are their mentors. Meaningful mentorship can be the foundation for developing a willingness to adapt and the ability to learn among young fisheries professionals. To enhance and sustain performance at each professional stage, mentors should consider the career goals, skill gaps, and experiences of individuals and develop customized approaches for gaining necessary skills. The selection of mentors by students and young professionals is a key to future adaptability and survival.

Mentors

Students and young professionals need to seek out mentors among wizened pack leaders. Mentorship of young fisheries professionals often begins with their university experience, especially among students within university programs where original research is a component of degree requirements. Students should find mentors who will dedicate one-on-one time for discussions about the relevance of their research and encourage them to think deeply about the application of science. Students should also seek advisors, committee members, and other mentors who reinforce the application of the scientific method and the differences among science, management, or advocacy. They also need mentors who will aid them in identifying gaps in their programs of study and are invested in their professional successes.

We often think that the obvious skills are those most important to survival. For coyotes, these are hunting skills; for fisheries professionals, these involve scientific methodology. However, a coyote pup that has already gained hunting skills still needs to develop social skills to survive. Similarly, a young fisheries professional who is proficient in scientific methods still needs to develop communication and cooperation skills in order to succeed. Tools to effectively use science-based education within the real world are ever more important in the work environment, but means to learn and practice these “soft skills” tend to be minimal components of curricula within science or natural resources departments at universities. Students who are science oriented need to identify mentors who can direct them to courses and experiences that will contribute to their social skills. Fortunately, several universities are developing curricula to assist students to develop skills in team building, problem solving, and conflict resolution, but formal courses are probably not enough. Mentors can fill this gap by providing, encouraging, or even requiring participation in these areas.

Young coyotes strive to become the dominant pup in their litter. The skills necessary to move up the ladder are not typically part of fisheries curricula or research specializations. The education of fisheries professionals should include experiences that prepare them to work with individuals from differing backgrounds and within group settings. Development of communication skills requires students and young professionals to go beyond their formal education and seek extracurricular, community, and professional activities to enhance these skills. For instance, activities to strengthen public speaking may include participation in Toastmasters International; presentations to student clubs, informal department seminars, local nongovernmental organizations (NGOs), or chapters of professional societies; or accepting leadership roles in community groups or professional organizations, such as a university's student subunit of the American Fisheries Society. Similarly, writing skills may be strengthened by writing for department, professional society, or NGO newsletters, blogs, or Web sites or by writing peer-reviewed material outside of thesis or dissertation research, such as editorial pieces for Fisheries. In these settings, mentors provide a valuable sounding board and source of constructive criticism to pups as they organize thoughts into outlines, prepare written drafts, and practice oral presentations.

Perhaps the most important, and maybe the most difficult, thing that a mentor can do to assist a student or young professional is to be available and open for questions, concerns, and discussions. If a coyote pup cannot find, or is intimidated by, its older pack members, it is unlikely to learn from them. Approachable, receptive mentors who are truly interested in a pup's success will have the greatest impact. Mentors who take the time to get to know their mentees as individuals will probably have a lasting influence on the future of their mentees as fisheries professionals. This relationship will promote the continuing development of a wide array of survival skills that are necessary for success.

Continuing Education

Coyotes continually learn from their experiences and adapt their behaviors in order to survive. Fisheries professionals must recognize that continual learning is needed to obtain the skills for productive careers. Involvement with continuing education can begin while students are still at universities. Even while working on their degrees, students can participate, as students or instructors, in online or professional-society-sponsored continuing education courses, outreach courses, or workshops. Experience with such courses can encourage young and mid-career professionals to continue to expand their skills upon completion of their university degrees. A challenge for students and university mentors is to distinguish skills that can be developed later in life from those that are more easily learned in a university environment. Once again, mentors can play a role by assisting students in identifying subjects that cannot be effectively learned by means of continuing education venues and ensuring that those subjects are accommodated within the graduate program. Courses that may fall into the latter category are those requiring higher-level mathematics, strong computer support, or highly technical laboratory techniques.

The Importance of Mentoring and Adaptability

Our experiences suggest that, like coyote survival, success as professionals in fisheries or related fields depends on adaptability. That means being broadly educated in the sciences, having flexible specializations within the realm of fisheries and aquatic sciences, and possessing sophisticated communication and social skills to address multidisciplinary problems and challenging situations. Students and young professionals need to seek mentors who will help them develop the ability to adapt to the ever-changing demands on fisheries professionals. We firmly believe that fisheries professionals must embrace lifelong learning not only to build technical knowledge, but also to enhance their critical thinking, communications, team building, problem solving, and conflict resolution skills.

There are, and will continue to be, parallel challenges for coyotes and fisheries professionals. Which coyotes (fisheries professionals) will survive? Those with the sufficient diversity to adapt to change. Which coyotes (fisheries professionals) will contribute to the long-term survival of the population (fisheries profession)? Those with the ability to adapt to an ever-changing environment. Established pack members (mentors) play a key role in developing the survival skills for pups (students) and juveniles (young professionals) to survive and maintain a viable population (corps of fisheries professionals) into the future. Mentoring and adaptability are critical to career success among fisheries professionals.

Biographies

Wayne A. Hubert began his career as an aquatic biologist with the Tennessee Valley Authority from 1972 to 1979. He spent three decades with the Cooperative Fish and Wildlife Research Units Program serving at Iowa State University and the University of Wyoming where he mentored graduate students and conducted research. He has been an active member of the American Fisheries Society, serving as president from 2010 to 2011, and remains professionally active through Hubert Fisheries Consulting, LLC.

Paula Guenther spent childhood free time on the lakes and rivers of national forests in the upper Midwest, Southeast, and West. After discovering limnology during her undergraduate work at the University of Wisconsin-Madison, she came west to study for a master's degree in water resource management at the University of Wyoming. She began her career with the U.S. Forest Service in 1989 and has worked in fisheries and wildlife biology, natural resource planning, and recreation on forests in the Rocky Mountains. She has also held positions with the Bureau of Land Management and The Nature Conservancy.

Diana Miller grew up on a guest ranch near Wapiti, Wyoming, playing in the mountains and streams of the national forest. She received her bachelor's degree in wildlife biology from Colorado State University and her master's degree in fisheries biology from the University of Wyoming. She has been working as a fisheries manager for the Wyoming Game and Fish Department in Jackson, Wyoming since 2008.

Mentoring from the Heart

Donald C. Jackson*
Department of Wildlife, Fisheries and Aquaculture
Post Office Box 9690, Mississippi State University, Mississippi 39762, USA

Key Points

- Mentors should by example challenge young professionals to integrate their core values as persons into their professional identity.
- Good mentors set the stage, equip young professionals with tools to succeed, and are there with and for the young professionals with words and actions of assurance, encouragement, and support.

Introduction

The explosion ripped into the armored vehicle on patrol in Iraq, killing the driver and shattering the left eardrum of the young soldier manning the machine gun on top of the vehicle. This event, one that changed his life forever, is still painfully vibrant in the student's eyes and his quiet but trembling voice as he shares his wartime story with me during breakfast at a local diner, trying to decide whether or not to quit university studies in a desperate quest to restructure his life.... Over the telephone, a baby is crying in the background as an exhausted young mother with a learning disability tells me why she must miss the exam scheduled for later that morning.... A student from an isolated rural community, with a black eye from raucous activity the night before, sits in a chair across the desk from me in my university office, confused, in a state of despair, envisioning his dreams slipping away from him because of academic and social failure in the vastness of the state's largest university.... The phone rings. The call is from the office of a U.S. senator inquiring about a student who has applied for a summer internship in Washington, D.C. that would be focused on natural resources legislation. The issues to be addressed are critical. Do I know the student well enough to make a recommendation: not just in terms of academics, but in terms of character and integrity?.... A "tough" young man drops by my office to share with me the news that he soon would be leaving the university to accept a professional position as a conservation officer with an agency in a neighboring state and to request (once again) that I not tell his classmates that he is the mystery poet whose poetry was among the selections I read to the fisheries management class that semester at the beginning of each lecture.... Another fisheries student drops by soon thereafter to tell me that he will be going to seminary (as had I) before pursuing further graduate studies in the sciences, that his undergraduate degree in fisheries solidified his commitment to good relationships between people and the Earth's resources, and that his growing understanding of the complexity and beauty of those relationships took him

* Corresponding author: djackson@cfr.msstate.edu

across the threshold of his decision.... A brilliant graduate student drifts red-faced through the halls, on the brink of collapse from days without sleeping and eating, convinced that his data and analyses are nothing but junk and that his pending thesis based on those data will be worthless, along with his life. I refused to allow him back into the classroom or into his office or laboratory until he can produce photographs of fish he must catch while angling alone on my private fishing pond located just outside of town.

What do these people have in common?

- They all graduated with degrees in natural resources.
- They all moved beautifully into new chapters of their lives and careers.
- They all have gifts to share as young professionals and are determined to do just that.

Formulating a decision to dedicate one's life to a specific field of endeavor, and staying the course, can be an extremely difficult and emotional process. For young professionals engaged in the sciences, and particularly those addressing natural resources, the challenges can be exacerbated because the required intellectual attributes for success in the field tend to be accompanied by a propensity for identifying and engaging an array of interests. Young professionals care deeply about life processes and tend to understand that the diversity of human experiences, not just the sciences, must somehow be incorporated into their professional identity. This can lead to consideration of alternative ways to focus their lives, and perhaps alternative careers, particularly when the demands of the natural resources fields get tough or are suffocated by demands from personal realms of their lives.

While such consideration can be and usually is healthy, and can lead to reinforcement of the career decision or revelation regarding new vistas of opportunity, it also can resurrect deep and painful questions and sometimes almost incapacitating doubts. Unless there is a mentor willing and able to engage the young professional in a timely and effective manner, there is risk that we will lose her or him (and the unique gifts she or he would bring into our realm) and that she or he will lose opportunity to meld with the beautiful currents that flow within the ranks of our professions in natural resources. There must be mentoring.

The Mentor's Framework

What is mentoring all about anyway? How do we go about it? It is certainly a complex affair but, ultimately, one of the more rewarding dimensions of who we are and what we do professionally and personally.

I believe the first step in mentoring is reflection back through the passages of the mentor's own pilgrimage. We need to be brutally honest with ourselves in this endeavor. Through such reflection, mentors remind themselves of, and subsequently are better able to share with young professionals, fundamental realities: stressful challenges in the world of professionalism in natural resources arenas are and will continue to be ongoing throughout careers, and the challenges are and always will be operative in concert with parallel challenges in personal dimensions of a professional's life.

During such sharing with young professionals, it is essential that mentors clearly articulate that careers are processes—that we never really arrive. Although mentors need to encourage young professionals to have goals, it is important to help young professionals understand that attainment of goals leads to new goals, and that it is the mosaic of these accomplishments that bring structure and meaning to our pilgrimages. Professionalism is (or should be, if the fit is right) a celebration of these processes. Accomplishments during the pilgrimage are beautiful stepping stones along a trail that never ends.

It is also important that young professionals be encouraged to understand that most mentors are motivated into mentoring as an important aspect of their jobs by the need to share, to give, to make sure that progress along the professional trail continues. The process of passing the flag is typically as inspirational to the mentor as it is to the young professional, for collectively they, the mentor and the young professional, are privileged to share hope and discern opportunity for the future of the profession—hope embodied in the freshness, vitality, energy, and dreams of youth in the fields, hope tempered and guided by those with experience long in the field, with opportunity limited only by their ability to engage joint ventures into creativity.

The primary responsibility for these endeavors, however, rests with the mentors, for mentors are the ones who, standing on firm foundations of established careers, are at the vantage points allowing them to see beyond the horizons and envision where the new professionals can take us, and share those visions. These visions of the future, once revealed, can be exciting to the young professionals and, when grasped and blended into the young professionals' own visions, can fill the mentor's life with incredible satisfaction and affirmation.

The Young Professional's Framework

The future envisioned by young professionals, while challenging and hopefully inspirational, can also be frightening. Young professionals need assurances from time to time that mentors, as established professionals, have faith in the young professional's abilities to succeed. Doing so requires that mentors have sensitivity to the young professional's individual limits of intellectual, physical, and emotional strengths. Objective, truthful appraisal of talent, ability, and accomplishment is absolutely essential. While encouragement is important, superficial praise rings hollow and should be avoided. Unearned praise can lead to a false sense of competency and, if carried to the extreme, ultimately to a fall and catastrophic failure in the young professional's career.

During the process of discerning how best to work with individual young professionals, it is important that mentors have an understanding of the issues and pressures that surround the young professionals. The social, economic, and technical frameworks that make up the young professionals' worlds may transcend the mentors' personal experiences.

For example, some young professionals are veterans of wars and may be carrying deep scars in a variety of manifestations. For mentors who have never known war, or perhaps have never served in the military or other forms of national service (e.g., AmeriCorps, Peace Corps, and VISTA), the demands of mentoring such young professionals can be daunting tasks. Yet, somehow, mentors must be able to relate, to find the common denominators of experience (e.g., stressful, perhaps life threatening field experiences; weather related emergencies; and cross-cultural challenges of international assignments) from which to draw when providing guidance to young professionals.

Insecurity regarding appropriate backgrounds and experiences can also challenge young professionals. Although drawn to natural resources professions and life with outdoor-oriented dimensions, many young professionals come into the mentor's realm from urban and suburban backgrounds where outdoor experiences and associated skill acquisition were difficult to obtain. Mentors can help young professionals overcome insecurity by providing them with settings, opportunity, and personalized instruction for developing competency and self-confidence in the wide array of skill sets necessary for their fields. In the realm of outdoor skills, I have found great enthusiasm among young professionals when I introduce them to such things as maintaining, hooking up, and backing trailers; maintaining and driving motor-powered boats; handling rowboats and canoes; dealing with emergency situations on the water (e.g., swamped boats, tired or distressed swimmers); using a wide assortment of fishing gears; driving tractors; cleaning fish (several ways); and caring for and using outdoor tools such as knives and chainsaws.

Integrating Personal Arenas

Many young professionals are deeply involved in developing relationships with people who may become life partners, or are recently married and/or have young children. Some are single parents. Some are dealing with illness, separation from loved ones, or loss. Military deployment disrupts many relationships.

Mentors must be sensitive to these situations and realize that young professionals are faced with incredible pressures, physically and emotionally. Sometimes all that young professionals need are assurances that the mentor understands and is willing to help in appropriate ways if the stresses become extreme. And, of course, there are always economic issues. An occasional helping hand on the economic side of life can make a real difference. How many of us during our student years or early professional lives received personal money from our professor or other mentor? I certainly did, and it was not a loan.

The Mentor as Colleague as Well as Counselor

Mentoring professionalism by example goes a long way. Honesty, frankness, and openness, coupled with the courage to listen and then to share your own story, can oftentimes calm a young professional's stormy seas. Although helpful and certainly recommended, it is not really necessary for mentors to have formal training in counseling. Young professionals simply need to know that someone cares enough to listen, and cares enough to reach out to them as a friend as well as a colleague.

If young professionals have a supportive, nurturing environment within which to work, they can generally handle the demands of their jobs/studies. It is subsequently the mentor's responsibility to provide this environment and, frankly, to do even more: to run occasional interference (discretely) so that the young professional can focus on getting the framework and foundation for a career firmly established.

Finally, and if at all possible, mentors should avoid abdication (kicking the can down the road) to professional counselors. Mentors should be, and typically are, the first points of contact with issues regarding academics, professionalism, and how these function within the framework and politics of institutions. However, mentors certainly need to recognize their limits and know when to seek additional assistance. Severe depression and anxiety; withdrawal from work, friends, and colleagues; substance abuse; and suicidal symptoms or talk must be advanced to other professionals who have the appropriate training, skills, and credentials to handle such situations.

Conclusion

Mentoring requires insight and courage, and frequently extreme humbleness, in order to provide meaningful perspectives, values, and reasons. The mentor's role becomes primarily one of helping young professionals through example, instruction, and encouragement, with the discernment and selection of trails. Mentors are the guides who light the way.

Mentoring is not just about helping young professionals with the acquisition of a set of skills. Mentoring also must incorporate and share with young professionals a particular mind set of connectedness with the rhythms of the earth and connectedness with the deeper currents of themselves as persons. In this process mentors become catalysts: they help reduce the amount of energy needed for the proper reaction to occur.

Good mentors set the stage, equip young professionals with tools to succeed, and are there with and for them with words and actions of assurance, encouragement, and support, working directly with them, and also behind the scenes, to make sure that the right doors are opened. The best mentors maintain rigor but pull up rather than beat down. They build bridges, not walls. But most importantly, mentors are there help young professionals understand that good professionalism reflects our humanity.

Acknowledgments

My sincere appreciation goes to my students, now scattered all around the world. It is they who mentored me. I am proud of them and the wonderful things that they are accomplishing in their careers.

Biography

Donald C. Jackson (American Fisheries Society president 2009–2010) is the Sharp distinguished professor of fisheries at Mississippi State University. A native of Arkansas, he received his B.S. and M.S. in zoology from the University of Arkansas and his Ph.D. in fisheries from Auburn University. He received seminary training from Lexington Theological Seminary (Kentucky) and was a U.S. Peace Corps volunteer (Malaysia). He loves writing and is the author of three books of poetry and outdoor essays (*Trails*, Strode Publishers, Huntsville, Alabama, 1984; *Tracks,* University Press of Mississippi, 2006; and *Wilder Ways*, University Press of Mississippi, 2012). He is devoted to working in leadership and service positions with the Girl Scouts and Boy Scouts and is the waterfront director and aquatics instructor at the scout's local summer camp. His passions are wing shooting, training bird dogs, and creek fishing.

Swift to Hear, Slow to Speak, Slow to Wrath

Jud Kratzer*
Vermont Fish and Wildlife Department
1229 Portland Street, Suite 201, St. Johnsbury, Vermont 05819, USA

Key Points

Three principles can help any fisheries biologist to effectively interact with disgruntled stakeholders.

- Be swift to hear: ask questions and listen intently.
- Be slow to speak: speak sparingly and respectfully, using questions rather than statements when possible.
- Be slow to wrath: always maintain composure, especially when the stakeholder does not.

Introduction

If you are like most fisheries biologists, you chose your career because you like working with fish, not because you like working with people. You definitely did not choose your career because you wanted to field complaints from disgruntled anglers or commercial fishers. I consider the fishing to be pretty good here in Vermont, where I serve as a fisheries biologist, and yet people still complain. "The best fishing in Vermont is in New Hampshire," they say. So the anglers in New Hampshire must be really happy, right? No, there the saying is, "The best fishing in New Hampshire is in Vermont (or Maine)." In truth, there are many happy anglers, and in most cases, the satisfied anglers are far more numerous than the dissatisfied ones, but angry anglers are much more motivated to contact their local fisheries biologist. When a disgruntled stakeholder does come to call, the interaction can be negative, frustrating, and wasteful or positive, exhilarating, and educational, both for the biologist and the stakeholder. The outcome depends largely on the skill and composure of the fisheries biologist, who can see any interaction with a stakeholder as an opportunity to inform and be informed. My experience in Vermont, where arguing is a sport, has demonstrated the value of three simple principles when dealing with disgruntled anglers (or coworkers or friends or family members). These principles are swift to hear, slow to speak, and slow to wrath.

* Corresponding author: jud.kratzer@state.vt.us

Swift to Hear

> The fundamental cause of trouble in the world is that the stupid are cocksure while the intelligent are full of doubt.
> —Bertrand Russell

To be swift to hear and slow to speak requires good listening skills. Early in my career, it was easy to listen. I did not know (and still do not know) all the answers to anglers' questions. I did not know, for example, why we stopped stocking Rainbow Trout *Oncorhynchus mykiss* in Little Averill Lake. I did not know why the Salem Lake Walleye *Sander vitreus* fishery had (supposedly) collapsed. All I could do was listen, ask questions, and listen some more. Often, the conversation would start out with an impassioned, and sometimes belligerent, rant by the angler about my agency's mismanagement of some fisheries resource. I have found that most rants follow the same general format: "I have been fishing (insert water body) for 20 years (or 30 or more), and the fishing for (insert species) is not what it used to be. The Vermont Fish and Wildlife Department is to blame." I learned early on that, if I just respectfully listened, asked questions, and respectfully listened some more, the situation became much less tense and the conversation often ended on a positive, even friendly note. Listening and asking questions was also a great way to learn. Of course, there were many half-truths and much conjecture, but I did glean some valuable lessons from most of my interactions with upset anglers. I also learned after the conversations by researching the questions that I was not able to answer.

Some clarification about what I mean by listening is in order. Listening means more than not speaking. It means giving the speaker your undivided attention rather than thinking about a witty comeback, or even a polite response, that you will deliver as soon as the other person is done talking. A true listener listens with eyes, as well as ears. The true listener's body language communicates that the speaker's thoughts and values are important. Asking follow-up questions is a particularly effective way to demonstrate a genuine interest in the other person's views. Listening can be hard work and requires energy and focus to take in, process, and store what the other person is saying.

Slow to Speak

> He who speaks without modesty will find it difficult to make his words good.
> —Confucius

> In a gentle way, you can shake the world.
> —Mahatma Gandhi

Asking questions and listening will often reveal an opportunity to inform the stakeholder on some topic. Now, it is time for the fisheries biologist to speak. Stakeholders often do not have all the facts, and even if they do, they may have used those facts to come to the wrong conclusions. An important duty of fisheries biologists is to educate the public about fisheries biology and management. We can see any interaction with an angler, even an angry, complaining one, as an opportunity to inform the public. In many cases, the stakeholder has an honest question or an obvious need for some piece of information. If you have that piece of information, you can share it. If not, it is okay to say so, go back to the office, and do some more research.

Questions, rather than statements, can also be used very effectively in fulfilling our commission to inform the public. If the stakeholder is wrong, it is usually due to a lack of information or sound reasoning, and the right questions can cause the stakeholder's arguments to self-destruct or at least make the stakeholder more open to what you have to say. The following is a contrived example that illustrates how questions can be used to respectfully educate a stakeholder.

Angler: You guys are not stocking enough fish! The fishing on the Passumpsic River was much better 20 years ago when you were stocking a lot more fish. Now, it's terrible. I have friends that won't even buy a fishing license anymore.

Biologist: Hmm, that's interesting. I do a bit of fishing on the Passumpsic, myself. What was the fishing like back then?

Angler: It was nothing to catch 20 trout a day, and there were some days when you just couldn't keep them off your line.

Biologist: Wow, that sounds like fun! What is a typical day of fishing like for you now?

Angler: Now, I'm lucky if I catch a fish.

Biologist: Well, that's too bad. What makes you think that stocking more fish will help?

Angler: It just makes sense. Fishing was better when you stocked more fish.

Biologist: Are you sure that all those fish that you were catching back then were stocked fish, or could they have been wild?

Angler: I don't know, but clearly if you stock more fish now, there will be more fish to catch.

Biologist: That's probably true, but where are we going to get those fish?

Angler: Doesn't the state have hatcheries?

Biologist: We do, but they are currently producing fish at full capacity or beyond. We can't raise any more fish, so if we were to stock more fish in the Passumpsic River we would have to take those fish from some other river, and that wouldn't be fair to the people that fish that river. We actually have a trout management plan that provides guidelines on when, where, and how many trout we stock so that we can allocate these valuable resources fairly and consistently.

Angler: Well, why don't you just build another hatchery?

Biologist: That's a very expensive proposition. Do you have any idea how much money goes into each trout we stock?

Angler: How much?

Biologist: About five dollars apiece.

Angler: Really? How many trout do you stock in a year?

Biologist: A little over a million statewide.

Angler: You guys need to stop stocking so many trout!

Of course, anglers are rarely convinced this quickly and decisively, but the example is useful nonetheless. Note that the biologist mostly just listened and asked questions. In fact, the biologist asked five questions before providing any information to the angler. One type of question that can be used in almost every interaction with stakeholders is "How did you come to that conclusion?" In the above example, the question was worded "What makes you think that stocking more fish will help?" This question may reveal that the stakeholder does not have good data to support his view and often provides the biologist with a great opportunity to provide real data and inform the stakeholder. Also note that the biologist attempted to relate to the angler by stating that he or she fishes the same river. In my free time, I try to fish as often as I can and in as many different locations as possible not only because I enjoy doing it, but also because I have learned that it is much easier to relate to an angler when I have fished the same water, especially if I have caught fish there. Finally, note that the biologist could have gone many different routes with this discussion. The angler has concluded that fishing is not what it used to be because too few fish are being stocked, but the biologist knows that there are several reasons why the angler might perceive fishing to be poor, such as fewer wild trout, habitat changes, or inaccurate memories of past fishing experiences. It is usually best to steer the discussion in the direction of your strongest data. It is fine to ask the question about wild trout, but without data on past and present wild trout abundances in the river, the biologist cannot effectively use that line of reasoning.

Slow to Wrath

> When the debate is lost, slander becomes the tool of the loser.
> —Socrates

Whether hearing or speaking, the fisheries biologist must always be slow to wrath, maintaining a cool, calm, professional demeanor, regardless of the stakeholder's behavior. The fisheries biologist is a paid representative of a larger agency, and his or her behavior reflects on the agency as a whole. Also, if the biologist becomes angry, he or she is likely to resort to raising his or her voice, interrupting the other person or otherwise trying to intimidate. In addition to being rude, resorting to these methods rather than sound reason suggests that the biologist's position is a weak one. The biologist should demonstrate confidence in his or her position by maintaining composure while asking questions and presenting facts in a respectful manner. The biologist should also aim to keep the stakeholder calm. An angry stakeholder is more likely to be emotional and defensive, but a calm stakeholder is more likely to listen to reason and learn something from the encounter. It is very important to remember that the stakeholder could be right, and a calm biologist is also more likely to learn from the conversation.

Conclusion

> [B]e swift to hear, slow to speak, slow to wrath.
> —Holy Bible, James 1:19

Learning to deal with angry stakeholders is probably not included in most fisheries curricula, but it is a valuable skill. Fortunately, anyone can practice being swift to hear, slow to speak, and slow to wrath in any conflict, whether it be with a family member, friend, or coworker, and while these three principles might not be very useful when working with fish, they can be of inestimable value when working with people. A fisheries biologist that masters these principles will be a better agency representative and a more effective educator of the public. These principles can also contribute to greater peace in professional and personal relationships. So, go start practicing now!

Acknowledgments

I thank the editors and peer reviewers for their insightful and helpful comments. Time spent in the writing of this vignette was paid for by the Federal Aid in Sportfish Restoration Act and Vermont fishing license sales. The views and insights represented herein are the author's and do not reflect any official position of the Vermont Fish and Wildlife Department.

Biography

Jud Kratzer studied fisheries science at Pennsylvania State University (M.S.) and Michigan State University (Ph.D.). He has worked as a fisheries biologist with the Vermont Fish and Wildlife Department since 2006. A Pennsylvania native, he is still adjusting to the fast-talking, blunt style of conversation practiced in New England, which has been an excellent proving ground for the swift to hear, slow to speak, and slow to wrath principles described above.

Sustainable Fisheries Is a Worldwide Objective: As a Fisheries Scientist, the World Is Your Oyster

MARGARET MARY MCBRIDE*
Ecosystem Processes Research Group, Institute of Marine Research
Post Office Box 1870, Nordnes, Bergen 5817, Norway

Key Points

- As a fisheries scientist, your skills are transportable and in demand wherever fisheries are conducted.
- New challenges present new opportunities, and chances to develop new skills and expertise.
- Stay conscious of new frontiers, and do not be afraid to knock on doors.
- To remain flexible, sometimes you have to go out on a limb; it is hard to earn a reward without taking a risk.

Introduction

Sustainable fishing is a worldwide objective. As a fisheries scientist, your skills are in demand wherever fisheries are conducted. During my 38-plus years as a fisheries scientist, I have enjoyed working at five different research institutions in three different countries on three different continents. The jobs all focused on very different sets of issues, problems, and concerns. I have learned valuable lessons along the way, and every day has been an adventure.

Fishing is one of the oldest human professions and has been an organized industry since the Middle Ages. It is a global enterprise, and today, seafood constitutes an important source of nutrition and animal protein for much of the world's population. The fisheries sector provides livelihoods and income, both directly and indirectly, for a significant share of the world's population. An estimated 70% of the earth's surface is covered with water, and roughly 150 countries throughout the world have coasts granting access to marine fisheries. Fish and fishery products are among the most traded food commodities worldwide, with trade volumes and values reaching new highs in 2011. This rising trend is expected to continue, with developing countries continuing to account for the bulk of world exports. While capture fisheries production remains stable, aquaculture production continues to expand and is set to remain one of the fastest-growing animal food-producing sectors. In the next decade, total production from both capture fisheries and aquaculture will exceed that of beef, pork, or poultry.

* Corresponding author: margaret.mary.mcbride@imr.no

Woods Hole: Getting One's Feet Wet

The international dimensions of fisheries science became apparent immediately after I began working in 1975 at the National Oceanic and Atmospheric Administration (NOAA Fisheries). NOAA Fisheries was established in 1885 to manage, conserve, and protect living marine resources and to serve the fishing industry by providing science-based assessments of the status of fishery resources. However, fishery science as a profession only began to emerge in the 1950s. The United States was an active member of the International Commission for Northwest Atlantic Fisheries (ICNAF)—among the first regional fisheries management bodies to be established that played a leading role in the assessment and management of fish stocks outside of national jurisdictions. The Woods Hole Laboratory was a key player in carrying out ICNAF's mission in the Northwest Atlantic. As such, there was a steady flow of research vessels from numerous countries docking at our piers to conduct joint ICNAF research surveys.

The existing policy was one of scientific exchange: a number of visiting scientists were placed onboard our NOAA vessels, and a number of NOAA scientists were placed onboard the visiting foreign vessels. Even so, I was amazed when—immediately after undergraduate studies, at the tender age of 22 and as green as grass—my first field assignment was a month-long survey onboard a Russian research vessel, the RV *Belagorsk*. Following this trajectory, my early career was colored with two- to three-week-long survey adventures onboard research vessels from Russia, West Germany, Poland, Portugal, Japan, France, Italy, and so forth. These were truly exciting times for a young NOAA scientist who had not traveled outside the States. Onboard these vessels you were exposed to different languages, different cultures, different cuisines, different customs, and different ways of interacting. Being thrown together onboard a ship for a few weeks presented the opportunity to share ideas, laugh out loud together, and establish friendships with researchers from other countries. In effect, the world became a little smaller in a very nice way. Important lessons here were that even long distances can be bridged, fishing sustainably is a shared objective, and people are transportable.

Learning the basics is important, and NOAA afforded great opportunities to do that. The Northeast Fisheries Science Center (NEFSC) conducts standardized research surveys from Cape Hatteras to the Scotian Shelf collecting fishery-independent data to determine the status of commercially valuable fishery resources to support sustainable fisheries management. I worked in the Population Dynamics Branch conducting assessments on stocks of Yellowtail Flounder *Limanda ferruginea* and Silver Hake *Merluccius bilinearis*. Early on, I was assigned the huge task of assembling time series of information assessing the status of commercially important fish stocks in the Northwest Atlantic falling under NEFSC jurisdiction. This effort gave me an overview of key concepts in fisheries science and resulted in the first issue of an ongoing series of NEFSC Status of the Stocks publications. Similar publications are now produced annually by NOAA Fisheries Science Centers nationwide. Working at the NEFSC taught me the tools of the trade. One central lesson was that growth overfishing—harvesting fish at an average size smaller than that needed to produce the maximum yield per recruit—is to be avoided. This underpins the importance of a second goal of fishery management: to avoid recruitment overfishing by not removing immature individuals from the stock before they can reproduce.

When undersized/unmarketable fish are captured, they typically are discarded at sea and do not survive the experience. Both landings and discards constitute total removals and should be incorporated into the stock assessment. However, data reporting amounts discarded are often unavailable, as was the case for Yellowtail Flounder. Under a NOAA career development program, I was able to pursue an advanced degree in Fisheries and Wildlife

Science at Oregon State University. Part of my graduate research was to develop a method using available data to estimate discard levels and total removals for the Yellowtail Flounder fishery. Using reported estimates of commercial landings, groundfish survey catch-at-length data, and bottom trawl catch-at-length selectivity curves, I was able to simulate expected total commercial catch-at-length. Around this time, a mutual friend introduced me to a scientist from Norway's Institute of Marine Research (IMR) with whom I discussed this common sense approach. As a result, I was invited to work at IMR for a year (1990–1991) as a visiting scientist to apply my method using data from Norway's Barents Sea Atlantic Cod *Gadus morhua* fishery. The United States and Norway are fellow NATO countries with a history of collaboration; NEFSC and IMR had also had earlier scientific exchanges. Therefore, I was able to secure a year of leave without pay to pursue this adventure. Fortunately, the work went very well and was published in a peer-reviewed journal. There were other quite serious consequences of this experience, which include that I now have been married to a Viking for more than 21 years, he and I have a 19-year-old son who holds dual citizenship, and after 17 years I was again invited to work at IMR where I have remained since 2007. But I digress; the career lesson here is that the most brilliant science may fall by the wayside if you do not communicate it. It is important to exchange ideas with other scientists and to document/publish your research findings.

Chesapeake Bay I: An Estuarine Extension

My husband is also a fisheries scientist, but not a U.S. citizen. Therefore, he could not become permanently employed by the U.S. Government. So, before we could join forces, significant compromises had to be made by both: I left my beloved Woods Hole to take a position at the U.S. Fish and Wildlife Service (USFWS) in Annapolis, Maryland (1993–1996), and my soon-to-be husband left his homeland after securing a position at the University of Maryland Chesapeake Biological Laboratory. In effect, we formed an international transportable fisheries team, and this became our modus operandi.

Working in the Chesapeake Bay was illuminating. It is part of the largest estuary in the United States, covering 64,299 square miles (166,534 square kilometers) and parts of six states. More than 150 rivers and streams drain into the bay, but its average depth is only 6.5 meters. Major fisheries are conducted for blue crabs *Callinectes sapidus*, Striped Bass *Morone saxatilis*, and historically for American oysters (also known as eastern oysters) *Crassostrea virginica*. More than 17.7 million people reside in the surrounding watershed. Water quality is compromised by negative anthropogenic inputs in combination its shallow depth and limited exchange with oceanic waters. Working with the USFWS on restoration of the Atlantic coastal Striped Bass was a lesson on the critical importance of estuaries to some coastal fisheries. The population ranged from the Gulf of Saint Lawrence to the Gulf of Mexico, but Chesapeake Bay is an important breeding and nursery area. A key element of the coastal Striped Bass restoration effort was to rebuild the stock through combining fisheries regulations with stock enhancement. Between 1985 and 1993, more than 9 million tagged hatchery-reared Striped Bass fingerlings were released into river and stream tributaries of the Chesapeake Bay system. The subsequent return of the coastal population to record high levels was considered a management success story.

Mozambique: An African Adventure

Call it wanderlust or call it madness, but the next chapter of this story takes place on the Mother Continent. My husband and I packed our bags and, with 22-month-old son in tow, headed to

Africa following an offer to work on an institution-building foreign-aid mission sponsored by the Norwegian Agency for Development Cooperation. We worked three-plus years at the Instituto Nacional de Investigação Pesqueira in Maputo, Mozambique. Being African-American myself, I had always wanted to work in Africa. To the contrary, the USFWS saw no advantage in my working for three years in Africa; I was unable to obtain LWOP (leave without pay) status. So, we jumped off the deep end and both resigned from permanent positions in the United States to pursue this African ambition, believing that it would happen either now or never!

Working in Mozambique was extraordinary. This country had endured a bitter history. Its Portuguese colonizers had neglected domestic development. This led to years of economic crisis after finally gaining independence in 1975. A struggle for dominance followed between independence government (FRELIMO) and resistance (RENAMO) political factions, which drew the country into a brutal civil war lasting more than 17 years. With the cease-fire in 1992, there came an end to the fighting and the opportunity to hold multiparty elections. The nation emerged in 1994 as a fragile new democracy determined to establish a free market economy.

At that time, it was considered one of the poorest countries in the world. Still today, almost 70% of Mozambique's 19 million people live in poverty despite an estimated growth in GDP of 7.5% for 2013. Yet, its eastern boundary extends 1,535 miles (2,470 kilometers) along the coast of the rich and productive Indian Ocean. Mozambique's fishery resources include shallow-water shrimp, deepwater crustaceans, scad, mackerel, and other demersal fish species. Much of Mozambique's fishery-related GDP results from the export of shrimp. In coastal regions, artisanal fisheries form the basis for subsistence and provide food security for local populations. Our charge was to help Mozambicans research and manage their own fisheries. Our objectives included helping design a survey of coastal artisanal fisheries and helping Mozambican scientists publish their research reports in English to reach a wider reader audience and secure ongoing financial support. It became my personal project to document how artisanal fisheries are an integral component of the coastal culture and to emphasize the importance of their sustainability. Featuring my husband's photos, I published a popular article in a Norwegian magazine called *Kyst*; an English translation of the article title is "Artisanal Fisheries in Mozambique: Timeless, Treasured, Threatened." Working in Mozambique illustrated for me something that is true for many coastal nations, that small-scale, artisanal fisheries are central to food security for local populations.

Chesapeake Bay II: Enter an Ecosystem Approach

We left Mozambique in 1999 and returned to Maryland. So begins Part 2 of lessons learned from the Chesapeake Bay. The mix of skills gained from professional experiences in the Northwest Atlantic, the Northeast Atlantic, the Chesapeake Bay, and the Indian Ocean helped me land a research position at the NOAA Chesapeake Bay Office, where I co-chaired a panel of 17 regional experts to develop the first fisheries ecosystem plan (FEP). Challenges to management of Chesapeake Bay fisheries arise from two major factors: (1) the surrounding watershed is densely populated, which leads to excessive nutrient (nitrogen and phosphorus) loading and other forms of pollution; and (2) many of the economically and ecologically important fishery species spend critical life stages outside of the bay. Only 5 of the 14 most valuable fishery species in the bay are year-round residents. These species cannot be entirely managed or protected within the watershed-to-mouth-of-the-bay management unit. There also are challenges to manage fisheries for species that cross state jurisdictions; none of the bay states can, by itself, effectively conserve its living marine resources or protect the interests of its citizens. Each bay state must work with the others and the federal government to achieve sustainable fisheries. It was not easy to carry this effort to completion due to limited funding. But it was gratifying that the FEP was well received and formally adopted by the Chesapeake Bay Program to guide its restoration efforts. It also was published in 2006 as a book, *Fisheries Ecosystem Planning for Chesapeake Bay*, by the American Fisheries Society.

Working to develop the FEP clearly illustrated that overfishing is not the only threat to sustainable fisheries. Pollution, degradation of viable fish habitats, and anthropogenic climate change also increasingly undermine the potential for healthy, productive, and sustainable fisheries. Today, more than 50% of the world's population lives in coastal areas; that percentage is expected to rise to 75% by 2025. Increased coastal development will likely be accompanied by increased pollution from industrial and sewage treatment plants and increased nonpoint-source pollution from freshwater runoff, precipitation, atmospheric deposition, drainage, seepage, or hydrologic modification. Maintaining viable fish habitat in view of coastal marine ecosystems in decline is an emerging issue that will likely have societal, political, and personal lifestyle implications. One key lesson from the Chesapeake Bay (Part II) is that it will be crucial for an ecosystem approach to marine resource management to effectively communicate scientific knowledge and ecosystem principles to the general public, political legislators, and other social policy/decision makers.

Norway: A Holistic National Approach

The year 2007 seemed a good time for our then 13-year-old son to discover his Norwegian roots. When I married the Viking, I anticipated a return with him to live in his country at some point. Fortunately, we both were welcomed back to work at IMR. So once again tickets were purchased, our home was sold, and belongings were packed and shipped. We returned to Bergen in early June.

With a coastline extending over 15,600 miles (25,148 kilometers), Norway takes sustainable fisheries seriously. For centuries, fisheries—conducted in the Barents Sea–Lofoten area, the Norwegian Sea, and the North Sea—have been important to its culture and national economy. New technologies and increased fishing capacity present new challenges to ensuring sustainable fisheries. This became apparent when the Norwegian spring-spawning Atlantic Herring *Clupea harengus* collapsed in the late 1960s. Although overfishing continued for decades, sustained efforts to rebuild the stock, including the implementation of a range of new management tools, gradually led to a coherent Norwegian policy to prevent overfishing and secure long-term sustainability.

During the 1960s and 1970s, fisheries policy objectives were predicated on the understanding that long-term ecological sustainability is not only an objective, but also a prerequisite for fisheries to be profitable without subsidies. Sustainable fisheries would ensure equitable wages for those working in the fishing industry, ensuring living conditions similar to those of other sectors of the strong Norwegian economy. Today, it is recognized that the fisheries sector on its own is no longer able to maintain settlement and employment in coastal communities. However, within the constraints of ecological sustainability and profitability, the fisheries sector will continue to contribute to meeting national objectives to develop and maintain settlements and provide employment in coastal communities, remain sufficiently profitable to secure wages and living conditions similar to those of other sectors of the Norwegian economy, and not deplete fish stocks.

Achieving these ambitious and socially democratic objectives required building a framework through close collaboration between scientists, fishery managers, and stakeholders.

Implementing policies at the national level required compromise and adapting behaviors to new realities. Major policies adopted included

- Limiting access;
- Ending subsidies;
- Annually distributing/allocating fishing opportunities with stakeholder participation;
- Reducing overcapacity through scrapping and license aggregation;
- Introducing a system of real-time closures (1984) to protect juvenile fish;
- Imposing a ban on discard at sea for Atlantic Cod and Haddock *Melanogrammus aeglefinus* (1987) that is currently in effect for most species;
- Requiring mandatory sorting grids for shrimp trawls (1991) and groundfish trawls (1997); and
- Setting total allowable catch levels through rational decision-making procedures using sustainable harvest control rules.

Norway has also adopted a holistic ecosystem approach to management of living marine resources. Integrated ocean management plans have been developed to present the overall framework for both existing and new activities in Norwegian waters and to facilitate the coexistence of different industries, particularly fisheries, maritime transport/shipping, and petroleum. The aim is to establish ecosystem-based management of these activities in relevant sea areas and to manage activities so that combined environmental pressures will not threaten the sustainability of ecosystems or the fisheries they support.

There are many lessons to be learned here on the Arctic frontier, but one is already quite apparent: meeting the objective of sustainable fisheries requires closely coordinated international cooperation, developing and setting national policies that support meeting this objective, and sound science-based management advice.

My research interests at IMR have expanded to include the effects of climate change on marine ecosystems in the Arctic and Antarctic polar regions and how climate variability and global change will impact future productivity and sustainability of fisheries resources in both regions. Arctic and Antarctic marine systems have in common high latitudes, large seasonal changes in light levels, cold air and sea temperatures, and sea ice. In other ways, however, they are strikingly different, including their geological structure, ice stability, and food web structure. Both regions contain very rapidly warming areas and a number of climate impacts have already been reported, and there are dramatic projections for future changes. The combined effects of a changing climate on oceanographic processes and food web dynamics are likely to influence future fisheries in Arctic and Antarctic regions in very different ways. This will have implications for the nature and availability of marine fishery resources, as well as on global food security. At the Arctic Frontiers conference in January 2013, I gave a talk entitled "Krill, Climate, and Contrasting Future Scenarios for Arctic and Antarctic Fisheries." My coauthors and I reviewed the published literature and have written an article accepted for publication in a thematic issue of the *ICES Journal of Marine Science* entitled "Marine Harvest in the Arctic." This is fascinating stuff, and the article is a good read. Publication will be both in print and online, so keep a lookout for it.

Closing Thoughts

Different world regions will have different sets of challenges to achieve sustainable fisheries. Each must take steps to address the multiple factors involved to achieve this objective. Providing science-based management advice is a part of the solution; thus, fisheries scientists have a critical role to play. This is an exciting and extremely multidisciplinary field with many niches to be filled. On any given working day at the NEFSC Woods Hole Laboratory, I might interact with coworkers who were aquarium keepers, biologists, chemists, computer programmers, cooks, economists, engineers, graphic artists, lawyers, librarians, modelers, administrative assistants, physicists, school teachers, scientific editors, scuba divers, ship captains, statisticians, taxonomists, oceanographers, you name it. Being a fisheries scientist can also take you to some pretty interesting and desirable places throughout the world. The past 38 years of my life will attest to that. Even now that I am stationed in Norway, I continue to have very interesting work travels, thus far to countries in Asia, Europe, and North America.

Having entered the field just after undergraduate studies, I did my share of grunt work and had to walk before I could run. But, I enjoyed the scenery along the way. Meeting each new challenge has taught new skills and presented new opportunities. Do not be afraid to knock on doors; people recognize and appreciate personal initiative and are usually more than willing to help. Staying open and conscious of new frontiers has, thus far, afforded me interesting and enriching experiences living and working on issues in the northwest Atlantic Ocean, the Northeast Atlantic, the Chesapeake Bay, the Indian Ocean, and the Arctic. These experiences have also expanded my professional network; currently, I am collaborating on exciting and very different projects with scientists located in Australia, Canada, France, Mozambique, Norway, the United Kingdom, and the United States.

To remain flexible, however, sometimes you have to go out on a limb. It is hard to earn a reward without taking a risk. And, it is a challenge to relocate when you have two careers and a family to consider. Prerequisites are a spirit of adventure, a desire to travel and experience different cultures, and agreement by all involved that the change would be good. Once the decision is made, jobs must be found and schools identified. This is a real process, and there is no single best way to proceed. However, there are a number of international organizations dealing with fisheries research and management that post job vacancies on their Web sites: FAO (Food and Agriculture Organization of the United Nations), ICES (International Council for the Exploration of the Sea), PICES (the North Pacific Marine Science Organization), and so forth.

Communication is a key. If young fisheries professionals choose to work in a country where English is not the first language, learning the national language might not be essential. But, making the effort to learn that language will help to ingratiate you with your colleagues, enhance your ability to contribute to meeting the institute's objectives, and make your stay more enjoyable and rewarding.

Biography

Margaret M. McBride is a senior scientist at the Institute of Marine Research (IMR) in Bergen, Norway, with more than 38 years of broad international experience at research institutions in the United States, Norway, and Mozambique. Her early career was as a fisheries biologist in the Population Dynamics Branch at NOAA's Woods Hole Laboratory. She also worked at the U.S. Fish and Wildlife Service in Annapolis, Maryland on the successful restoration effort for Atlantic coastal Striped Bass. As visiting scientist at the Institute of Marine Research in Bergen, Norway, she applied a method she had developed to estimate total removals from Norway's Barents Sea Atlantic Cod fishery. She dedicated three years to the Norwegian Foreign Service at the Fisheries Research Institute in Maputo, Mozambique. Later, she worked at the NOAA Chesapeake Bay Office where, as co-chair of a panel of regional scientists, she led development of a fisheries ecosystem plan for the bay. At IMR, she works within the Ecosystem Processes Research Group on issues related to ecosystem-based research and management. She believes that the human footprint on our planet should be minimized and that an ecosystem approach is needed for sustainable management of fishery resources.

Recommended Readings

Anyanova, E. 2008. Rescuing the inexhaustible... (the issue of fisheries subsidies in the international trade policy). Journal of International Commercial Law and Technology [online serial] 3:147–156.

Chesapeake Bay Fisheries Ecosystem Advisory Panel (National Oceanic and Atmospheric Administration Chesapeake Bay Office). 2006. Fisheries ecosystem planning for Chesapeake Bay. American Fisheries Society, Trends in Fisheries Science and Management 3, Bethesda, Maryland.

Food and Agriculture Organization of the United Nations. 2012. The state of world fisheries and aquaculture. FAO Fisheries and Aquaculture Department. Available: www.fao.org/docrep/016/i2727e/i2727e.pdf. (January 2014).

Gullestad, P., A. Aglen, A. Bjordal, G. Blom, S. Johansen, J. Krog, O. A. Misund, and J. Røttingen. 2013. Changing attitudes 1970–2012: evolution of the Norwegian management framework to prevent overfishing and to secure long-term sustainability. ICES Journal of Marine Science [online serial]. DOI: 10.1093/icesjms/fst094.

Maryland Fishery Resources Office. 2013. Striped Bass *Morone saxatilis.* U.S. Fish and Wildlife Service. Available: www.fws.gov/northeast/marylandfisheries/Fish%20Facts/striped%20bass%20facts.html. (January 2014).

McBride, M. 2006. Developing ecosystem-based fishery management plans. Pages 355–363 *in* Chesapeake Bay Fisheries Ecosystem Advisory Panel (National Oceanic and Atmospheric Administration Chesapeake Bay Office). Fisheries ecosystem planning for Chesapeake Bay. American Fisheries Society, Trends in Fisheries Management 3, Bethesda, Maryland.

McBride, M. 2006. Managed fisheries of the Chesapeake Bay. Pages 13–79 *in* Chesapeake Bay Fisheries Ecosystem Advisory Panel (National Oceanic and Atmospheric Administration Chesapeake Bay Office). Fisheries ecosystem planning for Chesapeake Bay. American Fisheries Society, Trends in Fisheries Management 3, Bethesda, Maryland.

McBride, M. 2006. Pathways to fisheries ecosystem plan implementation. Pages 365–368 *in* Chesapeake Bay Fisheries Ecosystem Advisory Panel (National Oceanic and Atmospheric Administration Chesapeake Bay Office). Fisheries ecosystem planning for Chesapeake Bay. American Fisheries Society, Trends in Fisheries Management 3, Bethesda, Maryland.

McBride, M. M., and B. E. Brown. 1980. The status of the marine fishery resources of the northeastern United States. NOAA Technical Memorandum NMFS–F/NEC-5.

McBride, M. M., and A. Fotland. 1996. Estimation of unreported catch in a commercial trawl fishery. Journal of Northwest Atlantic Fishery Science 18:31–4.

McBride, M. M., P. Dalpadado, K. F. Drinkwater, O. R. Godø, A. J. Hobday, A. B. Hollowed, T. Kristiansen, E. J. Murphy, P. H. Ressler, and S. Subbey. 2014. Krill, climate, and contrasting future scenarios for Arctic and Antarctic fisheries. ICES Journal of Marine Science. DOI:10.1093/icesjms/fsu002.

McBride, M. M., S. Subbey, and P. Baloi. 2011. Tradisjonelle småskalafiskerier i Mosambik: tidløs, tjenlig, truet. [Artisanal fisheries in Mozambique: timeless, treasured, threatened.] Kystmagasinet 8, Bergen, Norway.

National Marine Fisheries Service. 1999. Ecosystem-based fishery management: a report to Congress by the Ecosystems Principles Advisory Panel. National Marine Fisheries Service, Silver Spring, Maryland.

National Oceanic and Atmospheric Administration. 2002. White water to blue water initiative. Available: www.publicaffairs.noaa.gov/worldsummit/blueandwhitewater.html. (January 2014).

Northeast Fisheries Science Center. 2011. The historical development of fisheries science and management. Available: www.nefsc.noaa.gov/history/stories/fsh_sci_history1.html#a. (January 2014).

Richards, R. A., and P. J. Rago. 1999. A case history of effective fishery management: Chesapeake Bay Striped Bass. North American Journal of Fisheries Management 19:356–375.
Sherman, K., and A. M. Duda. 1999. An ecosystem approach to global assessment and management of coastal waters. Marine Ecology Progress Series 190:271–287.

When Opportunity Knocks, Don't Blink

Edward L. Mills* and James M. Watkins
Cornell Biological Field Station
Department of Natural Resources, Cornell University
900 Shackelton Point Road, Bridgeport, New York 13030, USA

Key Points

- Guidance of mentors is the foundation of a successful career path and impacts everything we do—research, creative activity, teaching, public engagement, and meeting day-to-day challenges.
- Prepare to travel many paths in your career through available opportunities and emerging issues.
- When opportunity knocks, don't blink.

Introduction

Careers are often built upon being at the right place at the right time and networking with key people. The 1970s was a time in U.S. history when the Vietnam War was waning, Americans waited in gas lines due to the Arab oil embargo, and jobs for newly graduated Ph.Ds. were scarce. Consequently, career building was a challenge during these times, and one learned very quickly that when opportunity knocks, don't blink.

Hard work, diligence, dedication, and an appreciation of the natural world are key building blocks for a career in the natural sciences. These traits and the willingness to take on new challenges can spell success no matter whether you are a student or a seasoned professional. When key people come your way in life, make sure that you look to them for strength, leadership, innovativeness, creativity, and humor. Engage yourself with mentors and take time to ponder what they are telling you. Give back to society and to future generations when possible. And for young professionals seeking a successful career, it must be repeated, when opportunity knocks, don't blink.

Undergraduate and Graduate Education: Guided by Mentors

The guidance of mentors sets the foundation of a career path and impacts everything we do as students and professionals. Mentors help shape a person's view of the world, philosophy, and understanding of the past. Mentors also help create and identify new opportunities.

Opportunity knocked early as an undergraduate biology major for Edward L. Mills (ELM) in the late 1960s at the State University of New York (SUNY) Geneseo, a small education and liberal arts school nestled along the ridge of the Genesee River valley plain in

* Corresponding author: elm5@cornell.edu

central New York State. Here, ELM met his loving wife, Lois, and Dr. Herman Forest, an aquatic ecologist and professor who introduced him to the amazing world of aquatic ecology and limnology. North America's freshwater lakes were choked in pollution from nutrient overfertilization and toxic chemicals at this time. High quality water was a key to life and a healthy environment, and the emerging public consciousness demanded that something be done.

The undergraduate years at Geneseo cemented ELM's appreciation for the natural world and provided a sound education and preparation for the next stage in life. Deciding whether to enter the job market or move on to graduate school was a pivotal question. With the help of undergraduate professors, ELM decided to pursue graduate studies and applied to several schools, settling on an M.S. program at prestigious Cornell University in 1969. Dr. Ray T. Oglesby, professor in the Department of Natural Resources, and Dr. Raymond Loehr, professor in the Department of Agricultural Engineering, had teamed up to provide assistantships with a focus on agricultural wastes and their effect on the environment through a training grant for new graduate students, and ELM was a proud recipient of an assistantship award.

Cornell opened up a new and exciting world of knowledge and learning. The late 1960s was a tumultuous period in United States history with the Vietnam War and all the antiwar sentiment, especially at universities like Cornell where there were numerous student protests against the war. Dr. Oglesby accepted three graduate students in 1969, two of which later voluntarily entered military service due to the fact that their draft lottery numbers were low. The Vietnam War complicated the lives of many students and their careers. ELM was one of the lucky ones, having completed an M.S. in 1972 and continuing on at Cornell for a Ph.D. under the tutelage of Dr. Oglesby.

ELM's graduate program at Cornell University provided valuable learning and networking opportunities, as well as the ability to design a unique graduate program consistent with one's professional and career goals. The latter led to independent and creative thinking, problem solving, and the ability to branch out and explore other disciplines, all critical elements that contributed to a successful career path. Cornell is an esteemed research university so it became readily apparent that graduate students needed to publish research so a broader community could benefit from one's latest findings. Last, Cornell taught the importance of networking—an essential ingredient for developing one's career by providing numerous opportunities to interact with fellow students and colleagues, scientists, policy makers, managers, specialists, and the public.

Cornell had some outstanding professors, and one soon learned the importance of absorbing as much knowledge as possible from these individuals. One of those influential mentors was Limnology Professor Gene Likens. Dr. Likens researched the Hubbard Brook Experimental Forest in the White Mountains of New Hampshire. An early finding of the Hubbard Brook study was that the rainfall in the region was highly acidic, and this finding led to one of the first scientific studies linking acid rain to air pollution. Dr. Likens' research revealed the critical link between ecosystem function and land-use practices and the importance of long-term ecological studies in assessing the impact of human activities on aquatic and terrestrial ecosystems. In addition, he taught the importance of science in guiding policy makers and the inclusion of the public in solving large-scale environmental issues. These concepts developed by Dr. Likens set the cornerstone of ELM's scientific career and provided the framework for designing and implementing long-term ecological studies of large lake ecosystems.

Next Step: Networking to Land the First Job

When opportunity knocks, don't blink. In the spring of 1974, ELM was introduced to Dr. John Forney by Professor Oglesby on the main Cornell campus. John was the first director of the Cornell Biological Field Station and a well-respected scientist and fish ecologist and considered a hall of famer by many professionals in the fish world. John sought a young research scientist with interests in limnology to work at the biological station located on the south shore of Oneida Lake. The position was funded for one year and could only be extended if additional funding became available. The interview was straightforward—use your skills and knowledge by teaching a one-day course on identification of aquatic organisms to undergraduates in a summer field course. In the fall of 1974, Dr. Forney offered ELM the job. For ELM, there was no blinking as it was very exciting to accept the position as a new Ph.D. graduate seeking employment. Now, with his foot in the door, this one-year position would grow and evolve for 34 years!

Dr. Forney's research focused on fish interactions in Oneida Lake, New York. These studies targeted interactions between Walleye *Sander vitreus*, Yellow Perch *Perca flavescens*, and zooplankton and were a first step toward a comprehensive understanding of how species' interactions impact community structure of freshwater lakes. ELM's research revealed a better understanding of the lower food web that included both zooplankton and phytoplankton. Our initial collaboration in the mid-1970s focused on the role of zooplankton in the energetics, growth, and food consumption by juvenile fish in Oneida Lake. These studies led to insights into mechanisms controlling larval fish abundance, including how interspecific competition and predation interact to regulate young fish recruitment. Likewise, we learned that predation by fish can extend to lower trophic levels. These early studies were pivotal in documenting that young fish could impact a lake's zooplankton community structure and that density-dependent effects of young fish wiping out their prey could indirectly impact their own growth and recruitment. And so the one year job opportunity with Dr. Forney to explore predation by young plankton eating fish would lead to further collaborative opportunities.

Engaging with colleagues at conferences, seminars, and other networking opportunities at Cornell confirmed that collaborations are essential and can lead to important scientific discoveries. ELM met Dr. John Confer, a zooplankton ecologist and colleague in the Department of Biology at nearby Ithaca College, at a Limnology and Oceanography Society meeting in Newfoundland where ELM presented findings linking predation by young fish and zooplankton. John Confer, intrigued by these findings, submitted a collaborative proposal with ELM to the National Science Foundation (NSF). True elation broke out when they were notified by NSF that the proposal had been funded, a big break for a young scientist (ELM) as it led to further collaborative opportunities. The NSF funded research focused on how feeding behavior and prey selectivity by young Yellow Perch influenced their growth and, ultimately young fish growth. The results challenged optimal foraging theory and showed that during their early life history, young fish select prey smaller than could otherwise be consumed, and by doing so, they consume prey with a higher surface-to-volume ratio, leading to greater conversion of food to tissue and growth. Consequently, years in which young Yellow Perch had abundant supplies of intermediate-sized zooplankton, namely *Daphnia*, were years in which fish growth increased, leading to better recruitment of juvenile and adult stocks.

Developing a Career Linked to a Long-Term Ecological Study of Oneida Lake

Long-term studies require a lifetime of dedication, passion, and hard work. Long-term studies of ecological systems are vital if we are to attain an accurate perception of ecological dynamics, understand the mechanisms that influence these dynamics, and differentiate the influence of human activity from natural variability. Looking back, the opportunity to work with a brilliant scientist like Dr. Forney and to contribute to the knowledge and understanding of a large freshwater lake ecosystem in Oneida Lake for more than three decades was both rewarding and monumental. John's passion for the Oneida Lake system and its fishery, his desire and dedication to scientific excellence, and his willingness to mentor ELM as a young scientist provided the ingredients for the development of a long-term ecological study of a large freshwater ecosystem in Oneida Lake.

Comprehensive long-term ecological studies like those on Oneida Lake now provide clues on the impact of such environmental issues as global climate change and biological invasions. Consequently, with the introduction of the zebra mussel *Dreissena polymorpha* to Oneida Lake in the early 1990s, the need to ask new research questions and develop new hypotheses soon became apparent. We hypothesized that primary production would decline as a result of mussel grazing in this shallow lake. Contrary to prediction, primary production did not decline despite increased water clarity and decreased algal biomass. The key to the absence of a decline in whole water column primary productivity was the compensating effect of increased water clarity resulting in photosynthetic activity penetrating deeper in the water column. Consequently, despite the order of magnitude increase in grazing rates and associated decrease in algal biomass, open-water production from algae to fish did not decrease in association with zebra mussels—a surprise indeed.

Expanding the Influence of Academic Research through Agency Partnerships

Partnering with state agency biologists in New York provided an opportunity to apply findings from the early zooplankton–fish studies on Oneida Lake to a broader array of lakes. It was through ongoing collaborations that John Forney had with agency biologists that ELM was able to establish new partnerships. One New York State biologist, Mr. Albert Schiavone, provided a unique opportunity to explore zooplankton–fish interactions in the Indian River lakes of northern New York, lakes in the same geographic setting but with different fish communities. The idea of melding fisheries and limnology led to interlake comparisons to assess the use of zooplankton as an index of fish community structure and ultimately as a basis for developing fish-management strategies. The research resulted in a publication in the *North American Journal of Fish Management* and was recognized by the American Fisheries Society as the most significant paper in the journal in 1983. The idea of using zooplankton as an indexing tool was further assessed for a broad spectrum of freshwater lakes. The published research was runner-up as the best paper in the *North American Journal of Fish Management* in 1987. Zooplankton–fish research continued into the early 1990s when studies were designed to assess the timing and movement of planktivorous-feeding Alewife *Alosa pseudoharengus* into the nearshore waters of Lake Ontario in the springtime. While these studies received recognition from the scientific community, a significant byproduct of the work was that the concept and ideas of using zooplankton as an indicator tool was accepted by fishery biologists at state and national levels. Now, fishery biologists began thinking about fish as part of a food web in their approach to fish management. The results of this research also inspired development of long-term data sets on zooplankton in the mid-1980s by state and federal agencies in the lower Great Lakes, namely Lakes Ontario and Erie.

Opportunities and Emerging Issues: North America's Great Lakes and Biological Invasions

One important lesson for professionals is the following: prepare to travel many paths in your career through available opportunities and emerging issues. The onset of such a lesson began after ELM met Dr. William Taylor (professor of fisheries, Department of Fisheries and Wildlife, Michigan State University) at the annual Limnology and Oceanography Society meeting held in St. Johns, Newfoundland. While this meeting brought us together because of mutual research interests, Dr. Taylor would later become a valued colleague, and this connection would lead to future collaborative research opportunities in the Great Lakes on an important emerging issue–the role of biological invasions. Through Dr. Taylor's efforts and his position as chair of the Great Lakes Fishery Commission's Board of Technical Experts (BOTE), Dr. Joseph Leach (Ontario Ministry of Natural Resources) and ELM became new members of BOTE and were charged with assessing the historical significance and importance of invasive species in the Great Lakes, surprisingly unknown information on one of the largest sources of freshwater in the world at the time (late 1980s). Initial studies were focused on patterns of invasion, entry vectors, and ecological impacts. Results of this research indicated that as human activity increased in the Great Lakes watershed, the rate of biological invasions increased. In fact, almost one-third of the organisms introduced into the Great Lakes (since 1959) coincided with the opening of the St. Lawrence Seaway. Two significant peer-reviewed publications resulted from this work. The first publication was entitled "Exotic Species in the Great Lakes: A History of Biotic Crises and Anthropogenic Introductions," which appeared in the *Journal of Great Lakes Research* and the second was entitled "Exotic Species and the Integrity of the Great Lakes: Lessons from the Past," which appeared in *BioScience*. Both publications were instrumental in the development of ship ballast water policy in the Great Lakes and led to the only ship ballast water law in the world. The opportunity that was presented to ELM by Dr. Taylor resulted in numerous collaborations, opportunities, and significant policy outcomes—an excellent example of when opportunity knocks, don't blink.

Mentoring from the Viewpoint of a Mentee

Reading his mentor's vignette reminded James M. Watkins (JMW) that however established a senior scientist can appear late in a career, attaining that position depended on a series of several unpredictable events, relationships, hard work, and a little luck. JMW first met ELM in 2005 while applying for a Lake Ontario data analyst position at the Cornell Biological Field Station. JMW had received a master's degree from Oregon State University several years prior but had been living a nomadic life teaching oceanography on sailing ships with the Sea Education Association. His only and most recent visit to the rural station was on a Cornell undergraduate limnology class field trip in 1988. JMW was looking to put down roots on land, but interesting opportunities were few. Fortunately, opportunity knocked. ELM offered JMW the Cornell position, and JMW did not blink. JMW was on his way to working on the lower food web of Lake Ontario and pursuing a career path in Great Lakes ecology.

At JMW's job interview, ELM asked in passing if JMW had considered going back to graduate school. JMW occasionally had but was unsure who his advisor or what his thesis project would be. After JMW started the position, ELM, Dr. Lars Rudstam, and assistant Kristen Holeck quickly introduced JMW to Lake Ontario as well as to a broad American–Canadian network of academic and agency scientists and policy makers devoted to its research and integrity. JMW found his oceanography background to mesh well with the study of large lakes and had the freedom to pursue several of his research interests. From this one-year study of the lower food web in Lake Ontario, JMW developed an important thesis project—understanding the recent decline of the native amphipod *Diporeia* in the Great Lakes. Funding was tight, but through a series of Integrative Graduate Education and Research Traineeship fellowships and research and teaching fellowships, JMW was able to achieve his goal of a doctorate in 2011.

ELM retired in 2009 but was a stalwart supporter of the renewal of JMW's academic path as his academic advisor. ELM's positive nature and greeting always brought JMW up from any bad news. He never hesitated to introduce JMW to any of his long list of collaborators. Most importantly, from day one, he encouraged JMW to seek external funding through grant writing. This new skill has served JMW well as a "young" professional. Last, he constantly reminded JMW to clearly separate work from home with his wide range of hobbies and interests with his wife, Lois. Although retired from the field station, he continues to use his emeritus status to actively lead local lake associations and federal invasive species committees. And, of course, his growing grandkids take front and center. (Thank you, Ed!)

Reflections, Recognition, and Relationships

Professional careers evolve and are shaped in many ways through mentorship and collaboration. As illustrated in this vignette, key individuals and networking can lead to opportunities that shape one's scientific and professional career. Scientific discovery evolves and scientists are only recognized for brief windows in time. You may be relishing the limelight today for an important scientific breakthrough and discovery, but that that will fade as new discoveries are made. Consequently, good mentors stress timely publication of results so that findings are not only shared with a broader community, but also researchers are duly recognized for their efforts.

Scientists are often so consumed by their work that they let the world pass them by. Take advantage of opportunities that bring you to new places and present new challenges. And the importance of meeting new people and building new relationships cannot be overstated. Take up a hobby or two that fosters networking with new and different people. Yes, today there are smart phones and all sorts of technology that allow for networking and communication opportunities. The most effective communication, however, continues to be written and oral. Practice your communication skills often and opportunities will abound. Face time with colleagues and fellow scientists is so critically important and can have significant consequences not only for developing new paths of scientific discovery, but also for creating new professional and career opportunities.

Give back to society when and where you can…become a mentor yourself. It is especially rewarding to see students utilize their skills and lessons learned and fledge into the world to become mentors themselves. Mentoring is a life-long journey—perhaps the best way to give back to society in a very meaningful way. The guidance of mentors is the foundation of a successful career path and impacts everything we do—research, discovery, creative activity, teaching, public engagement, and meeting day-to-day challenges. Good mentors challenge their mentees to problem solve, think critically, and strive for excellence. Last, the world of science and discovery is exciting and challenging, so when opportunity knocks, don't blink as you climb the professional ladder to success and make significant contributes to society. Your mentors are depending on you.

Acknowledgments

The authors wish to thank Cornell University and members of the Cornell community for providing our career path foundation as aquatic ecologists and for strengthening our resolve for initiative, integrity, and excellence. ELM wishes to thank the many students, support staffs, mentors, and colleagues that contributed so much to a successful research program and career. ELM also wishes to thank Cornell University, the Charles Brown endowment, and New York State for their dedication and support of long-term ecological studies, particularly those on Oneida Lake. JMW thanks ELM for his mentorship as supervisor and thesis advisor in developing him as a Great Lakes researcher.

Biographies

Dr. Edward L. Mills is currently emeritus professor in the Department of Natural Resources, College of Agriculture and Life Sciences at Cornell University. He became director of the Cornell Biological Field Station located on the south shore of Oneida Lake in 1995 following the retirement of the station's first director, Dr. John Forney. His love for Oneida Lake continues as he is currently serving as both director and president of the Oneida Lake Association, a large environmental organization organized in 1945 to protect the natural resources of this wonderful water body.

Dr. James M. Watkins is currently a postdoctoral research associate at the Cornell Biological Field Station. His research focus on the Great Lakes follows in the footsteps of Dr. Mills and current director Lars Rudstam. He is coordinating Cornell's recent collaboration with the U.S. Environmental Protection Agency's Great Lakes Program Office to monitor zooplankton, benthos, and chlorophyll for all five Great Lakes. He is also involved in the 2013 binational intensive sampling effort for Lake Ontario.

So You Say You Love Fish

MICHAEL P. NELSON*
Department of Forest Ecosystems and Society, Oregon State University
321 Richardson Hall, Corvallis, Oregon 97331, USA

JOHN A. VUCETICH
School of Forest Resources and Environmental Science, Michigan Technological University
1400 Townsend Drive, Houghton, Michigan 49931, USA

KATHLEEN DEAN MOORE
The School of History, Philosophy, and Religion, Oregon State University
322 Milam Hall, Corvallis, Oregon 97331, USA

Key Points

- Conservation requires both an ability and a willingness, both scientific information and ethics.
- Scientists can contribute importantly to the kind of understanding that leads people more directly to right action.

The rope lies in the cellar for years,
coiled, stinking of the sea and the fish
that once lived in the sea and the sweat
of the man who wishes he could save one
strand of the world from unraveling.
—Alison Hawthorne Deming, from the poem "Rope"

The scientific community has been relatively ineffective in conveying this message of planetary change to our society, whose collective choices propel us along this path. As scientists, we are trained to avoid speaking in ways that touch people's souls.
—F. Stuart Chapin, III

People become natural resource scientists and professionals usually, we believe, because they possess a deep-seated desire to protect and restore natural systems and populations. We see this desire in the conservation scientists with whom we have collaborated and in the students we have taught over many years. But why? Why do people choose a science career in order to protect the animals and populations they study? Perhaps because at some level they believe in the direct connection between what people know and what people do. That is, people who become scientists in this field believe that conservation science is directly

* Corresponding author: mpnelson@oregonstate.edu

linked to conservation, or so it is assumed. Perhaps aspiring conservation scientists believe that if scientists only knew how the mercury from burning coal affects fish populations, if citizens only knew how their use of fossil fuels was impacting the climate, or how their land use practices were directly linked to the extinction of bird species, we would collectively mend our ways, we would cease these harmful practices, we would become conservation-minded and activated.

This belief even has a name. Social scientists refer to it as the "information deficit model" (IDM) of behavior change. Within the realm of conservation, this model would imply that conservation (as a behavior) is most fundamentally limited by information. And information comes from science. The message could scarcely be clearer: the path to conservation is through information, the kind gathered by the sciences.

The assumed relationship between information and action is embodied in the current debate over anthropogenic climate change. As the public fails to respond adequately to this reality, scientists redouble their efforts to spread the message, speak with a single voice, speak with greater volume and force and skill.

But, here is the painful problem: the IDM does not reflect reality. As social scientists have been telling us for decades, the IDM does not explain how the world works, does not represent how behavior really changes. There is no, there can be no, necessary and linear connection between facts about nature, or facts about changes in nature, or even facts about how it is that people value nature, and the conservation of nature. We do not act simply because we know. We do not care for something simply because we understand how. The discovery that Canadian oil sand mining operations release toxins into freshwater ecosystems (Kurek et al. 2013) does not, by itself, imply the cessation of Canadian oil sands mining. The fact that Brook Trout *Salvelinus fontinalis* populations will almost certainly succumb to the affects of anthropogenic climate change does not, by itself, mean that we ought to abandon them to that fate. To arrive at a prescription for action requires us to consider facts together with deeply held values.

Several kinds of evidence show us that information and actions are not related in this way. Knowledge alone does not necessarily lead to right action because the health of the Earth's species and populations and ecosystems is not improving even as conservation science grows. On the contrary, we seem to cause greater harm even as information and knowledge increases. Moreover, social scientists have repeatedly demonstrated, for the past 40 years, how our basic beliefs about the world are not necessarily reflected in our actions. As one review paper puts it, there is a well-known "gap between the possession of environmental knowledge and environmental awareness, and displaying pro-environmental behavior" (Kollmuss and Agyeman 2002).

We may even find evidence against the IDM in our own actions, yours and ours. Think about the relationship of knowledge, attitudes, and actions in your own life. At this point in your career, you know a great deal about fish, more than nearly any other person in the world. Do you still sometimes treat fish and the environments they depend upon in ways of which you are not always proud? Why? Is it because you have not yet acquired enough facts about fish? Do you really believe there is some magical fact about fish that you have not yet discovered, that when you do you will begin to treat fish better?

When you are in a restaurant, perhaps you refrain from ordering fish you know to be harvested unsustainably. Maybe you would remove other species from your diet if you learned they too were being fished unsustainably. This is laudable, but think harder. Do you, for example, contribute more than you should to water pollution? If so, is this because you do not know that water pollution is a problem for fish populations? Likely you already know quite a bit about that. Is it possible that your actions are limited by not having yet learned how to love fish enough? Of course you love fish. But do you love them enough to pollute less? If not, the challenge is not the acquisition of more facts, but learning to love more. The challenge is learning how to do that and then teaching others how to do that.

This disconnect between "knowledge about" and "love of" is in large part an extension of a logical mistake. Consider the logic we use to reach conservation policy decisions of any type. Any argument arriving at a conclusion about what we ought to do must have two premises. The first presents the facts of the matter. These facts are delivered to us by science. The second premise presents the values and principles at stake, the culture's collective moral wisdom about what is just and good and fair. Consider this argument, that some might point to, as an example:

> Premise 1. Fish are capable of experiencing pain and suffering caused by catch-and-release fishing [or electrofishing].
>
> Premise 2. We should not cause unnecessary pain and suffering.
>
> Conclusion: We should refrain from catch-and-release fishing [or electrofishing] unless it is absolutely necessary.

The first premise is an empirical claim (Braithwaite 2010). It might be disputed, but it falls within the knowledge domain of science. But, the second premise is critical. If it is true, the conclusion is true. And this second premise is not the realm of science, but rather the realm of the humanities, art, poetry, and other disciplines charged with understanding, shaping, and conveying our collective values. Only when we combine facts and values can we arrive at a conclusion we would recognize as conservation (such as concluded above for these two premises).

In a nutshell, this is our mistake. We mistakenly take an important element of conservation for conservation itself. We mistakenly jump from the premise that good conservation science is critical for conservation to the conclusion that good conservation science will somehow deliver conservation. We forget that we conserve not only because we can, but also because we want to, that conservation is dependent upon both an ability and a willingness.

We believe all of this is both true and a tragic indictment of so much of our current conservation efforts. So the critical question for fisheries scientists and professionals is this: how can scientists, without leaving behind the scientific world in which they are credentialed, competent, and confident, contribute to the kind of understanding that leads people more directly to right action (see Box 1)?

The answer begins by recognizing that knowledge comes in two kinds. One kind of knowledge helps us do things in the world—knowledge to help us conserve nature, knowledge to restore damage we have caused nature, and knowledge to live sustainably. However, the knowledge that helps us do good things can also be used for the most disgraceful endeavors, to live unsustainably and to unnecessarily exploit others, be they human or otherwise. Some of our most unhealthy relationships with nature are fueled by knowledge of how nature works (or our belief that we know how nature works).

We might use our knowledge about wildlife habitat selection to work to save a species from extinction, and we might use that same knowledge to more efficiently eradicate a species. It is not our knowledge but our attitude that determines whether we use knowledge to do right or wrong, good or bad.

Box 1. Science and Ethics Dancing Together

In November of 2011, a group of environmental scholars—scientists, philosophers, writers, poets, and religion scholars—gathered at the H. J. Andrews Experimental Forest in the Oregon Cascades, a designated long-term ecological research site, to write a new, ecologically inspired ethic for our time. Note the interplay between science and ethics, between what we know about the world empirically and what we value. We offer this as one example of collaboration between science and ethics. Here is some of what they wrote:

> Humanity is called to imagine an ethic that not only acknowledges, but emulates, the ways by which life thrives on Earth. How do we act, when we truly understand that we live in complete dependence on an Earth that is interconnected, interdependent, finite, and resilient?...
>
> The questions of our time are thus: What is our best current understanding of the nature of the world? What does that understanding tell us about how we might create a concordance between ecological and moral principles, and thus imagine an ethic that is of, rather than against, the Earth?...
>
> The necessity of achieving a concordance between ecological and moral principles, and the new ethic born of this necessity, calls into question far more than we might think. It calls us to question our current capitalist economic systems, our educational systems, our food production systems, our systems of land use and ownership. It calls us to re-examine what it means to be happy, and what it means to be smart. This will not be easy. But new futures are continuously being imagined and tested, resulting in new social and ecological possibilities. This questioning will release the power and beauty of the human imagination to create more collaborative economies, more mindful ways of living, more deeply felt arts, and more inclusive processes that acknowledge the ways of life of all beings. In this sheltering home, we can begin to restore both the natural world and the human spirit.

The complete text and list of authors of "The Blue River Declaration" can be found online at http://springcreek.oregonstate.edu/documents/BlueRiverDeclaraton.2012.pdf.

Knowledge that can change our attitude about nature is the second, arguably more important, kind of knowledge. It is also the kind of knowledge that we spend less time and effort developing. Think about knowledge that makes you go, "Wow!" Wow, that's so beautifully complicated… wow, look how magnificently nuanced… wow, how astonishingly connected. Wow: to be held in a state of wonder about nature. It would seem exceedingly difficult to intentionally abuse nature while being held by its wonder. How can you do anything but care for nature, while astonished by its beauty, complexity, and interrelatedness?

How does one go about conducting wonderment-generating science? It is largely characterized by audience and purpose. In regular science, the audience is often just the dozen or so other scientists that might read the technical paper describing your discovery. For wonderment-generating science, the audience is much broader. It is any segment of the general public—elementary or secondary school children, adults at the public library or senior center, or whatever. In other words, the audience for wonderment-generating science is your fellow human beings, the people who are funding your salary and your research and the people counting on you to show them how we ought to relate to nature. The message is to convince this audience, convince them beyond a shadow of a doubt, why you love or are in awe of something in nature such as fish. Explain how it is that your life story brought you to this point. You have to explain why you love fish at every level in the hierarchy of life—fish genes, individual fish, fish populations, and the role of fish in the ecosystems they inhabit. You should know that you are better qualified for this task than virtually every other person on the planet. No one else can take your place. If you are unwilling or unable to communicate that love for fish, how can anyone else be expected to love fish? And this is not a one-time exercise. This is your life-long vocation: to become better and better at sharing with others why you love fish. Done effectively, this love becomes infectious. It will become the foundation for why we should all love fish, or anything else you that has wowed you.

Conducting wonderment-generating science requires two strategic skills. One skill is learning how to pursue wonderment-generating science in a world that is mostly focused on science aimed at learning to control nature. Certainly, there are always constraints, but within those constraints there is always a great deal of freedom, and it is too easy to forget that. Your challenge is to ensure your research is maximizing the opportunities to illustrate why we should love fish. If that ambition is ever on your mind when you are developing and conducting your research, you will achieve the objective. You also have to be prepared to conclude that some research is, quite simply, not worth conducting, even if someone is willing to pay for it.

The second skill involves messaging. Rather sadly, we, as scientists, have some serious inhibitions about sharing our love of nature. This is a skill that requires cultivation. Interestingly, people seem to begin loving nature without further provocation when they are presented with wonderment-generating narratives about nature, narratives that build on themes like interconnectedness, contingency, complexity, and empathy. If you learn how to communicate your research and love of fish as a story built on those themes, then you will be doing the most important work you could ever possibly be called to do.

Before engaging wonderment-generating science, you have to ask yourself an important question. Do I really love fish, and if so, why? Do I love them mainly for the selfish interest, to satisfy my own curiosity about how they work or because I love to work outside near the water? Do I love fish mainly because they are what have made me successful and admired by colleagues? Do I love fish mainly because fish are so important to human welfare? Curiosity, professional success, and human welfare—that is all fine, but it misses the mark. Do you love fish—fish genes, individual fish, fish populations, and their role in ecosystems—because they are wonderful creatures that deserve your love. Unless you and your peers scream "yes" to this last question, unless we can all hear the message of love echoing from harbors and river valleys, then rest assured, fish are doomed.

Acknowledgments

The ideas in this essay did not develop in a vacuum. We thank colleagues Rolf O. Peterson, Leah M. Vucetich, Joseph K. Bump, Jeremy Bruskotter, Paul Paquet, Meredith Gore, Frank

Moore, Robin Kimmerer, Charles Goodrich, Fred Swanson, Julia Jones, Carly Lettero, and many others who travel this path with us.

Biographies

Michael P. Nelson is the Ruth H. Spaniol chair of renewable resources; professor of environmental ethics and philosophy; the lead principal investigator of the H. J. Andrews Long-Term Ecological Research Program; a senior fellow of the Spring Creek Project for Idea, Nature, and the Written Word; a co-founder and co-director of the Conservation Ethics Group; and the philosopher-in-residence of the Isle Royale Wolf-Moose Predator-Prey Project. He is in the Department of Forest Ecosystems and Society at Oregon State University in Corvallis, Oregon.

John A. Vucetich is an associate professor of animal ecology and the co-director of the Isle Royale Wolf-Moose Predator-Prey Project (the longest continuous predator–prey study in the world), in the School of Forest Resources and Environmental Sciences at Michigan Technological University in Houghton, Michigan. He is also the co-founder and co-director of the Conservation Ethics Group.

Kathleen Dean Moore is a university distinguished professor of philosophy and a senior fellow (and director emeritus) of the Spring Creek Project for Ideas, Nature, and the Written Word in the School of History, Philosophy, and Religion at Oregon State University in Corvallis, Oregon. She is also an award-winning essayist. Her latest books include *Wild Comfort: The Solace of Nature* and (with Michael P. Nelson) *Moral Ground: Ethical Action for a Planet in Peril.*

Let's Play Two: Optimism Makes All Things Possible

LARRY A. NIELSEN* AND GRETCHEN L. STOKES
Department of Forestry and Environmental Resources, North Carolina State University
Campus Box 8008, Raleigh, North Carolina 27695, USA

Key Points

- Optimism is a key ingredient for a successful fisheries career, especially for rising leaders facing ever-growing and complex challenges.
- Reality, humility, and a firm grasp on risks and uncertainties are necessary to temper optimism.
- Fisheries professionals need to focus on turning obstacles into possibilities if they want a chance at changing the world.

Ernie Banks and Uncle Joel: Our Heroes

Ernie Banks always has a smile on his face. Many pessimists would say he has no reason to smile, but Ernie knows better. Ernie Banks played shortstop and later first base for the Chicago Cubs from 1953 to 1971. The Cubs were a perennially losing team during Ernie's career, but that never dampened his enthusiasm—or his success. He hit 512 home runs, played on 11 all-star teams, was twice named the National League's Most Valuable Player, and was inducted into the Baseball Hall of Fame in 1977. He was a perennial optimist. Throughout Ernie's career, baseball teams often played two games, one right after the other, on Sundays or to make up for a rain-out. Most players groaned at the exhaustion brought on by an 18-inning day, but not Ernie Banks. When asked how he felt about any upcoming game, he would flash his mile-wide smile and say, "It's a great day for baseball. Let's play two!"

Uncle Joel could rival Ernie Banks in optimism. A natural athlete, he plays basketball, swims, and skis with the best of them. He has been a beloved Scout leader for decades. He carried the Olympic torch for a leg of its journey across the United States in 1996. But another quality elevates his optimism above that of other athletes and leaders—he is also a paraplegic. Despite his paralysis, Uncle Joel has taken on and conquered every obstacle in his way, obstacles that would have discouraged most of us. He does it all with never a complaint or a negative word. He used his condition as the basis for a new career path, and he now teaches others in similar situations about how to return to the activities they love most.

Ernie Banks and Uncle Joel were our childhood heroes. One of us (LAN), growing up on the northwest side of Chicago, was born and bred to be a Cubs fan, a habit of faith that continues to this day. The other (GLS) got her first glimpse of lifelong optimism from Uncle Joel, learning as a young woman to complain less, appreciate more, and seize opportunities

* Corresponding author: larry_nielsen@ncsu.edu

in every aspect of life. As a consequence, we have both become unrequited optimists—a trait we believe is essential for success.

Behaviorists will argue about what separates humans from other animals, but one certain difference is that between action and reaction, humans engage the abstract process of choice. We choose how to react. And our assertion is that the right choice for success is always optimism. An optimist sees opportunities where a pessimist sees roadblocks. An optimist takes the initiative, grabs the controls, and steps on the gas while a pessimist takes a backseat, waiting until the easy or simple solution becomes apparent. Surrounded on many sides by pessimism and apathy, the optimist will often find a way to make a positive difference. The optimist is a can-do person, willing to tackle challenges that would scare away most people. An optimist will sometimes fail and also often reach only partway to a lofty goal, but more importantly, an optimist will sometimes succeed and will always eagerly anticipate the next opportunity, the one that just might change the world.

The Making of *Fisheries Techniques*

I (LAN) credit optimism for an outcome that almost every fisheries student and professional has held in his or her hands—the American Fisheries Society's (AFS) textbook *Fisheries Techniques.* The fisheries profession had needed a new techniques book for at least a generation, but the daunting prospect of writing such a book scared every pessimistic author away. A techniques book, after all, requires descriptions of tools and best practices from headwater stream to estuary, from electroshocker to bottom trawl, from shellfish to sturgeon. No one is a master of all these dimensions of techniques, and no one had been ready to throw a career at this need.

Throughout our graduate programs at the University of Missouri, my colleague, Dave Johnson, and I discussed this need and optimistically daydreamed about taking on the challenge. When I was elected president of the adolescent AFS Education Section while still an adolescent professor at Virginia Tech, I saw my chance. Dave and I presented a plan to our AFS faculty peers and negotiated through a spirited discussion of who, what, where, when, and how. In the end, our enthusiasm won, but I am sure our more senior colleagues left the meeting thinking they had heard the first and last of this harebrained scheme. But we are optimists, and we persisted. With the bravado that only naiveté can sustain, we dove into our roles as co-editors. We created an outline, a strategy, and a timeline. We purposefully assembled a team of authors who were not necessarily the world's experts on each topic but who were mostly mid-career teachers—professionals who knew the need for the book and could be counted on to write clearly and confidently for the intended reader, an undergraduate student in fisheries. We had to replace a few authors along the way—ones who disappointed us by letting obstacles get in their way—but mostly our band of optimists held together. Dave and I knew nothing about publishing a book, so we did it all the hard way, down to selecting a printer and even the paper and ink colors (examine the first edition closely and you will notice that it is printed with dark brown ink on off-white paper).

In early 1983, about 18 months after we started, the book rolled off the press, a testament to the power of optimism. Now in its third edition, *Fisheries Techniques* has sold more than 50,000 copies, outpacing any similar book in our profession's history. The impact, however, extends far beyond the book's iconic red covers. We knew that other textbooks were also needed, so we insisted that a portion of the price of each copy would establish a fund to help the AFS Education Section jump-start its next textbook project. Two more books followed, *Fish Biology* and *Fisheries Management*, produced by equally optimistic editors who took on and conquered an equally big and worthwhile challenge.

Two Voyages to Remember

As a young adult, one of my (GLS) pivotal leadership experiences took place aboard the SSTV *Gremlin*, a 38-foot sloop equipped with provisions for a weeklong Sea Scout Advanced Leadership training course. I loved the sea, and learning to handle a ship seemed like a good step towards a fisheries career. I stepped aboard full of optimism, well prepared and eager for the leadership challenges that would come, or so I thought. July 4, 2008 would challenge all of that.

Five of us, Sea Scouts from across the country, had gathered together to captain a boat through Galveston Bay. Each of us was assigned to take command for one day. My day fell on our final leg at sea. Feeling confident and skilled, I began with salutes and instructions for getting underway. Then reality stepped in. Dodging 60-ton shipping vessels in the Houston Ship Channel, miscommunication among crew members, and urgent calls to a nonchalant drawbridge operator were just a preview. The real show came in the force of tropical storm conditions bearing down on Galveston Bay and our ship.

As gale force winds blew lawn chairs and other debris past our bow, a leader's worst fear came true: one of my crew members toppled over from a blow to the back from the boom, a metal beam notorious for injuries. She lay on the deck, nearly paralyzed, unable to continue her duties, let alone move to safety. The injury required an immediate decision to return to shore and the associated navigational issues, all resting in my hands. I credit optimism and perseverance for helping me stay cool through the inevitable chaos and stand up to the test as I plotted a proactive approach that brought us to shore promptly and safely. Despite the fully saturated crew, a shortened cruise, and a trip to the emergency room, the day proved the power of positive thinking and led me to the top of my class of candidates.

My second voyage was not on a ship but in small villages in Vietnam and Cambodia. An independent study opportunity took me to Southeast Asia to study fisheries. While there, I interviewed a collection of subsistence fishermen, aquaculture managers, fish processors, and commercial operators, all of whom remarked about the dramatic reduction in size and abundance of fish stocks. Meanwhile, the public officials who set and enforced fishing regulations claimed that no such changes were occurring, seeming to look the other way in the face of both data and fishermen's testimony.

I was both frustrated and humbled to work alongside the men pulling their seine nets in the floating villages or the workers laying their small fish on racks to dry. While some people might have despaired over the crushing difficulty of improving this situation, I felt differently. Partially because of the example of Uncle Joel, partially because of my previous experiences, and partially because of my innate optimism, I felt an intensified motivation to create a positive change. Although I personally could not increase stock size, halt the building of dams, stop the illegal poaching of giant catfish, terminate sewage discharge in the lakes, or put more fish on the table, what I could do was learn, ask, and synthesize to better understand these issues. I am more motivated than ever because I know that by gaining an education and striving to become a leader in international fisheries management, I will be able to better support the livelihoods of these people. I cannot help but let my optimism prevail, knowing that this could be my opportunity to change the world.

The Trouble with Optimism

We would be bad optimists if we did not discuss the downsides of optimism. We offer two caveats to temper our virtuous portrayal of optimism. First, the optimist must acknowledge that every idea has downsides and risks. The ditches along the road to success are littered with innovators and entrepreneurs who failed to consider the possible negative consequences of an action.

I (LAN) chose to resign as Provost at North Carolina State University (NC State) because of a failure to anticipate downstream risks. I had hired the First Lady of North Carolina, who was then a faculty member at another state school, to create and implement a new speaker series on campus, which became known as the Millennium Seminars. The seminars were extremely popular, bringing world-class speakers to campus four times each year, and the First Lady herself was equally popular. Four years later, however, questions about how she had been hired caused a controversy in the state and on campus that just would not go away. I had never anticipated that a successful academic program might turn into a political liability; I had never anticipated that any decisions I might make would have political consequences. Now, I know that hiring a political figure, even when done appropriately and with universal support, always brings risk. A little more forethought by my advisors and a little less naiveté by me might have prevented a regrettable outcome, an outcome that no one wanted but eventually became inevitable.

I (GLS) have found optimism leading to overly ambitious actions when forming a new student sustainability organization. In my first year at NC State, I spear-headed a sustainability raft race that showed great potential; four groups made rafts from recycled materials and sailed them across a local lake, racing for a victorious win and a front-page story in the school's newspaper. We had a successful event, and we were all encouraged by the possibility of what the race could become in the future. The next year, we tried to scale up this event to become the greatest tradition at NC State. We asked many groups to build rafts as a major part of their annual activities, but they had other priorities already in place. We looked for sponsors throughout the university and community, but they had ongoing commitments and shrinking budgets. We tried to assemble an expanded leadership team and work crews, but few were able to drop other activities to join ours. This raft race was just too new, too unproven to bypass all the other programs and opportunities at the university. In the end, the attempt to scale up the activity in one big step became too much for the original leaders, as we ran out of time and energy ourselves. We fell short of our goal and had to settle back on the same small race we had before. As graduation loomed, we also had to pass our leadership in the race to the next generation of students, and we worried about their ability and interest to sustain the race, ironically in a sustainability organization. We realized that the optimism of one or two is not always enough. Had we been more patient and realistic, setting up achievable milestones for gradual expansion, our second year might have been a strong step for a successful future; if so, we might have passed the race down with more fuel instead of leaving it half empty. (We are happy to report, however, that our misstep was not fatal; a new group of optimists succeeded us, and the race continues today.)

Second, the optimist must avoid dogged loyalty to an idea that just is not going to work. Michael Eisner, former CEO of Disney and an avowed idea man, has observed that only about one idea in ten is a good idea, and only about one good idea in ten is practical. So, only one Pinocchio in a hundred is going to become a real boy—something the optimist must understand. The effective optimist must listen to bad news from his (her) peers and advisors when they give thumbs down to an idea. The successful optimist must be sufficiently self-aware to discard an idea that is dead on arrival.

As an avowed "idea man" for my entire career, I (LAN) have created and then abandoned many more ideas than I have ever implemented. I have planned for a campus in South Korea, a new medical school, a fisheries and wildlife museum, an environmental science degree, an endowment fund to support teaching, and a weekly doughnut time to bring faculty and students together. None of these have worked. Maybe they were bad ideas, maybe they were just impractical, but I gave them up willingly when it was clear they would not work.

Optimism and Fisheries

It is true that too often we see a grim reality: pollution continuing into the estuary, the latest assessment of an overharvested fish stock, or a conversation with a discouraged and worn-out natural resource manager. Fish and other aquatic organisms hide from us in a watery world that we barely understand. They roam over vast distances well beyond any one person's or one jurisdiction's control. The resources are mostly public, so everyone wants a say in how they are managed. Just about every characteristic of fisheries screams, "This is too hard."

Fisheries management is risky. Risky situations are ones in which the probabilities of certain outcomes are unknown. Is the stock fully exploited or could it produce more? Will these introduced fish spawn or not (and do we want them to or not)? Is 10 parts per million of a chemical safe or not? Making risky decisions takes courage and good judgment, qualities that may be lacking in a pessimist who shuns the responsibility for a decision gone wrong or in an optimist who has not learned to temper enthusiasm with reality.

Fortunately, our profession and society as a whole have made substantial progress in addressing risk over the past generation. Decision analysis, risk assessment, and the study of so-called "wicked" problems are tools that can be used to help us reduce the uncertainty surrounding our decisions. We recommend that every prospective fisheries professional become acquainted with these tools in order to better prepare for the time when they will be called to lead.

Despite these wicked problems and risky situations, an intelligent and persevering leader will use these obstacles as motivation to move forward, knowing that they have the power to create positive change in the world, the power to act before they must react. The tool of optimism is relevant to fisheries every day, but especially on those days when it looks the grimmest or foggiest. Faced with satisfying the different demands of aggressive fisheries stakeholders, a frustrated pessimist might wave the white flag of surrender. An optimist, however, might design a collaborative management process that brings battling stakeholders to a common table where they learn to understand and appreciate each other and work together for the nobler goal of restoring the fishery, no matter who gets the benefit. A pessimistic might give up trying to sample water quality throughout a watershed because there are too many sites and not enough time. An optimistic, however, might create a science project for schools in the watershed that has middle-school students collecting and reporting water quality data—which, coincidentally, may land some new recruits for the fisheries profession.

A successful career in fisheries needs optimism just as much as Ernie Banks and Uncle Joel do. There really is power in positive thinking: the power to inspire, motivate, encourage, and lead. The fisheries glass is neither half empty nor half full but rather twice as big, abundant with opportunity, rich with inspiration, and awaiting the extraordinary leaders of tomorrow. If you can strike out today and come back tomorrow ready to "play two," you are likely to be carrying the Olympic torch of fisheries one day.

Biographies

Larry A. Nielsen is professor of natural resources at North Carolina State University (NC State). He previously worked at Pennsylvania State University and Virginia Tech, in positions as a faculty member, department head, and school director. At NC State, he has been dean, provost, and, now, a faculty member again. He has degrees from the Universities of Illinois, Missouri, and Cornell. He is author, co-author, or co-editor of five books. He is an honorary member of AFS and was the society's president during 1990–1991. He teaches the principles of conservation of natural resources, an optimistic subject that matches well with his personality.

Gretchen L. Stokes is a recent graduate in fisheries, wildlife, and conservation biology at NC State. She is the recipient of Udall, Hollings, and NC State's Park Scholarships and a member of Phi Beta Kappa. At NC State, she served as the vice-president of the Student Fisheries Society, founder of the Sustainability Club, and ambassador for the College of Natural Resources. Her education has been enriched by learning experiences in Puerto Rico, Equatorial Guinea, Cambodia, Vietnam, and the Pacific Northwest. She plans to attend graduate school, followed by a career in fisheries ecology or international fisheries management, knowing that her optimism will propel her forward.

Curiosity Comes Naturally, but Three Other C's Must Be Learned—the Earlier the Better

Steven G. Pueppke*
Michigan State University
446 West Circle Drive, East Lansing, Michigan 48824, USA

Key Points

- Curiosity comes naturally to most scientists. Do not allow it to wither away as you mature and become comfortable in your career.
- No matter what you do as a scientist, your work always fits within a broader framework. The more you take the big picture into account, the greater your effectiveness. This is *contextualization*, the first of the learned C's.
- You can expand your perspective as a scientist if you work with others who see the world differently from you. Partnerships enhance your impact and allow you to have more fun doing your job. *Collaboration* is the second of the learned C's.
- Good work is necessary, but it is not sufficient. You also have to be able to inform others of what you do, inspiring them in the process. *Communication* is the third of the learned C's.

Introduction

A career in water might seem an unlikely choice for a farm kid from arid North Dakota, but I knew about food production and was determined to become a scientist. Like many of my generation, I was drawn to the concept of research solutions that might actually benefit humanity. And so, I would devote my life to erasing hunger with food harvested from the seas. Such were my convictions in 1971, when I was as sure of my future as any college senior can be. But, it would prove too good to be true.

When contemplating graduate school, I hedged my bets, applying to further my education not just in marine biology, but in two other disciplines that I found interesting. The graduate oceanographers invited me to join them, as did the botanists, but my application to the environmental science faculty at Cornell University was rejected. And so they sent it to the Department of Plant Pathology, which I had hastily scribbled onto the application form as my second choice. Soon my mailbox contained a letter asking if I was serious about this unknown discipline. Cornell had offered US$3,200 a year—real money in those days—and so I accepted their research assistantship and renounced my aspirations for a watery career.

* Corresponding author: pueppke@anr.msu.edu

I have occasionally wondered about the outcome, had the 21-year-old persisted with his initial dream. But I have taught myself to face forward, and I have learned that scientists often stumble into satisfying careers after having made shortsighted decisions early on. The rest of my story, like that of so many scientists, is one of restlessness and inquisitiveness tamed. Curiosity, a characteristic shared by most researchers, comes naturally, and universities know how to nurture those of us who wonder why.

Young people who gravitate to science have no trouble finding campus mentors to shape and mold their inquisitive nature, so that when fully formed, curiosity can be released to seek out interesting problems in need of solutions. That is certainly what happened to me, and I spent the first decades of my career pleasantly holed up in the laboratory, content to unravel the secrets that allow microorganisms to deliver nitrogen to crop plants.

I got to thinking about curiosity recently as I participated in a workshop with two small groups of scientists, each representing a different university. The goal was to share information and eventually team up to solve bioenergy problems. The workshop had been planned with great care; most of the scientists did not know one another, and all had made time in their busy schedules to attend. Participants were given 30 minutes to summarize their research, with plenty of time for questions and discussion.

Here is how the routine played out. Everyone comes into the room, opens his or her laptop, and begins to check e-mail. Upon being introduced, the first speaker closes her laptop, moves to the front, and speaks to the audience, which, with one or two exceptions, is by now surfing the Web. At the end of the talk and after applause, questions are asked, after which the speaker returns to her e-mail. Meanwhile, the second speaker is closing his laptop and proceeding to give his talk to an audience (including now the first speaker) that, again with one or two exceptions, is intent on searching the Internet to complete other unrelated tasks. And on it goes until the very end, when all proclaim that they have enjoyed the interaction greatly and commit to developing joint research projects.

There is a moral to what happened in my workshop, and you have probably concluded that it has to do with modern wireless communications. The Internet was there as temptation, but the real issue is how scientists go about relating their interests to the work of others. Deliberately seeking to link your research to that of others when you are presenting to the audience is half the story; you must also actively listen to others to see how your work might connect to theirs. If you understand these relationships, you have mastered contextualization. You have learned to punch holes through the boundaries between what you do and what they do, and you are inviting knowledge and ideas to flow unimpeded in both directions.

Contextualization

Contextualization is the first C that needs to be learned early, but it is routinely ignored and rarely encouraged by mentors as part of the learning process. In pursuit of scientific depth, barriers are too often erected to confine students to narrowly defined segments of the real world. Yet in reality, if you are unraveling the population dynamics of a single fish species, you will do best if you understand the species as part of the aquatic community. You may be dissecting the function a single protein, but in the real world, this protein floats in a cell as part of some organism's metabolic system. And if, like me, you are interested in a single nutrient, you had better appreciate how it fits into the ecosystem.

Nitrogen fascinated me, but I was not serious about viewing my work in a broader context. I would justify my research by pointing out that a pound of nitrogen delivered to crop plants by microorganisms is a pound of nitrogen that does not have to be purchased and applied to the soil as fertilizer. Glib statements of this sort punctuated my seminars and justified my applications for research funding. One day a prying dean asked me to give him an estimate of the economic benefits of my research on nitrogen. He actually wanted me to get out of my corner in the laboratory and talk to a social scientist, and I resisted. As a biologist, I did not want to think too deeply about anything beyond what I investigated with my own hands. And so I erected boundaries between what I did and the approaches of my colleagues, and I fortified them—in the process greatly frustrating the dean, whose inquiries I managed to evade.

Like my interaction with the dean, the attendees at the workshop were defending perimeters of their own making. Content within their enclosures, they simply tuned out discussion of other aspects of bioenergy and became seduced by the siren call of the Internet. Too often I see glazed, tuned-out eyes when the viewer ought to be deep in analysis of what is being said. Molecular biologists wield amazing tools to improve the environment yet are too often dismissive of scientists who do field work. Field biologists overlook the opportunity to anchor the work of molecular biologists to organismic phenomena. Freshwater biologists underestimate what they can learn from saltwater systems. Saltwater biologists reciprocate. And more generally, natural resources experts concentrate on "their species" as if it is the only one in the ecosystem.

Wouldn't it be exciting if curiosity endured and all scientists were eager to have their minds stretched? They would sit right up at workshops and ponder the relationship between what they really understand and the surrounding world that appears fuzzy and even inhospitable—a place where they are not necessarily the expert and where their tools are of little use. They would unabashedly ask questions, and as mentors, they would insist that younger colleagues start stretching early. A few might even begin to tiptoe beyond the fringes of what they know.

Collaboration

Tiptoeing often leads to collaboration, the second C that must be learned. Collaboration sometimes starts out badly—when you do it in the classroom, it is called cheating. And as young scientists step into career pathways, they are usually encouraged to toil in relative isolation. Still unproven, they are expected to demonstrate that they can generate ideas, gather data, and complete the analysis by themselves. Collaboration is taken as evidence of a paucity of intellect, of depending too much on others—a softer, gentler kind of cheating. So it is no wonder that many young researchers learn to avoid even partial reliance on others. It is the wise mentor who points out early on that collaboration enables scientists to build on their curiosity and expand their perspective, forging deep, working relationships with others who see the world differently.

Some of these interactions can be very deep, particularly if you are intent on confronting the big issues that bedevil society these days. Many of these problems relate to natural resources (think climate change), and they inevitably resist solutions offered from a single perspective. But collaboration can be unsettling to biologists like me, who have traditionally approached problems like carpenters with hammers. Armed with a technology or a special technique, or perhaps proud to have generated a breakthrough idea, we are on the lookout for matching problems—nails that can be pounded in. But, complex issues are not like nails, and so we must learn to shoulder toolkits and not just clutch a single tool.

The difference between tools and toolkits became clear recently in discussions about renewable energy in my state of Michigan. It is heavily forested, and we know that only about one-third of the annual accumulation of woody biomass is removed. We also know that as wood is taken away for pulp or lumber, biomass in the form of tops and branches is left behind. Could we retrieve some of these materials and convert them into bioenergy without doing environmental harm?

Biologists and engineers assure us that this is possible. We could literally have our cake and eat it too, selectively harvesting and selling wood for bioenergy yet retaining the ecosystem services that make natural areas so valuable to the state's citizens. But social scientists give us a different story. They remind us that the memory of clear-cutting and associated environmental exploitation is still fresh, even though it had run its course by the end of the 19th century, long before most of our grandparents were born. More sobering from the standpoint of economic development, they have conducted surveys and know that most private forest landowners in Michigan prize nature and are not motivated by profit.

These dynamics often surface in the fisheries world, too. Biologists are certain that fish can be harvested sustainably, providing jobs and other economic benefits. But people, often those living closest to the resource, are not so sure, and they may resist. When confronted with these complex issues like this, we need the whole toolkit, which includes biologists and engineers, who are like hardware developers. They offer tools in the form of technologies, techniques, and breakthrough scientific ideas. But a complete toolkit also includes social scientists, who are more like software developers. Their tools rely on an understanding of how human skills, motivations, and beliefs come into play as the hardware is put to use. Hardware and software are both necessary, especially for complex issues involving resource management, but neither alone is sufficient–we must have a complete toolkit, and that will require collaboration.

Communication

The third C that should be learned early is communication, the ability to organize and formulate ideas and then explain yourself so that others (who are likely curious but simply do not understand) come away inspired. Like collaboration, communication can start out badly. Junior researchers are routinely forced to adopt wooden prose as they draft their first manuscripts, and they are often encouraged to embrace a stiff, facts-only speaking style. These bad habits can be unlearned, but a more fundamental problem often lurks nearby: many scientists really do not like to write or speak. They see no inconsistency in decoupling generation of knowledge from the process of sharing the good news. Sadly, far too many students are encouraged to become passionate experimentalists yet are allowed to remain begrudging communicators.

Uninhibited and, for better or worse, at ease with the pen, I was different, but I started out with a lethal combination: overconfidence and a poor appreciation for standards. I learned my lesson early, and it was painful. As an undergraduate, I did independent research and entered a national awards competition. My accomplishments were to be judged on the basis of a 15-minute speech in front of an audience of more senior scientists. Confident of my rhetorical skills, I was indignant to learn that I would have to practice my talk in front of a gathering of seemingly ancient professors. One of them, a highly respected senior scientist, tore me apart. He was nice about it, but that is what he did. Black and blue, I had the sense to recast the entire presentation, ultimately winning the second-place prize. Wounds healed, and more than 30 years later, I met this professor, who was by then retired and had long forgotten me. It was a pleasure to give him tardy thanks for a valuable and timely lesson in the value of communication.

I got more lumps when, as an apprentice graduate student, I crafted my first serious research paper. Confident in my abilities, I submitted the final draft to a more senior student, who did it to me again—tearing me apart—and just a few days before the due date. This was more than 40 years ago, but I still wince when I recall the tone of his voice when he asked if I seriously thought that something like my paper would please my mentors. And so now with wounds reopened, I resolved to thoroughly rewrite my creation and received a grade of A on it. This was another early lesson learned about the importance not just of oral, but also written communication.

Narratives, both written and oral, add power to data. Every one of us has been impressed by evocative and sizzling science, and we all have been captivated by speakers who can tickle concepts so that they spring to life. But, we have also suffered through those unbearable words, "I know you can't read this slide, but." And we have endured pages and pages of scientific doggerel as we vainly search for the point of it all. I was once on a panel that reviewed a grant proposal that was bloated in this way. On and on the investigator wandered, with phrases in a tiny font and congealed paragraphs that bulged out toward the margins of each sheet. The reviewers did manage to excavate his research idea, and it was worth funding. We gave him some money, but we wrote him a note of warning: If you ever again glob up your ideas in such sloppily framed verbiage, we are not even going to read your proposal much less fund the work.

None of us wants to be in this investigator's position, having good ideas that for lack of clear articulation remain concealed and underappreciated. We need to guide our junior colleagues, not just to get by, but to employ communication skillfully as a tool to make science more meaningful and relevant. It is the last of the trio of C's that must be learned early.

All three C's are in short supply in science, and although they are assets that extend the naturally occurring C—curiosity—these C's can slip away unrecognized. Too often, they fail to take hold, not just in students, but over the entire course of professional careers. Yet, contextualization, collaboration, and communication are decisive. Scientific insight and impact can be slowed, even extinguished, unless researchers learn to place their work into a broader context, draw meaningfully upon the skills and insights of others, and communicate what they are doing.

Biography

Steven G. Pueppke was awarded a B.S. degree from Michigan State University in 1971. He failed to realize his early dream of becoming an oceanographer, but he does not regret having obtained a Ph.D. degree in plant pathology from Cornell University—or the opportunity to do research on plant–microbe interactions. He has held academic positions at universities in Florida, Missouri, and Illinois and spent two years as a guest professor at universities in Switzerland and Germany. He currently is associate vice president for research and graduate studies at Michigan State University, where he takes advantage every day of the three C's. He thanks all of his early mentors who helped him learn to appreciate their advice.

Facing Reality and Overcoming Adversity

Jerry L. Rasmussen*
U.S. Fish and Wildlife Service (Retired)
4 Cobblestone Lane, Le Claire, Iowa 52753, USA

Key Points

- At one time or another, you will face adversities in your professional life.
- Being determined in your mission and believing in yourself will help you overcome these adversities.
- Developing good communication skills will help you convey what you believe in.
- Developing strong professional relationships with your peers will provide you with the support network needed during these challenging events.

Introduction

Many of us pursue a career in natural resource management thinking we are going to escape to a life in the outdoors far from the hectic mainstream battles and confrontations of everyday life. But, it is not long before we realize that the natural resources we had hoped to escape to are being lost. If we wish to protect these resources, we have to face reality and return to the mainstream where we find ourselves facing people, organizations, and agencies that seem to care little about the things we cherish the most. Our mission then becomes one of confronting them in order to eliminate, reduce, or mitigate the impacts of their projects.

Our university degrees have given us the tools we need to understand basic biology and ecology but have usually given us little training and experience with regard to human nature, negotiations, and politics. Our employers sometimes provide us with limited training in these areas, but most of what we need has to come from within our spirit, from our own gradually acquired experience, and through collaboration with our peers.

Real Life Example 1: Master Plan for the Management of the Upper Mississippi River System

In the early 1980s, I represented the U.S. Department of the Interior as chair of a committee assigned to prepare environmental documents for a master plan for management of the Upper Mississippi River System (UMRS). Significant navigation expansions (i.e., new and larger locks and dams) had been proposed. Our job was to evaluate the impacts of such expansion and then propose measures that could mitigate any projected impacts on

* Corresponding author: ijrivers@aol.com

the environment and recreation opportunities. Working with an interdisciplinary committee from five UMRS states and two private entities, I developed strong bonds with my committee members. As team leader, I encouraged individual input from my peers. We developed timely and complete reports and documents, we brought in outside scientific expertise and developed strong scientifically based recommendations, and, last but not least, we socialized together. Despite being significantly outnumbered by representatives of the U.S. Army Corps of Engineers (the Corps) and navigation interests, subjected to severe budgetary constraints, faced with continuous scheduling changes, and intimidated at several levels by adversaries who threatened and caused the firing or reassignment of some of our members (including forcing the turnover of committee chairmanship three separate times), we succeeded in gaining support for a US$300 million, 10-year Upper Mississippi River (UMR) environmental management program (EMP). A program of this size and scope was unheard of at the time. Our accomplishment would not have been possible without (1) belief in ourselves and in one another, (2) determination and hard work, (3) timely and accurate reporting, and (4) unity as a group, both professionally and socially.

Lessons

- When working in defense of natural resource issues, you will usually find that you are outfunded and outstaffed by your adversaries. To succeed, you will need to culture strong relationships and support from sister agency employees who will likely be in the same predicament as you.
- Some of these relationships will come naturally through interactions and roles at various meetings. Others can be developed through involvement in outside activities and professional societies. These societies can speak out on issues that your state or agency often cannot.
- A very important way to enhance camaraderie among your peers is through social interactions, which oftentimes present themselves during social hours at the end of a day's meetings. The bonds that can be developed between colleagues through social interaction are surprising. The same can be true for your relationships with the staffs of the adversarial groups you may have been arguing with all day. Do not overlook these opportunities.

Real Life Example 1, Continued

After the master plan was completed (in the mid-1980s), I assisted my state colleagues in getting the EMP authorized and funded by Congress, even though my own agency, the U.S. Fish & Wildlife Service (FWS), demonstrated little interest. In fact, I was told later that, at one point, my superiors were ready to settle for a mere US$50,000 in mitigation funding. When the $300 million EMP was finally authorized, all funding came through the Corps and the program was broken down into two components: (1) habitat projects, and (2) a long-term resource monitoring and analysis program (LTRMP). The Corps took charge of the habitat component and looked to the FWS to handle the LTRMP. Because of my long-term commitment and belief in the EMP, the FWS asked me to apply for the position of LTRMP program manager. But, even though this was an FWS position, the Corps' would not allow me to lead the LTRMP because of their antagonism toward the master plan team that I had chaired. Since they held the LTRMP purse strings, they were able to pressure the FWS into recruiting someone more to their liking. I then became assistant program manager, essentially leading the program from behind. Corps officials then created many administrative obstacles, including requiring that I continually develop different planning scenarios and "what if" reports, which drained both time and money. Our funding was also reduced through significant overhead charges and diversion of funds to something the Corps called "savings and slippage." U.S. Fish and Wildlife Service administrators were equally guilty of funding diversions, including charging 38% overhead to the entire LTRMP project and diverting $500,000 to purchase a camera fitted for a regional airplane that was committed to taking blueprint-quality photos of wetlands being inventoried for the farm program, which had nothing to do with the LTRMP. The states themselves also charged much smaller overhead fees, and when all was said and done, the LTRMP operating budget was reduced to about 25% of the total appropriated in the early years. I had established a series of five state-led field stations dependent on program funding, and it was essential that I keep the states partners informed of funding shortfalls. So, I prepared a chart for LTRMP participants showing where all of the money was going. This information found its way back to the sponsoring congressman who then addressed funding issues with both FWS and Corps administrators. This, of course, did not sit well with certain federal officials, and at the end of the LTRMP's second year a new FWS LTRMP program manager was brought in, the LTRMP was restructured, my position as assistant program manager was eliminated, and I was reassigned to a different program on the Missouri River. But, after I had been gone for a year or two, the new LTRMP program manager and other FWS officials in the regional office apologized to me because they found through their own experience that I was right in exposing the funding problems. Also, because of my honesty with the states and because of our collective interest in and ownership of the EMP, it survived, despite the initial lack of federal interest. Twenty-two years later, and after a name change or two, the program that we created remains one of the nation's leading ecological programs with an ongoing accumulated budget well in excess of $1 billion. Because of our dedication, hard work, and professional bonding, my colleagues and I changed the way natural resource management was and continues to be done on the UMR.

Lessons

- Teamwork and team bonding can result in significant achievements.
- Sometimes, however, even though you succeed in your mission, your employer may not reward you. In fact, the opposite can be true.
- But, if you are on the right side of the issue, are true to yourself, and maintain your professional integrity, others will notice, and in the long run you will be recognized for your work.

Real Life Example 2: Mississippi Interstate Cooperative Resource Association

After leaving the LTRMP but before arriving in Missouri at my new assignment, I was recruited by a state colleague to coordinate a newly formed interstate group called the Mississippi Interstate Cooperative Resource Association (MICRA). This 28-state group was formed to address the needs of interjurisdictional fisheries over the entire Mississippi River Basin (MRB), and the states were in need of a federal coordinator. My reputation with my colleagues and my dedication to the resource paid off. This was a much better assignment than the one I had been reassigned to and came with a much higher profile. My state colleagues had taken care of me.

I spent the next 10 years coordinating and helping develop MICRA programs before I was faced with another major controversy—what to do about spreading Asian carp (i.e., Grass Carp *Ctenopharyngodon idella*, Bighead Carp *Hypophthalmichthys nobilis*, Silver Carp *H. molitrix*, and Black Carp *Mylopharyngodon piceus*) populations. In one instance in the late 1990s, when backwaters of a UMR national wildlife refuge were drained by seasonal drought, 97% of the fish left stranded and killed were Asian carp (Bighead Carp and Silver Carp). Also, when I traveled to southern Illinois to help film a documentary on Asian carp, I was amazed at the numbers we observed, including thousands of 15–17-inch Silver Carp, some of which jumped right into the boat with us. MICRA's members needed to be made aware of the growing problem and its potential for spread throughout the MRB, so I documented my concerns in articles prepared for MICRA's newsletter, *River Crossings*.

Then, in early 2000, MICRA formally requested that the FWS list Black Carp as an invasive species under the federal Lacey Act. Such action would prohibit transport and sale of the species across state lines. MICRA also asked that the FWS work with private and state entities to eliminate all remaining Black Carp that currently existed in the United States. It was felt that the numbers of the species present in captivity were still small enough to control and that if they escaped captivity and established wild populations, already threatened native mussel and shellfish populations would be at great risk.

By August 2000, MICRA had exchanged letters with the director of the FWS regarding the issue, and being an FWS employee my involvement was becoming more and more sensitive. But because of my past experiences and my knowledge of the issue, I was confident in myself and knew I had strong support from the states. Besides, my responsibility for supporting MICRA (as assigned by the FWS) and preparing *River Crossings* remained unchanged. So, for the summer issue of *River Crossings*, I included information provided informally by Arkansas biologists that they had first observed Asian carp populations in their state in river reaches near fish farms and that the species had spread outward from there across the state. Fish farms as a major pathway for the spread of Asian carp thus became widely publicized. This raised immediate controversy and resulted in one fish farmer, a state finance executive, and a congresswoman (all from Arkansas) traveling to Washington, D.C. to meet with the FWS Director. There, they complained about the *River Crossings* articles and, since I was an FWS employee, accused me of conflict of interest regarding MICRA's request to list Black Carp. This was during President Bill Clinton's Administration, and I was immediately ordered by the FWS Director to drop everything related to MICRA and report to the regional office in the Twin Cities, Minnesota, on the following day. Readers should note that before becoming President, Mr. Clinton had been Governor of Arkansas, so the state's fish farmers had a close connection to the White House.

I was informed that I, a 24-year FWS veteran, was being reassigned from my MICRA duties to that of a fishery technician. The environmental advocacy group Public Employees for Environmental Responsibility (PEER) learned of the situation and tried to get involved but informed me that since there was no reduction in pay, legal action was not feasible because any judge would likely refuse to hear my case. That left me with little recourse other than to accept the penalty. On the positive side, I received significant accolades and moral support from my peers in the states and elsewhere, and I was confident enough in myself and my position on the issue that it would only be a matter of time before my situation was corrected. In the meantime, I made the best of it, in part by preparing and editing a revision of the "Upper Mississippi River Conservation Committee Fisheries Compendium," a document summarizing UMR fisheries information, that I had first prepared in 1979.

But MICRA, left without the day-to-day leadership of a coordinator, struggled to survive, and *River Crossings* was not published for the following 20 months. Fish farming advocates had thus won a temporary victory. But during those 20 months the administration in Washington, D.C. changed, as did FWS leadership at both the Washington and regional offices. My new superiors recognized the significance of the Asian carp problem, felt that I had done the right thing, and wanted to use my talents to a better end. So, in 2001 I was asked to prepare a white paper on the threat of Asian carp to the Great Lakes and the MRB, and was later assigned to make numerous oral presentations on the Asian carp threat at meetings in cities all around the Great Lakes. As a result, I became somewhat of a celebrity for the Great Lakes Asian carp controversy (which continues to this day). The picture included with this paper was circulated widely in the press—my frown in the picture apparently reflected the sentiments of many people regarding the Asian carp issue (Figure 1).

I returned to my MICRA position in mid-2002. Later that year, when the need arose to establish a panel to address invasive species issues in the MRB under the National Invasive Species Task Force, MICRA's Invasive Species Committee took over that role, with me serving as coordinator of both MICRA and the federally recognized MRB Panel on Invasive Species.

FIGURE 1. The author holding a 26-pound Bighead Carp taken in 2001 from the Illinois River, approximately 30 miles downstream from the canals connecting it to Lake Michigan.

Lessons

- People will take drastic action to shut off a source of information when they fear it may affect their economic interests.
- People in leadership positions will likely then take whatever counter action is necessary to pacify political interests in order to protect themselves.
- Sometimes a price has to be paid, but when you are right and your convictions are firm and true, others will notice and wrongs will be righted.
- So, when in a difficult situation, trust your instincts, be determined, and most of all, maintain your professional integrity.

Conclusion

What these examples show is that natural resource management can be a contact sport, and not everyone may be suited to it. Those who are not may want to seek employment in academia or in one of the many nongovernmental organizations active in environmental issues. But if you (1) are prepared and on the right side of an issue, (2) are determined and believe in yourself, (3) maintain good communication skills, and (4) maintain strong professional relationships with your peers, you can succeed—even against powerful foes. I believe a natural resource manager is on the right side of an issue if his or her actions are

- driven by doing what is in the best interest of the natural resource;
- driven by doing what is necessary to protect the public interest or common good; but
- not driven by actions taken solely for personal gain, promotion, or profit.

Had I taken the option to go with the flow and become just another bureaucrat, perhaps I could have ended up in a higher paying job within the bureaucracy. But to do so, I would have had to compromise my principles and I would have lost myself along the way. As it is, I feel my career has made a difference, and I believe the nation's fish and wildlife resources are better off because of it. I encourage others to follow my example, pursue their own paths, and stand up for the critters and for what is right when the opportunity presents itself in their careers.

Acknowledgments

The author would like to thank William Taylor and Abigail Lynch, Michigan State University, for their encouragement, guidance, and patience in developing this paper. He would also like to thank Richard Sparks, Illinois Natural History Survey (retired); Tom Gengerke, Iowa Department of Natural Resources (retired); and Mark Laustrup, U.S. Geological Survey (retired) for their critical review of this paper. All three of the reviewers are former colleagues of the author who shared portions of his professional journey described, in part, by this paper.

Biography

Jerry Rasmussen graduated from Iowa State University with a B.S. in fish and wildlife biology in 1968 and from Colorado State University with an M.S. in fisheries science in 1971. His career included working for two states, a private consulting firm, and two federal agencies. He retired in 2008 after a 32-year career with the FWS that included an assignment to the White House Floodplain Management Review Committee in the aftermath of the 1993 floods. He won many awards and accolades from his peers over his 37-year professional career, including the National Honor Roll Award from the Izaak Walton League of America in 1986 and the American Fisheries Society's first President's Fishery Conservation Award in 1995. Today, he continues to contribute by serving as a private consultant.

Today's Fisheries Challenges Require Creative, Interdisciplinary Problem Solving

Kyle S. Van Houtan*
Pacific Islands Fisheries Science Center, NOAA Fisheries
1601 Kapiolani Boulevard, Suite 1000, Honolulu, Hawaii 96814, USA
and
Nicholas School of the Environment and Earth Sciences, Duke University
Box 90328, North Carolina 27708, USA

Eric C. Schwaab
National Aquarium
501 East Pratt Street, Baltimore, Maryland 21202 USA

Key Points

- Natural resource challenges require interdisciplinary analyses to solve scientific problems and improve management practices.
- Interdisciplinary thinking requires diverse experiences and continual education throughout one's career.

Introduction

Long-term solutions to today's most pressing fisheries challenges require an increasing understanding of complex system interactions. Ecosystem-based management that links diverse data streams and disciplines has been a focus of fishery scientists for decades (Christensen et al. 1996), yet ecosystems are not the only systems important to fisheries management. For real and lasting solutions, social and economic systems deserve just as much attention and study as natural systems. Here, we share our personal experiences understanding complex problems in marine science and management. In doing so, we illustrate how stepping outside our respective disciplines and thinking about system dynamics at a range of scales in both science and management can shed light on, and offer new solutions to, challenges faced by natural resource practitioners.

Despite a need for strategic, systemic view of the natural resources challenges we face, we too often maintain a narrow approach to our vocation. Pimm (1991) pointed this out when he wrote, 20 years ago, that "very little ecology deals with any processes that last more than a few years, involve more than a handful of species, and cover an area of more than a few hectares." But, the problems extend beyond science, and they are exacerbated in both university and professional settings. Academic and natural resource institutions, in an attempt to generate experts of renown, often require intense focus on specific disciplin-

* Corresponding author: kyle.vanhoutan@noaa.gov

ary topics while avoiding training that is more interdisciplinary, holistic, and practical (Van Houtan 2006). Spurred on by litigation, management agencies require simple, immediate results for complicated problems. The perhaps ironic result is that the mindset required to understand and manage something as complex as an ecosystem is constrained. However, the issues of taxonomic and spatiotemporal scales present in fisheries management may only approach the complexities of social and economic systems. So in addition to scientific frontiers, equally important opportunities exist to incorporate systems approaches on the management side—in addressing the organizational, political and economic challenges to sustainable management of fisheries and other aquatic resources.

Catch Shares: Melding Science, Economics, and Sociology

For decades, fishery managers have struggled with boom and bust cycles in commercial fisheries. Managers have leveraged long-term monitoring and stock assessments into closed seasons, closure areas, catch limits, and other tools to achieve conservation outcomes. In common-access fisheries, achieving compliance with such regulations often ran against the short-term economic motivation of individual fishermen, cooperatives, and political will. Even the most conservation-minded fishermen operate in systems that often disproportionately award individual actions that run counter to the rules applied across the fishery. By operating outside of agreed-upon rules, individual actors can accrue greater immediate benefit while spreading the harmful effects across all of the fishery participants. Conversely, any individual, voluntary conservation action could be diluted by the actions of others, thus discouraging individual conservation actions.

This is the classic argument of the tragedy of the commons (Hardin 1968). Seeing its very real effects motivated me (ECS) to look for ways to integrate economics with fisheries management. From my position at the Maryland Department of Natural Resources Fisheries Service, I took a month-long fellowship with economists and other resource managers at the Kinship Conservation Institute hosted by the Property and Environment Research Center in Montana. Working closely with a diverse group of economists and resource managers, and drawing from my own experience with traditional fisheries management, I developed a blueprint for a "catch share" system for the then-depleted Chesapeake Bay blue crab *Callinectes sapidus* fishery. Though it was not implemented at that time, fishermen and research managers continue to work toward a quota-based system and are now exploring alternatives to traditional restrictions based primarily on gear limitations, seasons, and workday restrictions. The Chesapeake Bay blue crab population has rebounded, and the fishery and coastal fishing communities are benefitting substantially.

In 2010, I brought this experience to the National Marine Fisheries Service (NMFS) to help rebuild fisheries and the livelihoods and communities that depend on them. Recent changes to U.S. federal fisheries laws (NMFS 2006) placed significant new requirements to curtail overfishing and rebuild depleted stocks in federally managed fisheries. That change in the law led to scientifically based catch limits and a renewed emphasis on the use of rights-based fishery management systems. These catch share systems depend at their core on the allocation of shares of an annual quota to individual or small groups of fishermen. By allocating individual shares up front, these systems better align the economic motivations of individual fishermen with the broader conservation goals set for the fishery, essentially avoiding some of the pitfalls of the tragedy of the commons.

Following this renewed emphasis in federal law, NMFS, in concert with various regional fishery management councils, has supported the development of a number of new catch share systems in federal fisheries. Under these new management systems, the agency is already seeing better alignment of broad conservation goals with the economic incentives at work in various fisheries, leading to improved economic performance, stronger support for scientifically based catch limits, and increased rebuilding progress. During 2011, the first year of operation of the Pacific groundfish trawl fishery under a new catch share management system, discards of unwanted species dropped dramatically, accelerating rebuilding of depleted species in the fishery. In the Gulf of Mexico, commercial Red Snapper *Lutjanus campechanus* fishermen advocated a shift to an individual quota system that has eliminated a derby fishery in which fishermen raced to catch a portion of a shared quota during a short season. Under the new system, fishermen have a year-long season and increased dockside values while meeting quotas that are supporting a rebuilding stock. While these results seem to speak for themselves, it is only through sound application of economic theory, incorporation of experiences from fisheries in other parts of the world, and transfer of that knowledge in work with local fishing communities that transformative change occurs.

The March of Atoms to the Sea

Unlike most of my peers in marine turtle ecology, I (KSV) am still a relative newcomer to the field. I am a population ecologist and I approach scientific and management problems through interdisciplinary analyses. My undergraduate research training began in biogeochemistry. I worked in laboratories with an astounding breadth of scale. Some researchers chronicled the impacts of acid deposition to Shenandoah Valley streams; others studied the global nitrogen cycle. While I did not gravitate towards flow charts as they can seem too engineered and sterile, I recall looking at simple diagrams of nitrogen flowing from one compartment to another and being fascinated by the idea, as well as the complexities within each box itself. Despite their abstractness, the boxes and arrows touched upon my first-hand experiences growing up on an estuary—playing in streams and marshes amid the flow of life—and stirred my imagination.

These connections came together crisply while reading the epilogue to a biogeochemistry textbook (Likens and Bormann 1995) where the authors had the good sense to reference that the quantitative descriptions in their text were essentially foretold by a Midwestern park ranger a half century earlier. In *A Sand County Almanac*, Aldo Leopold tells the story of a hypothetical compound X as it journeys through the life of a prairie ecosystem:

> Between each of his excursions through the biota, X lay in the soil and was carried by the rains, inch by inch, downhill. Living plants retarded the wash by impounding atoms; dead plants by locking them to their decayed tissues. Animals ate the plants and carried them briefly uphill or downhill… No animal was aware that the altitude of his death was more important than his manner of dying. Thus a fox caught a gopher in a meadow, carrying X uphill to his bed on the brow of a ledge, where an eagle laid him low… An Indian eventually inherited the eagle's plumes, and with them propitiated the Fates... It did not occur to him that they might be busy casting dice against gravity; that mice and men, soils and songs, might be merely ways to retard the march of atoms to the sea." (Leopold 1949:107–108)

The accessible narrative grabbed my attention, but the underlying connections in trophic dynamics and nutrient cycling were inspiring. I thought Leopold was a brilliant observer of nature, and it motivated me to improve my ecological knowledge. I then embarked on a decade-long journey of field research and graduate schooling that led me across the world. I worked on coral reefs in Belize, sustainable forestry in the Rocky Mountains, bird behavior in the Amazon, and protected area planning in Southern Africa. I made it a habit to read the personal biographies of important scientists and great intellectual quests (Darwin 1958; Beckmann 1971; Rhodes 1986; Anton 2000) and even the philosophy of science (Lakatos 1978; Ziman 1984). I spent many solitary hours in remote locations observing wildlife, taking detailed notes, and designing figures for manuscripts. I developed my data analysis skills, learning how to find answers in large data sets and use data others considered uninteresting. In essence, I mapped a scientific process onto my natural curiosity.

These skills came together when I began investigating a sea turtle disease shortly after I arrived in my present post in Honolulu. Since the 1980s, a tumor disease has been plaguing green sea turtles around the globe, but particularly in Hawaii. Though many hypotheses had come and gone, its ultimate cause was unknown. The largest advances at the time were that it was linked to a herpesvirus and that it was anecdotally observed in impaired waterways. Due to its remote location and physical geography, Hawaii was a special place to understand this problem. It was also an opportune setting as the local NMFS office maintained a 30-year database of diseased sea turtle records. From these data, I began to map the occurrence of the disease across the islands and pinpoint hotspots and locations that were disease-free. I printed out color-coded maps, hung them on the wall by my desk, and stared at them for weeks. One day, I showed them to a colleague and asked her what they might be, playfully not telling her what they were. She said they looked like the distribution of an invasive species of marine macroalgae.

Recalling my undergraduate biogeochemistry and Leopold's "march of atoms to the sea," I started to think of ways to account for nutrient runoff to Hawaii's coastal waters. Though I already knew that elevated environmental nitrogen can profoundly affect ecosystems and promote invasive species, I did not yet comprehend how it might promote tumors. Using a suite of available land-use databases, I used geographic information systems and spatial modeling to create a nitrogen footprint map that strikingly resembled the disease maps I previously developed. Connecting the two maps did not take long. After immersing myself in the literature, I identified an important overlap between turtle ecology, plant physiology, and herpes virology. Macroalgae flourish in nutrient-rich coastal waters, but they store environmental nitrogen in their tissues in a particular amino acid—arginine—that also regulates the herpes life cycle (Van Houtan et al. 2010). Because green turtles are herbivores, feeding on the arginine-rich macroalgae could fuel the herpesvirus that causes the debilitating tumors. It was an important new idea that came through a series of data mining, modeling, and a familiarity with diverse disciplines. Just as much, it required a willingness to step outside of my immediate field of expertise, to make new colleagues, and to humble myself to learn new material. However, I may have never conceived of its possibility if I did not spend significant time studying biogeochemistry as an undergraduate.

Conclusion

Though one of us is a marine resource manager and the other a population ecologist, our experiences have similar conclusions regarding the need to reach outside of traditional disciplinary boxes to find real solutions to our scientific, management, and policy challenges. As our stories attest, through broad systems perspectives and cross disciplinary observation and action, we can achieve creative new solutions and a path to lasting success. True solutions align a diversity of scientific disciplines and engage social and economic systems toward sustainable outcomes. Diverse experiences, continual training, and the desire to learn are important attributes for leaders today and for generations to come.

Biographies

Dr. Kyle Van Houtan leads the Marine Turtle Assessment Program with NOAA Fisheries and is an adjunct associate professor in the Nicholas School of the Environment and Earth Sciences at Duke University. His research and teaching address the problem of extinction and combine field ecology, quantitative modeling, ethics, and theology. Currently focusing on marine population dynamics and climate change, his research has been featured in *Nature*, *Science*, *The New York Times*, *National Geographic*, *New Scientist*, and *The Guardian*. Dr. Van Houtan was recently recognized by the White House as a recipient of the 2011 Presidential Early Career Award for Scientists and Engineers, the highest honor given by the federal government to scientists in the early stages of their careers. He and his family live in Honolulu.

Eric Schwaab is the senior vice president and chief conservation officer at the National Aquarium in Baltimore. From 2010 to 2013, he was the assistant administrator for fisheries at the National Oceanic and Atmospheric Administration. In that capacity, he directed the National Marine Fisheries Service, the principal federal agency that protects and preserves the nation's living marine resources through scientific research, fisheries management, law enforcement, and habitat conservation. Mr. Schwaab has a long history of innovative natural resource management and leadership at state, federal, and nonprofit organizations, including more than 24 years at the Maryland Department of Natural Resources where he served as the agency's deputy secretary, fisheries service director, forest service director, and director of wildlife, forest and heritage service, and as a park ranger and natural resources police officer.

References

Anton, T. 2000. Bold science: seven scientists who are changing our world. Freeman, New York.

Beckmann, P. 1971. A history of π (pi). The Golem Press, Boulder, Colorado.

Christensen, N. L., A. M. Bartuska, J. H. Brown, S. Carpenter, C. D'Antonio, R. Francis, J. F. Franklin, J. A. MacMahon, R. F. Noss, D. J. Parsons, C. H. Peterson, M. G. Turner, and R. G. Woodman. 1996. The report of the Ecological Society of America committee on the scientific basis for ecosystem management. Ecological Applications 6:665–691.

Darwin, C. 1958. The autobiography of Charles Darwin 1809–1882. W. W. Norton & Company, New York.

Hardin, G. 1968. The tragedy of the commons. Science 162:1243–1248.

Lakatos, I. 1978. The methodology of scientific research programmes. Cambridge University Press, Cambridge, UK.

Leopold, A. 1949. A sand country almanac. Oxford University Press, New York.

Likens, G. E., and F. H. Bormann. 1995. Biogeochemistry of a forest ecosystem. Springer-Verlag, New York.

NMFS (National Marine Fisheries Service). 2006. Magnuson-Stevens Fishery Conservation and Management Reauthorization Act (PL 109–479). U.S. Government Printing Office, Washington, DC.

Pimm, S. L. 1991. The balance of nature? University of Chicago Press, Chicago.

Rhodes, R. 1986. The making of the atomic bomb. Touchstone, New York.

Van Houtan, K. S. 2006. Conservation as virtue: a scientific and social process for conservation ethics. Conservation Biology 20:1367–1372.

Van Houtan, K. S., S. K. Hargrove, and G. H. Balazs. 2010. Land use, macroalgae, and a tumor-forming disease in marine turtles PLoS ONE 5(9):e12900.

Ziman, J. 1984. An introduction to science studies. Cambridge University Press, Cambridge, UK.

Expand Your Horizons: The Importance of Stepping Out of Your Comfort Zone

So-Jung Youn* and William W. Taylor
Center for Systems Integration and Sustainability
Department of Fisheries and Wildlife, Michigan State University
115 Manly Miles Building, East Lansing, Michigan 48823, USA

C. Paola Ferreri
Department of Ecosystem Science and Management, Pennsylvania State University
408 Forest Resources Building, University Park, Pennsylvania 16802, USA

Nancy J. Léonard
Northwest Power and Conservation Council
851 S.W. Sixth Avenue, Suite 1100, Portland, Oregon 97204, USA

Amy Fingerle
Department of Aquaculture and Fish Biology, Hólar University College
Háeyri 1, Sauðárkrókur IS-550, Iceland

Key Points

- Breaking out of your comfort zone can provide you with personal and professional growth that makes the challenge worthwhile.
- You can break your intellectual boundaries by pursuing new opportunities, engaging in experiential activities, and finding great mentors that care about your personal and professional well-being.

If you do what you've always done, you'll get done what you've always gotten.
—Anthony Robbins

To make a huge success, a scientist has to be prepared to get into deep trouble.
—James D. Watson

Being Successful Often Requires Taking Risks

Beginning in childhood, we develop a self-image rooted in what we are good at and what we like to do. These ideas are often interconnected; oftentimes, we like to do what we are good at doing! One's self-image becomes more defined through the course of pursuing a particular educational path and participating in extracurricular activities. Once we have completed our undergraduate studies, we often define ourselves by the title of our degrees. By saying

* Corresponding author: younsoju@msu.edu

"I am a biologist" or "I am a fisheries scientist," you may feel that you are essentially stating that you are not any number of other things that your fellow students might have pursued. Further, a bad experience during a class presentation may have you thinking "I am not a good public speaker" while never having cast a line to catch a fish may have you convinced "I am not a fisherman."

These boundaries, often defined as your comfort zone, may severely limit your willingness to pursue new personal and professional opportunities. Quite naturally, we become comfortable working in a certain area with a particular set of skills, sometimes without even realizing it. Sooner or later, however, problems will likely come along that challenge us to step out of that comfort zone in order to adequately solve a knotty problem. We (the authors of this vignette) have all had an experience of pushing our personal and professional boundaries.

Although breaking out of your comfort zone can be difficult and frightening, taking this step will help you to grow as a person and greatly expand the contribution you can make to the fisheries profession and society in general. As an emerging professional, we encourage you to take risks and pursue new opportunities in order to break out of your comfort zone to grow as a person and a professional. This may be difficult to do, but without taking risks, neither science nor you can grow in knowledge or impact. While you attempt to do new things, realize that you will need help, so get a great mentor and remember that failure is a stepping stone to success, not an end in and of itself.

Breaking Out of Our Comfort Zones

Bill: I learned early on of the need to take risks if I wanted to be a valued member of a team. I grew up as the last of four children. Because my family moved often, we were fairly close-knit and relied on each other for everything, from camaraderie to safety. Being so much younger, I always struggled to keep up with my siblings, physically, emotionally, and intellectually. I often discovered that no matter how hard I tried, there just were some things I could not accomplish (at least not at the age I was), but knowing this did not stop me! If I did not keep up, I would be left behind—which was unacceptable, in addition to being frightening.

But, what could I do to make sure that I was not left behind? I had to learn how to be a valued member of the team, and the only way I could do this was to break out of my comfort zone of being compliant with their expectations of whom I was or should be and allow my unique talents, including persistence and a sense of adventure, to assist team Taylor in our pursuits. For example, when I was five years old, a storm surprised us on a family boat outing in the Adirondacks. The waves rolled over the side of the boat and we all realized this was not going to be an easy trip to shore. My sense of adventure and enjoyment in entertaining my family quickly kicked in. To calm my siblings, I began singing "Davy Crockett" and swinging a rope above my head. I turned the scary boat ride through crashing waves into a memorable team Taylor adventure that we often recount and share to this day.

Once I figured out that the best way to be valued was to be myself, I did all I could to grow and hone my abilities, becoming an ever more respected (or at least accepted) member of the team. I learned that I could not grow or add value to the team if I was afraid to risk, and failure to me was not an option. Consequently, I chose not to be afraid of taking (hopefully) calculated risks. I also found that, even when the risks resulted in unintended consequences, I had learned how to pick myself up, often with the help of others, and try again. The next time, I was better prepared, both experientially and intellectually. I found that taking risks, regardless of the outcome, always had rewards and these experiences have shaped my life view and my mentoring abilities.

So-Jung: For most of my life I was comfortable being a follower and would never have described myself as a leader. I had no desire to hold leadership positions and believed that I did not have the ability to be a good leader. During my sophomore year of undergrad, however, I unexpectedly found myself in the president's position for a club, with which I was involved, two weeks before the club's major annual show. The unexpected resignation of the former president, combined with the lack of returning club members, resulted in the executive board unanimously voting me in as president. Holding an executive position was something I had no inclination to do, but with time running short, there were no other options besides cancelling the show entirely. Needless to say, I was being pushed outside of my comfort zone. I had to adapt quickly from being a follower to a leader. I realized that I could not let the club members down, so I rose to the challenge and worked hard to both pull off that show and lead the club successfully for the rest of the year.

Due to this opportunity, for which I would never have volunteered on my own initiative, I discovered that I do, in fact, have the potential and the ability to become a leader. Knowing that I successfully helped coordinate this show, as well as all the fundraisers, speakers, and club events I organized that year, gave me the confidence to take advantage of new opportunities and also made me more eager to take proactive, leadership roles in other appropriate projects.

Nancy: Often when we think of comfort zones, we envision our skills and knowledge, such as being good in math, but sometimes our comfort zone may involve something more intrinsic to ourselves. For me, I find it challenging to engage in large group discussions. Early on in my career, if I found myself stuck in a group environment, I preferred to remain quiet. This reaction was not due to my lack of skills or knowledge, but rather to my hearing impairment. Although I have developed quite a knack for reading speakers' lips to compensate for my hearing loss, I know that there is a 50% chance that I may not accurately understand the conversation. Having guessed wrongly a few times about what was being said and having the embarrassing experience of contributing something that had no relevance to the topic at hand ultimately made me want to hide and disengage during group discussions in any forum. I really wanted to avoid a reoccurrence of such missteps, as you can well imagine. Obviously, this strategy was limiting my ability to contribute to and advance within my profession, but it felt safer than suffering the embarrassment I had felt by not fully understanding what was being discussed.

When I began working with Bill, he helped me realized that I was unfairly limiting myself by hiding my impediment. Instead of keeping my hearing impediment to myself and trying to deal with these limitations by staying in the shadows, Bill encouraged me to overcome my hesitations and inform others about my hearing loss. This resulted in the unexpected outcome that I can now usually find someone involved in the group that is willing to keep me on track when I start missing the essence of a conversation and by having those in the group accommodate my needs by looking at me more as they speak and not covering up their mouths with their hands or papers or microphones so that I can read their lips. Overcoming this limitation by stepping out of my comfort zone and involving the groups with which I work has helped me contribute more and has made me a more valuable participant. Though this has not solved all my challenges, I am more comfortable speaking up to indicate that I need help to accurately hear what is being said in group discussions.

Paola: I started my Ph.D. program in fisheries and wildlife with a background in general biology. Although my M.S. research focused on Sea Lamprey *Petromyzon marinus* production and stream habitat, I lacked the classical fisheries and wildlife background that most of the students in Bill's laboratory had at the time. After the first year of my program, Bill decided that I should mentor two undergraduate women who were interested in pursuing fisheries and wildlife as a major. The thought of having to supervise someone was terrifying to me at the time. How could I give these two young women a worthwhile experience when I felt that I myself did not know that much about the field? What if I made mistakes that caused them to change their minds about the major or about pursuing higher education in general? How could I be trusted to guide the development of these two young women while I was still developing myself?

With Bill's help, I developed a plan to get them involved in some field and research experiences for several months, and with great trepidation, I stepped from the role of student and mentee into the role of mentor. It was really hard at first; I second-guessed myself every time I interacted with these students. Was I doing the right thing? I did not feel qualified to be supervising anyone. As time passed, I found that it was actually a lot of fun and very rewarding to work with these young women. It was still uncomfortable for me at times because I was still learning myself and many times did not know the answers to their questions or even how to best approach a problem we were facing with the research. In the end, I realized that it was okay for me to not always know the answer. In fact, through working together to solve problems, they learned that hard work, perseverance, and team work is how to make progress on research. Now that I am mid-career, I look back on that experience and realize that stepping out of my role as a student at the time and stretching myself to try the mentoring role really helped me grow both personally and professionally. I learned that I enjoyed teaching and mentoring so much that this experience out of my comfort zone changed my career trajectory.

Amy: When I was younger, travel caused me a great deal of anxiety. I would arrive in a new location confronted by worries about the things I had forgotten to pack, the friends I would miss, and the difficulties of navigating in an unfamiliar city. Deep down, I knew that I genuinely enjoyed exploring new places, but I struggled to overcome the challenges that went along with being away from home. When it came time to decide where to attend university, I picked the logical, but also safe, choice in my hometown. I sensed I was missing out on something, however, when I realized that I could see my preschool from the window of my dorm room. I was not sure that I would ever become truly independent if I continued to live, work, and study in the same place where I had learned to walk. Although college was ultimately an exciting and productive time for me, I still wondered about the person I could be if I did the one thing that scared me most: move far from home. If I could overcome my fears and broaden my horizons in such a literal sense, what other challenges would become easier by comparison?

I took action on these feelings by applying for graduate school abroad, and a few years later, I found myself sitting on a streambank in northern Iceland, observing juvenile salmonids as field research for my master's degree. I had finally mustered up the courage to leave my familiar surroundings and call another country home. Pursuing a graduate degree in another country presents its own unique challenges, but this path has forced me to grow both professionally and personally in a way and at a pace that would not have been possible otherwise. Making a fool of myself while attempting to speak Icelandic has encouraged me to be braver in communicating the results of my research. Navigating through immigration bureaucracy has shown me that every problem has an answer, no matter how roundabout the solution may be. I began to break out of my comfort zone by deciding to move abroad, but the truth is that almost every day I encounter a situation that forces me to push my intellectual and emotional boundaries. What I have learned is that the more frequently I do things that challenge me, the easier it becomes to face new challenges. By consistently stepping out of my comfort zone, I have created a new normal for myself, so that what used to make me uncomfortable is now just a part of my daily existence. Ultimately, the things I have learned as a result of pursuing new opportunities have demonstrated that any discomforts experienced along the way are far less painful than never stepping outside the confines of my fears, and far more rewarding.

How Can You Break Out of Your Comfort Zone?

As we have all learned, it is very easy to remain within your comfort zone, but doing so can severely limit your personal and professional growth, which in turn can limit your potential contribution to advancing the field of fisheries. The profession cannot grow without people who are willing to take risks and expand the boundaries of the field itself. Taking the risk to break out of your comfort zone will be a very scary but ultimately rewarding endeavor! As an emerging professional, you are in a good position to seize new opportunities. People generally expect you to take risks and forgive mistakes as you learn how to contribute to the field. To expand your horizons and step outside of your comfort zone, consider

- Pursuing leadership experiences within professional societies (such as the American Fisheries Society).
- Exploring how decisions are made through an internship or fellowship within the legislative or executive branches of government.
- Interning or volunteering in a field outside of your area of expertise.
- Engaging with a state, federal, or tribal agency; a hatchery; a legal group; or a nongovernmental organization.
- Presenting a seminar at your university, as well as engaging in public meetings held by agencies, commissions, and nongovernmental organizations.

As you move forward in your career, it can be tempting to focus only on your area of expertise using the skills in which you excel. But, take a risk and get involved in areas and activities about which you are less knowledgeable! This will open up your world to new people, connections, and opportunities. You may discover that by expanding into new areas, you can draw connections and generate new ideas. As you begin to expand your horizon and break out of your comfort zone, remember that you do not need to do this alone. In fact, your trusted mentors, mentees, and colleagues will often be the ones to reveal a boundary you were not aware you were hiding behind and help you expand past it.

Biographies

So-Jung Youn is an M.S. and Ph.D. student under Dr. Taylor in the Department of Fisheries and Wildlife and Center for Systems Integration and Sustainability at Michigan State University. She is interested in studying the role of inland fisheries in local food and economic security, particularly in Asia. So-Jung received her B.S. in biology, with a minor in management and organizational design, from the College of William and Mary.

William W. Taylor is a university distinguished professor in global fisheries systems in the Department of Fisheries and Wildlife and the Center for Systems Integration and Sustainability at Michigan State University. He has been active in the American Fisheries Society throughout his career, serving as president of the society in 1997–1998. He currently holds a U.S. Presidential appointment as a U.S. Commissioner (alternate) for the Great Lakes Fishery Commission.

C. Paola Ferreri is an associate professor of fisheries management in the Department of Ecosystem Science and Management at Pennsylvania State University. She is interested in understanding the link between habitat, management actions, and sustainable fisheries.

Nancy J. Léonard is the fish, wildlife and ecosystem monitoring and evaluation manager with the Northwest Power and Conservation Council. Her work experience in multi-jurisdictional fisheries governance with the Department of Fisheries and Wildlife at Michigan State University, and her work with the Great Lakes Fishery Commission give her a strong professional background working with diverse interests towards common goals in fish recovery. She is active with the American Fisheries Society (AFS), currently serving as the scientific program co-chair for the 2015 AFS annual meeting. She has an M.S. in biology from Carleton, an M.S. in water resources from Miami University, and a Ph.D. in fisheries and wildlife from Michigan State University.

Amy Fingerle is a master's student in the Department of Aquaculture and Fish Biology at Hólar University College, Iceland, studying the behavior of stream-dwelling Arctic Char *Salvelinus alpinus*. Prior to graduate school, she earned a B.S. from the Program in the Environment at the University of Michigan and worked as a research assistant for the USGS Great Lakes Science Center and the Partnership for Interdisciplinary Studies of Coastal Oceans (PISCO) at Oregon State University.

Diversity to Appreciate

Succeeding as a Nontraditional Graduate Student: Building the Right Support Network

ROBIN L. DEBRUYNE*
Cornell Biological Field Station
Department of Natural Resources, Cornell University
900 Shackelton Point Road, Bridgeport, New York 13030, USA

EDWARD F. ROSEMAN
U.S. Geological Survey Great Lakes Science Center
1451 Green Road, Ann Arbor, Michigan 48105, USA

Key Points

- Nontraditional graduate students bring distinct perspectives and reflections, which meld their passions and interests in academics, research, and their home life and family.
- Success of nontraditional graduate students may depend on the support networks and flexibility provided by institutions, advisors, mentors, and employers.

Schedules, routines, and support networks. That is what allows the entire day (and week and semester) to make sense and flow together, from one to the next. From when I (RLD) *get up, to when my daughter goes to bed, and beyond. Trial and error has taught me that following our schedules and routines brings a day with fewer tantrums, better eating, and, most importantly, better sleeping than the unscheduled day. One of the best scheduled events occurs every few days: my mid-afternoon coffee break. My time to relax with a friend; moan a little about life, school, and research; and recharge for the final push of productivity until 4:30 pm. And today I'm not missing that!*

We meet outside the building and walk to the library and into the coffee shop. We get our drink and take a seat in the large foyer of the library, being overwhelmed by the hustle and bustle of the undergraduates and fellow academics scurrying around us. The background noise is almost music, which relaxes me in the comfortable din. Then, we can begin catching up. We talk about anything interesting from the day, research triumphs or failures, plans for the weekend, upcoming field season preparations. I know my friend needs this time away from her desk as much as I do, and that makes me feel better, knowing that her life can be just as crazy without a kid as mine is with one.

Then she asks, "Are you going to the 4:30 pm seminar and reception tonight for the invited speaker?"

Am I? Well, no. That's not really possible this week. Even though the first solution would be to have my spouse watch my daughter, he has a job and other obligations that prohibit this. This week my husband is off in the Adirondack Mountains, two and a half hours away, conducting fall fisheries assessments in small remote lakes. Because we are both fisheries professionals, I don't begrudge him for

* Corresponding author: rdebruyne@usgs.gov

his frequent and extended absences. It's just part of the job and one of our shared passions. He always gives me 100% support in my research endeavor, so I give the same to him in return. A babysitter? Not likely. All of the people I know without kids, and, more importantly, those that my daughter knows, will want to attend the early evening seminar and reception. Bring her with me? Definitely not. My own networking would not be enhanced by following a toddler around to be sure she doesn't break anything.

But luckily for me, I know my friend doesn't need convincing regarding all the reasons to explain my absence at the gathering. I just reply, "I can't. My husband is sampling this week."

"That's too bad. I'll let you know how it goes."

"Thanks." I'm so lucky she understands. Others I work with don't always understand. But how can I fault them? I didn't understand when I was the traditional graduate student, straight out of undergraduate school beginning my master's degree. I tried to be sympathetic to those with spouses, children, other jobs, or obligations that weren't school or research related, but I didn't understand. I had no idea what it was like for them. I thought of my project, my academic work, and my planned social outings. I would go to seminars and the receptions and wonder why those other students weren't there. Of course, my lack of understanding was simply a byproduct of my own ignorance. Now that I have these other factors influencing my daily routines and considerations, I completely understand what it was like (generally) for those nontraditional graduate students. And I empathize with their need for more flexibility in their schedules, why they have such organized days, and why they weren't typically involved in last minute plans. Their (and now my) days are in part predetermined to maximize their time with their partners and/or their children. The flexibility afforded to us by our advisors and employers is something that we greatly appreciate.

"Did Little Miss get over that fever she had last week?" my friend asks. She's referring to the three day fever my daughter had at the end of the week. That fever had me out of work for two days, including both of my teaching assistant sessions that the professor and other teaching assistant graciously covered for me. The support I have received from my advisors and employers has been second to none.

"Yes. It was from a common virus in kids, roseola. Now that she's had it, she shouldn't get it again. I wasn't able to come to work because I couldn't take her to daycare with the fever. She's been back this week, but I'm still catching up." I look at the big clock above the doors into the new library. "Well, it looks like I should be heading back to the office."

"I'm ready to go. I have to finish some analyses before the seminar," she responds. We get up from the little silver table, push in our chairs, and head to the corridor toward the outside. We chat a little more before she heads into the department's main building to her office and I head to the pale yellow building that was once a chicken coop but is now a storage building with some graduate student offices on the top floor. I push through some more grading until 4:30 pm when I walk to my car and head to daycare.

****

Our experiences as nontraditional graduate students (NTGSs; defined as those who did not move into graduate school directly from undergraduate matriculation or who may be older, married, divorced, have children, have full or part-time jobs, be veterans of the military or Peace Corps, or a member of cultural or ethnic minority groups) were more variable and complex compared to the experiences of traditional graduate students (RLD was both). We were both NTGSs but had different circumstances that led to this classification. We encountered different barriers and took advantage of different support resources. But one thing we both experienced, that in no small way led to our success, was flexibility and understanding from our advisors, employers, and from our peers. Flexibility, not in terms of workload, responsibility, or expectations, but in terms of work schedule, support structures, and openness to considering different problem solutions, is what is most helpful to relieving stress for the NTGS.

The wide-ranging backgrounds, experiences, and priorities of NTGSs suggest that they have very diverse needs and stressors that can become obstacles for their success. How can students, advisors, university professionals, and employers provide support to meet those needs, or at least help to minimize or eliminate obstacles and stressors for this growing proportion of the graduate student body and provide the tools to succeed?

Retention and success of any student, traditional or nontraditional, begins with recognition of the vast diversity in backgrounds, experiences, priorities, and goals of the students and integrating this diversity into the goals of the academic institution and individual academic departments. Based on our experiences, success begins with developing personal priorities and commitment that allows for a delicate balance between family, education, career, health maintenance, and social networking. As evidenced in the scenario above, commitment to our families is always the top priority, and having a network of mentors, colleagues, and friends who understand, accept, and support this commitment relieves stressors associated with challenges such as marriage and child rearing in an academic setting. For example, upon arrival at graduate school, we were pleased to be greeted by department administrators with a list of local childcare facilities, playgrounds, and museums and to be introduced to other students who were also spouses and parents of young children. This early interaction provided an immediate sense of belonging and bonding that was the start of an important valuable social network that has lasted well beyond graduate school. Departments or graduate student associations can compile lists of resources by appointing a staff member or student to search local listings, query staff and students, and distribute this information at orientation and on departmental Web sites.

Working in an academic community that valued family as a priority and nurtured NTGSs was also instrumental in our academic success. Our mentors and advisors made efforts to be accessible at times that fit our schedules, and they provided an open ear for discussions on topics relevant to family life as well as coursework and research. Mentors and advisors also provided direction and a workload that was realistic and considered the unique situation and experiences of the NTGS. Many professors were accepting to the one or two times when all options were played out and we were forced to bring our well-behaved children to a lecture. They also provided flexibility in attendance and deadlines when family priorities could not be avoided. Our gratitude for their acceptance of our circumstance and flexibility is unending. The compassion shown by our mentors to provide flexibility when life happened was and still is reciprocated by us to our mentors, mentees, and coworkers. When our mentors' plans changed unexpectedly, we accommodated them anyway possible, from moving meetings, cutting meetings short, or going through an analysis via e-mail because their schedules were busy. There were times when the weekend was the best, or only, time a meeting could occur. If you are an NTGS needing additional flexibility, speak up; work with your advisor or employer to find an amicable solution. Advisors and employers can improve their preparedness for working with NTGSs by attending workshops and seminars and by reading books and scholarly papers on the subjects of mentoring and advising nontraditional students. Also, making the topic of advising and mentoring NTGSs a part of faculty meetings (even only once per year) or having the departmental graduate studies coordinator distribute tips on mentoring NTGSs can encourage discussions among advisors to share lessons and resources across the department.

In addition to the support rendered by our mentors, the academic community also provided many unique resources catered to NTGSs. These often included campus child care, family counseling, career services, Peace Corps and military veteran's services, tutoring, and social activities that were scheduled with families in mind. Information on many of these services was available on the university's Web site as well as through the departmental counseling and advising office. The departmental (or university) graduate student organization also played a positive role by recognizing that a majority of graduate students had families and sponsored family-friendly social activities. Universities have many of these programs and services free of charge to students.

Student veterans are a growing segment of the student body. Nearly one million veterans, service members, and their family members have made use of the Post-9/11 GI Bill since it was made available in 2009. As a military veteran, I (EFR) found comfort and solace interacting with other student veterans. Our shared experiences in the military provided a unique bond and an atmosphere where we could discuss personal and scholastic issues that we felt awkward discussing with nonveterans. Off-campus resources such as the local American Legion provided collegiality and friendship with American Legion members and other student veterans. These social interactions provided another sense of belonging and appreciation for military experiences and service and helped me and other student veterans make the transition to civilian life.

Several scholarship and award opportunities are also available to nontraditional students. These opportunities often require an application essay but prove economically worthwhile by providing extra financial support for research, living, and child care expenses, as well as recognition for academic and research excellence. A valuable link to some of these scholarship opportunities can be found at the Association for Non-Traditional Students in Higher Education Web site (www.antshe.org/), as well as from your department and university career counseling offices.

Last, and by far most important for us to remain motivated and inspired to pursue higher education, was the support and understanding we received from our spouses, children, parents, and other members of our families. Our successes were achieved as a team effort where our families shared in sacrificing time and finances so that we could accomplish our educational and career goals. And now, we make sure our families know our appreciation as we share with them the rewards, satisfaction, and pride in our accomplishments.

Conclusion

As fisheries professionals, we are a passionate group. We care deeply about the natural resources around us and strive to secure this experience for generations to come. NTGSs bring distinctly different perspectives and reflection, which meld their passions and interests in academics, research, and their home and family life. We are able to visualize the values for multiple and diverse beneficiaries almost simultaneously and want to share the positive experiences and places we visit, rehabilitate, or restore with those multiple user groups, such as our children. As more students engage in graduate education later in life, the instances of NTGSs will increase, as will the diversity and size of the fisheries profession. As such, earning success in higher education will require special consideration by the student, their mentors and professors, and other students. If you are an NTGS, recognize the strengths you bring as an NTGS and work with the challenges of your circumstances. Gather information about the support networks (academic, financial, emotional) in your area and use them to the fullest extent. If you are an advisor or employer of NTGSs, know that any support you provide (e.g., information about local childcare, parks, museums, libraries, family-friendly activities, scholarships, and support groups) is greatly appreciated. And finally, whether you are a NTGS or a NTGS mentor, peer, or coworker, remember everyone needs flexibility when life happens and share your solutions to these situations, if possible.

Useful Resources for Nontraditional Students and Mentors

- Association for NonTraditional Students in Higher Education (www.antshe.org)
- National Association of Graduate and Professional Students (www.nagps.org)
- Student Veterans of America (www.studentveterans.org)
- Resources at other universities and organizations (www.rackham.umich.edu/faculty_staff/information_for_programs/academic_success/mentoring_advising/)
- University and departmental academic advising, counseling, cultural, and career centers
- University and departmental graduate student associations
- Fellow students and faculty

Acknowledgments

The authors would like to thank our families, advisors, mentors, employers, coworkers, and peers who have supported them through their graduate and professional careers. This vignette is dedicated to their brothers, Lee Alan DeBruyne, Jr. and William John Roseman, who are no longer able to celebrate life's accomplishments with them. The authors are grateful to Ellen George and three anonymous reviewers for comments that improved this manuscript. This is Contribution 1770 of the U.S. Geological Survey Great Lakes Science Center.

Biographies

Robin L. DeBruyne is a Ph.D. candidate from the Cornell Biological Field Station in Bridgeport, New York (Cornell University) and a contractor with the U.S. Geological Survey Great Lakes Science Center, in Ann Arbor, Michigan. Her dissertation research examines percid-cormorant interactions on Oneida Lake, New York. She received a master's degree (2006) from Central Michigan University in Mt. Pleasant, Michigan, examining Lake Whitefish *Coregonus clupeaformis* population dynamics. She is married with two children.

Edward F. Roseman is a fishery research biologist for the U.S. Geological Survey Great Lakes Science Center where he investigates deepwater fishery ecology and early life history dynamics. He served in the U.S. Army for four years, stationed in the United States and Germany. He received a master's degree (1997) and Ph. D. (2000) from Michigan State University in East Lansing, Michigan, studying physical and biological factors influencing year-class strength of Walleye *Sander vitreus* in western Lake Erie. He has one son currently serving in the U.S. Navy.

Understanding Generational Differences in the Workplace

KELLY F. MILLENBAH*
College of Agriculture and Natural Resources, Michigan State University
Justin S. Morrill Hall of Agriculture, 446 West Circle Drive, Room 121
East Lansing, Michigan 48824, USA

BJØRN H. K. WOLTER
State of Alaska, Department of Education & Early Development
Post Office Box 110500, Juneau, Alaska 99811, USA

Key Points

- Each generation has its own experiences that influence the way its members think, act, and interact.
- Be aware of your own generational expectations, opinions, and experiences, and recognize that they might not be the same as other generations.

I (Millenbah) was ecstatic when my parents announced that they were moving from a PC to a Mac. Finally! My Mac-versed sibling could field all of the calls from my parents about the almost daily computer induced dramas: This program is not working! What's a URL? What do you mean a computer can get a virus? I would spend countless hours attending to my parents as they attempted to navigate the world of computers, something that seemed so effortless and commonplace to me and yet so completely and utterly challenging and frightening to them. I would lose patience as I walked them through how to copy and paste for what seemed like the 100th time. My parents are highly educated individuals and held high level positions in each of their past careers so it is not that they are incapable of learning or understanding computers. This interaction between my parents and me got me thinking. Am I really any different than they are? My own children look at me with distain with my own apparent lack of computer savvy. "Hey, I am on Facebook!" I boast. Little did I know that to be really computer-literate I needed to be a tweeter and active on Tmblr, all while multi-tasking via instant messaging, text messaging, and playing Halo on my Xbox. So, am I really that different from my parents? How can three generations of family members be so drastically diverse in the way we communicate, live, and survive?

Today's workplace is complex, filled with individuals of different generations all with their own expectations, motivating forces, experiences, and opinions. The differences among generations can result in a suite of challenges, as well as opportunities, for solving pressing natural resources issues. Before we can be effective in advancing our collective work on the conservation of natural resources, we need to better understand what defines, describes, and motivates workers in the field and how we are influenced by our generation.

* Corresponding author: millenba@msu.edu

What Characteristics Describe Generations in the Workplace Today?

About 15% of today's workforce is comprised of the veteran or traditionalist generation (born between 1925 and 1945). This generation grew up during the Great Depression and World War II; thus, many of their behaviors and decisions are based on those experiences. They are called traditionalists because this generation has an affinity for protecting traditional values such as civic pride, loyalty, and living within one's means. They respect authority and expect respect for a job well done. Veterans tend to be hard workers who stick to instructions given, follow the chain of command, do not complain, are quiet, and believe a good paycheck requires hard work. Most in this generation have been with the same employer since they first entered the workforce.

The baby boomers (~46% of today's total workforce, born 1946–1964) are products of the post-World War II boom in births, and the sheer size of this generation gives them unprecedented clout in determining workplace policy and atmosphere. In their youth, this generation was known for its liberal and idealistic views and counterculture distrust of institutions and authority. Boomers grew up in an era of civil rights movement, empowerment, and emphasis on diversity, which is likely tied to how they have been so instrumental in winning many workplace victories so many take for granted today. Members of this generation frequently prefer to be leaders rather than followers, which may lead to confrontation. As they have aged, boomers have become more conservative and focused on financial success. They are often deemed the inventors of the 60-hour workweek and are described as living to work.

Generation X (born ~1965–1980) represents 30% of today's workforce, and they are the children of the baby boomers. Gen Xers were the first generation to arrive home after school to an empty house, as often both parents worked outside the home. Their parents expected them to be as committed to their jobs as they were. As a result, this generation is one of the most overanalyzed, hypercriticized (often by their parents) sections of society, and they are often quite harshly referred to as slackers. Many Gen Xers disagree with this sentiment and argue that they are much more independent than either their parents or their children. Unlike their parents, they work to live rather than live to work, only wanting to be in the office if there is something important to do. Gen Xers know and understand technology (and want to use it), may have lots of career interests and paths, demand individuality, and enjoy multi-tasking. Gen Xers grew up in an era of corporate downsizing in the 1980s and 1990s where they watched family members and friends lose their jobs in the name of corporate profit. Because of this, they are perceived as prioritizing personal growth and independence over workplace fidelity, and perhaps labeled as jaded or cynical. As a result, Gen Xers are more likely to have both multiple careers and jobs than previous generations.

The Millennial Generation (born ~1980–2000) is the most socially and culturally diverse generation our nation has ever seen. Their representation in the professional workforce is growing but affected by the reluctance of baby boomers to retire. The following adjectives have been used to describe Millennials: special, sheltered, confident, pressured, team-oriented, conventional, and achieving. They have been plugged into technology since they were babies and expect to use it constantly in the workplace, are accustomed to highly scheduled days, expect to multi-task in most aspects of their life, and will have six to eight careers in their lifetime. Boring or menial tasks are not favored by Millennials, and they tend to seek out jobs where their creativity is most important, noticed, and rewarded. They often have unrealistically high expectations of success and reward, and while a good paying job is important to Millennials, so is a favorable work environment—they are willing to trade one for the other.

Post Gens (born after 2000) are the newest generation to be described, and few have entered the workforce yet. However, the Post Gens have already experienced a wealth of generation-shaping events (e.g., evolving and shifting foreign policies, a more global perspective of the world) that will influence the way they live and work. While data are still being generated about this generation, early information suggests that this generation will be "the most contributory, collaborative, and perhaps engaged generation in our memory" (Meet the PostGens 2011). How they will influence the workforce remains to be seen.

Strategies for Working across Generations

Given the differences we have detailed across current workforce generations, we have some recommendations for how to facilitate work based on our own experiences and experiences shared by others we know.

1. Veterans/traditionalists—Veterans have a strong work ethic and the apprenticeship model of learning is ingrained in them. This makes them excellent mentors and coaches. Take the time to sit down and listen to this generation. Their wealth of information and experiences can be instrumental in not repeating mistakes of the past. Because they are digital immigrants (interacting with technology is something that has been learned rather something they were born into), they can be skeptical of technology multi-taskers and may assume that those who engage in multi-tasking are wasting time rather than working. Veterans will complete tasks on time and will expect those around them to follow suit without complaint or effusive praise when a task is completed.
2. Baby boomers—Personal connections are important for baby boomers. Baby boomers are fond of the saying "We've always done it this way," but they will accept a challenge if offered. They are readers, thinkers, and analyzers. Baby boomers make good leaders and want to work in groups. However, many in this generation are uncomfortable managing conflict and can be more focused on process rather than results. Boomers are the original workaholics. They expect colleagues around them to be willing to work the same long, hard hours they are willing to invest to get the job done and to dedicate time to follow-up, documenting, and organizing. Boomers can be judgmental of others whom they perceive as not being equally dedicated. It is the boomers who are responsible for labeling Gen Xers as "slackers" because they work to live rather than live to work. Similarly, they strongly believe that one earns privileges rather than having a right to them, without experiences to back up that right.
3. Gen Xers—Unlike the baby boomers, Gen Xers work as a means to an end. They tend to be impatient with traditional processes and are willing to log long hours on the job, but only if the work is relevant and meaningful to get to the desired end product. Focus is key for Gen Xers—objectives and goals need to be crystal clear to get them to buy in. This generation is fundamentally different from boomers in that they are focused on product, not process. Gen Xers do not subscribe to the idea of a 40-hour workweek, perceiving downtime in the office as wasted time. They are highly independent, which should be capitalized on. Gen Xers can be placed in charge of diverse projects with little supervision and be counted on to deliver. Balancing work and personal responsibilities is important to this generation. As a result, they enjoy working in short bursts and at flexible times to allow connections to their family and activities outside of the work-

place. Gen Xers prefer communications that are succinct and to the point. Because they tend to opt for technology-related communication strategies over in-person connections, they can be perceived as unfriendly by coworkers of older generations. This is not the "watercooler" generation; however, structured opportunities to socialize within the office can bear fruit with Gen Xers. Hierarchy and protocol is not overly important to this generation. While Gen Xers value feedback, they are not afraid to offer feedback in return. This generation thrives on being "in the know" and can be offended if they are left out of important decisions or feel uninformed.

4. Millennials—Millennials are grounded in wanting to make a difference while also having fun. Many Millennials are willing to give up a larger paycheck to work in areas that connect with their passion. If they do not feel like they can influence the end product in a position, they will seek out other positions rather than adjust their expectations of success on the job. Millennials are digital natives and they prefer to use short, quick modes of communication such as texting; for this generation, e-mail is used less and less. Millennials demand flexible work atmospheres but require many different types of learning experiences to help build their workplace knowledge repertoire. Problem-solving abilities and autonomous work habits are not as well developed in Millennials as in Gen Xers; however, they do work well in groups, perhaps more so than any previous generation. Success of such groups depends on identifying a core leader capable of motivating and nurturing the group. Millennials are highly respectful of a diverse workplace and expect to engage with diverse groups. They value rules but have no problem breaking them if they get in the way of their goals. Millennials should be cautious of overconfidence in their knowledge base. Many tend to overestimate their own abilities and set unrealistic goals for what they can accomplish and deliver. Accolades are important for this generation, even in situations where other generations may feel it is unwarranted. Consistent and purposeful supervision is expected and needed for this generation.

So, Do Generations Have Anything in Common?

Working across generations can be challenging. What is most important to recognize is that differences among generations are not right or wrong; rather generations are simply different. However, there are many similarities that do resonate across all generations, and creating work environments that support those similarities can be critical to workplace success. Zemke et al. (2000) identify six important similarities that can tie generations together:

- Work is a vehicle for personal fulfillment and satisfaction, not just a paycheck.
- Workplace culture is important.
- Being trusted to get the job done is the number-one factor that defines job satisfaction.
- Everyone needs to feel valued by his or her employer to be happy on the job.
- Flexibility in the workplace is valued.
- Career development is the most valued form of recognition, even more so than pay raises and enhanced titles.

Working across these generations will be an important strategy for successfully conserving and protecting our vital natural resources. What traits and characteristics do you bring to the table, and how will you reach out to other generations to find common ground for advancing our most important natural resources issues? We recommend the following for each generation to consider as you move forward in the workplace:

- Baby boomers: Accept that different does not have to mean worse, and understand that following generations are just as dedicated to the work at hand but approach it from a different perspective.
- Gen Xers: Try to avoid the persecution complex that can come from being caught between two large generations, and focus on maximizing the coming years as you will come into your own in the workplace.
- Millennials: Understand that previous generations often have the weight of experience on their side, and be willing to adapt to the situation at hand. It may be expedient to explain why you are approaching your work the way you do so that others realize that different approaches do not necessarily indicate a lack of focus or attention to detail.

Biographies

Kelly F. Millenbah is the associate dean and director for academic and student affairs in the College of Agriculture and Natural Resources at Michigan State University (MSU). She is also a faculty member in the Department of Fisheries and Wildlife. Millenbah is a scholar of natural resources education with a focus on teaching and learning within the classroom, curricular design and development, and recruitment and retention of students in the sciences. She earned her B.A. degree in biology from Ripon College, Ripon, Wisconsin, and her M.S. and Ph.D. in fisheries and wildlife at MSU.

Bjørn H. K. Wolter is the science supervisor for K–12 education in the state of Alaska. He is also an adjunct faculty member with the Department of Fisheries and Wildlife at MSU and the University of Alaska-Fairbanks. Wolter holds a Ph.D. in postsecondary science education from MSU and M.S. and B.S. degrees in ecology from Western Washington University. His research focuses on the scholarship of teaching and learning in undergraduate sciences, student transitional experiences to college, and building scientific literacy. Together, Wolter and Millenbah have coauthored numerous articles on natural resources education.

References

Meet the PostGens. 2011. The Tru Report (July [special edition]):3–11.
Zemke, R., C. Raines, and B. Filipczak. 2000. Generations at work: managing the clash of veterans, boomers, Xers, and nexters in your workplace. Amacon, New York.

Recommended Readings

Howe, N., and W. Strauss. 2000. Millennials tising: the next great generation. Vintage Books, New York.
Strand, M., M. S. Johnson, and J. F. Climer. 2009. Surviving inside Congress. Congressional Institute, Alexandria, Virginia.

Can We Really Have It All?

Jessica Mistak*
Michigan Department of Natural Resources, Fisheries Division
6833 US Highway 2, 41, and M-35, Gladstone, Michigan 49837, USA

Key Points

- The balance of work and life will change throughout a career.
- Work–life balance applies to everyone and is different for everyone.
- Work–life balance can be improved at any point in a career by creating a coalition of support, taking steps to ensure fulfillment, and adjusting work schedules.

Introduction

Not so long ago, it was easier to leave work at the office. Now, the boundaries have been blurred by improvements in technology. With so many people having a smart phone or Internet access at home, who has not checked his or her voicemail or e-mail after work? This fact when added to the downturn in the United States' economy, and therefore a shrinking workforce, often equals more work being expected of each individual employee. The work–life balance slope is slippery and those who are resting at the bottom of that slope will be the first to say they do not know how they got there. As with most mountains, there are ways to clamber back up again; however, the simplest route is to avoid the negative effects of work–life imbalance and thus the need to climb upwards.

There are many things one can do to remain engaged and fulfilled—both personally and professionally—throughout a career. The most important thing to remember is work to live, not live to work. If one needs help in living this ideal, there are many people in fisheries science who serve as great examples. Our colleagues can be counted among a diverse cadre, including supportive partners, devoted parents, spiritual leaders, community leaders, coaches, athletes, travel enthusiasts, foodies, artists, and devotees, and users of natural resources. I encourage everyone to look around and find people with similar life experiences who can share their stories of successes and missteps in balancing their career with their personal life.

Once you begin your career, the phrase "work–life balance" will undoubtedly be tossed around. While most probably know what it means, few will spend much time thinking about it initially. On one hand, you have a job that you had to work very hard to get (and, congratulations for successfully navigating your way into the competitive field of fisheries!), so life should be simple, right? Well, there may be little time for celebration if personal time is spent thinking about such things as buying a house and paying back student loans. You may also be torn between working extra hard to prove yourself as an employee and not quite being ready to let go of your more carefree days as a student. Please let me welcome you

* Corresponding author: mistakj@michigan.gov

to the beginning of your new career by sharing some lessons learned in balancing the pull between personal and professional life.

Embrace the Mess

What Does Work–Life Balance Mean?

In its simplest definition, work–life balance means having enough time for work and enough time to have a life. This is easier said than done because, as I have discovered, work–life balance continuously changes throughout one's career. Personal demands are different whether single, a newlywed, a parent, a caregiver for aging parents, or a soon-to-be retiree. When there is little to tie one down, it is easier to work longer hours or travel more for work. Once additional factors are thrown in that affect one's personal life, for example entering into a committed relationship or having children, additional work and travel must be carefully considered or may not be possible at all. As an example, a female colleague and I were lamenting on how difficult it was to work at home after we became mothers and she replied that we "turn into pumpkins," meaning that we basically become useless as employees after our work hours end . . . and, no, we do not get to dance with Prince Charming at the ball first! This situation is much different compared to the early days of my career when I would often bring work home with me.

Work–life balance also means different things for different people, analogous to how some people can function on little sleep while others need much more sleep to be coherent. For example, I know of colleagues who put in over 40 h a week at work, play an active role in their child's school, manage a farm, and come to work looking as though they just spent the day at the spa. Other colleagues put in their 40-h week and live in an apartment so they can spend as much time as possible fishing. I also know of people who seemingly spend all of their waking hours at work. There is no one size fits all for work–life balance.

It Is Going to Be Messy.

Nora Ephron (1996) offered advice on having both a career and family in her Wellesley commencement speech by saying,

> It will be a little messy, but embrace the mess. It will be complicated, but rejoice in the complications. It will not be anything like what you think it will be like, but surprises are good for you. And don't be frightened: you can always change your mind. I know: I've had four careers and three husbands.

Although the statement was directed towards women, it is just as important—and difficult—for men to maintain a balance amid the chaos of life. The same goes for people with and without children.

For me, balancing work and life equals being adaptable. There are times when priority work assignments will mean putting in extra hours. There are also times when I may have pressing personal needs that take me away from work, such as a sick child or pet. I am fortunate to have a supportive spouse who can pick up the slack this creates at home, and I do the same in return when needed. Things really get messy, however, when my spouse and I are concurrently putting in extra hours or need to travel away from home. Messy times are when things that used to be a priority, like having a clean house, home-cooked meals, or a weed-free vegetable garden, get pushed to the back burner and instead I focus on what really matters–making sure my family knows they are loved (while we sit on the dog hair-covered couch, eat frozen pizza, and discuss whose turn it is to run the rototiller). As Cathie Black, former chairman and president of Hearst Magazines said, "You can love your job, but your job will not love you back." Do not expect to find a perfect balance between work and personal life on the first try. Instead, determine what is important right now, and appreciate that your definition of important will be different than that of others and will change over time.

The Myth

Look up the phrase "have it all" and you are bound to find just as much information on how to achieve it all as you are on the impossibility or myth of achieving it all. I, along with my messy house and weed-ridden gardens, agree with Alain de Botton (British writer, January 3, 2013) who said, "There is no such thing as work–life balance. Everything worth fighting for unbalances your life." Yet, even though it is arguably impossible to truly achieve work–life balance, there are ways in which some equilibrium can be established.

One of my valued mentors, retired Michigan Department of Natural Resources Fisheries Division Chief Kelley Smith, recommended the following eight actions to balance life with work:

1. Set goals, and therefore expectations, for home and work.
2. Follow your own core values and principles; do not let others set those for you.
3. Put life (family) first.
4. Learn to say the most powerful word in the English language—NO!
5. Decide what matters and what does not, and then focus on the important.
6. Have faith and trust in your fellow employees, and learn to delegate.
7. When at home, truly be at home.
8. Take time off, and truly make it time away from work.

Some of the ways in which these actions can be accomplished include creating a coalition of support, taking steps to ensure fulfillment, and adjusting work schedules.

Find Supportive Individuals, Both at Home and at Work

Be sure to actively seek out support, especially during periods of high stress at home or at work. This coalition of support will often evolve organically, although sometimes it will need to be sought out. My first recommendation is to find trusted allies. These trusted allies could be mentors, friends, family, or coworkers who understand what you are going through and appreciate the value of an enriched personal life. For example, I have been able to coordinate with coworkers to take turns attending evening meetings; that way, no one person is away from home for extended periods. Likewise, my coworkers also stepped up to attend meetings requiring travel during my first year as a parent, allowing me more time at home. In turn, I do the same for my coworkers when their schedules are tight. This is such an appreciated act of kindness because no one, whether single or married, child or childless, career newbie or career veteran, wants to constantly be away from home. I can also not say enough about the role of the American Fisheries Society (AFS) in building a coalition of support. Through involvement at the student subunit, chapter, division, or society level, one will meet colleagues in AFS who share similar passions and help build a sense of professional fulfillment. These colleagues energize us professionally and, most importantly, provide support throughout our careers.

Ironically, support often comes from where you least expect it rather than from where you expect it most. Even though fisheries science is a male-dominated profession, I have experienced nothing but support from my male colleagues through every changing phase of my personal life and career; yet I have met a few women in fisheries who were not supportive of my decision to be a parent. My best explanation for this is that the women who truly paved (and continue to pave) the way for females in fisheries often made tough decisions to be deemed credible by their peers, including working harder during regular hours and taking on unpopular assignments. I believe these women wanted the best for me and their lack of support was, in a sense, them saying that they did not want my personal decisions to interfere with my professional credibility. Thankfully, most women in fisheries, including all of those who I have gotten to know through AFS, want to pay it forward and will go out of their way to help others, both females and males, in their career development. With a strong and diverse coalition of support, it will be easier to handle life's bumps in the road—and the rare naysayers—and decide what is in your best interest.

Pursue Mental, Physical, and Spiritual Fulfillment

This means finding ways to enjoy your home life, yourself, and your job. For me, this involves continual engagement at work and growth as an employee, with the caveat that this engagement and growth fit within the four corners of my regular work schedule instead of adding extra hours to my already busy week. Sometimes professional fulfillment takes some creativity, especially during lean budgets when there is little to no funding for formal training. Growth options in the fisheries profession always exist, however, and things I have done include increasing my level of involvement with AFS and taking part in mentoring and cross-training opportunities.

Being physically and spiritually fulfilled means taking care of yourself, whether it is through eating well, exercising, attending religious functions, taking vacations, or simply spending time with friends. By far, one of the best things I do to make sure I am personally fulfilled is take on hobbies ranging from fly fishing to crocheting to marathon running. These hobbies have changed, evolved, and ebbed and flowed over time depending on where I lived and my personal situation; yet I have always had passionate interests outside of school or work. My advice is to sign up for that rock-climbing class, take a hunter safety education course, try growing tomatoes, sit through a cake-decorating class, be a tourist, or sign up for a 5k race. It's never too late to try something new! By having interests outside of work, it is much easier to leave work behind when going home, and conversely, by having interesting work, it is much easier to go to work each day.

Set Your Own Schedule

In her book *Lean In: Women, Work, and the Will to Lead*, Sheryl Sandberg (2013) discusses an important lesson she learned early in her career—that it is up to us as individuals to decide how much work we are going to put in each day, how many nights we are willing to spend away from home, and where we will draw the line on what we are willing to do. She goes on to say that employers will always have assignments and demands on our time, and as employees, it is our obligation to set expectations for a sustainable workload over the long term. Most employers offer tools to help, including flexible work arrangements, and these tools will ultimately allow more freedom for your personal schedule by improving efficiency while at work. Examples of flexible work arrangements include working an alternative schedule, such as starting earlier or working later than typical hours, or working a compressed work week such as four 10-hour days instead of five 8-hour days. Other examples include working part-time, flexible scheduling, job sharing, or telecommuting. For me, it seems like I have tried almost every flexible working arrangement possible. Certain schedules have worked better at certain points in my career. For instance, some jobs or assignments lent themselves well to working at home while others required active in-person dialogue. It previously made sense to work long hours so I could get every Friday off, while now I would rather work an 8-hour day so I can have additional time with my family in the evening. The bottom line is that fulfillment and balance can be improved by taking control of our work schedule.

Success

Sheryl Sandberg (2013) also eloquently said "If I had to embrace a definition of success, it would be that success is making the best choices we can . . . and accepting them." Deliberate choices concerning personal and work obligations will help us be a better and more successful partner or spouse, parent, and thus employee. Stress, fatigue, and unsustainable expectations by your employer that are often the result of spending an inordinate amount of time at work can also be avoided by making the best choices we can and accepting them. Keep in mind that to be truly successful at maintaining a work–life balance, individual situations will need to be periodically reassessed and changed, whether this means finding additional support, actively taking steps to become more personally and professionally fulfilled, or altering work arrangements. By managing our time and relying on others when needed, we will be happier, healthier, and more productive as employees.

Acknowledgments

I give credit to all women who showed me that it was conceivable to be well respected in the natural resources profession while also being a wife, mother, hunter, angler, and passionate addict of the outdoors. These women include my first mentor back in high school, Dianne Kolodziejski, a resource manager with the U.S. Army Corps of Engineers, and my mentor during my first years as an emerging professional, Jan Fenske, who was the first female fisheries biologist hired by the Michigan Department of Natural Resources. I am also eternally grateful for my husband whose love and support enables me to enthusiastically embrace the mess and for my daughter who delightfully adds to the mess.

Biography

Jessica Mistak is the Northern Lake Michigan fisheries supervisor for the Michigan Department of Natural Resources, where she is responsible for the administration and implementation of inland and Great Lakes fisheries management programs within a multi-watershed area of the Lake Michigan basin. Previously, as a senior fisheries biologist, Jessica provided technical and policy assistance to the department on a variety of aquatic habitat related issues, including mining, dam removal, fish passage, stream restoration, and hydropower licensing. She holds a B.S. in natural resources from The Ohio State University and an M.S. in fisheries and wildlife from Michigan State University. Jessica is also a senior fellow in the Environmental Leadership Program, is involved with the American Fisheries Society as its constitutional consultant, and has previously served as president of the Michigan Chapter and president of the North Central Division. In her free time, Jessica enjoys spending time with her family, trail running, gardening, and catching fish of all sizes.

References

Ephron, N. 1996. Commencement address. Wellesley College, Wellesley, Massachusetts.
Sandberg, S. 2013. Lean in: women, work, and the will to lead. Knopf, New York.

Reconnecting People to Their Natural Environment

CHRISTINE M. MOFFITT*
U.S. Geological Survey, Idaho Cooperative Fish and Wildlife Research Unit
University of Idaho, 875 Perimeter Drive, MS 1141, Moscow, Idaho, 83844 USA

ZACHARY L. PENNEY
Idaho Cooperative Fish and Wildlife Research Unit, Department of Fish and Wildlife Sciences
University of Idaho, 875 Perimeter Drive, MS 1141, Moscow, Idaho 83844 USA

LUBIA CAJAS-CANO
Environmental Science Program
International Programs Office and Idaho Cooperative Fish and Wildlife Research Unit
University of Idaho, 875 Perimeter Drive, MS 1141, Moscow, Idaho 83844, USA

Key Points

- It is our responsibility as natural resource professionals to build stronger and more relevant ways to connect the public and ourselves with the importance of local, regional, and global aquatic resources.
- Creative and relevant outreach efforts must engage people that are physically or mentally separated from the natural environment to understand the critical components that support all of us.
- Measures that allocate the quantity of resources used in the production, distribution, and consumption of products are excellent tools for understanding the complexity of individual personal decisions.

Introduction

Smartphones, social networks, and countless search engines provide information about the surrounding world with the simple push of a button, click of the mouse, or touch of the screen. Ironically, many of us are mentally and physically disconnected from the natural environment around us. Nowadays, it is more common to have a TV, a cell phone, or a computer at home than to have a vegetable garden, yet even simulated farming is available on the Web (i.e., FarmVille). How can we increase the public awareness of the global and local human connections to the environment that provides our food, water, and air? Effecting these changes will require committed professionals working locally, regionally, and internationally.

For young students studying in urban settings, careers in science are often perceived as related to medicine or human health. I (CMM) was a Sputnik generation student in grade school when the United States rose to challenge Russia's lead in the space race. Even with

* Corresponding author: cmoffitt@uidaho.edu

that national science challenge, my junior and high school exposure to science in suburban Los Angeles involved the bright, likely college-bound kids taking a biology class with an emphasis on human health and molecular biology. Physical sciences, including astronomy, were occasionally included as an option. The emphasis on human health was likely because it was the most visible connection with science in the urban environment. My love of animals was interpreted by my college-educated parents to be a career in veterinary medicine, again a health-related field. They and most others in my community were ignorant of the opportunities for natural resource studies, and no one had an idea of careers in natural resource management or ecology.

Reflecting on my own beginnings, and the serendipity of ending up in fisheries resource education, I am especially concerned about the critical need to bring into all our communities information about and training in aquatic and other natural resources. These challenges are international and multicultural. When our civilizations and communities were simple and more locally place-bound, understanding the natural environment was part of the survival package. If you did not prepare for and exceed the capacity of the natural environment, you migrated or you did not survive. With increased ease of mobility for human and natural resources, and the wealth of large corporations, the environmental costs of development and commodity production are less obvious. Today, natural resources come to us, rather than us to them. Most of our foods are not grown locally, although there is increasing interest in local foods in some communities. Most people have no idea how their food and other goods are obtained, or the pollution associated with their production and distribution. The consequences of production of all commodities are not well incorporated into general understanding and decision making.

I pose three broad questions to all our young professionals. How do we as a society incorporate the larger context of place, the region, and earth into our own lives and the greater public in meaningful ways? Can we create a sense of community and use modern tools to help us make these connections to the greater public? Can we provide a challenge for young citizens to be informed and aware of the importance of their natural world to provide sustainable water, food, and air to breathe?

I asked two graduate students with diverse backgrounds to address these disconnections and provide their perspectives on how to educate and reconnect humans back to the natural environment.

Perspective from Zachary Penney, an Indigenous Graduate Student

Many indigenous peoples able to retain ties to their ancestral homelands feel that there is no clear separation between themselves and the natural environment. The health and well-being of one's body is ultimately determined by the health and state of the environment they belong to. As a Nimiipuu (Nez Perce tribal member), I was raised with this philosophy and taught to practice it from a very young age through hunting, fishing, root gathering, and camping trips with my family. Listening to oral traditions told by Nimiipuu elders was especially vital in establishing my connection to the landscape. These traditions taught not only moral lessons and values, but were also used to describe the origin and traits of organisms, as well as the formation of various geological features of the local landscape. As a result, these stories established a strong sense of belonging and responsibility to the place I lived, as well as relationality (a connection to the *act* of doing with your ancestors, as well as with the stories, power of place, and spirituality of the act). This sense of place was continually reinforced by physically interacting with my environment, whether it be dipnetting salmon and hunting with my dad and brother (Figure 1) or digging camas with my grandmother. These experiences required the use of all five senses (sight, hearing, touch, smell, and taste), which in my opinion, is a critical link being lost to many people living apart or segregating themselves from the natural world.

Despite my strong cultural connection to the Nimiipuu homeland, my transition to the study of fisheries and natural resources was not an easy one. Like many young students, I found it hard to relate many scientific and mathematical disciplines to the world that existed outside the classroom. The very idea of natural resource management or stewardship implies that we (humans) must keep dominion over the environment and that natural resources are simply a commodity to be sold or used for recreation. Culturally, these ideas are contradictory to what I learned growing up. Fortunately, my aversion to science changed following a high school internship with the Nez Perce Tribe on projects focused on the restoration and protection of various fisheries. Salmon restoration is a particularly important goal to the Nimiipuu because salmon not only provide nourishment, they act as a symbol, providing a sense of comfort that our world is healthy much in the same way keystone species are viewed by modern ecology. I was able to clearly see and appreciate the relationship of science and math to the world I belonged to through my experience as a high school intern. Thus, it was this experience that helped me make this connection and solidified my choice to pursue a career in fisheries.

FIGURE 1. Aaron Penney, Nez Perce tribal member, dipnetting salmon in Idaho's middle fork of the Salmon River. Photo by Ermie Whitman.

Having not grown up in an urban area, I cannot claim that people raised in rural areas are any more connected to the natural environment than people raised in urban settings. Today, even some of the most remote rural areas are equipped with Internet access and cellular service. With our increased use of technology for information, communication, and entertainment, we have become increasingly reliant on only two of our senses: sight and hearing. At the same time, many of these same technologies allow us (humans) to search, find, and disseminate information at an astounding rate, which undoubtedly has provided more opportunities for people to educate themselves about the natural environment beyond their own backyard. However, reading about elk hunting in a magazine or observing how a Nez Perce dipnets salmon via computer offers only a limited glimpse of the actual experience. These physical elements cannot be substituted by a book, game, phone application (or App), TV show, or movie. Therefore, I believe that experiences engaging all five senses are an essential component to helping reconnect people to the natural world.

While it is easy to assert that people need to go outdoors and rekindle their physical connection to the natural world, this statement is both idealistic and aimless. As fisheries and natural resource professionals, we have a special opportunity to help establish and foster this strong sense of place and connection to the landscape among people of all ages. If people are not able to see the value or relevance of the natural environment beyond what they learn from a textbook, phone, TV, or computer, many will only see the world as a distant observer without making the connection between their own health and happiness and the state of the environment. Like the elders who took time to sit and talk to me, we must be willing to take the time and transfer our knowledge to students of all ages in a language they can understand. Considering that not everyone can simply venture out and experience the natural environment, we must be creative in how we engage people into understanding how the natural environment works from "unnatural" settings (e.g., classroom, conference centers, and public forum).

Various forms of outreach are already used by tribal, state, provincial, and federal agencies, but an increase in hands-on experiences that engage all five senses will be highly beneficial. From an educational standpoint, programs such as "Fish in the Classroom" provide students with actual hands-on learning experiences in regards to fish biology, life history, and the importance of a clean, healthy environment for their fish. As an example, elementary students at the Lapwai School District in Idaho raised steelhead *Oncorhynchus mykiss* derived from local hatchery broodstock from the eyed egg to fry stage (March to May). The students were responsible for feeding, cleaning, and maintaining the proper rearing environment for their fish (e.g., temperature, dissolved oxygen, and pH) in preparation for release into a local stream containing a recovering population of steelhead. Throughout the rearing process, tribal (Nez Perce), state (Idaho Fish and Game), and federal (U.S. Fish and Wildlife Service) scientists, as well as tribal elders, shared expertise and helped students connect their own existence to the juvenile steelhead they were raising (Figure 2). In the end, students at Lapwai Elementary released 98 fry out of an original 100 steelhead eggs and told the incredible odds their steelhead would need to endure to survive. Although this lesson occurred primarily within the classroom, these students were intellectually and emotionally invested into the well-being of their fish and began to understand that there were numerous factors influencing their survival. Programs providing experiences such as this can serve as an important tool into helping connect people to the natural environment regardless of the environment they come from.

FIGURE 2. Justin Petersen, a Nez Perce tribal fisheries employee, teaches Lapwai elementary students about healthy streambeds and macro and microinvertebrates. Justin worked with the fifth graders on a weekly basis and helped to monitor the steelhead fry. Photo by D'lisa Penney-Pinkham.

The physical experience of learning, especially in relation to the natural world, is important in reconnecting people to the world they live in. Connections, relationality, respect, and reciprocity cannot be made if students do not have an experience that is culturally responsive and relevant. In the case of education, there is no such thing as too early or too late. With time, the technology we use will continue to change; however, the need to educate and remind people about their connection to the natural world will always remain.

Perspective from Lubia Cajas-Cano, a Hispanic Graduate Student

I was born and raised in Guatemala City. Guatemala is a Central American country slightly smaller than the state of Tennessee. Because of its rich cultural heritage and dynamic landscape, my country is known as "the land and heart of the Mayan people" or "the country of the eternal spring." Culturally, Guatemala is comprised by a mix of Amerindian-Hispanic population and numerous indigenous peoples (mainly of Mayan heritage). Guatemala's natural environment is shaped by highlands, volcanoes, rainforest, and exotic flora and fauna (e.g., ceibas, orchids, jaguars, and resplendent quetzals). Guatemala has abundant water resources (freshwater and both the Atlantic and Pacific oceans) that sustain indigenous populations. Historically, these populations have worked with nature rather than exploiting it; however, their livelihoods are now more dependent on commercial economies. Unfortunately, the need for more food and increased agriculture has contributed to changes in traditional uses of the land and deterioration of the environment fueled by lack of awareness of how to create sustainability, poor law enforcement, and limited infrastructure.

Increased urbanization surrounding many of our freshwater ecosystems has resulted in a decrease in water quality and increase in eutrophication. In many areas, there are no regulations for water use or wastewater management, thus no way to enforce the sustainable use or treatment of water. Many of the freshwater lakes, such as Amatitlán, Atitlán (Figure 3), and Peten Itza lakes, receive urban and industrial wastewater from the surrounding areas without any treatment.

My father, a veterinarian parasitologist who specialized in cattle, was among my influential mentors. He understood the connections between clean water and food safety for human and environmental health. His knowledge and guidance helped me develop a personal passion for protecting water and the environment. My interest expanded to learn about ecosystem services and conservation of water.

One of the best approaches for protecting our water resources is to help people realize that our livelihood is closely related to the quality of the environment in which we live. I teach methods for calculating water footprints for different human activities. Water footprinting is a tool used to estimate the volume of freshwater consumed and polluted during

FIGURE 3. Lake Atitlán, Guatemala, viewed from a lake overlook. The lake is surrounded by volcanoes and small towns. Photo by Victor Cajas.

the entire life cycle (from cradle to grave) of any product or activity, similar to a carbon footprint. I conduct workshops about environmental protection for Central American exchange students and their teachers. In one of the lessons, I give away T-shirts with logos promoting environmental sustainability, but then I explain to participants that each T-shirt has a water footprint of 22,000 liters. It is a big surprise for the students to learn the impact of their consumer decisions on the environment.

Clothing choices are minimal compared to the large footprint of agricultural food production. In the United States, it takes an average of 2,000 liters of water to produce 1 kilogram of soybeans, and 16,000 liters for 1 kilogram of edible boneless beef. Learning and understanding about water footprints can lead to improved choices of consumption and encourage mitigation of pollution. I encourage my readers to think of water as money (as an example); how much and how/where would you spend a certain amount of water that is allocated to you in a certain period of time (daily, or in your lifetime). These tools will help regulators and those managing water resources locally and regional to improve their management of water resources. Some countries are taking water footprints into account in promoting certain human activities. For example, the Kingdom of Jordan has prioritized their imports and local productions in order to protect their water resources. Using the water footprinting tool helps regulators and those managing water resources to improve their use and better water management.

It is important for everyone to understand the embedded water and the water pollution involved during the production chain of all goods and foods. Use of bottled drinking water may be needed in some locations when water systems are not available or safe, but use of bottled water for convenience or prestige is not needed. Bottled water must be transported, and the bottles manufactured and then disposed of.

Regardless of the country or ecosystem, engaging humans to make changes in their nature stewardship is difficult, especially if many are not so inclined or there are not feasible alternatives. Environmental programs related to water use and natural resources management are needed at all levels in the education to enhance the people's understanding of how they affect the environment and how their choices will affect their future health and quality of life. The use of the water footprint is just one of the tools that can be used to help establish this connection and can be used at many levels of sophistication. In Guatemala, environmental education has been successful when there is an environmentally friendly alternative or choice for people to use. I am highly supportive of educating people about the power of individual and community choice on the natural environment quality, but a key element remains to provide people with the tools and knowledge to make this education effective.

Conclusion

Education is central to engaging students in careers in fisheries and natural resources, but as professionals we must be willing to go the extra step to help students and the public make these connections. Although various technologies can interfere with direct connection to the natural environment, some can serve a role in education, and the ease of collecting data from across the globe improves our ability to grasp a world context. By enhancing the awareness of the connections from local to global with tools that include experiential education, stories from elders, and water footprinting tools, we will aid understand of the linkages between our environment, our personal health, and our prospects for the future.

Acknowledgments

We thank the following for their thoughtful review of earlier drafts of this manuscript: Frances Gelwick, Texas A&M University; Jay Hesse, Department of Fisheries Resources Management Nez Perce Tribe; Lonnie Gonsalves, NOAA National Ocean Service, and two other anonymous reviewers for improvements to the manuscript. We thank Abigail Lynch, Nancy Léonard, and Bill Taylor for their synthesis and oversight for all essays in the book. Any use of trade, firm, or product names is for descriptive purposes only and does not imply endorsement by the U.S. Government.

Biographies

Christine Moffitt is a U.S. Geological Survey scientist and professor at the University of Idaho. She has been engaged in science education in the United States for more than 45 years.

Zachary L. Penney is a Ph.D. candidate in fish and wildlife sciences at the University of Idaho and a member of the Nimiipuu or Nez Perce Tribe. He has educational and research experience with indigenous tribes in Alaska, Washington, Oregon, Idaho, and British Columbia.

Lubia Cajas-Cano is a Ph.D. candidate in environmental science at the University of Idaho. She is a native of Guatemala and has teaching and educational experience with students from Central America, the Caribbean, and Idaho.

Changing the Game: Multidimensional Mentoring and Partnerships in the Recruitment of Underrepresented Students in Fisheries

Stacy A. C. Nelson*, Ernie F. Hain, and Brett M. Hartis
Department of Forestry and Environmental Resources and the Center for Earth Observation North Carolina State University, Campus Box 7106, Raleigh, North Carolina, USA

Ashanti Johnson
Office of the Provost, Division of Faculty Affairs, University of Texas at Arlington Box 19128, 701 South Nedderman Drive, Suite 300, Arlington, Texas 76019, USA

Key Points

- Our emerging professionals face a shifting economic and demographic landscape that will ultimately reshape future science missions and public dollar funding priorities.
- In order to ensure we are able to effectively manage the priorities of our natural resources, we must ensure that the next generation of professionals arises from a more diverse and interconnected system of resources and academic experiences.
- As our nation continues to diversify, a comprehensive solution must involve attracting the brightest and most highly motivated students to the fisheries and related natural resources fields that are representative of all the communities we serve.

Introduction

Historically, the American education system has been regarded as one of the greatest academic institutions in the world. It has attracted young scholars from developed and developing nations to experience the renowned American melting pot. The role of our institutions of higher education is ever more crucial today as the United States faces growing competition from other nations in science, technology, commerce, capitalism, and education. Often these other nations are less racially diverse than the United States, which is now regarded as one of the most racially diverse countries in the world. Thus, as the United States continues to become increasingly multiethnic, multiracial, and multicultural, our colleges and universities have a unique requirement to ensure that future generations of leaders remain nationally and globally competitive. Our institutions must expand access to research and education opportunities for future leaders who are just as racially and culturally diverse as

* Corresponding author: sanelso2@ncsu.edu

the nation they serve. In this chapter, we present a broad overview of the challenges faced in attracting and recruiting historically underrepresented minorities in the fisheries sciences at graduate levels, as well as highlight some of the personal experiences of coauthor Stacy Nelson. Finally, we provide insight into a new program that develops valued partnerships with minority-serving intuitions to attract, recruit, and retain students in the discipline.

Recruitment Challenges of Underrepresented Students

The System

Our nation's higher education system continues to attract more ethnically and culturally diverse students from around the world. However, certain academic disciplines still struggle to recruit and retain students from our domestic, historically underserved populations. According to the National Science Foundation's 2013 report "Women, Minorities, and Persons with Disabilities in Science and Engineering," underrepresented minorities in the United States made up approximately 20% of the science and engineering graduate-level degrees awarded in 2010. While these numbers show an increasing trend in the number of master's and Ph.D. degrees over the past two decades, domestic underrepresented minorities, which are described as African-Americans, Asian/Pacific Islanders, Hispanic, and Alaska/Native Americans, account for only 8% of the master's degrees and 5% of the Ph.D. degrees awarded in 2010 in the earth, atmospheric, and ocean science category (NSF 2013).

Careers in fisheries and natural resources have been recognized as emerging fields in the United States, as well as on the global job markets. However, as the popularity of these disciplines grow among students and programs in the majority-serving institutions, this trend is not typically seen in institutions that predominantly serve our domestic underrepresented students groups: historically black colleges and universities (HBCU), Hispanic-serving institutions, and tribal college and university systems. In fact, minority-serving institutions still remain one of the primary sources of baccalaureate and postbaccalaureate education for our nation's underrepresented and underserved communities in science, technology, engineering, and mathematics (U.S. Commission on Civil Rights 2010).

The Tradition

Disparity among the varieties of disciplines offered at minority-serving versus majority-serving institutions may be a result of a number of different factors. For example, most minority-serving institutions originate from a rich history steeped in the custom of providing more practical curricula that meets the traditional educational needs of the communities they serve. For example, several HBCU institutions have historically had outstanding programs in medicine, nursing, engineering, and law. Additionally, a lack of historical familiarity with discipline-specific research, career, training, and teaching potentials, such as those available in fisheries, may further complicate the ability to expand nontraditional curricula in these schools to include courses, like those in the fisheries sciences.

The Future

Too often, academic institutions, whether minority- or majority-serving, when faced with dwindling budgets, are saddled with a one-size-fits-all approach that usually requires cutting back on all but the most essential programs. However, one component of strategic positioning that a discipline-specific field such as fisheries must contend with is ensuring increased access and representation of the best and brightest student talent from historically underrepresented and underserved groups. The very survival of those fields of science that have been underrepresented by nonmajority groups becomes more critical as the nation experiences a demographic shift toward a majority-minority by the middle of the 21st century. This demographic evolution will surely have resounding implications on our nation's political orientation, the development of its future workforce, private and citizen engagements, and its overall economic structure.

The emerging demographic composition of our country will influence the way we approach everything from economics to education, navigate cultural sensitivities, and prioritize the state and federal funding that has remained the impetus for the technological discovery. In fact, it has been the development of these resources that has made our nation's educational and research systems among the most sought after in the world. To maintain a competitive edge, it is essential that the emerging stewards of the discipline-specific fields of science and technology are as racially and ethnically diverse as the demographics they serve.

Attracting, Recruiting, and Retaining—A Personal Experience

As a coauthor of this chapter, I, Stacy Nelson, began my academic experience in higher education at a moderately sized, hometown, urban HBCU in the Deep South. Like so many kids that grew up watching Mutual of Omaha's *Wild Kingdom* with Marlin Perkins every Sunday night and the *Undersea World of Jacques Cousteau* television specials, from an early age I was drawn to the aquatic sciences. This fantasy may have even been even stranger for a young child growing up in Jackson, Mississippi, with a limited exposure to natural resources, having never recreationally fished, hunted, or gone hiking or camping. I could only rely on television programs and books to introduce me to the natural environment outside of the inner city.

Entering a traditionally black college in the mid-1980s, I found few opportunities to pursue a broad range of courses in the aquatic sciences. An even more foreign topic was trying to explore an opportunity to pursue a major in fisheries or marine science. In fact, I was often strongly encouraged by advisors and peers to follow the prescribed degrees paths of general biology, general chemistry, or general biology premedicine. After all, these were the only opportunities with any kind of historical track record within the school's program. Fortunately, by applying to different summer internship programs and receiving encouragement from a faculty advisor, I was able to spend every summer in different programs during my undergraduate years, including a summer at sea aboard the National Oceanic and Atmospheric Administration's ship *Oregon II*, a summer field program at the Gulf Coast Research Laboratory, a summer research experience at the Duke University Marine Laboratory, and multiple summer research experiences at the Virginia Institute of Marine Sciences (VIMS).

The internship experiences required me to go against the culturally familiar and what was academically comfortable, toward opportunities to move into environments where I often found myself alone, racially. In fact, in some programs I was the only student participant of any race. Also, I often encountered preparedness barriers, whether self-imposed or externally imposed (sometimes blatant, sometimes furtive), that often made me second-guess whether or not I had made a sensible decision in pursuing culturally less-traditional fields of study. In fact, pairing very young and inexperienced undergraduates with some of the top researchers in their fields alone is often enough to shake a student's confidence. Certain experiences, be they racial, cultural, or academic, definitely shook mine. However, this was also coupled with an elementary school diagnosis that I had a reading disability. Determined not to let this diagnosis be the determining factor of what I would be able to achieve, I knew that I had to study harder, study longer, take double notes, and rewrite them often to maintain my academic college scholarship and get accepted to graduate school in a field of science where African Americans have been historically underrepresented.

I entered VIMS to pursue my master's degree in the early-1990s. Although, my previous internships had somewhat prepared me for adjusting to an environment where I would be one of very few African Americans, I still found the budding metropolis of Gloucester Point, Virginia to be very different from the culturally very comfortable experiences at my undergraduate institution. Also, I was still battling my perceived disability, in which I felt I had to overstudy to complete the simplest of tasks. These insecurities were created by my own perceptions of being drastically underprepared academically, especially coming from a general science undergraduate program and going into one of the highest-regarded marine science programs in the world. I would often hear comments from fellow graduate students that came from other highly regarded academic programs and had background majors in marine science that "this program is hard. I've never even had to study before." Needless to say, despite feeling like I needed to do everything I could to combat my early-age reading disability diagnosis, as well as keeping up with the graduate-level studies and research, I often felt overwhelmed by the tremendous amount of work that getting through the degree program required of me.

After graduating from VIMS, teaching for a couple years at another Deep South, urban HBCU, as well as a research stint at the NASA Stennis Space Center, I began a Ph.D. program in fisheries sciences at Michigan State University (MSU). For me, the urban-esque setting of East Lansing/Lansing provided the cultural comfort I was not able to find in early to mid-1990s Gloucester Point, Virginia. Being a part of the larger university community also helped me to overcome the diversity isolation I had felt in previous programs. However, I still had to overcome my reading ability insecurities. I had largely kept this issue to myself since early grade school. I had found a coping mechanism that worked: work hard, do not sleep, no excuses! This strategy had always given people on the outside the perception that I was highly productive. However, to keep up this level of productivity, I was quickly burning myself out.

While still working on my degree at MSU, Dr. Bill Taylor, who recruited me and frequently checked in on my progress, wanted to know why I had not yet scheduled my comprehensive exams. At that time, I was literally burning the candle at all ends: courses, research, papers, conferences, workshops, committees, and so forth. In fact, the best sleep I got was in seminar classes! I explained I just needed to wait until I had enough of the time I needed to adequately prepare. Bill was not buying it. In fact, after deeper conversations, I finally had to come clean and explain that I clung to several insecurities, including past academic experiences and my reading disability. Bill, again, was not convinced. He immediately picked up the phone and arranged several sessions with a counseling/reading therapist. Despite my protests of needing more time, needing to implement the strategy I have always used and was comfortable with, and my inability to afford healthcare (much less therapy!), it was settled. I was scheduling my comprehensive exams and going to therapy (free of charge).

Although the techniques I learned through therapy are techniques I still have to use to this day roughly 12 years after completing my degree, it was one of the most defining points in my career. Often, minority students face the same cultural, preparedness, and personal insecurity challenges when entering into fields of science in which they have been historically underrepresented. These challenges require efforts of champions within the field to say, believe it or not your experiences are not too far different than my own and we have steps in place to make sure that these challenges will not represent a barrier to your success in our program. I think this is a huge lesson learned, even for me currently serving in the professorate. The lesson is that a lot of students have a certain amount of baggage, be it cultural, racial, or emotional. However, the true courage of a program, and the program's leadership, is one that is willing to recognize this diversity and provide opportunities for each student to perform at his or her best. A tremendous opportunity still exists in exposing more students from historically underrepresented, underserved, and economically disadvantaged backgrounds if we are truly of the belief that it is our responsibility, as educators, to make sure that everyone has a voice at the table in deciding the future of our natural resources.

Attracting, Recruiting, and Retaining—A New Approach

Many majority-serving institutions have tried to combat the challenges of finding and recruiting talented students from historically underrepresented and underserved groups through established research experiences for undergraduates (REU) programs. Often these programs are very effective. Most operate by establishing a one-way transfer of the most talented students from the typically smaller or minority-serving institutions. However, many REUs may be limited, due to funding cycle or timing constraints, in creating a broader partnership that not only affects the targeted student participants and majority-serving institution, but also provides a capacity to affect the institutions that sponsor the students. A new strategy that has the potential to circumvent the limited opportunities to build capacity at the student's home institution, which is common with many REUs, is the development of innovative and collaborative approaches that are geared towards increasing the engagement of underserved groups through active partnerships. Ideal partnerships should be inclusive of a host of partners on many institutional levels: federal research and land management agencies, local and state agencies, and majority- and minority-serving colleges and universities. This multidimensional approach creates a win–win–win scenario that is aimed at developing relationships among all of the institutions involved, making it easier to share resources and expertise, increasing institutional capacities at less-well-resourced institutions, and exposing our next generation of researchers to a more-connected environment of achievement that greatly mimics the global culture in which we must all live and work.

One such program is currently operating as a pilot program at North Carolina State University and Michigan State University. This program partners with multiple University of Puerto Rico campuses (Rio Piedras, Humacao, Mayaguez, and Aguadilla) and the U.S. Department of Agriculture Forest Service (USFS). The focus of the program is on using newly recruited graduate students from minority-serving institutions as a conduit for developing programmatic linkages between all partners. The graduate students are recruited as a cohort to the majority-serving schools by a multi-institutional panel of faculty and advisory committees that include faculty from both the majority- and minority-serving schools. Graduate student research projects are structured to allow research to take place on both majority- and minority-serving school campuses. Additional active engagement between partners include advisement or committee participation by local and state natural resources agencies, as well as opportunities for internships and future employment within the natural resources agencies, such as the U.S. Forest Service.

This level of engagement is also beneficial for the student participants in that it attempts to circumvent some of the challenges involved in recruiting and training talented student from historically underrepresented and underserved groups. For example, issues of tradition are lessened because the students involved also have faculty mentors and committee members that has been selected from their home institutions. This mentor helps to design the student's research project, as well as help to attract other agency participants. The program is also designed to give student participants a more inclusive, team-oriented experience while completing course and laboratory work at majority-serving institution, in that students are selected in cohorts with schedules that allow them to start and complete their respective degrees at the same time. Also, recruiting students in a cohort helps to minimize feelings of isolation and, over time, attempts to establish a critical mass of students within a program with like cultural backgrounds.

The unique nature of this program also has the added benefit of allowing the leveraging of shared resources (i.e., instruction, laboratory space, field sites, project budgets, sampling and analysis equipment, professional expertise, etc.) among all partners involved. Thus, this program not only provides the student participants with the traditional major professor/laboratory-team network, but also an expansive network partnership that, to date, has included the following:

- Department of Forestry and Environmental Resources, North Carolina State University
- Department of Fisheries and Wildlife, Michigan State University
- Department of Marine, Earth and Atmospheric Sciences, North Carolina State University
- Department of Ecology, University of Puerto Rico at Rio Piedras
- Department of Biological Sciences, University of Puerto Rico at Humacao
- Department of Agronomy and Soils, University of Puerto Rico at Mayaguez
- Vida Marina Center for Coastal Restoration and Conservation, University of Puerto Rico at Aguadilla
- Center for Marine Science and Technology, North Carolina State University
- USFS International Institute of Tropical Forestry
- USFS Southern Research Station, Eastern Forest Environmental Threat Assessment Center
- Commonwealth of Puerto Rico Department of Natural and Environmental Resources
- U.S. Fish and Wildlife Service
- U.S. Geological Survey

Finally, to combat the controversial concerns that advanced degrees may be preparing students for jobs that do not exist (see Shea 2013), this program is tightly integrated with the sponsoring agency's workforce diversity objectives. Through this objective, the USFS is committed to building a skilled and diverse, multilingual and multicultural workforce that reflects the future demographic projections of the U.S. Census populations and the civilian labor force. This program also complements agency recruitment initiatives to train student participants in areas that effectively replace needed skills that have been identified in the U.S. Forest Service Human Resources Management workforce and succession plan and recruitment strategy. These efforts provide a level of institutional support for historically underserved student groups that may be less familiar with career opportunities in the fisheries and natural resources-related fields. They also serve to develop a lasting network that puts the students involved directly in a working relationship with future employers, stakeholders, and community-centered research needs. This level of active engagement retains student partners by providing direct lines of communication and close association with agency professionals as they matriculate through the program.

Final Comments

The personal experiences described here provide a broad view into some of the perspectives students from historically underserved communities may hold. The proposed approach represents one method to increase our ability to recruit students from historically underrepresented and underserved groups into the fisheries sciences. Also, this program aims to ensure that the next generation of professionals arises from a more interconnected system of resources. Regardless of the academic pedigree and institutional challenges, emerging professionals are, in fact, facing a shifting landscape that will ultimately reshape future science missions. Mission priorities will be contingent upon the needs of the communities they represent and those communities' perceived relevancy of the required expenditure of public dollars that are necessary for the natural resource professionals to perform. Despite competing with a shrinking resource base from traditional sources, a comprehensive solution must involve attracting the brightest and most highly motivated students to the fisheries and related natural resources fields. However, it is also critical to realize that, with this nation becoming one of the fasting-growing, ethnically diverse countries in the world, sound and sustainable research, management, and policy decisions must broadly connect with the community they serve. Effectively meeting this goal requires inclusion of both traditional and nontraditional groups in the preparation of careers in both traditional and nontraditional natural resources and fisheries employment opportunities. It is true that many students today, despite their ethnic, racial, or economic backgrounds, have faced significant challenges and, given these challenges, are nothing less than future stars in the making. Academic programs serving these fields remain the great keepers. Admissions committees that only clamor for the students with perfect SAT, GRE, and GPA scores may miss the opportunity to expose their programs to a truly diverse and rich academic experience. Leadership within these programs must have the courage to stand up to bureaucracy and red tape to recognize that this diversity is important and to provide opportunities for each student to perform at his or her best and ensure that everyone has access and a voice at the table in deciding the future of our natural resources.

Acknowledgments

The authors would like to thank the editing staff of this text (William "Bill" Taylor, Nancy Léonard, and Abby Lynch) and the additional associated reviewers whose comments significantly improved this manuscript. Funding for the USDA Forest Service Capacity Building program was provided through the Southern Research Station FS Agreement No. CA-11330101, with additional support and program advisement provided by Cheryl Jefferson, Kier Klepzig, Felipe Sanchez, Robert Ragos, and Jimmy Reaves. Additionally, the authors would like to acknowledge the bold leadership of several mentors that have provided support, direction, and the occasional swift kick where it counts when necessary. These role models have left an indelible impression on each of our lives and continue to impact everything that we do, how we value ourselves, and how we forge new relationships.

Biographies

Stacy Nelson is an associate professor at North Carolina State University within the Department of Forestry and Environmental Resources and Center for Earth Observation. He received a master's degree from the College of William and Mary's School of Marine Science, the Virginia Institute of Marine Science (VIMS), and a Ph.D. from Michigan State University's Department of Fisheries and Wildlife. His research revolves around the use of geospatial technologies to address questions of land-cover change and the impact this change has on aquatic communities at the regional and local scales.

Ernie Hain is a research fellow with the U.S. Department of the Interior Southeastern Science Climate Center at North Carolina State University and a recent Ph.D. recipient in the Fisheries, Wildlife, and Conservation Biology Program. His research combines techniques from hydrologic modeling, geographic information systems, and remote sensing, population ecology, and molecular ecology to investigate how native species respond and adapt to a changing environment. He is especially interested in how changing habitats, flow regimes, and community structures can affect the demographic rates and population connectivity, and thus persistence, of native species.

Brett Hartis is an aquatics extension associate at North Carolina State University, with responsibilities in aquatic invasive species management. His research interests include coupling traditional mapping applications with geospatial technology, including remote sensing. Dr. Hartis is a recent Ph.D. recipient in fisheries, wildlife, and conservation biology at North Carolina State University. He received a master's degree in the same program. He received a degree in biology from East Carolina University with a concentration in coastal resources management. Prior to accepting his current position with the university, he worked as a research assistant, quantifying aquatic vegetation.

Dr. Ashanti Johnson is the assistant vice provost for faculty recruitment and associate professor of environmental sciences at the University of Texas at Arlington, and also the executive director of the Institute for Broadening Participation. Her areas of research specialization include (1) aquatic radiogeochemistry, (2) professional development of students and early career professionals, and (3) STEM diversity-focused initiatives. In 1999, she became the first African American to earn a doctorate in oceanography from Texas A&M University. She has received numerous honors and awards, including a PAESMEM award at the White House in recognition of her mentoring, professional development, and diversity-related activities.

References

NSF (National Science Foundation). 2013. Women, minorities, and persons with disabilities in science and engineering. NSF, Special Report NSF 13-304, Arlington, Virginia.

Shea, A. A. 2013. For graduate science programs, it's time to get real. The Chronicle of Higher Education (March 14).

U.S. Commission on Civil Rights. 2010. The educational effectiveness of historically black colleges and universities. U.S. Commission on Civil Rights Publications Office, Washington, D.C.

Openness to the Unexpected: Our Pathways to Careers in a Federal Research Laboratory

Kurt R. Newman*, David B. Bunnell, and Darryl W. Hondorp
U.S. Geological Survey Great Lakes Science Center
1451 Green Road, Ann Arbor, Michigan 48105, USA

Key Points

- Emerging fisheries professionals should be open to opportunities that may differ from their initial expectations.
- To achieve professional success, set high expectations.
- A strong and supportive team environment is invaluable.

Introduction

Many fisheries professionals may not be in the jobs they originally envisioned for themselves when they began their undergraduate studies. Rather, their current positions could be the result of unexpected, opportunistic, or perhaps even lucky open doors that led them down an unexpected path. In many cases, a mentor helped facilitate the unforeseen trajectory. We offer three unique stories about joining a federal fisheries research laboratory, from the perspective of a scientist, a joint manager-scientist, and a manager. We also use our various experiences to form recommendations that should help the next generation of fisheries professionals succeed in any stop along their journeys.

The Scientist

My entry into the fisheries profession surprised me. I always enjoyed spending time outdoors (e.g., hiking, camping, boating, and fishing), but through my undergraduate years, I always wanted to be a high school biology teacher. My first research experience occurred during my senior year in college. I now recognize how fortunate I was to have two fisheries scientists in the biology department of my small liberal arts school who opened my eyes to research and guided me towards M.S. assistantship opportunities in the fisheries field. The more research I conducted, the more I wanted to make it a larger component of my job. I still remember camping along the Chattooga River in South Carolina following radio-tagged Brown Trout *Salmo trutta* and being amazed that I was actually earning a master's degree in such a beautiful natural setting.

After completing my Ph.D., I still envisioned myself as a teacher, but on a college campus where I could teach, conduct research, and mentor graduate students. Like most new graduates anxious for their first real job, I could not afford to be choosy and applied for jobs

* Corresponding author: knewman@usgs.gov

outside of academia. Today, I realize how fortunate I am to have landed at a federal research laboratory. Initially, I accepted the offer because of its tremendous upsides: smart scientists I admired, the ability to work in the Laurentian Great Lakes, access to large research vessels that I would have no responsibility to maintain, and proximity to well-respected research universities. The position offered me a great opportunity to translate what I had learned in eight years of graduate school for the betterment of critical fishery resources. I also believed all the upsides of the job would allow me to keep the door open to academia should I decide that was a preferred career path. Little did I know that I would ultimately be able to incorporate some aspects of an academic career into my research position. Over time, I have acquired adjunct status at nearby universities, taught a fisheries management seminar (in the evenings), and have had the opportunity to mentor graduate students and postdoctoral scientists. I should emphasize that some federal research laboratories may not offer opportunities for the same level of academic involvement. My experience demonstrates, however, that an initial openness to nonacademic positions allowed me to grow into a job that supported my passion for research and to develop additional opportunities for teaching and mentorship.

Beyond being open to unexpected career opportunities, I offer additional recommendations to help guide new fisheries professionals on their journeys. First, when you have the opportunity to mentor, I believe it is important to set high expectations and to even provide opportunities for your students or technicians to achieve beyond what normally would be expected. For example, my Ph.D. advisor's expectation that I oversee a several hundred-thousand dollar research project spoke of his belief in my abilities and his expectation for me to not only conduct top-tier science, but also to manage its logistics. In my current position, I try to offer technicians and students the opportunity to do work beyond their 40-hour week technical duties or theses. In many instances, this extra research opportunity has led to stand-alone, publishable research papers. The bottom line is that an environment of high expectations can lead to higher productivity and accomplishments for all.

Second, I do not think that the importance of regular meetings to facilitate communication and intellectual growth can be overstated. As a graduate student, I was a part of weekly meetings where technicians, graduate students, and faculty members were expected to be equal contributors. Although I realized that the decision-making power was unequal within that room, my mentors created an environment where each of us felt our opinions mattered. In my current position, I try to schedule frequent (sometimes weekly) meetings to discuss research progress and monthly meetings to discuss journal articles that help us all better understand the scientific underpinnings of our daily research activities.

Finally, I believe it is important to take time to celebrate accomplishments. When my M.S. and Ph.D. advisors hosted parties in their homes, it helped forge the bonds of our team. I also remember the thoughtful handmade gifts from my M.S. advisor. In my current position, I have frequently opened my home for gatherings. Similarly, I have tried to set celebratory milestones (e.g., completion of sample processing, end of a long field season) that create additional opportunities for morale-building and merriment. In the end, creating a culture of celebrating success pays large dividends for the morale of individuals and for the team.

The Manager-Scientist

My professional journey to a manager-scientist position with a federal agency illustrates that opportunity often will figure more prominently in your decision to pursue or to accept a given position than will education and training. Originally, I wanted to be a professor at a research university, but after more than a decade of preparation, first as a graduate student and then as a postdoctoral research scientist with two separate organizations, I still had not found a good fit or been offered a position as a college or university professor. Driven largely by a need to secure more stable employment (my wife was six months pregnant with our son at the time), I applied for, and eventually was offered, my current post as an assistant program manager and fishery scientist at a federal laboratory. In this position, I supervise a team of junior-level scientists, co-coordinate our agency's fishery research vessel fleet, and apply science and research skills to a wide range of project management tasks.

The decision to remain open to new opportunities has been profitable for me personally, as well as for my employer. For example, while managing a variety of different projects, I have developed new skills in leadership and in team management. I've also discovered that I possess natural talents in employee supervision and diplomacy that in some cases are more critical to successful personnel and project management than are my technical skills. At the same time, I have been able to use my past training as a researcher to more effectively organize personnel and resources to support my agency's science mission.

The characteristics that have influenced my approach to management and science stem from interactions with many amazing and talented individuals across a spectrum of position types. You need not have one specific person that you envision as your main mentor in order to have a successful and fulfilling fisheries career. Below, I recount interactions with those I consider as my mentors and what each taught me about what it means to be a mentor and to execute my current role in a way that benefits my supervisors, subordinates, and peers in the fishery profession.

Gladys (not her real name) taught me the importance of being grateful for the opportunity to be a fishery professional. Gladys is a collaborator on a project that requires us to collect fish during the spring in a region where spring temperatures may be anywhere from 30°F to 80°F. Every single day, on the boat ride to our sampling sites, Gladys looks at me and says, "Can you believe we get paid to do this kind of work?" I readily agree on the sunny, warm days, but Gladys makes this remark just as often on dark, cold mornings when her face is frozen and her hands raw from pulling fish out of bitterly cold water. Gladys is genuinely grateful to be a fishery professional, and her attitude reminds me that it is a privilege to research and manage an important natural resource. Inspired by Gladys, I make an effort to be grateful both for the work I do and for the opportunity to interact with many dedicated scientists, resource managers, and policy makers. This is particularly important for me on days when my plate is full of tasks that seem difficult or mundane.

And then there is Ralph (again, not his real name). Ralph taught me the importance of pursuing excellence in all aspects of my work. Ralph was a former supervisor with whom I had very little in common except for the fact that we were both fishery scientists. Ralph was highly successful by most professional standards, but I found him to be demanding, unnecessarily dogged in defense of his opinions (particularly when they were at odds with mine), and outright abrasive at times. We had (and still have, to my knowledge) completely different worldviews, and I confess that towards the end of my tenure, I often did not like Ralph very much. That said, Ralph was excellent at his work and also excellent at challenging younger scientists like me to identify and ask the most relevant and important scientific questions. Ironically, some of the most celebrated times in my career were when Ralph complimented my work because his praise was saved only for recognition of excellent work. Despite my differences with Ralph, I respected him and continue to respect Ralph for his insistence on quality in every aspect of the job. I have found that my supervisors appreciate my efforts to produce excellent work regardless of whether my efforts are directed towards science or towards improving laboratory processes.

Shelby (not his real name) taught me that the necessary companion to excellence is the ability to encourage your fellow professionals, particularly when the outcomes of their efforts are far from excellent. During one day in the field, I was using an expensive pinger-and-hydrophone system to track the depth of a net towed behind our vessel. At some point, the hydrophone stopped detecting the pinger, and upon inspection, I discovered that thanks to the vessel's propellers, I now had a hydrophone cable but no hydrophone. This event abruptly ended the research trip and predestined what I expected would be a very uncomfortable conversation with my then advisor, Shelby. I was terrified, especially since I had been his student for less than six months. Upon hearing my tale and profuse apologies, Shelby smiled and said, "Now that you've broken something expensive, you can relax and get some work done." Shelby's response taught me not to walk on eggshells when it comes to scientific exploration and that minor disasters will occur no matter how careful our planning and preparation. I have found that this is no less true in managing programs or supervising employees. Recognition of this truth also permits me more freedom and enjoyment in my work than I would otherwise experience.

The Manager

The ideas of a career in fisheries or mentorship were unimaginable to a nine-year old kid fishing an Indiana farm pond, but in some ways that was the beginning of both. I am not sure my uncle understood how frightened I was when he showed me how to bait a hook and pointed me in the right direction to the back forty. Or if he understood the pressure I felt when he said "Don't forget to bring back dinner!" Or if he knew how great it was sharing my adventure with the rest of the family that night. I do know that the foundation for lessons I would continue to learn over the twists and turns of my career were set that day. I have had to be willing to grasp opportunities that were not part of my plan or comfort zone as a student, researcher, and ultimately leader at a federal research laboratory. I have had to embrace high expectations from my peers, my own supervisors, and those I supervise with an eye on success. And, I have had to learn how to make a genuine and meaningful connection with others along the way.

My path to becoming part of the executive leadership at a federal research laboratory has been the result of hard work, unexpected opportunities, and relationships that I have developed in the fisheries profession. Leaving my job at a local television station and returning to school after a long hiatus felt like a big step into a completely unknown future. I had a passion for fishing but no real idea of where an education or career in fisheries would lead me. I was lucky enough to meet someone who did early on and quickly found myself in a university fisheries lab working side by side with a group of high-performing graduate students who themselves have gone on to very successful careers in the field. Those students accepted me as part of their team and challenged me on a regular basis to learn more than what I was being exposed to in the classroom. They gave me opportunities to support their graduate projects in a variety of roles, trusted me to lead elements of that research when it made sense, and expected me to deliver my very best. We were always ready to help each other succeed. We were a team, accountable to each other. Key to our success was the leadership of our professor who valued developing real relationships with each other as much as our knowledge about fisheries. Those connections we made have persisted for more than

25 years, and I have either sought out or tried to recreate that team environment wherever I have worked.

I have been lucky to work in strong team environments throughout most of my fisheries career, no matter what the focus of my efforts were at the time. That has been gratifying in many ways and has helped me suit up and show up on days when I was not feeling particularly good about what I was facing. Defining what a gratifying career in fisheries might look like is clearly an individual matter, but I have picked up a few basic ideas since my days in the university lab that I hope will be useful. First, keep an eye out for new challenges. Do not be afraid to step outside your comfort zone and risk taking a new path. There are many different careers available in fisheries, and often, the greatest adventure is born out of opportunity rather than careful planning. Leaving the job I had for seven years to pursue an education in fisheries was a giant leap into the unknown. My first job as a researcher for a state fisheries agency was in a field very much outside the expertise I had developed in graduate school. Moving from researcher to supervisor and eventually into a leadership role with the federal research laboratory involved taking risks each time. Careful consideration of each opportunity that presents itself and the challenges it brings can be exciting and might just be the first step to other opportunities around the corner.

Second, work hard. I know this seems like a basic rule we learned as kids, but my experience has proven to me over and again that this is not a universally held truth. Set your personal expectations high and do everything you can to exceed them. Look for jobs where employees have high expectations of themselves and each other. There will be a lot of positive energy in those organizations, and nothing opens up new opportunities like hard work.

Finally, learn to make real connections with others. Be open to listening and learning about the people you work with. Try to understand who they are before you pass judgment on why they did something. Then, be ready to share ideas and work towards solutions of common problems together. Those connections and relationships often are the foundation for successful teams and can be deeply satisfying throughout your career.

Conclusion

Each of us still remembers when that first hook was set and we realized that fisheries could be a profession. Since that time, we have each been open to new opportunities, even if they did not provide the path of least resistance or the path that fit our preconceived notions. In our experiences, these new opportunities led to fulfilling fisheries positions. Although we shared a diversity of recommendations in our own vignettes, they included common themes. We each spoke about the value of high expectations. We described the importance of creating strong teams, where members cooperatively work hard while encouraging one another, effectively communicate, develop meaningful relationships, and celebrate accomplishments.

We also recognize that the courage to step towards the unexpected path was assisted by mentors that helped form us along the way. Relationships with mentors are generally developed in school or in the workplace, but new professionals should also be aware of the networking opportunities (and potential mentoring relationships) available at regional and national American Fisheries Society meetings. It is not easy to introduce yourself to that person whose research or management skills you admire, but we have found fisheries professionals to be a friendly bunch (especially if the initial conversation begins with a compliment!). Making connections outside your local sphere can pay big dividends down the road—it may even lead to unexpected opportunities!

Acknowledgments

This article is Contribution 1823 of the U.S. Geological Survey Great Lakes Science Center.

Biographies

Kurt Newman grew up in Michigan playing on the shores of the Great Lakes. He received his B.S. from Michigan State University where he was advised by Dr. William Taylor. He went on to work with Dr. Andrew Dolloff to earn his M.S. from Virginia Tech. He returned to Michigan State and completed his Ph.D. under Drs. Daniel Hayes and William Taylor. He spent 10-years working as a researcher, as a unit supervisor, and in executive leadership with the Michigan Department of Natural Resources before becoming the Western Basin Ecosystems Branch chief for the U.S. Geological Survey Great Lakes Science Center.

David “Bo” Bunnell is a proud Kentuckian who has found the Great Lakes so easy to love. He received his B.S. from Centre College (Kentucky) where he was advised by Drs. Christine and Mike Barton. He worked with Dr. Jeff Isely to earn his M.S. from Clemson University (South Carolina). He completed his Ph.D. at Ohio State University under Dr. Roy Stein. He also had great postdoctoral research experiences with Dr. Tim Johnson (Ontario Ministry of Natural Resources), Mr. Carey Knight (Ohio Department of Natural Resources), and Dr. Tom Miller (Chesapeake Biological Laboratory, University of Maryland).

Darryl W. Hondorp is a Michigan State Spartan (B.S. in fisheries management 1995) who lives and works in the shadow of the University of Michigan where he received his Ph.D. in aquatic ecosystems ecology in 2006. He commends his coauthors as individuals that not only teach excellence and leadership, but also demonstrate these qualities.

Mentoring Minorities for More Effective Fisheries Management and Conservation

MAMIE PARKER
Ma Parker & Associates, LLC
45788 Shagbark Terrace, Dulles, Virginia 20166, USA

DANA M. INFANTE*
Department of Fisheries and Wildlife, Michigan State University
Manly Miles Building, 1405 South Harrison Road, Suite 318
East Lansing, Michigan 48823, USA

Key Points

- Mentoring minorities can help encourage and retain passionate individuals with novel ideas and perspectives and, in turn, increase diversity and talent in the field of fisheries management and conservation.
- Empowering individuals who may have new insights and strategies for solving problems can ultimately support more effective management of fish, fisheries, and aquatic habitats.

Introduction

The importance of biodiversity is well established in ecological contexts. E. O. Wilson (1992) states, "We should judge every scrap of biodiversity as priceless while we learn to use it and come to understand what it means to humanity." Diversity within the workforce is also widely acknowledged as important by our society. Many of us have been taught this in school, and the value of diversity is typically enforced through workplace training programs. Few would disagree with the notion that a diverse work force can provide benefits to collaboration and problem solving, an idea that holds true within the field of fisheries management and conservation. In our field, dominated historically by white males, more women and racial minorities in the past few years have brought new perspectives and strategies that may have been previously underrepresented. Such perspectives may contribute to problem solving, foster greater degrees of innovation, and lead to new insights that can allow for greater success towards a shared goal: improving conservation and management of fish, fisheries, and their aquatic habitats. In support of these outcomes, we believe that mentoring minorities is a critical component necessary to increase diversity and retention of underrepresented groups in natural resources. Good mentors can help minorities to better understand and navigate group dynamics, develop important connections, and feel comfortable with their uniqueness.

* Corresponding author: infanted@anr.msu.edu

Despite these benefits, and despite efforts to foster greater diversity in natural resources, minorities are still missing in vital spaces and places in our profession. In fact, fewer women and less racial diversity can be found in fields including fisheries management and conservation than in our population as a whole. The Bureau of Labor Statistics reports for 2013 that of the 964,000 individuals working in farming, fishing, and forestry occupations, 21.7% were women and 5.6% were black or African American (Bureau of Labor Statistics 2013). Also, of the 94,000 individuals working as environmental scientists (a broad grouping of employed individuals that includes geoscientists), 27.9% were women and only 4% were African American (Bureau of Labor Statistics 2013). While these statistics capture wider trends than those represented solely by our field, they help to frame our understanding of gender and racial differences in related sectors of the workforce. Further, they suggest that minorities seeking careers in fields with such disparities may not only face a lack of peers within their professional community, but also a lack of mentors who may have experienced the particular challenges that they will face. This disparity may begin for minorities within academia, where they receive training for their chosen careers. As reported by the National Education Association in 2011, for full-time faculty at postsecondary institutions who reported their race/ethnicity, 79% of individuals identified as white, 6% as black, and 4% as Hispanic (National Center for Education Statistics 2013). Mirroring this trend, while 57.4% and 62.6% of individuals receiving B.S. or M.S. degrees, respectively, were women in the 2009–2010 academic year, only 10.3% and 12.5% were black while 8.8% and 7.1% were Hispanic (Aud et al. 2012). Similar biases in race and gender may be found in natural resource management agency offices around the country, and they may also be observed when attending professional meetings and conferences. Such imbalances in the composition of their professional communities may contribute to perceptions that minorities are outsiders; that we may not fit in; and that our ideas, which may be different from those held by the majority, may be inappropriate for reasons other than their merit. Further, it is important to note that as these perceptions may not be held by members of the majority, they may still be experienced by the minority, contributing to their feeling like an outsider and leading to a complex situation where the minority feels challenged to express their views and unconfident in their abilities.

The notion of being an outsider, however, is not unique to being female or a racial minority. Other factors, including sexual orientation, socioeconomic background, spiritual views, and disabilities, can contribute to this perception. And independent of these factors, many of us have felt this way at points in our careers. In some situations, we have been the only researcher charged with working with a group of managers or the public, or perhaps the only manager in a room with policy makers. We have been confronted with cultural norms, institutional practices, and personality traits that may be foreign to us, setting us all up as outsiders at points, affecting our confidence, and eroding our ability to work collectivity as a group. Yet, this is where good mentoring can play an important role for all of us. A good mentor can help an individual overcome this sense of isolation, allowing the mentee to function and even thrive in such settings.

The goal of this vignette is to present some of the benefits that good mentoring can offer to those who may be in a minority position within the field of fisheries management and conservation. As authors of this vignette, we draw from personal experiences to highlight examples of how this can occur and acknowledge that good mentoring may provide many other benefits that are not described here. We emphasize how a good mentor can help a mentee to understand the dynamics of a situation, initiate and develop important relationships, and sharpen confidence in his or her abilities and in expressing his or her individuality. In sharing these personal examples, we highlight the importance of mentoring at key stages of our careers and share the difference that mentoring made to us. Mentoring can also mean the difference to others, and keeping minorities encouraged, motivated, and confident allows us to work more effectively towards a goal that many of us share in our field: managing and conserving fisheries resources into the future.

A Good Mentor Can Help a Person Better Understand the Dynamics of a Situation

Dana Infante: *As a new researcher, I was charged with a project intended to become part of a national effort to promote protection and conservation of fish habitat throughout the nation (formerly the National Fish Habitat Action Plan [NFHAP] and currently the National Fish Habitat Partnership [NFHP]). In 2007, I attended my first meeting with an oversight committee comprised of researchers and managers from around the country. Committee members (who were nearly all male) were selected based on their previous experiences in the field; I knew some of them by their reputations and through published works. I looked forward to the opportunity to interact with these professionals, and I hoped to make a good impression. I spent a lot of time preparing for the meeting and anticipated that I would be reporting on the status of my research and answering questions related to data being used and approaches being taken. I also expected to be given feedback on the work and was open to making changes to the work to improve it.*

While I felt prepared and calm as the meeting began, those feelings were replaced by confusion and anxiety. Committee members were not shy about expressing dissatisfaction with my ideas or those of other members, and the discussion was heated at times. Besides raising questions on project details, members challenged the overall utility of the project as well as the likelihood that it could even be completed on time. Points were raised about funding for the effort and the conversation included several requests from members that I could not incorporate because they deviated from established goals. At one point, three members were simultaneously yelling over one another. I will always remember freezing during that moment—I had no idea what to say or how to handle their concerns amid the conflict. At the end of the meeting, these same individuals left together, apparently comfortable with each other. This was especially confusing because they were arguing with each other a few hours before and yet they were obviously on good terms. While some members congratulated me on the success of the meeting and the proposed work, I was certain that they were just trying to make me feel better. I left the meeting feeling dejected. Between the conflicts, requests that could not be addressed, and members' doubts on the success of project, I was certain that I had failed. While I felt confident that I could meet the analytical expectations for the project, I did not know if I could manage the dynamics of another meeting and, more importantly, if I was the right person to gain the trust and respect of the committee members for this important effort. I linked the challenges to the newness of my position (certainly a factor), but I also wondered what else about me personally contributed to the manner in which the criticisms were raised. I wondered if I should have been more assertive in making points or tried to insert myself into the conversations of these senior managers and researchers more directly. Right or wrong, I wondered if the committee viewed me as less competent than they would if I were male. Regardless, I was concerned that my role in the project would compromise its chances for success, and I contemplated stepping away from the effort.

Following the meeting, I shared my concerns with an individual who has since become a trusted mentor. I suggested that the lead on this effort be passed to somebody else and that I could instead play a supportive background role. Fortunately, he convinced me that this was not a good idea. He then provided his insider view of the meeting, helping me to understand much of what had occurred. He explained some of the interests of committee members and pressures that they were facing from their respective agencies, along with some of the history among members. It was a completely new view into the situation. These insights resulted from his years of experience working with these people and were invaluable in helping me understand what happened. This allowed me to adjust elements of my research to address needs of some members, but perhaps more importantly to me personally, his insights revealed sources of tension and dissatisfaction that were not directly related to my work or to me. With this understanding, I had the confidence that I needed to continue.

This meeting occurred seven years ago, and a lot has happened since that time, including the culmination of that initial effort, which now serves as the basis for future work that I am engaged with for NFHP and for other agencies. I have also gotten to know past and current members of the committee and have a fuller appreciation for their commitment to the effort. I also want to share that my experience with NFHP has enforced my own commitment to conducting applied research; if I had stepped away from the effort, I may not have realized this. In telling this story, I wish to highlight the importance of that conversation with my mentor and thank him for not letting me quit! His insights helped to minimize my doubts about myself, provided me with some much-needed confidence, and helped me to realize my life's passion.

One trait of a good mentor includes their perspective that comes from time working within the field. An experienced mentor may be aware of agency or institutional characteristics or personality traits of individuals gleaned from years of experience that would not be transparent to a newcomer or that may be harder to grasp by a minority. Following a meeting, for example, where the mentee may have felt largely unsuccessful in expressing his or her views or having his or her opinions heard, a good mentor may be able to lend insight into a situation that allows the mentee to better understand the dynamics of what occurred and what led to certain outcomes. By sharing this knowledge and aiding mentees in developing their own richer perspective of the situation, mentors provide them with greater awareness that can benefit not only how mentees perceive situations in the future, but also in how they take action. This greater awareness can help to dispel feelings of self-doubt and lend confidence to mentees, giving them an insider's view and reducing feelings of being an outsider, often experienced by minorities.

A Good Mentor Can Remind a Person of His or Her Strengths and Help Him or Her to Become Comfortable with His or Her Individuality

Mamie Parker: *Living in the space of gratitude, I am grateful for the mentors and the network of individuals that helped me to see that while I was different and an outsider, my ideas and perspectives could make a difference in fisheries management and conservation. They always encouraged me to focus on my positive attributes and experiences when it was so easy for me to see the negatives, the rejections, and the loneliness of being the only minority female. Looking back, I know that it was the many long talks and walks with my mentors that kept me in the game. In fact, early in my career at the U.S. Fish and Wildlife Service (Service), when I was often frightened, stuck, or confused, I reflected on the memorable teachings of my first mentor, Ma Parker, my mother—the sharecropper, the maid, and the avid angler. She helped me the most in maximizing my human potential. It was her actions and words of encouragement that first helped me in the sixties when our small town of Wilmot, Arkansas was forced to integrate the public school system. As one of the first little black girls to attend the all-white school in this segregated community, I was ignored, not included in activities, and often told that this was simply because I did not look like everyone else. I realized for the first time that being different was not always good. I learned that being different was very difficult. It did not get easier over time. In fact, when I was in third grade, I recall coming home and crying, describing my feeling of isolation and desire to quit. My first mentor's actions and encouraging words helped me cope with being different. These were certainly times to see these unpleasant, or lack of, encounters as times to learn, listen, and leave those uncomfortable negative thoughts.*

While I listened and learned a lot from my first mentor, it took me years to actually leave the negative thoughts behind and realize that being different does not mean that you cannot make a difference. As a young girl in that elementary school and during most of my young adult life in the fisheries management and conservation community, I learned to survive but struggled with the greater challenge of feeling truly confident that I belonged. I often was reminded of this when my great ideas were repeated by a man and suddenly it became a profound statement that moved things forward. It was still necessary to respond to questions regarding whether you had enough years in the Service to justify the position you held. My gender and race were still an impediment to my success. More often than not, I wanted to leave the profession and go to places where I felt more comfortable and accepted. My mentors gave me the support that I needed to stay in the profession. I realized that fisheries management and the conservation community needed the silent voices of minorities in urban communities, and that encouraged me to inspire other minorities and women to not quit, even when it was difficult being different.

In fact, my mentors, including Hannibal Bolton, another minority biologist in the Service, helped me with this struggle. While serving as assistant director for Fisheries and Habitat Conservation, I encountered many unpleasant situations. He was a great mentor as I struggled to lead the Service's Fisheries Program. We were working very hard to gain more internal support and budget requests in the Service for NFHAP. This great partnership would bring the various entities in the fisheries community together to protect, restore, and enhance fish habitat. Powerful things were happening outside the Service within the fisheries community to move this plan forward. However, we were facing roadblocks and I lost sight of a positive outcome. My negative thoughts and bruised ego paralyzed me at times. Hannibal and other mentors, in addition to the Fisheries Management Team, had many heart-to-heart conversations with me when I was about to give in and give up. When I really lost the confidence, mentors pushed me to find and use that astounding creativity and productivity that kept me from quitting during those first dark days in my elementary school. Mentors shared their resources and contacts in the White House and on Capitol Hill to help us get over the hurdles in the Department of the Interior. More importantly, when I cried on my mentors' shoulders and picked their brains, my mentors gave me a safe environment while teaching me that the hard times are learning times and are not times to give up. We learned from each other by sharing our own failures and taking risks. Eventually, we overcame many challenges and got the much-needed internal support for NFHAP. I celebrate my mentors that continued to support NFHAP for years after I left the Service and am excited that recently the U. S. Senate Committee on Environment and Public Works passed the National Fish Habitat Conservation Act (S. 2080) authorizing the plan. This is another proud moment for my mentors and many others.

Mentors like mine can really help future minorities in fisheries management and conservation. Minorities have more confidence and take more risks when they are encouraged by mentors and know that being different is difficult but necessary in the fisheries management and conservation community. The future of fisheries should focus some efforts on making people with different experiences, different backgrounds, and different ideas and approaches feel accepted and welcomed. Mentors can make a difference by encouraging minorities and women with the words that Christopher Robin told his bear, Winnie the Pooh: "You are braver than you believe, stronger than you seem, and smarter than you think."

In the field of fisheries management and conservation, successful outcomes are not always immediate. Sometimes we wait months or years for results of our efforts to transpire. We may not often hear that we are doing a good job, and we are often left to interpret outcomes of meetings and exchanges on our own, with no recognition or indication that our work has been on track. For all of us, this can be discouraging, leaving one's confidence on shaky ground. However, for a minority, this may lead to additional challenges. As a minority, challenges of the work at hand may be interpreted as being linked to the minority's differences from other members of a working group or team, leading the minority to attribute unwelcome outcomes to differences in his or herself versus simply the nature of the work. A series of such self-assessments may lead a person to feeling overly critical about his or her abilities and erode his or her confidence. Further, with one's confidence shaken, a person may be much less willing to take risks or to raise ideas that may seem outside of the conventional focus. A person may be inclined to keep novel ideas under wraps, may be reluctant to suggest new courses of actions, and may be less willing to contribute to group dynamics as a whole. Good mentors can help us assess situations more objectively. They can help us see when and how we have been successful and also how our unique perspectives can be shared and valued. In fact, many mentors have sent minorities and women back with encouraging words and feedback. They send us back to look for the good list in each of us.

Working towards Our Shared Goal

Being a minority in our field, even being new to our field, can be compared to being a pioneer. To settle the western United States, many people forged ahead into unmapped territories, without knowing exactly what was waiting for them and unsure of all the skills that they would need to survive. Armed with their will and spirit, their success depended on luck in some cases, trial and error in others, and certainly, chances of success were improved with the assistance of guides who could teach them needed skills, help them navigate through unmapped territories, and keep their spirits up when they felt discouraged. Guides aided pioneers in their efforts, just as good mentors can help to ensure the success of minorities or others they mentor. And just as guides and pioneers shared the same ultimate goal, to explore and settle the West, we believe that those of us working for fisheries management and conservation also share a goal.

We are all aware of the myriad challenges facing fisheries today, with examples including climate change, human land uses, spread of invasives, and overfishing. We likely all acknowledge that addressing these challenges effectively is critical for the long-term sustainability of fisheries and their habitats. We propose that despite differences in race, gender, or any number of other factors in our professional community, we have all chosen careers in this field because we share a desire to promote sustainable management and protection of aquatic resources. Mentoring minorities can ensure that passionate, committed, and creative individuals remain in our field, and encouraging and empowering the next generation of researchers and managers can only improve our chances for more effective fisheries management and conservation.

Acknowledgments

We wish to thank the many individuals who have served as mentors to us during our careers. For Mamie, these special people include her family and first mentor, her mother, the late Cora Parker, along with her teachers, administrative assistants, supervisors, coworkers, colleagues, employees, and women in conservation that reached down and raised her self-esteem, self awareness, and spirits. For Dana, thanks goes to Dr. William W. Taylor for believing in her enough to help her believe in herself and achieve more than she ever thought possible. She also wishes to thank Drs. Lizhu Wang, J. David Allan, Jack Liu, and Joan Rose; members of the Aquatic Landscape Ecology Lab; and her family. By sharing time, knowledge, and insights, and sometimes by just listening, these special people have helped to keep us confident, encouraged, and motivated.

Biographies

Mamie A. Parker worked as a biologist and senior executive at the U.S. Fish and Wildlife Service (Service) for nearly 30 years. She attended the University of Arkansas and University of Wisconsin and worked throughout the United States on efforts including the Partners for Fish and Wildlife program. She made history when promoted into the Senior Executive Service as regional director of 13 northeastern states headquartered in Massachusetts—the first African American in this position in the agency's 135-year history. As the agency's assistant director and the country's head of Fisheries, she was instrumental in developing the National Fish Habitat Action Plan. She was honored to receive the Ira Gabrielson Leadership Award for most outstanding leader and mentor in the Service and also received the Presidential Rank Meritorious Service Award. An environmental consultant, facilitator, and coach, she is founder and president of Ma Parker and Associates near Washington, D.C.

Dana M. Infante is currently an associate professor in the Department of Fisheries and Wildlife at Michigan State University (MSU). She has been part of the department for more than 7 years and previously worked as a postdoctoral research associate for the Institute for Fisheries Research in Ann Arbor after receiving her Ph.D. from the University of Michigan. Many of Dana's research efforts have occurred over large spatial extents and are highly applied, including characterizing influences of landscape factors on river systems and performing ecological assessments. Since joining MSU, she has led an effort to conduct a national assessment of stream fish habitats for the National Fish Habitat Partnership (NFHP). In 2011, she and members of her laboratory were acknowledged for scientific achievement in support of fish habitat conservation by the NFHP Board for completion of the initial national assessment of fish habitats in 2010.

References

Aud, S.,W. Hussar, F. Johnson, G. Kena, E. Roth, E. Manning, X. Wang, J. Zhang, and L. Notter. 2012. The condition of education 2012 (NCES 2012-045). U.S. Department of Education, National Center for Education Statistics, NCES 2012-045, Washington, D.C. Available: http://nces.ed.gov/pubsearch/pubsinfo.asp?pubid=2012045/ (May 2014).

Bureau of Labor Statistics. 2013. Household data annual averages. Table 11: employment by detailed occupation, sex, race, and Hispanic ethnicity. Bureau of Labor Statistics, Current Population Survey. Available: www.bls.gov/cps/cpsaat11.pdf. (April 2014).

National Center for Education Statistics. 2013. Fast facts: race/ethnicity of college faculty. Available: https://nces.ed.gov/fastfacts/display.asp?id=61. (April 2014).

Wilson, E. O. 1992. The diversity of life. W. W. Norton and Company, New York.

We Are All in This Together: Capitalizing on Individual Abilities for Collective Benefit

William W. Taylor*
Center for Systems Integration and Sustainability
Department of Fisheries and Wildlife, Michigan State University
115 Manly Miles Building, East Lansing, Michigan 48823, USA

Nancy J. Léonard
Northwest Power and Conservation Council
851 S.W. Sixth Avenue, Suite 1100, Portland, Oregon 97204, USA

So-Jung Youn
Center for Systems Integration and Sustainability
Department of Fisheries and Wildlife, Michigan State University
115 Manly Miles Building, East Lansing, Michigan 48823, USA

C. Paola Ferreri
Department of Ecosystem Science and Management, Penn State University
408 Forest Resources Building, University Park, Pennsylvania 16802, USA

Key Points

- Recognize that we are all passionate about different things and consequently motivated differently to learn and excel.
- Take the time to discover and appreciate the unique abilities and passion of others because they will enrich your personal and professional experiences and advance your overall learning.
- Seek and provide opportunities that will contribute to discovering, learning, and strengthening abilities in others and yourself.

We are like pieces in a jigsaw puzzle. We are all unique, and have our own special place in the puzzle of the universe. Without each of us, the puzzle is incomplete.
—Rod Williams, musician

Introduction

We have all learned, over time, through our daily interactions with a wide variety of people that we have different abilities, natural talents, or skills that we have gained over years of practice. As you expand your professional network, you should challenge yourself to see

* Corresponding author: taylorw@anr.msu.edu

how the abilities of others can complement yours, especially when these abilities may be very different from what you think would benefit your current project. We use the word "challenge" because taking the time to learn and appreciate other people's abilities is not often an easy task. Sometimes, you may think that certain people cannot contribute to achieving your objectives. But take the time to discover and appreciate the true value of their potential contribution and you will come to learn why Stephen R. Covey stated that "strength lies in differences, not in similarities."

Mentoring is my (Bill's) way to challenge myself and grow personally and professionally. Every day I spend with a mentee, I learn and expand my own skills and knowledge base by exploring what my mentees are passionate about and what abilities they possess. By being mentored by my diverse mentees, I have learned that the fisheries profession benefits by uniting people with different interests and abilities. Together, we can provide input that strengthens our work and makes it relevant to more people.

Appreciating Others for Their Abilities

I have not always appreciated the value of my mentees, especially when their passions and abilities did not match my ideas of what would be most beneficial to a fishery professional. My upbringing and education originally led me to believe that, as a mentor, I had a mission to produce professional progeny who specialized in the same area of expertise as me, inland fisheries ecology and population dynamics. During a particularly difficult set of experiences with a few of my students who did not share my passion for population dynamics, I came to the realization that I had been guiding my students all wrong. There are many different paths to success and many different ways to contribute to the fisheries profession. There was no one correct path, nor was it productive to attempt to force my mentees along my path. The more I pushed my mentees to go down a path that did not fit them, the unhappier we were with each other and the more our relationships suffered. Nancy was one of my mentees that reinforced that I needed to focus on her interests and abilities instead of what I thought she should learn and be able to do.

When I first met Nancy, she had a diverse background ranging from raptors and insects to public outreach, with a focus on animal behavior. I thought that she lacked background in fish morphology and identification so I suggested that she take classes in those areas. She was not pleased with this suggestion and argued that she already had more than enough experience in identifying species and she had no interest in spending six hours a week in an ichthyology course. What I needed to recognize and accept was that Nancy was interested in a career that would require behavior and policy, not fish identification skills. I could have forced her to take courses that I had enjoyed and found beneficial. But, catching myself, I realized she did not need to invest more than the minimum amount of time in this area of study. Nancy was heading down a different career path than mine. So, instead of arguing and making both of us miserable, we compromised. She learned the basics of what I thought would be beneficial for her related to fish identification skills and we devoted more time in finding more relevant but nontraditional fisheries management classes that would augment her passions and abilities (e.g., social network analysis classes). Over the years, I have learned that attempting to force my mentees to fit my image of the ideal fisheries professional is counterproductive. Having learned this lesson, I have been able to be a better mentor to Nancy and many others.

Passion in the Pursuit

The reason it took me a while to learn to value people's different passions and abilities was that I had forgotten that we are all different. No two of us are the same and no amount of standardized curricula or work experience can ever make us identical. I learned that I could, however, celebrate and enhance my mentees' individual abilities, whether these were in communication, sociology, or areas more similar to mine, such as fish population dynamics. I had to learn to accept that the colloquialism "don't ever try to teach a pig how to sing; it annoys the pig and wastes your time" was correct. Clearly, I needed to step back and think about ways to help people be successful, not just mold them into my own idealized perspective of who they should be, an idealized clone of myself!

My job was to find the passion and abilities of each of my students, help augment these abilities, and provide tools to make them even better. Once I figured this out, with the help of my mentors and mentees, I developed a new found respect for individual diversity. I discovered that my laboratory and my research program were enhanced and strengthened by embracing the different passions and abilities of my mentees. My research program expanded from fish population dynamics to involve many new areas of expertise, such as policy, governance, outreach, and social network analysis. Based on these positive experiences, I am thrilled when students, such as So-Jung, whose background focused on business and invertebrate ecology rather than a traditional fisheries and wildlife undergraduate degree, consider me as a potential advisor for their graduate degree. When So-Jung was looking for a graduate program, she received a lot of guidance to focus her search on her current area of expertise, cultivated by what she had accomplished during her undergraduate degree. Luckily for So-Jung, she did not give up easily; she had a sense of her passion and wanted the opportunity to try and see if she could develop the needed abilities to combine her business and ecology backgrounds. Even luckier for me, So-Jung kept looking for a mentor and advisor that could provide her with those opportunities. Now, we are working together to develop graduate experiences that link her passion and abilities to fisheries so she will be a uniquely valuable fisheries professional. This journey of discovery is always the most exciting time for me when I begin working with a new student, as it pushes one to get out of a box and think creatively about the future!

In retrospect, it seems the lesson that I took so long to learn is simple, but unfortunately the traditional university curriculum and the type of work rewarded to achieve a degree or tenure encourages us to confine ourselves to a predefined idea of what a fisheries professional should learn and master. This traditional and narrow focus drives mentors, like I used to be, to train fisheries professionals in the same skill sets; every mentee should be a proficient population modeler as well as an ecologist and a terrific communicator with fisheries stakeholders!

Interestingly, all we need to do is reflect on our own lives to realize the fallacy of this view. None of us can do everything, but by working with a team and putting people into areas that allow them to effectively use their abilities and talents, anything is possible! We can, as a team, be proficient fish-population modelers and understand fish ecology and effectively communicate the findings from our model to diverse stakeholders.

Be Patient and Give People a Chance

As I long ago learned, and aim to teach all of my students, the best team you can work with is a diverse group who each value one another's abilities and knowledge because the synergy of the group will leverage each individual's abilities to yield an outcome that far exceeds expectations of any individual alone. To teach this important lesson, I have always felt that a firsthand experience was the best way. Consequently, I provide opportunities for my mentees to work in situations that force them to struggle and learn how to work with people that have passions and abilities that differ greatly from their own. For one of my mentees, Paola, that opportunity arose when I encouraged her to assist me with the arrangements for the American Fisheries Society's (AFS) annual meeting that was held in Detroit one year.

As Paola advanced in her Ph.D. studies, she was becoming increasingly focused on her academic pursuits and isolated into a very small universe of like-minded students who had the same priorities and motivations as her. Working on the arrangements committee for the AFS annual meeting, consequently, was an eye-opening experience for her. She had to work with people representing a wide array of backgrounds and interests, including agency administrators and biologists, other university professors and students, AFS staff, staff from the convention bureau, and hotel management staff. Each person brought different interests and skills to the group; for example, some people were interested in the budget and finances related to the meeting while others were concerned about the hospitality aspects. In the beginning of the planning process, Paola became frustrated when she wanted the group to focus on the program for the meeting–the theme, the plenary, and the technical sessions–because she believed that these were the most important parts of a professional meeting. Early on, she was exasperated by the time spent discussing other items like the budget or the types of signs that would be used during the meeting because she truly thought that if we just focused on the program, the meeting would be great.

As the planning process unfolded, Paola began to realize that each person had an important contribution to the overall quality of the meeting, and in many cases they brought ideas of which she would not have thought. In the end, the meeting was a successful one because everyone worked together and brought their individual strengths to the group. After all, the program we planned could have been a great one, but who would have been able to find their way around the convention center if there had not been someone paying attention to the size and type of signs we used?

We all still find ourselves sometimes thinking that someone may not be a valuable contributor but, now based on experience with past successes, catch ourselves, take a deep breath, and focus on what he or she could do and how this can be leveraged to improve the outcome of our joint efforts. To succeed, you must put aside your perception of how things should be, try to be objective to what can be, and figure out how those unique abilities and passions can be nurtured and used to contribute to the success of our profession. After all, individual differences are what add spice to life in general.

The Whole Is Greater than the Sum of Its Parts

What we learned from our experiences as individuals, graduate students, professors, and colleagues is that it is important to take the time to invest in discovering peoples' abilities and passions. Recognizing how these can contribute to your efforts and the profession will enrich your personal and professional experiences and advance your overall learning and effectiveness. By being willing to invest in this journey of discovery, you will find yourself exposed to new ideas, perspectives, tools, and even topics that you may never have considered as being a facet of your own area of expertise. It is equally important to recognize that everyone can be valuable colleagues as long as we focus on their abilities and do not force them to do something that does not align with their passion or skill sets.

By investing the time needed to understand others, you will not only learn about them as individuals, you will also be exposed to new ideas and experiences that will help you grow and become a better person. This hard work will not only reward you with the most important gift of having a network of people with whom to share successes and struggles, but will also make you a better mentor and professional. So remember, as you travel along your exciting and fast-paced fisheries career journey, take the time and challenge yourself to appreciate one another's abilities, especially when these are very different from your own, as the rewards will greatly justify the energy you invest in accomplishing this challenge.

Acknowledgments

We greatly appreciate the thoughtful and constructive comments from our anonymous reviewers and Abby Lynch.

Biographies

William W. Taylor is a university distinguished professor in global fisheries systems in the Department of Fisheries and Wildlife and the Center for Systems Integration and Sustainability at Michigan State University. He has been active in the American Fisheries Society throughout his career, serving as president of the society in 1997–1998. He currently holds a U.S. Presidential appointment as a U.S. Commissioner (alternate) for the Great Lakes Fishery Commission.

Nancy J. Léonard is the fish, wildlife and ecosystem monitoring and evaluation manager with the Northwest Power and Conservation Council. Her work experience in multijurisdictional fisheries governance with the Department of Fisheries and Wildlife at Michigan State University and her work with the Great Lakes Fishery Commission give her a strong professional background working with diverse interests towards common goals in fish recovery. She is active with AFS, currently serving as the Scientific Program co-chair for the 2015 AFS Annual Meeting. She has an M.S. in biology from Carleton, M.Env.S. concentration in water resources from Miami University, and a Ph.D. in fisheries and wildlife from Michigan State University.

So-Jung Youn is an M.S. and Ph.D. student under Dr. Taylor in the Department of Fisheries and Wildlife and Center for Systems Integration and Sustainability at Michigan State University. She is interested in studying the role of inland fisheries in local food and economic security, particularly in Asia. So-Jung received her B.S. in biology, with a minor in management and organizational design, from the College of William and Mary.

C. Paola Ferreri is an associate professor of fisheries management in the Department of Ecosystem Science and Management at Pennsylvania State University. She is interested in understanding the link between habitat, management actions, and sustainable fisheries.

Skills to Develop

Not Fish, Not Meat: Some Guidance on How to Study Fisheries from an Interdisciplinary Perspective

ROBERT ARLINGHAUS*
Leibniz-Institute of Freshwater Ecology and Inland Fisheries
Müggelseedamm 310, Berlin 12587, Germany
and
Humboldt-Universität zu Berlin
Philippstrasse 13, Haus 7, Berlin 10115, Germany

LEN M. HUNT
Centre for Northern Forest Ecosystem Research, Ontario Ministry of Natural Resources
955 Oliver Road, Thunder Bay, Ontario P7B 5E1, Canada

JOHN R. POST
Department of Biological Sciences, University of Calgary
2500 University Drive NW, Calgary, Alberta T2N 1N4, Canada

MICHEAL S. ALLEN
Department of Fisheries and Aquatic Sciences, University of Florida
7922 NW 71st Street, Post Office Box 110600, Gainesville, Florida 32653, USA

Key Points

When designing an interdisciplinary project,

- Develop a solid disciplinary foundation before becoming an interdisciplinary scientist.
- Choose the right project leader as knowledge broker.
- Employ the right mix of people.
- Conceptualize the problem to be addressed with the whole interdisciplinary team.
- Plan the integration at the onset of the project.

Introduction

Fisheries can best be viewed and understood from a systems perspective, which is defined as a web of interrelated and interacting ecological, biophysical, social, economic, and cultural components. Unfortunately, reductionist approaches focused on single-species fisheries biology as a discipline have long dominated fisheries science. Consequently, many well-

* Corresponding author: arlinghaus@igb-berlin.de

intended fisheries-management actions have failed to meet their objectives, either because of unexpected human responses or because of complex ecological dynamics. To address the resulting implementation uncertainty, scholars have increasingly asked for research programs that study the implications of management actions throughout the whole coupled social-ecological system. To achieve this aim, interdisciplinary science and the integration of disparate knowledge sources is needed, something that few graduate programs in fisheries specifically focus on.

A key assumption of this essay is that the simplification of key feedback processes and a general lack of integration of the natural and social components of fisheries may lead to system responses that are often characterized by high social and economic costs. To avoid such costs, we need a better understanding of the type and function of cross-scale and nonlinear feedbacks among the human and environmental subsystems because these feedbacks determine how fisheries as systems respond to disturbances and management interventions. We are convinced that the greatest breakthroughs in capture fisheries science wait at the interface of the social and ecological components of fisheries. Here, we offer some advice for the aspiring fisheries professional on how to develop a successful interdisciplinary agenda (see Box 1 for terminological clarification).

Before listing our advice, a disclaimer is in order: interdisciplinary projects in fisheries are no panacea, and in many cases it is just fine to work from single disciplines. For example, if the task is to estimate the current stock size for a purely scientific, or a theoretical, purpose, a quantitative stock assessment project that analyses abundance and catch-at-age data works well and is appropriate. Or, if the task is to learn how the broader angling public in a region feels about an existing fisheries regulation, a survey-based project based on probabilistic sampling conducted by a social scientist knowledgeable with the particular fishery system is a perfectly suitable approach. However, we can also think of many situations where an interdisciplinary research approach would be superior. Think about situations of marine spatial planning where multiple stakeholders, coastal zones, and transboundary fish stocks are involved. Or consider developing a holistic analysis of the impact of harvest regulations or other policies on ecosystems and fishing communities in a landscape of freshwater fisheries. Surely, integrating the ecological, evolutionary, and human dimensions of fisheries may be fruitful to solve these and related complex situations where ecological and social systems strongly interact though cross-scale interactions and feedbacks. Here is our (entirely subjective) list of recommendations that should help researchers enjoy the many advantages and mitigate any potential disadvantages of an interdisciplinary research path in fisheries.

Develop a Solid Disciplinary Foundation before Becoming an Interdisciplinary Scientist

Deeply entrenched disciplinarity is thought to be a barrier to interdisciplinary collaboration. However, some level of specialization in a given subject is needed to develop the foundation for basing future interdisciplinary projects. Hence, preparing oneself for interdisciplinary work involves attaining specialized depth in a given subject through a dedicated M.S. or Ph.D. program. Often, in the fisheries profession, such programs will be fisheries or applied ecology programs. However, as one specializes, one must maintain a broad interest, read widely (e.g., human dimension of fisheries, natural resource economics), and possibly also take some interdisciplinary courses to receive an appreciation for the multitude of approaches that exist to tackle a given problem. Recommending additional classes to a student enrolled in a busy graduate program is no trivial matter and may even mean extending the studies by one or

Box 1. Some Semantics on the "...disciplinarities"

There is wide variation in what is understood as interdisciplinary and transdisciplinary research, which should be separated from multidisciplinary research approaches. *Multidisciplinarity* refers to the study of an object such as a fishery through the lenses of multiple isolated scientific disciplines. For example, when a fisheries biologist and a human dimension researcher work side by side in the same management agency on the same fishery, each with his or her own research question, conceptual framing, and methodological toolbox, and with little attempt to integrate findings to solve a common research objective, one would talk about multidisciplinarity. *Interdisciplinarity* differs from multidisciplinarity in some important ways. Most importantly, research problems and questions are answered using methods, frameworks, and concepts from at least two separate schools of thought. In a prototypical interdisciplinary project, scholars from at least two disciplines would work together in an integrated fashion to answer common research objectives. For example, a bioeconomic model to help identify an economically suitable management action would demand the integration of a behavioral model of the fisher, a fish population model, and associated evaluation criteria, and hence be forced to use theories, variables, concepts, and models from different disciplines, such as economics, fisheries ecology, and operation research, to answer the research questions. Finally, *transdisciplinarity* is interdisciplinary research that substantially integrates the world of action into the knowledge generation and integration process. Here, stakeholders and practitioners are part of the scientific knowledge generation process and may be involved in framing the problem, collecting data and interpreting results, or in all of this; hence, the suffix "trans." A special form of transdisciplinary research is action research where the research process is conducted in sites and areas used and managed by communities and in close collaboration by researchers and practitioners. Transdisciplinary research of all variants aims at democratizing research through deliberate involvement of stakeholders to increase capacity building, ownership of results, and knowledge transfer to solve local and regional sustainability issues. One example of inter- and transdisciplinary fisheries research is a German research project called Stocked Fish (www.besatz-fisch.de) led by the first author of this article. In this project, principles of sustainable fish stocking in German angling clubs were derived using jointly conducted fish stocking experiments that took place in the club's waters and that were planned, conducted, and evaluated by researchers and angling club heads in joint teams.

two semesters. However, for interested students, this investment will usually pay off. Assisted by an appropriate interdisciplinary mentor, it is important to identify what classes outside the own narrow discipline would be worth taking and what literature to consult. We recommend that a motivated student carefully choose mentors and advisors that are themselves broadly interested and that have a proven record (grants, papers) of successful interdisciplinary work. Also, some fisheries programs have produced more interdisciplinary output than others, and hence, the M.S. or Ph.D. fisheries program to pursue may also constitute a decisive choice. Any resulting foundation of depth and breadth can then provide the raw material for facilitating the branching into interdisciplinary endeavors.

There are three reasons for why one needs both depth and breath before engaging in interdisciplinary work. First, any interdisciplinary project needs methods developed in a specific field, hence methodological depth. Second, to foster interdisciplinary projects and to build teams one needs a basic knowledge of jargon and methods used in alternative relevant disciplines (i.e., scientific breath). Finally, on a more practical level, many of the more traditional faculties emphasize specialized knowledge of some sort in their hiring processes despite the appreciation and increasing value attached to interdisciplinary interest and expertise. Therefore, there are real risks to a traditional career path for those who become interdisciplinary researchers too early on (e.g., at the masters level). Hiring committees at very traditional disciplinary departments, and even in multidisciplinary ones where you apply to a position demanding a specific methodological toolbox (e.g., fisheries stock assessment within a natural resource management unit), might disfavor your application with the simple argument "this person is neither fish nor meat." This statement means that she or he has no deep understanding in any school of thought and cannot bring any specialized knowledge into the program. This assumption might actually be false, but often the perception of the committee members matters. Hence, scientific depth might be important to safeguard tenure and promotion.

Such critical assessment was levelled on some authors of the present essay, even when applying at prestigious interdisciplinary schools. Even there, the question was asked "What approaches and methods do you bring to the table that no one else currently does in our unit? What in-depth disciplinary course can you teach?" The first author was even given "friendly" career advice to start conducting "true" fisheries research (meaning population dynamics of exploited fish), after finishing a Ph.D. in the human dimensions of fisheries. Apparently, fisheries biology was perceived as the only valid fisheries science discipline by some leading fisheries professionals in Germany. However, the first author had completed an aquatic ecology-based fisheries degree before branching out into the then unfamiliar domain of the human dimensions of fisheries. It is, of course, possible to learn the foundation of other disciplines in the period of a Ph.D., such as the human dimensions of fisheries, and then return to fisheries biology or to branch out. However, not every hiring committee is prepared to think that way. Therefore, interdisciplinary fisheries researchers have to be prepared to compete with disciplinary scholars during the chase of tenure.

Choose the Right Project Leader as Knowledge Broker

To facilitate true integration, rhetorically strong knowledge brokers as facilitators and integrators are needed. These brokers are people who are well read in multiple disciplines; they can help translate disciplinary jargon and provide the necessary kit for interdisciplinary teams. These peoples have the expertise for problem conceptualization, are able to run effective meetings, and are good motivators of team members. Although the leaders of most projects often involve tenured senior scientists, this might not be the case. Catalysts of interdisciplinary work usually have other qualities that are not contingent on age or experience in the science community. Basically, the leader of interdisciplinary teams has to think outside (all of) the narrow specialized boxes and be able to conceptualize in a holistic systems perspective. Leaders of interdisciplinary projects must feel excitement when they open a social science journal and find a paper about angler behavior, yet the same person must equally feel excitement when reading a paper about the genetic impacts of stocking or any other fisheries ecological theme. The key innovation is bringing thoughts together that have been developed in isolation. Often, the same general concepts are developed and applied in different disciplines. The problem is that these same concepts often have different labels. Careful reading can afford opportunities to see the similarities in concepts across disciplines. For example, ideal free distribution theory from behavioral ecology offers the same predictions in behavioral economics when the fitness function of the (human) predator is replaced by the utility function from economics. In such cases, theory developed in ecology and in economics can be merged and predictions tested once the homology of thought is identified among disciplines. The leader would then have the role of helping the disciplinary team members appreciate the complementarity of the various approaches (i.e. facilitating cooperation among economists and biologists leading to the formulation of frameworks, research questions, hypotheses, and methodological approaches that can only be solved from an interdisciplinary perspective and that help solving the sustainability issue). The very same team leader must over time also accept that she or he might sometimes feel bereft of a true disciplinary home. Symptoms of success include subscription to listservs of seemingly nonoverlapping research domains, membership in unrelated scientific communities, and travel to conferences that do not share a single common attendee other than oneself! This success usually involves abandoning the security of a true disciplinary home and choosing instead to feel excitement through the enrichment of intellectual lives from the experience of multiple homes.

Employ the Right Mix of People

Interdisciplinary projects usually involve a range of expertise and competencies. It hugely pays off to choose the right mix of people. Often, scientists are brought into interdisciplinary teams for the particular expertise they know best. However, this overlooks the importance of interpersonal skills, intellectual openness, and curiosity, which is equally or even more important if interdisciplinarity is to succeed because the best expertise might be unavailable to the interdisciplinary project if the person is not willing to sit down with others from other disciplines and develop a joint problem conceptualization. Usually, you do not want to include principal investigators who are known to only enjoy disciplinary research outputs, however excellent these people are, unless they promise to contribute a very particular method and expertise that nobody else is able to bring to the table. Members of interdisciplinary teams must also be patient when training young scholars in novel, unfamiliar theories and methods and be willing to integrate findings to solve the sustainability issue at hand. Otherwise, one risks interdisciplinary projects developing into multidisciplinary ones where the integration of knowledge bases is not achieved at the end. The first author of this paper has had this experience in the first interdisciplinary project that he guided. In the so-called Adaptfish program (www.adaptfish.igb-berlin.de), the goal was to study the adaptive dynamics of recreational fisheries from local to regional scales by linking local-level angler decision making to broad-scale governance and institutional dynamics. Although the project was intended to develop an interdisciplinary endeavor, it ended as a multidisciplinary project in which team members (usually Ph.D. students) developed their own disciplinary research approaches, publishing in disciplinary journals and receiving their Ph.Ds. in disciplinary fields. It was only after the official end of the four-year project that the first truly interdisciplinary research products were developed, but these products were only achieved with a small subset of team members who had developed integrative research questions and had invented novel modelling techniques to reap the benefits of integration and cross-disciplinary cooperation.

Take Your Time and Conceptualize the Problem with the Whole Interdisciplinary Team

Expect interdisciplinary work to take substantially more time than discipline-specific projects to develop common grounds and terminology among team members. It is important to be prepared in order to avoid frustration with some unavoidable time lags. You need the time and resources to invest in team building, problem conceptualization, and reading diverse literatures. One should plan at least a year of interactions, including a couple of excellent meeting (whose organization is the task of the above-mentioned knowledge broker), to reach common ground in interdisciplinary teams. If multiple disciplines are involved, make sure the team agrees, understands, and commits to common research questions. It is our experience that it helps to develop concepts that serve as bridges among disciplines and to develop a glossary of terms and definitions. Concept mapping exercises can help to conceptualize the system under study and to reveal the hidden perceptions and assumptions of all team members. For example, studying the issue of fish stocking from interdisciplinary lenses involves identifying critical components (concepts), feedback, and interactions within the ecological system (e.g., genes, phenotypes, and species) and among the ecological, social, governance, and policy systems. Developing maps of relevant concepts, relations, and interactions using mapping exercises will expose the team to the complexity of the interaction web and help nail down the most important feedbacks for the project to address. All team members, even those with the most diverse backgrounds, must ultimately agree with the small set of joint research questions and the general methodological approach to be taken that emerge from these exercises. Such consensus is not easy, but an early focus on this it will pay dividends as the project unfolds. Communication must regularly occur throughout the project to keep all involved in the research results and to maintain mutual understanding. This communication can best be achieved by agreeing on a research framework in which all commonly agreed specific research questions are embedded and all contribute to the overarching research goal. Regular meetings about preliminary research findings keep the subteams informed, involved, and motivated, and this helps the final integration of research results. For example, if the overall research goal is to understand the sustainability of fish stocking, subquestions may deal with how stocked fish interact with wild fish or how anglers respond to stocking. Answering these subquestions, using disciplinary or interdisciplinary approaches, is needed as intermediate steps before the final integration and answering of the overarching research problem can take place. It is important to keep the whole team engaged in enjoying the intermediate successes, which in some cases might embark changes to research directions.

Plan the Integration at the Onset of the Project

Successful interdisciplinary projects (i.e., those that help solving the chosen sustainability problem) are based on a joint problem conceptualization by all team members that are then decomposed into smaller research questions, whose answers help to solve the overarching sustainability issue. The approach to integration of the smaller-scale research results must be planned a priori. Questions to be answered are as follows: Which social and ecological data could be easily integrated and which cannot? What collection methods and models will best facilitate data integration? When and which data are needed for integrated model building? Who in the team is willing and able to integrate and synthesize? Will joint products such as publications, reports, and presentations be generated that provide evidence of the integration? Who will be the authors and who should be the audience for the products? It is our experience that while many people are broadly interested in integrating social and natural science information, often people develop disciplinary interests as projects unfold and have difficulty in (or even deeply rooted resentment towards) integrating the disparate knowledge in the end. Part of this dilemma is caused by specific reward systems in various disciplines. For example, economists often are rewarded for sole-authored papers, whereas such papers will be the exception in interdisciplinary projects. Hence, it makes sense to think through the research products from the onset and to agree on deliverables and strategies to fulfil the integrative demand and manage expectations.

Closing Thoughts

As in other areas of natural resource use, substantial institutional, organizational, and academic hurdles have to be overcome when one attempts to integrate the natural and social sciences in fisheries. When these hurdles have finally been cleared, however, huge payoffs await. Well-executed interdisciplinary projects offer many rewards such as a more holistic system understanding that supports management recommendations, which are robust to irreducible uncertainties. Academically, interdisciplinary science is also lots of fun. There are also various downsides to these complex projects, such as the need for considerable time investments into capacity building for learning new specialized terminology and for managing teams of diverse expertise and competencies. Moreover, interdisciplinary research is not always appreciated in hiring processes and, hence, may turn into a disadvantage for the young scholar when applying for tenure in strictly disciplinary schools and faculties. Also, interdisciplinary journals sometimes suffer lower status in more traditional scientific subcommunities, although this evaluation is changing. In fact, some well-respected multidisciplinary journals such as *Proceedings of the National Academy of Sciences of the United States of America* have special sections that are specifically tailored towards high-quality interdisciplinary research output in relation to natural resource use problems (the section called “Sustainability Science”). Nevertheless, in many organizations there remain important disincentives to collaboration across disciplines and faculties. Despite these challenges, we predict that the need for interdisciplinary studies will increase, rather than decrease, particularly in applied research fields such as capture fisheries, simply because sustainability problems are very difficult to be solved by other modes of research. Many challenges lie ahead of us, and, as senior scientists, we are looking at the training of a new generation of fisheries scholars to join us in our quest for integrated discoveries in capture fisheries. Welcome!

Acknowledgments

We thank three reviewers for excellent input that helped sharpen our message. R. A. thanks the Gottfried-Wilhelm-Leibniz-Community for funding the interdisciplinary Adaptfish project (www.adaptfish.igb-berlin.de). R. A. also thanks the German Federal Ministry for Education and Research (BMBF) for funding of the transdisciplinary project Besatzfisch (www.besatzfisch.de) in the Program for Social-Ecological Research (grant no. 01UU0907).

Biographies

Robert Arlinghaus is professor of integrative fisheries management at Humboldt-University of Berlin, Germany and fisheries scientist at the Leibniz-institute of Freshwater Ecology and Inland Fisheries. He has been pursuing interdisciplinary recreational fisheries science since

2004. Arlinghaus is most interested in understanding how anglers interact with fisheries resources and how the feedback processes work. In 2011–2012, Arlinghaus and collaborators led the development of the United Nations guidelines for sustainable recreational fisheries on a global scale. He is recipient of the Award of Excellence in Fisheries Management by the American Fisheries Society and the Medal of the Fisheries Society of the British Isles.

Len Hunt is a human dimensions of natural resource management research scientist with the Ontario Ministry of Natural Resources. He is most interested in human dimensions studies that help to address the uncertainty associated with implementing different management actions. He has increasingly worked and published on the development and application of integrated recreational fisheries (and other resource management) models that explicitly connect changing resource, managerial, and other social conditions to human behavior and connect human behavior to changes in ecological and social conditions.

John Post is professor of ecology and evolution in the Department of Biological Sciences, University of Calgary, Canada. He is interested in growth, survival, and population dynamics of freshwater fishes and sustainable harvest of recreational fisheries. He and his students use experiments, models, and landscape-scale adaptive management approaches to assess sustainable management approaches. Ongoing research focuses on effective integration of fisheries biology, human dimensions, and management using adaptive experimental management approaches in freshwater recreational fisheries.

Micheal Allen is a professor at the University of Florida whose work has focused on population dynamics and ecology of fishes. Allen has evaluated fisheries management strategies for recreational fisheries in lakes, reservoirs, and marine environments. His research uses a combination of field studies, experiments, and computer models to explore how management strategies (e.g., harvest regulations, habitat restoration, and stock enhancement) can improve recreational fisheries. Allen joined the faculty at the University of Florida in 1997 and is currently investigating fisheries ecology and management problems in the USA and around the world.

Make a Science of Communication

ELIZABETH L. BEARD*
American Fisheries Society
5410 Grosvenor Lane, Bethesda, Maryland 20814, USA

SAMANTHA M. WILSON
4440K CTTC, Carleton University
1125 Colonel By Drive, Ottawa, Ontario K1S 5B6, Canada

Key Points

- Science can only have an impact if it is understood.
- Formal scientific communication requires preparation, thoroughness, and a willingness to seek and accept feedback from a variety of sources ranging from your peers to your mentors to anonymous reviewers.
- Informal communication can be just as crucial for your long-term career, so practice your storytelling skills and learn to refine your message for different audiences.

Introduction

Communication often seems like an afterthought to scientists, who often have a reputation as being more focused on practicing science than communicating. Many fisheries scientists, particularly students and young professionals, are intimidated by a variety of forms of communication. Relax! Humans are born communicators, and when you think about how many times a day you communicate in any form (speaking, writing, and nonverbally), communication may seem like a more natural part of your work. Note that we do not say "an extension of your work"; communication is something that should be incorporated into your research from the very beginning. We encourage young professionals to use many forms of communication to disseminate their research or to discuss their work and their excitement about science.

Here, we consider two types of formal scientific communication young professionals may be trying for the first time—peer-reviewed papers and presentations—along with more informal types of communication like public outreach and education. We asked several young professionals to share their early experiences in communication, and we use those examples to illustrate some larger principles of scientific communication for the young professional.

The Write Stuff

In a world of publish or perish, emerging professionals need to quickly assimilate knowledge and learn to organize and communicate through scientific publications. Students and

* Corresponding author: bbeard@fisheries.org

young professionals are faced with developing a new writing style and critical thinking skills, as well as understanding and responding to editorial and reviewer comments. Many new authors are intimidated by the writing process and find their first publication difficult and potentially frustrating, especially for some increasingly competitive journals.

Multiple resources are available for students to learn the writing process (see Recommended Resources). The simplest way to begin writing may be to find a few papers with a layout and writing style you like and use them as examples. Other ways include taking university courses on writing scientific papers or consulting books on writing and style guides from organizations like the American Fisheries Society. Although no one will reject a strong scientific paper over a few misplaced commas or incorrect citations, many such mistakes begin to call the author's thoroughness, dedication, and credibility into question. Young professionals especially should take care to submit polished work, since they do not have the benefit of an established reputation in their field.

Ninety percent of completing a scientific publication occurs before you type your first sentence, according to Hilary Meyer of the U.S. Fish and Wildlife Service. She encourages all students to participate in review groups to help develop critical thinking skills and become more familiar with published works. Participation in review groups can clarify what makes a good paper and help students learn from mistakes of other authors. If you do not have a review group, start your own or ask a professor to direct you to one.

"Don't scoff at literature review groups!" Meyer says. "You may think you have better things to do as a graduate or undergraduate student, but it will certainly improve your writing and your critical thinking skills."

Academic advisors, coauthors, and experienced labmates are excellent resources to help with organization and with critical analyses of methods, statistics, and layout of your paper. Many students refuse to share their papers, attempting to get a perfect first draft before getting input. Although most people do not relish critique, the process can save many fruitless hours of rewriting or just being stuck. Being open, asking questions, and sharing problems throughout the writing process, even before the first draft is complete, is helpful for timely completion of publications.

"A well-thought-out manuscript can take time to organize, but it reads easier and it conveys your conclusions much more effectively than a poorly organized paper," Meyer continues. "I would strongly urge new writers to read as much as they can about how to write a scientific paper. There are so many nuances to remember, and it's so very different than the writing you do every day."

Making a schedule and taking your time can also help the task seem less overwhelming. One commonly recommended method for getting around writer's block is brainstorming by quickly writing as many key words, phrases, and ideas on a piece a paper as one can in a set period of time. Many scientists also often write their papers in reverse order, starting with the results section. Procrastinators can find focus by making an appointment with themselves, free of distractions such as cell phones or e-mail.

Courtney Saari, who recently received her master's degree in oceanography and coastal sciences with a focus on marine fisheries, advises, "Write a little bit each day because your thesis or manuscript won't write itself. Figure out what time of day you focus best and set aside that time each day, even if it is only for 30–60 minutes, to write. And when you write, try to focus on one or two paragraphs at a time, or one section of the manuscript, not the entire document. And finally, you do not always need to start writing at the beginning; it is okay to start with the methods or results section and finish the introduction later."

Once your paper has passed the copious rounds of revisions from all coauthors, it is ready to be submitted for publication. But is it ready to be published? Papers are submitted to two to four reviewers, depending on the journal, and reviewer responses can be frustrating and discouraging to many new authors. The best advice is to remain patient and positive. Remember that reviewers are usually unpaid volunteers, and those who take the time to make detailed comments are obviously quite interested in your work. Read through the comments once, give them time to sink in, and come back to them later as needed. Address as many comments as possible, but when you disagree with them, do not be afraid to justify why you decided not to make the changes. Editors will consider your comments—especially those that are scientifically justified, polite, and professionally made.

Nick Lapointe, a postdoctorate researcher at Carleton University, says, "Don't be afraid to communicate with editors over the phone or through e-mail, but remember to pick your battles, trust your gut, and don't be afraid to walk away. One of my papers was accepted in a higher impact factor publication after giving up publishing through another journal. So don't give up and don't let yourself get too frustrated. Many papers are rejected the first time. I have had papers that were rejected three times, but in the end they get published and they are always better after going through the review process."

Power in PowerPoint

Written communication may be the backbone of science, but verbal communication is its nervous system, providing instantaneous feedback and inputs of new information. The time-honored tradition of oral scientific presentation can be a crucial test of your work's strengths and weaknesses. Some even find that considering how to tell the story of their research as a presentation can provide helpful organizational clues that can in turn improve their written paper.

Although it may seem more intimidating than writing, presenting your work to your peers usually follows a natural progression that allows you to build confidence—beginning with presenting to your classmates and then to your scientific society student subunit, state or local chapter, regional division, and, finally, national society or even international scientific meetings. Posters, especially at meetings that include lively poster sessions, are a great way to learn how to concisely present your most important results and field questions about your work. When you move into oral presentations, follow the conventional wisdom about how many slides will work within your timeframe and how much text will fit on each slide while still being legible from the back of the room. Practice your timing and avoid reading your slides to your audience. Finally, watch presentations by seasoned professionals, take note of how they make their points, and compare what does and does not work well.

Courtney Saari practiced her presentation in front of her labmates and professor and received many good tips. But a lack of sleep combined with too much coffee and the dry meeting room air led to a serious case of the jitters on the day of her presentation.

"I continued to drink coffee up until my presentation and even during my presentation because my throat felt dry," Saari says. "What went wrong was the caffeine, and my nerves combined to make my heart beat extremely fast, make my hands shake, and cause me to talk very fast. For me, my presentation felt like it was going in slow motion and as if my heart was going to jump out of my chest. Unfortunately for the audience, I sped through my background slides and spent too much time talking about my graphs and not using the laser pointer because of my shaky hands."

Even when things do not go completely according to plan, presentations can still have a lasting impact. In Saari's case, her presentation sparked a former professor's interest and they had a great conversation over lunch about possible explanations for her results. Subsequent presentations were much more successful; she learned a great piece of advice about having a glass of cool water and taking a sip at natural pauses in the presentation. This pause also helps refocus the audience's attention on the speaker. Saari also learned to control the natural upward inflection in her voice at the end of sentences, which made her statements sound like questions. From observing presentations by other students and young professionals, she realized that often they were so excited to have found significant results that they neglected to allocate enough time to discuss the meaning of those results.

From Elevator Speeches to Open Houses

Formal communication such as publications and presentations, although vital for many careers, makes up only one small aspect of communication. One of the first encounters with informal professional communication many young professionals face is during their transition from students to teachers, whether they are educating undergraduates or even younger students, or members of the general public. Presenting sometimes very complicated scientific principles, data, and statistics to a nonscientific audience requires finesse. You must be able to present data in a simplified fashion so a lay audience can understand the results and their importance.

Emily Lescak, a master's student at the University of Alaska, was asked by her advisor to put together a workshop for a group of native Alaskan students. The goal of the workshop was for the students to learn how to extract DNA from a fin clip and then perform a polymerase chain reaction assay to sex genotype the fish. Along the way, the students would acquire basic understanding of the central dogma of biology and genetics.

"I realized that some of my verbal explanations were too cumbersome, confusing, and complex," Lescak says. "I wish that I had thought more about how to convey complicated information to my specific audience. I think I had become too accustomed to communicating with fellow scientists with whom I can use technical jargon without the fear of misunderstanding."

Lescak and her advisor developed a lecture component with animations to explain the role of the reagents and the significance of each step in a protocol. Otherwise, the many steps involved may have seemed arbitrary and unimportant. Sometimes visual communication can clarify a complex subject or task in a way that oral communication cannot. For example, think about if you had to read instructions on how to tie shoelaces rather than just watching someone demonstrate it.

"Animations and virtual laboratories that are done well can be great tools to teach students new protocols and the underpinnings of complicated processes," Lescak continues. "I also think that it is important to know the audience and explain new concepts using examples that they will understand."

Other popular ways to introduce new concepts include board games, infographics, and "a day in the life of" stories. Making a scientific story accessible to a general audience does not necessarily mean dumbing it down. Even complex concepts can be explained if you avoid jargon in favor of more relatable words—perhaps an invasive species is like a bully or a tailwater is like a washing machine.

Outside the classroom, communicating scientific concepts needs to be even faster and more simplified. Natalie Sopinka, a Ph.D. candidate at the University of British Columbia, tested her skills in telling a simple yet engaging story about her research when she joined a competition that challenged graduate students to summarize their theses in only three minutes. The competition, called Three Minute Thesis or 3MT, requires graduate students to develop a three-minute speech that summarizes their theses in a way that a lay audience could understand. With only a single PowerPoint slide to accompany the speech, this challenge forces students to be as concise as possible while also creatively telling a compelling story with their data.

"This contest forces you to think outside of the box," Sopinka says. "A lot of times in science we are told that we must follow the same layout—introduction, methods, results, and discussion—and no one ever tells you that it should tell a story. That was one of the things that we were told to do: to tell a story about your research in everyday language. It was fun because we were allowed to be more creative in discussing our research, and there is little room for creativity in the publishing realm."

Sopinka's single slide, featuring the open maw of a giant grizzly bear, developing embryos, and a majestic pair of spawning fish, clearly depicted the complex, magnificent, and stressful life of her study species, Sockeye Salmon *Oncorhynchus nerka.* She finished in first place both in the Faculty of Forestry and at University of British Columbia's campus-wide finals. Sopinka's experience with 3MT, which included three rounds of public speaking, significantly increased her confidence and helped her learn how to clearly explain her research to a lay audience.

Sopinka encourages emerging professionals to try to summarize their work in 30 seconds (or even just one sentence) so that when they meet their academic hero in the elevator at a conference, they will have a concise and interesting introduction to their field. She also recommends practicing presenting research to a general audience, such as friends and family.

"Feedback from friends and family helped me," Sopinka said. "My parents didn't even know what 'anthropogenic' meant. I really had to think differently about how to present my science."

For Sarah Wheeler, a joint doctoral student at San Diego State University and the University of California Davis, public outreach meant having the public in for a visit. She noted that the new marine laboratory at San Diego State was not well known, even on campus, especially when compared to the well-established Bodega Marine Laboratory at Davis. A university student group she helped co-found, the Marine Ecology and Biology Student Association, embarked on a mission to foster a sense of community, taking on projects such as creating a new Web site for the laboratory, along with a logo, e-mail list, and seminar series. But building awareness among the broader community beyond the campus required even more coordination and teamwork to host the laboratory's first-ever public open house.

Students organized interactive activities among eight laboratories, attracting more than 300 people, who in turn contributed more than US$1,700. Donated catering and raffle items enlivened the festivities, and the laboratory itself was transformed into a more visitor-friendly environment by adding educational posters about marine habitats and display tanks for live organisms.

"This was the first time that we got some feedback from faculty, who were just so proud to work there. They felt like they were giving back to the community," Wheeler says. "We've been talking a lot more to educators with experience in the teaching side.... What are our goals for our audience to learn when they come to the laboratory?"

Wheeler plans to give pre- and postactivity questionnaires during the next open house to measure communication goals for educating lab visitors about the impact of human actions on the environment. In fact, many types of fisheries research projects lend themselves to this kind of public outreach and education. By working with your organization's education or communication department from the beginning, you can achieve quantifiable results for your outreach and education efforts, with opportunities to publish results similar to your main research focus.

Fast Forward

In the social media age, today's technology is driving communication forward at an exponential rate. This provides a rare opportunity for younger professionals to more quickly establish themselves in their chosen field. When new forms of communication can blossom and become nearly ubiquitous in only a year or two, younger professionals may have an edge over some older professionals in taking advantage of these new outlets and understanding their cultural norms. However, these days everyone needs to be more flexible in their approach to communication and broaden the reach of their communication efforts. More and more, professional networking takes place through informal social networks, transforming these media from fun pastimes to ways to broaden your knowledge, raise your profile, promote your work, and correspond with colleagues (and best of all, they are still fun!). Our only caution would be not to confuse speed with haste and to use high professional standards in all of your communications.

A successful social media presence is built on interesting content, frequent posting, and a unique voice. Posts with visuals, tasteful humor, and the potential for interaction have the widest reach, but developing a larger following mostly takes hard work and consistency—qualities you want to demonstrate anyway.

Whether you are writing a 50-page paper or a 140-character tweet, remember who your audience is and how your story relates to them. Although the format of the stories has changed, the art of storytelling is still a crucial component for communication success. Fisheries science touches on environmental, economic, and social issues around the world, all of which can be woven into your story to show its relevance. If you want both your career and scientific projects to advance, building relevance through consistent communication is a vital key to success.

Acknowledgments

We would like to thank Nick Lapointe, Emily Lescak, Hilary Meyer, Courtney Saari, Natalie Sopinka, and Sarah Wheeler for volunteering to share their stories. We also thank Steve Cooke for helping to arrange an introduction so we could work together on this project and the reviewers and editors for their many thoughtful and helpful suggestions.

Biographies

Elizabeth Beard is the digital content and engagement strategist for the American Fisheries Society in Bethesda, Maryland.

Samantha Wilson recently received a master of science from the Fish Ecology and Conservation Ecology Laboratory at Carleton University in Ottawa, Canada.

Recommended Reading

American Fisheries Society. 2013. A guide to AFS publications style. American Fisheries Society, Bethesda, Maryland. Available: http://fisheries.org/docs/pub_stylefl.pdf. (January 2014).

Barrett, S. C. H. 2010. Some tips on how to succeed at graduate work and in writing papers for publication. Available: http://labs.eeb.utoronto.ca/barrett/pdf/How%20to%20write%20a%20paper.pdf. (January 2014).

Bik, H. M., and M. C. Goldstein. 2013. An introduction to social media for scientists. PLoS Biology 11(4): e1001535.

Bourne, P. 2005. Ten simple rules for getting published. PLoS Computational Biology [online serial] 1(5):e57.

Clapham, P. 2005. Publish or perish. BioScience 55:390–391.

Jennings, C. A., T. E. Lauer, and B. Vondracek, editors. 2012. Scientific communication for natural resource professionals. American Fisheries Society, Bethesda, Maryland.

Knuth, B. A. 2012. Expanding the reach of fisheries science and management through strategic social networking. Plenary presentation to the 2012 American Fisheries Society Annual Meeting, St. Paul, Minnesota (video). American Fisheries Society, Bethesda, Maryland. Available: http://vimeo.com/53687056. (January 2014).

Powell, K. 2010. Publish like a pro. Nature 467:873–875.

For even more resources, see the following Pinterest boards: www.pinterest.com/elbeard/scientific-communication/, www.pinterest.com/elbeard/social-media-tips/, and www.pinterest.com/elbeard/writing-tips/.

Creating Professional Networks for Successful Career Enhancement

T. DOUGLAS BEARD, JR.*
National Climate Change and Wildlife Science Center, U.S. Geological Survey
MS-400, 12201 Sunrise Valley Drive, Reston, Virginia 20192, USA

Key Points

- Successful professional networks can be a key component of career advancement.
- Creating these networks involves active engagement, volunteering, leadership, and dedication to your profession.

Introduction

Professional networks are a critical component of a successful career. Creating a professional network is an important task for emerging fishery professionals to accomplish. Professional networks not only lead to possible career advancement, but also create lifelong friendships that make working in the fisheries profession rewarding. I can point to multiple career opportunities (jobs, leadership in international organizations, research projects, and leading study abroad courses) that have been a result of growing a strong professional network. A key role for mentors is to help emerging professionals develop and nurture professional networks. One of the most rewarding approaches to ensuring a successful career is to join and become involved in your professional society by volunteering for work group activities; attending local, regional, and national meetings; developing and presenting symposia and papers; and serving as an elected official for various components of the society. Professional service has contributed greatly to my career success, and many of my closest friends and colleagues have resulted from my volunteer contributions to the American Fisheries Society (AFS).

My first AFS meeting was with the Central Pennsylvania Chapter of AFS. I had the good fortune of working on my M.S. degree with Dr. Bob Carline at Pennsylvania State University, and he was responsible for my initial involvement in these meetings. He encouraged me, very early on as a student, to develop a poster for the local chapter, outlining my proposed thesis project and using the chapter meeting as a way to solicit feedback on the research and, probably more importantly, to start building a local network that I could call on to help implement my project. My master's project required the involvement of multiple volunteers to help locate and identify trout redds though the entire length of a stream, something I could not have done on my own. By presenting a poster and participating in local chapter activities, I was able to explain my project in detail, find others interested in my research, and find volunteers to help on my project, allowing me to complete my degree in short order.

* Corresponding author: dbeard@usgs.gov

The nascent network I started building during my master's project taught me many lessons about engagement that I carried with me to my first professional job at the Wisconsin Department of Natural Resources (WIDNR). I made it a point to attend the Wisconsin Chapter of AFS meetings and to almost always give a science presentation to the local chapter. Engagement in the local chapter not only yielded important feedback on the science that I was working on, but also brought a sense of broader engagement with other fisheries professionals in the state of Wisconsin. In turn, this interaction helped me develop close friendships and partnerships that allowed me to be successful during my time with the WI DNR. For example, I was able to work with other fisheries professionals in the state to develop and publish a number of manuscripts on Walleye *Sander vitreus* management that not only advanced our understanding of the system, but also built trust and respect among the fisheries professionals working on this fishery within the state. Active engagement in local chapter activities also led to engagement at the North Central Division level and, eventually, leadership in the parent society.

Building Networks at Multiple AFS Levels

As I look back at my career, it is hard to explain how engagement at multiple levels of AFS led to professional career opportunities. Networks tend to grow if you are at all active in a professional society. However, I can point to my transition from the WIDNR to the U.S. Geological Survey as directly attributable to connections I made at AFS. I can point directly to my becoming president of the World Council of Fisheries Societies, which led to a six-week fellowship at the University of Tokyo as a result of the network I built through AFS involvement. Participation in multiple graduate committees across the nation is also a result of connections I have made through AFS. However, like many fisheries professionals, my attendance at AFS meetings over the years usually required active involvement in the meeting. I did that through three mechanisms: (1) development and teaching of continuing education courses, (2) development of symposia, and (3) submission of presented papers. All avenues were very rewarding, and certainly as an emerging professional, it is critical to submit papers for presentation at all levels of AFS meetings.

Presenting at conferences is the easiest and perhaps the quickest avenue to engage broad groups of like-minded professionals in your research, solicit feedback, and make a case for future employment (or building that next step in your own research program). Presenting at AFS meetings is incredibly critical for those that are job hunting or trying to build professional science networks. Presentations are the first step in generating name recognition; you never know which potential employer might be sitting in the audience or browsing through the abstracts. Giving presentations shows your leadership and engagement in science.

However, I believe moving beyond submission of papers and helping to organize components of an AFS meeting really allows you to get directly engaged in advancing the fisheries profession. Organizing activities allows you to identify and contact other fisheries professionals, giving you an opportunity to show your leadership skills. Often, being in a leadership role for an activity can be used to generate travel support for attendance at meetings, as agencies are very willing to invest in potential future leaders. I have organized components of AFS meetings through the creation of continuing education courses on the use of fisheries models and the use of microcomputers in fisheries (yes, there was a call for that in the early 1990s!) by participating in the organization of symposiums on fish habitat issues, inland fisheries, and climate change and fisheries, and I once took on the organization of the entire fisheries program for the Midwest Fish and Wildlife Conference. Moving beyond the submission of papers allows you to focus the science at a macro level and set the science agenda in a direction that provides broad interest, develops your professional network in new and interesting directions, and actively advances the science and management of the fisheries profession. Exploring areas beyond your expertise into broader scientific topics forces you to branch your network out to other areas of the profession, move outside your initial comfort zone, meet new professionals, and find out about other activities going on around the country. Additionally, if you have organized your symposium well, you can learn a lot of new information and science in an incredibly short period of time. For example, my leadership in the development of the Midwest Fish and Wildlife Conference fisheries program in 1997 really helped me focus my Ph.D. as curiosity about other aspects of ecological systems, outside my area of fisheries expertise, led to a deeper interest in social-ecological systems. Through my Ph.D. connections, I became engaged with the Resilience Alliance and, ultimately, the Millennium Ecosystem Assessment (to this day, a highlight of my career). Today, I am still engaged as the official U.S. science lead for the Intergovernmental Platform on Biodiversity and Ecosystem Services, all as a result of allowing the development of a fisheries program to lead me in new science directions and to new science connections. Nothing is more scientifically enriching than seeing the breadth and depth of fisheries science that is contributed at AFS meetings and using those ideas to inform your own career. Trying to duplicate the amount of information presented in one AFS parent society meeting would require a very long semester course in fisheries science!

Volunteering Helps Drive Network Development

AFS, like most professional scientific societies, is run largely by an elected leadership of volunteers. Although AFS is managed by an executive director, the governance decisions, the priorities of the society, and the way in which the society is organized and completes its work are all determined by an elected group of volunteers. Professional scientific organizations focused on resource management in the United States have small staffs largely dedicated to membership and publishing issues, but the lifeblood are the committees that advance a number of initiatives. In AFS, there are numerous committees ranging from membership (e.g., Membership Concerns Committee) to administration (e.g., Time and Place Committee) and science (e.g., Centrarchid Committee). My first volunteer activity was with the parent society of AFS in the early 1990s as part of the Membership Concerns Committee. In the early 1990s, I expressed my willingness to be a member of a national AFS committee, not knowing exactly what the expectations would be. I was asked to be on the Membership Concerns Committee, and most importantly, I followed through on the work that I was asked to contribute (i.e., I was not a silent committee member). Our committee goal was to put together a situational ethics survey of members to understand what AFS membership thought were ethical approaches to many issues (Membership Concerns Committee Situational Ethics Workgroup 1993). Working with other committee members, we designed the survey and structured the report in a way that allowed us to publish the results in *Fisheries*, a good thing for a new scientist. Of course, my first experience on an AFS committee was very positive, and I am not sure how many AFS committees, at all levels from local to national, I have been on over the years. I would like to believe that my early contributions to AFS were recognized in a way that led to further opportunities to build my professional network (i.e., become known as someone who would follow through). The key to being successful in building networks through volunteer activities is to actually participate, not volunteer to be on a committee and then disappear. Every one of these volunteer experiences allowed me to expand my professional network, learn about other components of the greater fisheries community, and generate information and activities that were beneficial to the members of AFS. It is easy to volunteer for AFS committee activities today: simply go to the AFS Web site and find the committee volunteer form and submit it. If you do not hear back from AFS in a timely fashion, send a note to the president or president elect letting them know your interest. Showing initiative will always help you get engaged in society activities.

Obviously, the first activity you do with AFS will likely not be running for an elected office, but if you have been actively engaged in the society, one day you will be asked to take on a leadership role. Only say yes to these opportunities if you truly have the time to devote to the duties of running, leading, and pushing forth the mission of a component of the society you are asked to lead. Absentee leadership can lead to a loss of respect and others in the society questioning your commitment, and may leave fellow professionals wondering if they can trust you to get the job done. Perhaps as important, it is not going to lead to more network connections. Be honest about your time commitments, especially when starting your career. Your initial focus should be on establishing your identity as a professional, and you should only volunteer for activities that directly overlap with your interests and abilities. As you progress in your career, you will be better able to expand your commitments and move beyond your initial volunteer interests. I have had the privilege to be president of two sections of AFS, the Computer User Section (now the Fisheries Information Technology Section; the connections here led to my career at the U.S. Geological Survey) and the International Fisheries Section (which lead to my engagement on the World Fisheries Council). Being elected to leadership of a committee, a chapter, a section, or the parent society demands time above and beyond that of just working on a committee activity or pulling together a symposium and generally comes about because your professional network recognizes your ongoing dedication and contribution to society activities. Elected leadership usually is for a two-to-five-year time period and requires you to work nights and weekends visualizing, implementing, and moving forward an agenda, and, perhaps the hardest thing, finding other volunteers to help you. You will be expected to work with other elected leaders on the governance of the society, actively participating and representing the individuals that are members of your component of AFS. As president of a section of the parent society, not only did I have to arrive at the parent society meeting two days early to sit through the governing board meetings, I was also expected to attend a two-day midyear governing board meeting. Governing board meetings are smaller, consisting of leaders from all over the nation, and are really a place to expand your network (as opposed to wandering the halls of the 1,500+ attendee AFS meeting hoping to meet someone).

However, just attending governing board meetings is not enough; you will be expected to arrive at these meetings fully informed about the decisions that board will make and have the ability to represent the members of your section and effectively engage in debate about the future of the society. Component leaders need to canvass section members on issues before the governing board, identify the position of the group they are representing, and work with other members of the leadership team in their part of the society to formulate their position on issues prior to the governing board meeting. You have to be able to listen to often-diverging opinions and work to find consensus positions for the issues at hand. All too often, I have sat at a governing board meeting exasperated by the lack of preparation by members of the board. How can you make informed decisions if you have not done the homework? Further, at some point during the AFS meeting, you will be expected to chair a meeting of your section, including development of an agenda, leading the membership through decisions and discussions and reporting on the decisions you have made on their behalf during the governing board meetings. Leadership positions are not to be taken lightly, and I encourage anyone that is interested in being a part of the elected leadership to attend governing board meetings prior to agreeing to run for an office. Get a feel for the decisions that are being made, the process by which the meeting runs, the issues that might face the board in the future, and what you might add to the discussion if you are elected. It was a privilege to be elected president of two sections, and I took it as an honor to represent others in the society. I still call on many members of the governing board of the sections I represented or the AFS governing board for help on everything for job recommendations (both giving and receiving), program management advice, and engagement in the program I currently run. Treat an elected opportunity as an important responsibility, not a chore, and you will be rewarded both personally and professionally for your volunteer efforts.

Professional Networks and Personal Friendships

I regularly present an invited talk to university students called "From Process to Function: Evolution of a Fisheries Biologist." Among the main themes of this talk is the huge role my engagement in AFS has had in my career success and the development of an incredibly rewarding set of friendships. I met my wife and my best man at AFS parent society meetings (not at the same one!), many of my lifelong friends are individuals I have met through attending and being involved in AFS, and I have managed to parlay my AFS activities into opportunities to learn about fisheries all over the world. My network of AFS professionals have helped me find new jobs and new opportunities for leadership at national and international levels and has even allowed me to meet both Emperor Akihito of Japan and the Prince of Wales (Prince Charles)—once-in-a-lifetime experiences I will never forget. And this was all achieved because I was actively engaged; I volunteered and followed through. I signed up to lead and expand my career horizons. So, sign up. Go to the AFS Web site, pick a committee that seems interesting, and volunteer. Then, follow through. Do the work. You just never know where it will lead you!

Acknowledgments

My career success would not have occurred without the wonderful mentorship of Bob Carline, Lee Kernen, Mike Hansen, Mike Staggs, Steve Carpenter, Bob Szaro, Bruce Jones, and Matt Larsen. I have been fortunate to have such wonderful mentors. My career would not have been as rewarding without my friendships with Dirk Miller, Don Pereira, Bill Taylor, Dale Burkett, Chris Goddard, Andy Loftus, Doug Austen, Craig Paukert, Jodi Whittier, Nancy Nate, Steve Cooke, Robert Arlinghaus, Ian Cowx, Gary Whelan, Glen Contreras, Shugo Watabe, Barb Knuth, Bob Gresswell, Stu Shipman, and numerous others from the fisheries profession. May there be many more aquarium socials in our future!

Biography

Douglas Beard is the director of the U.S. Geological Survey's National Climate Change and Wildlife Science Center. Doug has worked for the U.S. Geological Survey (USGS) for 10 years, starting with management of the fisheries and aquatic resources program. Prior to coming to USGS, he was a staff fisheries biologist for the WIDNR for 13 years. Doug is currently president of the World Council of Fisheries Societies, which consists of the world's major professional fisheries research organizations, and has been engaged in numerous AFS activities since 1987. Doug has authored or coauthored 37 peer-reviewed publications and coedited two books on fisheries.

References

Membership Concerns Committee Situational Ethics Workgroup. 1993. Should we eat those fish? A situational ethics survey of AFS members. Fisheries 18(2):19–23.

Interviewing Strategies and Tactics for Success

HENRY (RIQUE) CAMPA, III*
Department of Fisheries and Wildlife, Michigan State University
480 Wilson Road, 13 Natural Resources Building, East Lansing, Michigan 48824, USA

ALEXANDRA LOCHER
Biology Department, Grand Valley State University
212 Henry Hall, 1 Campus Drive, Allendale, Michigan 49401, USA

Key Points

Researching a potential employer and reviewing your materials will enable you to effectively communicate with diverse audiences during academic and nonacademic interviews.

- Plan and prepare for the interview by asking effective questions.
- Research and review your materials, communicating your qualifications to potential employers.
- Take care of yourself so you can be prepared for varied and unexpected experiences during the interview.

Introduction

When preparing for an interview, remember potential employers generally like to know three things about you: what are your professional goals, how will this position contribute to you meeting those goals, and, last, what will you do to contribute at the various levels of the institution, agency, nongovernment organization, or company. So, what can you do to have a successful interview? It all comes down to preparation and planning.

When should you start preparing for a potential job interview? You just submitted your materials a few days ago. How should you start preparing? To prepare, what materials should you investigate? Whom should you talk with about the position? What materials do you want to take to the interview if invited? Do these questions sound familiar? If so, do not worry as you are not alone. They are questions many job candidates ask or should be asking of themselves when applying for various positions.

If you have submitted your job application materials, in all likelihood you have already done some of the crucial work that will prepare you for the interview. You may have researched, prior to applying for a position, the institution, agency, or nongovernment organization offering the position, possibly by checking out its Web site. You likely dissected the position description and identified the mission of the potential employer and the primary re-

* Corresponding author: campa@msu.edu

sponsibilities associated with the position. Based on the position description, you may have modified your existing resume or CV and cover letter to emphasize your experiences and skills in relation to the position responsibilities. All the steps that you have taken to complete your application means you have already started preparing for a potential interview. Whew! Preparation is the key to having a successful job interview and usually requires a significant time commitment, so plan accordingly.

The goal of our chapter is to provide guidance on how you can prepare for a successful interview, whether you are pursuing a full-time academic or nonacademic position. We based our recommendations on our collective professional experiences from working in academic and nonacademic positions; serving on search committees; information gleaned from colleagues over the years who have worked in various types of academic institutions, agencies, and nongovernment organizations; and the work of others who have studied the professional development needs of individuals seeking professional positions as well as what employers want (see Recommended Readings).

Understanding the Interview Structure

What do we mean by the "interview structure"? Knowing the structure of an interview is crucial, especially if the position you are applying for is with an agency or nongovernment organization. Occasionally, these types of employers will have structured interviews, which means you may be interviewing for one to three hours with a search committee composed of multiple people (e.g., your potential supervisor, an upper-level administrator, and someone who would be a peer within the agency or organization). During this time, you might be asked a standard list of questions by the respective committee members and they will sit, listen, and take notes on your responses. Generally, you will not be asked to provide further details on your answers. In this format, there is usually minimal to no dialog between the candidate and the committee. These types of interviews can be intimidating if you are not ready for them—mainly because you do all or the majority of the talking. We do encourage applicants to take notes during these types of interviews so that they can address or clarify questions that were vague. Additionally, taking notes allows applicants to gather additional information about the nongovernment organization or agency that they may be able to ask about at the end of the interview.

For interviews with most academic institutions, whether at a research intensive institution, comprehensive university, or a liberal arts college, these will usually be full-day or multiple-day events requiring you to present a research seminar and/or teach a class and meet with numerous individual faculty members, administrators, groups of faculty, groups of students, and staff members. During these sessions, expect a significant amount of dialogue. Many meetings will be with individuals who are not from your discipline, which can be challenging. So being able to explain what you do, how you do it, and what your interests are to multiple types of academics will be important. For this reason, it is important to know whom you will be meeting during the interview. Do some research about their backgrounds and interests; they will certainly be researching you.

Ask the Right Questions Before the Interview

When you receive an invitation to the interview you need to know a priori what questions you will ask the potential employer to help you effectively prepare. It is also important to know what questions you should not ask. Remember, only ask those questions that you cannot find out through your own investigations. Otherwise, asking these questions may seem trivial to the person inviting you for the interview. Can you think of a few questions you might need to ask the Search Committee chairperson for an assistant professor of fish ecology position at a research intensive university or for a fisheries biologist position with a state agency? Below are some of the questions that could complement those on your list to help you prepare for a successful interview.

What will be the structure of the interview?

Will I receive a schedule (e.g., with a list of individuals with whom I will be meeting) for the interview? If the answer to this question is "No," we suggest that you probe to get at least a draft of your schedule (even without the list of individuals, this may not be finalized) so that you know where you are expected to be when.

Where will the interview take place, what time, what day, and is there a phone number I can get in case something comes up the day of the interview or while I am traveling (just in case you have car trouble, miss a connecting flight, get sick, etc.)*?*

How would you describe the composition of students attending the university or college and those within the department (e.g., proportion of community college transfer students, international versus domestic students, and number of students within the major) (for academic positions)*?*

Who would be my immediate supervisor in this position and how are departmental duties assigned among employees (asking this question may help clarify the hierarchical structure of the agency or organization)*?*

Asking effective questions when asked to interview for a position can help you accomplish three goals to help you secure the job of your dreams. First, asking questions can demonstrate your genuine interest in the position and institution/agency to your potential future employer. Second, asking these questions allows you to be proactive and will help determine how well your education and experience match the duties and responsibilities of the position. Last, these questions may help you assess whether or not you are interested in working in this position.

Review, Review, Review—Did We Mention Reviewing?

In addition to the work you did preparing your job application materials, it is wise to review (1) the position materials, (2) your application materials and work experiences, and (3) the findings from your research efforts. Reviewing these three packages of materials may provide you with more questions you would like to ask during the interview and bring up other questions that you need to research further. For example, if you were applying for an academic position, does the position description mention if it is an academic-year or annual-year position? Is it with a land grant institution? Do you know what a land grant institution is? Are you expected to bring in extramural funding or lead research experiences for undergraduates? Is it a tenured track position or fixed term? Or, if you are applying for an agency position, does the description mention if it is a supervisory position? Will you be employed under a union? With whom will you be collaborating on a day-to-day basis? Other biologists, stakeholders? What will be your primary responsibilities? These are just a few items that might appear in position descriptions that would need additional attention and research prior to your interview.

Last, review your preparation materials and review them often. Rique Campa was recently listening to a seminar by a candidate for an administrative position. The candidate was describing how his recent administrative accomplishments paralleled nicely with the mission of the academic institution in which he was hoping to work. Much to the surprise of the candidate, only days before the interview, the university had adopted a new mission and a corresponding set of new initiatives.

Reviewing the mission of your potential employer and who you will be meeting with during the interview is crucial, as your knowledge of the institution or agency will help guide how you will answer questions. For example, if you are interviewing for a primarily teaching-related position at a liberal arts college and a dean asks you to summarize why you are interested in the position, you probably do not want to answer the question by describing how you would like to continue doing research related to your dissertation. Instead, talk about your teaching interests, your experiences working with undergraduates on class projects, or your use of case studies to promote an active and collaborative learning environment. Knowing background, interests, and positions of the personnel you will be meeting during the interview demonstrates your knowledge of the expertise within the institution or agency and may prepare you well for answering questions related to how you might collaborate with other individuals.

Use Your Experiences

If a potential employer has decided to interview you, the search committee obviously saw something in your application that indicated you were qualified for the position. Maybe it was your educational background, perhaps your prior work experiences, or maybe your references gave you rave reviews. More than likely it was a combination of these and perhaps other factors.

Keep in mind that your application illustrated your experiences enough to land you the interview, but do not assume that everyone you will meet during the interview has carefully read your application materials. Therefore, practice quantifying and accurately describing your work experiences or providing demonstrable evidence from your experiences so that interviewers are convinced you are the right person for the position. For example, suppose one of the qualifications in the job description specified that the successful candidate "should have an interest in collaborating with university faculty and natural resource management agencies." While preparing for a likely interview question asking about your interest in collaboration, which of the following responses would make a greater impact?

"I am greatly interested in collaboration with various individuals because I feel that it could enhance the quality of my work and increase communication and work flow within the department."

"I agree that collaboration would be an important part of this position to facilitate communication and productivity. In fact, over the past year, I have collaborated with two researchers from *X* institution and an agency biologist on *Y* project. Our productivity resulted in the publication of a manuscript and development of an additional project."

The second response is much stronger, as it provides specific evidence from experiences showing that you are capable of meeting the qualification.

Do Not Experiment on the Big Day

If you do not drink espresso, starting the day of the interview with one is not a good idea. If you have never had sushi, you might not want to have it the night before the interview. If you usually run first thing in the morning to get you mentally and physically prepared for the day, then try to squeeze in a run before the interview. If you are usually in bed by 10:00 p.m., then try to go to sleep the night before your interview by 10.

We recall one instance during a faculty interview that turned detrimental for a candidate who changed her daily routine on the day of the interview. This individual normally was a breakfast eater but skipped breakfast to prepare for the day. When the candidate was about to begin her seminar, her face turned pale and she passed out on the floor. After regaining consciousness and adjusting her crooked glasses, she commented awkwardly that it probably was not a good idea to skip breakfast and dinner the evening before! Our point is, on the day of the interview do not experiment with new things. You need to take care of yourself and try to minimize distractions.

Be Prepared for Inappropriate Questions

Inevitably, during an interview, someone at some point will ask you an inappropriate/illegal question. These inappropriate questions may include topics such as kids, marital status, political or religious views, family descent, or questions related to medical history. These topics are considered inappropriate for discussion at interviews because of federal, state, and sometimes local laws. The inappropriate nature of the question may not be intentional; it may evolve from a conversation you were having. For instance, during an interview, one of the interviewers may mention how proud he is that his son hit a winning home run during last night's baseball game and then, informally, ask if you have any kids. Alternatively, the inappropriate questions may be intentional, as the interviewer is trying to glean additional background information about you. In either case, if or when asked an inappropriate or illegal question you have a few options. (1) You can briefly answer the question anyway, if you are comfortable doing so. (2) You can respond to the intent of the question. For instance, if you were asked whether or not your job decision depends on a spouse, you may reply that you feel you will make the best decision to fulfill your personal and professional goals. (3) You may choose to ignore the question and change the subject. (4) You may refuse to answer the question and indicate that the question is irrelevant to the specific requirements of the job.

Of course, there may be repercussions with any one of the responses, and as a job candidate, you must be ready to accept them. For example, providing or intentionally giving the interviewer personal information about children or a spouse may offer an opportunity to learn about daycare or potential employment for your spouse. Alternatively, the interviewer may interpret your personal situation as a potential distraction to your productivity. Your choice of responses may not provide the information requested by the interviewer, but the tone of your responses may indicate something about your personality and composure. Perhaps, if you answer with an irritated or arrogant tone, you may anger or embarrass a key person that may lead to you not being selected for the position. If a situation occurs where you are presented with an inappropriate or illegal question, evaluate the context of the question and decide whether the inappropriate nature was unintentional or deliberate and why, and then decide your course of action. Use the experience to help you judge whether this position is still the right one for you. Proceed with caution regarding what personal information you willingly provide during an interview.

Conclusions

Remember, a potential employer likes to hear three things during an interview: what your professional goals are; how this position will contribute to your meeting those goals; and what you will contribute at various levels of the institution, agency, nongovernment organization, or company. By remembering a few themes about successful interviewing, you should feel confident in securing the position of your dreams. Remember, your primary goal for the interview is to get a job offer! Researching the institution or agency will help you ask the right questions before the interview. Reviewing your application materials and the position requirements will help you identify and use your experiences as demonstrable evidence of your ideal fit for the position. Do not use the interview as an opportunity to try new espressos or daily routines. And finally, be prepared for inappropriate questions or potentially awkward situations that may arise during the interview. Your ability to remain collected and professional during these stressful times will increase your confidence and enhance your awareness about the job environment in which you may be working. Good luck!

Biographies

Henry (Rique) Campa, III is an associate dean in the Graduate School and a professor of wildlife ecology at Michigan State University. Teaching and research interests include wildlife–habitat relationships, effects of disturbances on wildlife and wildlife habitat, and the professional development of graduate students. His prior work experiences include positions in the Michigan Department of Natural Resources Wildlife Division and the U.S. Fish and Wildlife Service.

Alexandra Locher is an assistant professor of natural resources management in the Biology Department at Grand Valley State University. Teaching and research interests include wildlife–habitat relationships, responses of wildlife to forest management activities, and applications of remote sensing and geographic information systems to natural resources management. Ali's first faculty position was in the School of Forest Resources at the University of Arkansas at Monticello.

Recommended Readings

Fiske, P. S. 2001. Put your science to work. The take-charge career guide for scientists. American Geophysical Union. Washington, D.C.

Gray, P., and D. E. Drew. 2008. What they didn't teach you in graduate school. 199 helpful hints for success in your academic career. Stylus Publishing, LLC, Sterling, Virginia.

Howard Hughes Medical Institution and Burroughs Wellcome Fund. 2006. Making the right moves: a practical guide to scientific management for postdocs and new faculty, 2nd edition. Burroughs Wellcome Fund, Research Triangle Park, North Carolina and Howard Hughes Medical Institute, Chevy Chase, Maryland.

Influencing Your Agency's Thinking

ROBERT F. CARLINE*
U.S. Geological Survey (Retired)
123 Gibson Place, Port Matilda, Pennsylvania 16870, USA

DAVID A. LIEB
Pennsylvania Fish and Boat Commission and Western Pennsylvania Conservancy
450 Robinson Lane, Bellefonte, Pennsylvania 16823, USA

Key Points

- Influencing your agency's thinking will require considerable time and energy.
- New ideas are often met with skepticism.
- You must convince agency staff and stakeholders that your initiative is important.
- Be prepared to repeatedly tell your story using a variety of technical and non-technical media.

Introduction

Can a young biologist actually influence the thinking of an agency? One's first reaction might be "no way." But perhaps it is possible. Below, we each share an experience demonstrating how, despite setbacks and challenges, we were eventually able to influence our agency's thinking. We hope our experiences will prove useful to the next generation of agency biologists.

Bob's Story

My fisheries career began in 1967 after completing an M.S. in fisheries at Oregon State University and taking a job with the Wisconsin Department of Natural Resources Coldwater Research Unit. I was a project leader assigned to assess the biological responses to dredging of small, spring-fed ponds that supported wild Brook Trout *Salvelinus fontinalis* and Brown Trout *Salmo trutta.* It was a great research opportunity, and I was excited.

After about one year on the project, I decided that it would be really useful for the agency to do an economic analysis of this management program. After all, it was a costly program and the agency should be able to justify these expenditures on the basis of economic returns resulting from improved trout fisheries. While the need for this kind of analysis may have had some merit, my approach to convincing others was way off base.

Armed with one graduate course in economics, I began delving into the pertinent literature and sketched out a study plan. For some unknown reason, I decided to keep this idea to

* Corresponding author: m-bcarline@comcast.net

myself until I had developed it fully, rather than seeking counsel from my supervisor. This was the first big mistake. My strategy was to unveil my plan during the annual meeting of our fisheries research staff without forewarning anyone—the second big mistake. When my turn on the agenda arrived, I used a carefully prepared set of flip charts to explain the need for an economic study and how it would be done—with, of course, consultation from real economists.

In short, my presentation bombed. My fellow biologists and program administrators looked at me as if they were wondering what type of illegal substance I had been smoking. My recollection is that no one even hinted that my proposal had some merit. My ego was crushed, and I learned what it felt like to go home with one's tail between one's legs.

While trying to heal my battered ego, I began wondering if perhaps my failure was not due to a bad idea, but rather to an ill-conceived approach. I should have discussed the idea with my supervisor. If my supervisor had thought the idea lacked merit, I could have simply dropped it. Conversely, if he had agreed that the idea had merit, then I could have addressed any concerns he might have had and sought his advice on how best to sell it to the administration. Surely he had some experiences from introducing new programs to the administration.

My second realization was that natural resource management agencies do not like surprises, nor do they like new, somewhat unconventional ideas. New ideas are often met with skepticism and the usual responses: "that won't work" or "we tried that once when Truman was President."

After conferring with my supervisor, I could have met individually with a few administrators and introduced them to the proposal. My goal should have been to familiarize staff with the ideas, and, at the same time, seek their advice about how best to execute the proposal. By inserting some of their recommendations into the proposal, it would have been better thought out, and since it embodied some of their suggestions, it would have engendered their support.

A few years later, with the end of my research project in sight, I began to consider the best approach to convince agency administrators to adopt management practices that I was developing on the basis of my research on dredging and altering spring-fed trout ponds. This time, I conferred with my supervisor and explained my strategy, which was to ensure that all recommendations were familiar-sounding practices that administrators and managers had heard about for several years.

My research group met annually with agency managers to update them about the progress of our projects. These meetings were fruitful, but only a small proportion of the managers were in attendance. Our other mode of communication was a monthly research activity report that was circulated to all management offices in the state. Based on letters and phone calls, it was apparent that a substantial proportion of management biologists were reading these reports. My tactic was to more effectively use the monthly reports to communicate with management.

During my first few years on the job, I viewed monthly reports as a bothersome duty. I wrote them rather quickly and detailed the usual uneventful activities: how many benthic samples were collected, how many fish scales were read, and so forth. Despite the boring nature of these drab reports, folks asked questions.

During the last few years of the dredging project, I made sure that each monthly report included some significant finding and identified the management implication of that finding. Often, I would subtly suggest how management actions might change to capitalize on these findings. Most of the conclusions and suggestions were not earth-shattering, but together they constituted a rational management approach.

At the end of the dredging project, I made several presentations to management biologists and administrators and completed a final report. My findings, conclusions, and recommendations were well accepted without controversy. The most disappointing reactions I received were from folks who said, "We knew all of that stuff—nothing particularly new here." There was nothing new because I had been supplying them with the essential information for two years.

The tactic worked. The information that I transferred was received and assimilated. The only downside was that some biologists and administrators did not recognize my work as a new contribution. Nonetheless, most management personnel recognized the value of the work and largely adopted the recommendations.

Dave's Story

I was anxious to make a difference. I had just completed my Ph.D. and had discovered that exotic crayfishes were spreading rapidly across Pennsylvania. They were eliminating native species and damaging aquatic systems. Humans were responsible for their introduction and spread. It was clear to me that something had to be done. To my delight, a position with the Pennsylvania Fish and Boat Commission opened up, and I got the job. I was responsible for a new initiative focused on the management and conservation of Pennsylvania's aquatic invertebrates. Here was my chance to develop regulatory and management measures to slow the spread of exotic crayfishes in Pennsylvania and make a difference for the state's aquatic resources. But how was I going to accomplish that goal? How was I going to convince agency administrators and managers that stopping the spread of exotic crayfish should be a high priority for the agency?

As a newly minted Ph.D., my focus was on research and publishing that research in scientific journals. My initial plan reflected that mindset, and I set out to publish my research findings and management recommendations in a major scientific journal, naively assuming that successful publication would automatically result in action by the agency. Although my manuscripts were published and greeted with interest and enthusiasm by many at the agency, they did not result in regulatory and management action. Moreover, not everyone was convinced that stopping the spread of exotic crayfish should be a management priority of the agency. I had to do a better job of convincing agency personnel of the urgency of the situation. But what could I do?

I consulted with my supervisor and other upper-level administrators in the agency. They suggested that I formally present my findings and recommendations at the next meeting of the agency's commissioners, the policy-making body of the agency. I followed this advice and presented my findings to the commissioners and at a number of other meetings and workshops. My presentations were greeted with a good deal of interest, and although most attendees expressed support for my ideas, there was some skepticism. There were often questions and informal discussions following the presentations. Many of those questions were thought-provoking and forced me to explain my ideas more completely and provide more background than was possible during the actual talks. Presenting the same ideas repeatedly seemed redundant, and it was difficult to listen to myself say the same things over and over again; however, the talks seemed to be effective and my ideas regarding how to regulate and manage crayfish in Pennsylvania appeared to be gaining greater and greater traction within the agency.

Although I was getting closer to achieving my goal, there was still work to be done. I was asked to write an internal agency report formally explaining and justifying my ideas and recommendations. Admittedly, my initial thought was "This is unnecessary—my regulatory and management ideas are already available in the published literature." After more thought, however, I realized that my publications lacked the background needed to fully explain and justify my ideas. I thought back to the questions that I had been asked after my talks and realized that a report was, indeed, necessary. I completed the report, which was circulated among agency personnel and received favorable reviews. Having garnered considerable support within the agency, I was now ready to move forward with implementing my regulatory and management initiatives. I again consulted my supervisor and upper-level administrators.

Although they were also anxious to move forward, they reminded me that new regulatory proposals that lack public support are rarely accepted by the agency. There was also some concern within the agency that the bait industry might resist attempts to regulate crayfish in Pennsylvania. To address this concern, I conducted a survey of the state's bait shop owners. Most of the owners contacted expressed support for regulatory and management actions aimed at slowing the spread of exotic crayfish. I also wondered how Pennsylvania's anglers would view my ideas. Previous interactions with them indicated that, if informed, most anglers would support regulatory and management action, whereas anglers who were not aware of the state's exotic crayfish problem would be less likely to support such actions. The logical next step was to inform Pennsylvania's anglers about the state's exotic crayfish problem. Since most anglers do not read scientific literature, I would have to rely on the popular media. The question was how could I efficiently reach the state's nearly 1 million anglers?

Fortunately, I was afforded space for a short article about invasive species in the *2012 Pennsylvania Fishing Summary: Summary of Fishing Regulations and Laws* (Summary Book). In that article, I described Pennsylvania's exotic crayfish problem. I was excited about the opportunity because each of the state's 800,000+ licensed anglers is given a copy of the Summary Book. As an added benefit, after reading the Summary Book, a newspaper writer became interested in Pennsylvania's exotic crayfish problem and wrote an article about it. The article was circulated to more than 80,000 newspaper readers in the central part of the state. I also published an article about Pennsylvania's exotic crayfish problem in a popular fishing magazine and posted information on the agency's Web site, which is frequented by anglers. I was now satisfied that many of Pennsylvania's anglers had been informed of the state's exotic crayfish problem. Agency administrators were pleased that I had taken the time to reach out to the state's anglers and bait shop owners prior to formally proposing regulatory and management measures.

Although countless hours over the course of nearly two years were spent garnering agency support and educating anglers and bait shop owners, regulatory actions (laws) intended to slow the spread of exotic crayfish in Pennsylvania have been formally proposed and have made it to the highest levels of management. Their formal acceptance by the agency and subsequent addition to Pennsylvania law now seem likely.

Conclusions

Can a young biologist influence agency thinking? Yes, we believe you can, but do not expect it to happen overnight. Seek counsel from your supervisor and from other well-respected staff who understand the culture of the agency and have experience promulgating new ideas and initiatives. You may have to be imaginative in deciding upon the most effective approach. Do not be discouraged if your ideas are not instantly accepted; persistence and patience will be required.

Biographies

Robert Carline spent the first 10 years of his career as a coldwater research biologist with the Wisconsin Department of Natural Resources. He then served for seven years as assistant leader and leader of the Ohio Cooperative Fishery Research Unit, and he retired after 23 years as leader of the Pennsylvania Cooperative Fish and Wildlife Research Unit. He is a former editor of the *North American Journal of Fisheries Management* and former president of the American Fisheries Society.

David Lieb is an invertebrate zoologist with the Pennsylvania Fish & Boat Commission and Western Pennsylvania Conservancy and is responsible for the development and implementation of conservation, management, and regulatory initiatives that target Pennsylvania's aquatic invertebrates. Recent initiatives have focused on developing ways to slow the spread of exotic crayfishes in Pennsylvania. Lieb's recent publications have focused on the ecology, distribution, conservation, management, and genetics of crayfishes. Lieb holds a Ph.D. in ecology with a minor in statistics from The Pennsylvania State University and was previously employed by the Stroud Water Research Center and the Academy of Natural Sciences of Philadelphia.

"If I Know All the Science in the World, I'm Going to Change the World"—The Fisheries Scientist's Fallacy?

Amy Fingerle*
Department of Aquaculture and Fish Biology, Hólar University College
Háeyri 1, Sauðárkrókur IS-550, Iceland

William W. Taylor
Center for Systems Integration and Sustainability
Department of Fisheries and Wildlife, Michigan State University
115 Manly Miles Building, East Lansing, Michigan 48823, USA

Key Points

- Human values often override even the best science, so we need more than science to save the fish.
- In order to get fish "on the table" in policy and decision making, fisheries professionals must communicate the importance of fish to social well-being and prosperity.
- We can use business to communicate the value of sustainable fisheries by putting fish in economic terms.

Fish as an Environmental Indicator and Economic Commodity

I (Bill) grew up around freshwater in upstate New York (USA) at a time before many people cared about the environment. The consequences of our careless interactions with the land and water could be seen in the lakes I played in as a child. Lake Ontario, the lake closest to my home, was not a pleasant place to be around, with blue-green algae and dead Alewives *Alosa pseudoharengus* dominating the waterscape. I wondered how I could make a difference in improving our environment and the lives of those who depend on its quality. Unfortunately, early in my life, there was seemingly little interest in environmental issues. Economic prosperity and material abundance were paramount in our culture, and few actions were taken to care for the environment. All of that changed dramatically in the mid to late 1960s, a seminal time in the ecological history of the Great Lakes. The public saw lakes polluted, rivers on fire, piles of dead fish, and beaches closed. People realized that they needed to advocate for the environment and by so doing could improve the health of their children and the prosperity of their communities. This was a formative time for me. I was in high school and trying to decide what my future would be. All of a sudden it was clear: I cared about the environment, I cared about people, and I could have a career of linking the two together.

* Corresponding author: amy@holar.is

I chose to become, first and foremost, a biologist. I wanted to be a biologist in the way that my father was an engineer: I believed that learning more about the structure and function of a system would help to design a more sustainable future. Sustainability to me meant high-valued fisheries, no fish kills, and clean waters. I concentrated on fish in particular because I saw them as the great integrator of ecological systems, whose presence and production reflected the state of the physical, chemical, and biological environments in a watershed. Fish would allow me to better understand human impacts on the environment in order to increase societal health and provide solutions to environmental degradation. I was enthused—if I could save the fish, I could save mankind! An idealist notion, to be sure, but this was at the time of the great idealism that ultimately spawned the environmental movement in the United States.

As time progressed, however, I found myself increasingly disillusioned with my abilities to save fish and people. I became painfully aware that I was actually a chronicler of the demise of many of our fisheries and that my biological knowledge was not making much of a difference in the environmental policies we designed or in how we interacted with aquatic resources. I spent a long time thinking about why, despite our knowledge and data, fish were not being valued or saved. Jobs, money, safety, and public health drive people to make decisions, but most fisheries scientists were hesitant or unable to discuss fish in those terms, much less to advocate for them in the political arena. If we could discuss the value of fisheries in monetary units—its effects on food security, poverty alleviation, tourism, and trade, for example—people might listen. Fish would go from being biologically interesting to being socially and economically important. To do this, we had to understand fish in economic terms in order to demonstrate to society the value of fish and what we know about them.

I knew the importance of applying for grants and securing funding as an academic, but it took me longer to understand that agencies, the organizations that study and manage fisheries, are inherently businesses. Agencies manage a flow of money and information in order to deliver products—sustainable fisheries—that are valued by stakeholders. Early in my career, we were witnessing a decline in the ability of agencies to attend to their mandates because of inadequate funding. I began to think about why and how to get more funding. The answer was clear: we obtain funding by demonstrating value, and business could help us do this.

At this time, I began to actively work with business leaders and read the business literature in order to see what science might learn from the various disciplines (e.g., marketing, management, economics) that compose the study of business. It became apparent to me that to demonstrate the value of fisheries, we needed to create a management strategy by setting goals, developing clear objectives, and understanding how to market the value of what we know and the products that it produces, both market and nonmarket. Fisheries scientists were generally not thinking about how to develop a management plan that was inclusive of societal needs. But, from my point of view, having a piece of business in our thinking would be essential to achieving sustainable fisheries and healthy aquatic ecosystems. We had to sell the value of fish as both an economic commodity and an indicator of healthy, well-functioning environments! Once I understood this, I spent my career working together with scientists, policymakers, agencies, business leaders, and other stakeholders to better understand and communicate the value of fish and their habitats.

The Business of Environmental Stewardship

I (Amy) spent my first year of college as an economics major, scribbling down notes on the intricacies of supply and demand. As I progressed in my degree, however, I began to wonder why the discussions in my courses on topics such as globalization and environmental regulation were largely one-sided. The undertone I sensed was that anything that was good for business would harm the environment and that environmental protection was always bad for business. This to me was a false dichotomy—why could the two fields not learn from each other? Because of this, there did not seem to be room to discuss the kind of things I wanted to pursue with a business degree. The topics I was studying and the goals I wanted to pursue seemed mutually exclusive. I therefore decided to take a new direction. I changed my focus to environmental sciences, thinking that I would approach my goal of environmental protection in a more direct way. My lifelong love of the lakes and rivers I grew up around inspired me to take an aquatic ecology course. It was then that I decided to become an ecologist and wash my hands clean of what had frustrated me in my economics courses.

I graduated from college with a passion for environmental stewardship, but I was not convinced that all of my hours in a classroom and behind a microscope were going to effect any real change. They were a good place to start, but I had seen and read enough to know that people do not always pay attention to the "best available science," and sometimes for pretty good reasons. I came to understand and appreciate that social and economic needs, such as irrigation projects in impoverished agricultural areas, will sometimes trump ecological well-being in the form of free-flowing rivers and untouched wilderness. I also came to understand that scientists generally underestimated the importance of effectively communicating their results, and the value of those results, to those who most need to hear them. The questions in my head became "How we do we get science 'on the table' when policy valuation is being conducted? What is the true trade-off value when sacrificing fish and their habitat?"

Using Business to Communicate the Value of Sustainable Fisheries

In the words of the late astronomer Carl Sagan, "Science is more than a body of knowledge; it's a way of thinking." The same is true for business. Business is not inherently evil; it is how we use it. Amy turned away from business, believing it to be the antithesis of conservation and environmental well-being. What Bill learned over the course of his career, however, is that we need more than science to save the fish. This is true because human values often override even the best science. Economists and business people ultimately decide on the monetary value of the varied uses of water that our fish depend upon—shipping, hydropower, agriculture, human consumption—so we must also put fish in economic terms in order to link science to how society makes its decisions.

For example, imagine that a dam is proposed to be built in order to provide a community with a reliable supply of electricity. Planners of the dam will state its predicted capacity in kilowatt-hours of electricity, and they might show that the price per watt is lower than current forms of electricity generation. Politicians supporting the project may be able to state, in actual dollar amounts, how much money the dam would save constituents on their monthly energy bills and easily link the dam to social prosperity and well-being. In order to be considered alongside this competing use of freshwater, fisheries professionals must therefore demonstrate the value of the fish in the river—and their habitat—and what might be lost to society (both locally and globally) if such a project is realized.

Of course, determining the economic value of the creatures we study is easier said than done. For most of us, our education has prepared us to be biologists, not economists! We can, however, take a page from the business literature to help us get fish "on the table" in decision making. Businesses utilize a marketing plan to promote the goods and services they offer. This requires that businesses have a clear understanding of why and how their products are a unique contribution that will be valued by customers. Businesses do this by determining their target market: who will benefit from their products, and what will those benefits be? To communicate the value of fish, scientists must think in these terms!

Let us understand this by returning to that proposed dam. Our data may tell us that without careful design, construction of the dam will block the passage of salmon upstream and hinder their spawning efforts. But why should this matter to the policymakers and politicians, the people with the ability to give the go-ahead to the project? We must demonstrate how the salmon in that river are valued by the community and contribute to its economy. And not only are the salmon valued, but their habitats are as well: consider the recreational enthusiasts who enjoy kayaking in the free-flowing rapids and the nearby homeowners who appreciate the view of the lotic waterscape. We can put these values in monetary units by considering the dollars tourists spend in the local community for such things as lodging, dining, bait, and fishing licenses. Add to that the value of the fish to the subsistence and commercial fishers and the money paid for the salmon at the local market. Top it off with the nonuse value of the fish and their habitat, such as the increased home values near the river, the value people derive from knowing there are fish in those shadowy depths, and the value of saving an unimpounded river for our children and grandchildren. When you add it all up, those slimy, scale-covered ectotherms are actually worth quite a bit. Fish are more than just cool animals; they represent money, jobs, and health. We should tout their value!

Along with demonstrating the value of fish and their habitats, we must recognize and promote the unique contribution of fisheries scientists to social and ecological well-being. This is necessary for us to do in order to secure funding to study fish and their habitats. Here again, we can learn from business! It is essential for a business to distinguish itself from the competition by determining the main thing it provides that is of clear benefit to society; if it fails to do this, it may go bankrupt. The main things that fisheries scientists provide are the data necessary for politicians and agency leaders to make decisions that are informed by science. If we neglect to demonstrate the value of fish in economic terms, our scientific knowledge will likely not be seriously considered in relationship to other societal priorities.

Going Beyond Science for a Sustainable Future

It is an exciting but also a very challenging time to be a fisheries scientist! Continued funding cuts and increased demand on our water resources necessitate that we be innovative in our research and creative in our endeavors to secure adequate funding for sustainable fisheries. We are at a critical time in our ecological history; more than ever before, we as fisheries scientists must think beyond our laboratories, field sites, and peer-reviewed publications. We cannot sit idle at our desks, merely chronicling the decline of species that are essential for societal well-being in both the developed and developing world. As fisheries scientists, we have an obligation to communicate with the sectors that impact and depend upon productive, sustainable fisheries, as well as with the general public who may not fully realize the benefits they receive from healthy fisheries and habitats. We can know all we want to know about fisheries ecology, but if society does not value and use that knowledge, it will be irrelevant.

By coupling a marketing plan with well-formulated science, we can communicate the value of what we know and get fish "on the table" with other priorities for water resources. Fish will not always win in these decisions, but at least they will be seriously considered during policy discussions that impact wild fish and their habitats. This means that you need to have the courage to speak for the fish and the people who depend on them! Only by doing this will fish and humans have a chance for a sustainable, prosperous future.

Biographies

Amy Fingerle is a master's student in the Department of Aquaculture and Fish Biology at Hólar University College, Iceland, studying the behavior of stream-dwelling Arctic Char *Salvelinus alpinus.* Prior to graduate school, she earned a B.S. from the Program in the Environment at the University of Michigan and worked as a research assistant for the USGS Great Lakes Science Center and the Partnership for Interdisciplinary Studies of Coastal Oceans (PISCO) at Oregon State University.

William W. Taylor is university distinguished professor in global fisheries systems in the Department of Fisheries and Wildlife and the Center for Systems Integration and Sustainability at Michigan State University. He has been active in the American Fisheries Society throughout his career, serving as president of the society in 1997–1998. He currently holds a U.S. Presidential appointment as a U.S. commissioner (alternate) for the Great Lakes Fishery Commission.

How to Make a Difference When Fighting for Something You Love

Denny Grinold*
Fish N' Grin LLC
2738 North Grand River Avenue, Lansing, Michigan 48906, USA

Marissa Hammond
Center for Systems Integration and Sustainability,
Department of Fisheries and Wildlife, Michigan State University,
115 Manly Miles Building, East Lansing, Michigan 48823, USA

Key Points

- Use your abilities to form relationships and gain trust.
- Build trust and collaborate for effective fisheries management.

Developing a Passion

As a child, I (D. Grinold) spent hours fishing with my father, and as I grew up, I developed a passion for understanding motors, fish, and interactions among people. By the time I was 27 years old, I had a family and was running my own business. Regardless of how busy I became working at my auto repair shops, I always made time for fishing with my wife and children. In the late 1960s, Coho Salmon *Oncorhynchus kisutch* and Chinook Salmon *O. tshawytscha* were successfully stocked into Lake Michigan, and I fell more in love with fishing. When my children grew up, they began to lose interest in fishing, but I was not about to let that stop me from enjoying what I had grown to love so much, so I decided to become a charter boat captain. I knew this would allow me to fulfill my passion for fishing while also introducing others to the emerging and exciting Great Lakes fishery. When I became a charter boat captain, I never expected it would be something that would land me in the management arena fighting for the fishery I had grown to love. By 1986, the Chinook Salmon population was booming and the charter boat fleet was thriving. The industry became a valuable economic component of Michigan's tourism industry and served the livelihoods of many people, including me.

The introduction of Pacific salmon into Lake Michigan, and subsequently into the entire Great Lakes, may be one of the greatest fish stories of all time. By the 1940s, the Great Lakes were nearly void of any native commercial or sport fishing species, mainly due to overfishing and the invasion of Sea Lamprey *Petromyzon marinus.* In 1988, Chinook Salmon began washing up on beaches, catch rates plummeted, and the fishery collapsed. After developing a strong passion for Chinook Salmon fishing, I knew I could not give it up.

* Corresponding author: oldgrin@aol.com

Diagnosing the Problem

When the fishery collapsed, the mechanic in me wanted to diagnose and fix the problem. As a mechanic, I have had many opportunities to fix vehicles that need significant repairs. Often I may not know what is causing the problem right away, but I never give up until I figure out what that problem is and know exactly how to fix it. When I do fix the problem, I always return the vehicle to the owner in a condition that is better than before it entered my shop. When the Chinook Salmon fishery collapsed, I knew I had to figure out what was causing the problem and develop a solution that would make the fishery even greater than it had been my entire life, so generations following me could experience the love for this fishery that I had.

In the late 1980s, there was little knowledge available to scientists, managers, or fishermen about the life history, biological interactions, and ecological needs of Chinook Salmon in the Great Lakes. A working relationship was lacking between fishery managers, jurisdictions, and stakeholders, so many people accused the charter boat industry of overharvesting Chinook Salmon while others accused the Michigan Department of Natural Resources (MDNR) of not stocking the fish they claimed they had. In light of the collapse, the Lake Michigan Task Force was created to bring various stakeholders and managers together to determine what caused this dramatic downturn in the fishery. With the development of the task force, I recognized an opportunity to become involved in a way that would allow me to help diagnose and fix this problem.

Building Relationships

The point of this task force was to share information, build relationships with stakeholders, and learn more about the fishery. I had no idea what to expect during the first meeting, but I soon realized that people's assumptions regarding the cause of the collapse were based on what they thought they knew. Our first task force meetings were filled with a wide array of ideas and opinions. I had my own opinions and ideas, but because I was coming from the charter boat industry, an industry where blame for the collapse was being placed, I decided the most strategic move would be for me to listen and learn as much as possible before speaking up. This choice helped me to build a foundation for the trust and working relationships that were lacking at the time and would be necessary if I wanted to make a difference. I was convinced that the charter boats were not the cause for the collapse; because I had taken the time to listen and learn, I was able to develop trust among members of the task force. I proposed the idea that before imposing stiff regulations upon charter fishing vessels, we needed to determine what impact, if any, was caused by the charter boat industry. The task force agreed, so I worked with a Michigan legislator to craft legislation that required charter boats to report their catches. This piece of legislation would help determine the actual catch rates of charter boats and would eventually prove to be a valuable management tool for the MDNR Fisheries Division.

When the task force further investigated the Chinook Salmon collapse, they found that bacterial kidney disease (BKD) caused the massive die-off of salmon and, in turn, the collapse of the fishery. Fish tend to fend off most common pathogens, such as BKD, when in good health. However, a shortage of prey and an overabundance of Chinook Salmon rendered the fish more vulnerable than usual. After uncovering the cause of the collapse, state agencies and the general public backed off of the charter boat industry.

Although working relationships and trust had improved between all parties, collaboration between all stakeholders was not where it needed to be to solve the issue behind the cause of the collapse. As more discussion occurred, everyone learned more about the knowledge, expertise, and desires that each individual brought to our task force and the blinders that prevented us from realizing what was really going on were lifted. The need for more information came to the forefront.

Using Trust

Established through the task force, the working relationship between stakeholders and managers made proposing to manage the Chinook Salmon population much easier. In the late 1990s, the proposed solution to avoid another Chinook Salmon collapse was to cut stocking efforts in Lake Michigan. This created more tension because most people still had the idea that more fish in equals more fish out. But after discovering BKD and learning more about the ecological needs of Chinook Salmon and about their predator–prey interactions, scientists realized that this wasn't exactly the case, so, it became easier to justify why stocking numbers should be cut. Fishermen soon found that less meant more, especially when catch rates began to increase as a result of cuts in stocking efforts. Cutting the number of Chinook Salmon stocked into Lake Michigan was proposed again in 2006 and was even easier to implement because everyone was well informed. This type of management and interaction was not possible before the task force was created because working relationships and trust did not exist. There was no more pointing fingers or placing blame; everyone was now engaged in the decision-making process and everyone would be responsible for making sure the fishery did not collapse again. When you have this sort of collaboration between all parties involved, you can be more successful.

The success of the original Lake Michigan Task Force, now known as the Lake Michigan Citizens Fishery Advisory Committee (LMCFAC) and chaired by me for 20 of the 25 years of its existence, resulted in citizens fishery advisory committees for all of Michigan Great Lakes waters. Working off the original model, Michigan enjoys an enviable relationship with fishery stakeholders within the Great Lakes basin. Many of the unknowns surrounding the 1988 collapse have been identified; ongoing scientific research produces information that fisheries managers use to make decisions that have increased Chinook Salmon catch rates to numbers rivaling those of the mid-1980s.

Throughout this process, I had to maintain a limitless vision for what I wanted to accomplish and I had to take advantage of my abilities to establish relationships. I learned that my involvement could make a difference. The more involved I became, the more successful I was and the more satisfied I felt. However, because I chose to be so involved in this process, I had to put extra time and energy into the task force, on top of running my auto repair shops and spending time with my family. Being involved meant that I had to balance more responsibilities, but it was a choice I wanted to make because it meant that I could help find a solution to prevent another Chinook Salmon collapse, ensuring that future generations could experience an even greater fishery than my family, my clients, and I had.

Applying Lessons Learned

In recent years, an ever-changing ecosystem plagued by invasive species and an imbalance of predator–prey populations in Lake Michigan has become prominent. Because of ongoing research and collaboration, quicker recognition of the risk for another Chinook Salmon collapse has been possible. A multijurisdictional team was assembled and met for nearly a year to discuss various fish stocking scenarios. Because of my involvement with the task force, I was one of three Michigan stakeholders selected to participate with other stakeholders and managers from Lake Michigan's surrounding state jurisdictions. Using the same collaborative model developed by the task force, the decision was made to cut Chinook Salmon stocking in Lake Michigan by 50% starting in 2013 and to use a feedback policy to adjust stocking levels up or down as dictated by predator–prey abundance. The LMCFAC supported the strategy with some members even arguing to cut stocking of Chinook Salmon altogether. The trust and working relationships that were formed because of the original Lake Michigan Task Force in the late 1980s have grown stronger and have proved to be very important for management of Great Lakes fisheries more than 20 years later.

Our parents spend time teaching us how to build trust and working relationships with people, but we often forget how important those lessons can be when we have to fight for something we care about. As a professional, you have to remember to capitalize on your best qualities; for me, that has been my finesse, my ability to listen, and my curiosity to diagnose and fix problems. We each have qualities that set us apart or make us good at what we do, so it is important to use those qualities to your advantage. It is also important to become as involved as you can when you are fighting for something that you love. It takes extra time and energy on top of what you already have going on in your life, but it is incredibly important if you want to make a difference and have an influence on something that plays a significant role in your life. In the fisheries profession, viewpoints are different and often conflicting, so it is always important to take advantage of your abilities to help you build trust and working relationships between stakeholders, managers, scientists, and fishermen to make a difference and effectively manage fisheries.

Biographies

Denny Grinold was born and raised fishing in Michigan. He splits his time between Denny's Auto Diagnosis in Lansing and Fish N' Grin charter service out of Grand Haven. Past president of the Michigan Charter Boat Association and currently its state affairs officer, Grinold also is chair of the U.S. Committee of Advisors to the Great Lakes Fishery Commission, is chair of the Lake Michigan Citizens Fishery Advisory Committee, serves on the executive advisory board for Michigan Sea Grant, and is a commissioner on the Michigan State Waterways Commission. Denny recently received an Honorary Alumnus award from Michigan State University.

Marissa Hammond grew up in Wiscasset, Maine and has had a personal stake in the commercial lobster fishery throughout her life. In spring 2012, Marissa graduated from the University of New England with a B.S. in marine biology and environmental science. Marissa is currently a master's student in the Fisheries and Wildlife Department at Michigan State University and is studying the changes in population dynamics and subsequent management of Lake Whitefish *Coregonus clupeaformis* in the upper Great Lakes.

Resource Management in the Face of Uncertainty

DANIEL HAYES*
Department of Fisheries and Wildlife, Michigan State University
480 Wilson Road, 13 Natural Resources Building, East Lansing, Michigan 48824, USA

BRYAN BURROUGHS
Michigan Trout Unlimited
Post Office Box 442, Dewitt, Michigan 48820, USA

BRADLEY THOMPSON
U.S. Fish and Wildlife Service, Washington Fish and Wildlife Office
510 Desmond Drive SE, Suite 102, Lacey, Washington 98503, USA

Key Points

Uncertainty in the state of nature and how nature will react to management actions is unavoidable in natural resource management. Professionals in the field need to

- Develop some level of comfort working in an uncertain world,
- Refine their ability to effectively communicate their uncertainty, and
- Avoid or prevent using uncertainty as a stalling tactic to delay implementation of needed management actions.

"You can't regulate us! You don't know how many fish there are out there!"

This was a common theme from commercial fishermen while I (DH) was with the National Marine Fisheries Service in Woods Hole, Massachusetts, working on Haddock *Melanogrammus aeglefinus* as a stock assessment scientist. While it is true that there was considerable imprecision or uncertainty in our population estimates, the weight of evidence from fishery independent surveys and from the fishery itself clearly showed that Haddock was severely overfished. To convey this message, we stated, "there was no uncertainty in the overall status of haddock and more importantly, in the cause of their depleted state." In the politically charged and reactive environment of fishery management during the 1990s, the message from the scientific community did not get any traction with fishers or local congressmen and senators. This seriously limited how quickly actions could be taken to limit the fishery and the extent to which harvest reductions could be implemented.

* Corresponding author: hayesdan@msu.edu

"We're confident in our assessment! We are sure there won't be any negative impacts!"

This is something I've heard frequently from fishery managers while I (BB) have been with Trout Unlimited in Michigan working on stream trout management. With well over 18,000 miles of trout streams in the state (as well as 10,000+ inland lakes and the lion's share of Great Lakes waters to manage), basic information about fish population status or angling pressure on inland trout streams is often nonexistent, sparse, incomplete, or decades out of date. Managers are often asked to evaluate and recommend action on changes to (1) increase harvest via increased bag limits or lower minimum lengths or (2) reduce angling mortality to increase abundance, catch rates, or trophy fish opportunities. Managers often make recommendations for changes in regulations in the absence of basic information (e.g., fish population status, current angling pressure, and predicted angling pressure under changed regulations), often relying on outdated data or personal opinion as rationale. In some of these instances, managers have communicated these recommendations as "certain," not explicitly addressing or communicating any of the uncertainty involved. When this occurs, anglers on either side of a proposed regulation change quickly recognize the general level of uncertainty present and its potential ramifications. This failure to address uncertainty explicitly usually results in stakeholder community divisiveness and damage to the credibility and perceived impartiality of the fishery managers. When a fishery stakeholder sees a proposed decision contrary to their preference, they need to see compelling rationale for why that decision is being chosen. In the absence of this compelling rationale, a passionate stakeholder will be left to assume either a flawed or incomplete professional analysis was the result or that manager bias or political intervention is to blame. Both of these conclusions are damaging to the perceived integrity of the decision makers and the agency they represent. Uncertainty will be present, and it is best to explicitly recognize it and explain how it factors into the management decision at hand.

"I don't trust the model results! It's a black box and it relies on expert opinion for the inputs!"

When I (BT) started my career in fisheries science and management with the state of Washington, this was the most common refrain I heard from professional biologists collectively charged with prescribing Pacific salmon habitat restoration plans. The concern was consistent across teams of diverse yet equally passionate stakeholders that I assisted in many watersheds. Nearly all teams contracted the same consultant to use the same mechanistic model to predict salmon population responses to changing habitat conditions. The consultant's model structure was complex and required a large quantity of data to make predictions of salmon response. The information feeding the model had a lot of uncertainty associated with it, even though the model generated very specific, point estimate results. The specificity of the model's results when the input data were highly uncertain caused a lot of discomfort to the team members. Many of the teams viewed this specificity with skepticism and worried about the utility of the modeling results when choosing among salmon recovery strategies. For example, when the model predicted a proposed habitat action would result in an increase of 214 fish, was potential error ±10 fish or ±1,000 fish? In response, I collaborated with peers to develop a means for generating confidence intervals for many of the model outputs. We then applied this to predictions for each watershed and found that for most watersheds, the analysis indicated narrow enough confidence intervals for model outputs to greatly alleviate the original concerns regarding the model's utility in the salmon recovery planning process.

"What's a new biologist to do?"

These three case studies from across the United States highlight some common but important considerations for fishery managers and fishery scientists. *Accept that uncertainty is unavoidable in natural resource management; you need to develop a level of comfort working where you do not know exactly what the state of nature is or how it will react to management actions.* All evaluations of natural resource systems entail sampling, which provides only a statistical description of resource conditions and not a complete census. One of the unfortunate consequences is that gains in precision flatten out very quickly; it generally requires four times the effort and cost to halve the standard error or confidence around an estimate. "The truth–you can't afford the truth!" is a phrase I (DH) have often used while consulting on sampling plan design. Beyond the need to subsample fish populations and their habitats, many of the decisions we make are based on models, as highlighted in the case study regarding salmon in the Pacific Northwest. Model precision is generally limited by data inputs, but perhaps more importantly, the consequence of incorrect assumptions is that model predictions can be seriously biased. However, it is important to realize that whether the models used are formalized as mathematical abstractions or simply as a mental model, they are prone to these issues. Our main piece of advice is to run multiple models (both mathematical and conceptual) to better understand the potential variation due to model choice and consider this in the decision process.

A strategy for model building that has much appeal is to adopt an adaptive management approach. Adaptive management helps to deal with the issues we raise in a number of ways. First, it helps to clearly identify which models are explicitly being used in the decision-making process. Secondly, direct engagement with stakeholders helps them to understand the degree of uncertainty inherent in the process. Finally, critical uncertainties can be reduced by viewing management actions as experiments, whereby data are collected that help improve precision of the models.

Once you are comfortable (or at least accept) working in an uncertain world, *realize that you need to explicitly recognize uncertainty and communicate it clearly to the public while working with them to develop realistic expectations about reducing it further.* As the trout stream management case study highlights, the public and resource users are becoming increasingly conversant and experienced in natural resource management, and trying to put on an air of total certainty erodes the trust that is necessary in resource management. Lessons can be learned from how weather forecasts are communicated. Weather forecasts do not come across as it will or will not rain today, but rather as a chance of rain. It is then up to the listener to determine how to use that information. For example, most people would bring an umbrella along if there was a 90% or greater chance of rain. The number of people bringing an umbrella would be more mixed, however, if a 30% chance of rain was reported. Another lesson from the meteorology world is that despite the investments in capital and personnel toward weather forecasting, there are limits in terms of improved forecasting accuracy.

During communication with stakeholders, listen as well as present the technical information available. We observe that when information is provided at too detailed or too complex of a level, a common response is for people to mistrust the information provided. The same can result when too little information is provided. Thus, it is crucial to ensure that people are with you during the consideration of the evidence, and listening to their questions and concerns helps develop the co-learning environment that is needed for trust to develop.

Once you are comfortable with the uncertainty you face and have worked with stakeholder to develop the best collective understanding possible, *avoid hiding in the shadows of uncertainty and refine your ability to determine when others are using uncertainty to promote their own agendas.* We have all heard that before a decision is made, more research is required. Managers and stakeholders alike can use this phrase to support their preferred positions. To avoid the pitfalls of this, it is best to openly assess and communicate the existing uncertainty levels and their consequences for making a management decision in light of it, and for all involved in a fishery management decision-making process to consider together how much the proposed additional research will actually help in reducing uncertainty and whether the anticipated reduction would likely lead to different conclusions or management actions. It is difficult to avoid the problem of defaulting to a "more data" approach as it requires integrating how more information will add to the big picture view of the overall problem. The ability to do this comes with experience with this problem, but also with time spent understanding the specifics of the problem at hand. We encourage young professionals to frequently reflect on the question of "how would more data actually change what we do?" and accept that sometimes the answer is "not at all."

Conclusion

As in other facets of our lives, the first step toward addressing a problem is acknowledging its existence. Uncertainty is unavoidable in natural resource management, and given fiscal resources and the current state of our science, levels of uncertainty are often going to be larger than most of us would like. Like it or not, we need to make decisions despite the uncertainty we face. We encourage you to allocate more of your time and energy on developing management approaches that minimize the risks that an uncertain world poses than on complaining about how variable things are and how there is not enough money to sample sufficiently. As in other parts of our professional lives that are uncomfortable (e.g., speaking in front of a hostile group of stakeholders), thoughtful preparation helps greatly, but in the end, reflection on the experiences you have will help foster your professional growth most.

Biographies

Daniel Hayes has worked on fish populations and habitats in streams, lakes, and the Great Lakes, as well as marine systems. In his current position, he teaches and does research, but is also actively engaged in resource management through service on committees within the Michigan Department of Natural Resources. An avid outdoorsman, Dan recognizes what it is like to be a stakeholder in the complex and messy world of natural resource management.

Bryan Burroughs specializes in stream fisheries management, fish population dynamics, and fluvial geomorphology. As both a fishery scientist, and the executive director of Michigan Trout Unlimited, he is intimately involved with research, management, policy, advocacy, and on-the-ground stream restoration practice. The intersection of science and policy for fisheries management is a field he practices in daily and continually learns from. He is also an avid outdoorsman who relishes the diversity of experiences that angling provides.

Bradley Thompson has worn several professional hats as a research scientist, modeler, and policy decision maker for state and federal conservation agencies in the Pacific Northwest. His experience includes leading research teams and providing technical assistance to state, tribal, and international fisheries management bodies, as well as working with partners to deliver conservation programs that benefit native aquatic species. In his current position, he manages the U.S. Fish and Wildlife Service's Endangered Species Listing and Recovery programs for aquatic and terrestrial species in the state of Washington. Similar to his fellow coauthors, he is also an avid outdoorsman.

Carry a Big Net—Cast It Far and Wide

ROBERT M. HUGHES*
Amnis Opes Institute
112 Aspen Meadows Road #39, Driggs, Idaho 83422, USA
and
Amnis Opes Institute
2895 SE Glenn, Corvallis, Oregon 97333, USA

DANIEL J. MCGARVEY AND BIANCA DE FREITAS TERRA
Center for Environmental Studies, Virginia Commonwealth University
Trani Life Sciences Building, 1000 West Cary Street
Post Office Box 843050, Richmond, Virginia 23284, USA

Key Points

To have a rewarding and varied career with a positive effect on natural resource conservation, we recommend that you

- Dedicate yourself to developing new collaborations,
- Develop an eye for large-scale questions that necessitate large-scale collaborations and large-scale data sets, and
- Make the effort to learn a second language.

Introduction

This is a vignette of events that led three fisheries scientists from very different places to collaborate on transcontinental and intercontinental research projects involving large data sets that none of us could have collected or synthesized alone. How did this happen and what results have come of it?

Three Sinuous but Separate Paths

Robert Hughes (Bob): I grew up near a Michigan lake, where fishing was an important form of recreation and fish were a key component of the family diet. In this time, I began to understand the consequences that rapid population and economic growth can have on aquatic ecosystems. This catalyzed my decision to study biology and to earn an M.S. in resource planning and conservation from the University of Michigan, with the hope of protecting the natural resources I had come to value. I subsequently completed a Ph.D. in fisheries at Oregon State University and was hired as a contract research scientist at the U.S. Environmental Protection Agency (USEPA) lab in Corvallis, Oregon, where I worked for 30 years.

* Corresponding author: hughes.bob@amnisopes.com

Daniel McGarvey (Dan): I also grew up in the Midwest, where I spent much of my free time fishing in local streams and reservoirs or watching Sunday morning fishing shows. Studying biology and ecology as an undergraduate was therefore an easy decision for me. So was the decision to pursue an M.S. in fisheries at Pennsylvania State University. After completing my M.S., I moved to the Pacific Northwest and spent the next several years working as an industry consultant (aquatic bioassessment), traveling around the country to survey streams and rivers in a range of very different environments. These years were a wonderful opportunity to travel and learn, but I eventually decided that my career aspirations could only be met with a Ph.D. I felt this way for multiple reasons, but the truly decisive factor was simple curiosity: I wanted to know why the fish and invertebrates that I had spent the past several years collecting were so different in, say, the Willamette (Oregon), Susquehanna (Pennsylvania), and Pascagoula (Mississippi) River basins.

Bianca de Freitas Terra (Bianca): I am from Brazil, home to some of the planet's most biologically diverse biomes (e.g., the Amazon rainforest, the Cerrado, and the Atlantic Forest), its largest river (the Amazon), and its richest fish fauna. Although Brazil is a biologically and culturally rich nation, it is faced with ongoing conflicts between conservation and economic development. As a child, I witnessed the rapid degradation of the Atlantic Forest and decided to pursue a career in natural resource conservation and rehabilitation. I attended the Universidade Federal Rural do Rio de Janeiro (UFRRJ), where I entered the fish ecology program and began work on my first research project. As part of a team of undergraduate and graduate students, I surveyed streams and rivers throughout the Paraíba do Sul River basin. In this time, I came to know and appreciate the entire river system, including its many native fishes and the diverse people living along the river.

The Paths Converge

Bob and Dan

Bob: Our professional relationship began with a serendipitous encounter at the 2002 Annual Meeting of the North American Benthological Society (now the Society for Freshwater Science). Dan was presenting some of his previous bioassessment work and preparing to begin his Ph.D. studies in the Freshwater Interdisciplinary Sciences program at the University of Alabama. His recent exposure to rivers and streams throughout the United States had caused him to wonder why fish assemblages were so distinct in different parts of the country, and he was beginning to lay the groundwork for a dissertation on the ecology and biogeography of North American freshwater fishes.

Dan: Learning about my interests, Bob saw an opportunity for collaboration and told me about a large fish data set that was being compiled through the USEPA Environmental Monitoring and Assessment Program (EMAP). He suggested that the EMAP data might be ideally suited for the kinds of macroscale analyses I wanted to perform. Shortly thereafter, Bob agreed to serve as a member of my Ph.D. committee and I began a dissertation project that split my time between the Southeast and Pacific Northwest. Years later, Bob and I remain friends and his intuition regarding the usefulness of the EMAP fish data (and subsequent national-scale surveys) has proven to be correct; my research regularly incorporates a Southeast versus Pacific Northwest contrast (this is natural because the Southeast has the most diverse fish fauna in the United States, whereas the Northwest fauna is relatively depauperate), and the paper that I coauthored with Bob remains my most highly cited.

Bob and Bianca

Bob: I first met Bianca when she was an undergraduate at UFRRJ. Several years previous to this, I was approached by her (to-be) major professor after a talk I gave at a conference in Brazil. Her major professor invited me to come observe his work on the Paraíba do Sul River the next time I was in the country, and two years later I took him up on his offer. At that time, Bianca was part of a student crew working on the project and I enjoyed spending some time in the field with her and her classmates.

Bianca: I was initially hesitant to approach Bob because my English language skills were not as strong as I would have liked. I was also a bit intimated by his reputation; his expertise in aquatic bioassessment was considerable and my major professor clearly respected him. But Bob's dedication and personality motivated me to work even harder, particularly in my English studies, and I eventually overcame my fear of not being able to communicate with him. Looking back, I am glad that I did because he has become a wonderful friend and mentor. He even helped me secure a five-month residency in Corvallis, Oregon while I completed my Ph.D. studies. And I am now proud to say that we have successfully published multiple papers together, including a cover story in *Fisheries*!

Dan and Bianca

Dan: Bianca first came to my attention during an email exchange with Bob. I was filling him in on the latest developments in my new role as an assistant professor in the Center for Environmental Studies at Virginia Commonwealth University when he asked if I would be interested in working with a postdoc from Brazil. (Side note: you know you have found a great mentor when years later he or she still checks in periodically to see how you are doing.)

Bianca: Bob mentioned that he might know of a suitable postdoc mentor for me and put me in touch with Dan. After a couple of emails, Dan and I agreed that the kinds of biogeographical questions he was working on would be a good way for me to apply and expand my research skills, which also pertained to field-based biological assessment projects. For instance, we discussed the possibility of using hydrologic time-series data to build macroecological models of the richness of tropical, subtropical, and temperate fish assemblages. We quickly drafted a research proposal and submitted it to the Conselho Nacional de Desenvolvimento Científico e Tecnológico of Brazil (CNPq) Science without Borders program. Several months later, I learned that I had been awarded a postdoctoral fellowship and I began making preparations to move to Richmond, Virginia.

Our Proverbial Glue: Large-Scale Collaboration and Large-Scale Data

Bob: Working at the Corvallis USEPA lab, I was fortunate to have the opportunity to help design and conduct national-scale research on river and stream ecosystems. EMAP and its successor, the USEPA National Streams and Rivers Assessment, required close collaboration with a large number of agency and academic geographers, computer programmers, ecologists, chemists, hydrologists, statisticians, and aquatic biologists. Becoming an expert in all of these diverse but interrelated fields is not something that any one person can hope to accomplish, and so it was quickly evident to me that large-scale collaboration would be the key to success in large-scale surveys. Efforts to develop biological, sediment, and nutrient criteria for U.S. streams and rivers, which will (hopefully) lead to more sustainable regional and national land and water management practices, have gone mostly according to plan. But, the really fun part of this large, complex process—the part that I could not have anticipated in advance—has been the increased opportunities to work with talented collaborators, many of whom are from places that I had never been and might otherwise never visit. I now have ongoing projects in South America, Europe, and Asia, where collaborators are eager to apply the regional surveying and bioassessment lessons that I have learned here in the United States. And to top it all off, I get to work with and mentor some really bright, up-and-coming young scientists!

Dan: Let me start by pointing out that it is we "young scientists" who are lucky to be working with Bob. Since completing my Ph.D., I have often struggled to find data sets that are as useful for multifaceted, large-scale research projects as the EMAP family of biological surveys. Most of the time, large-scale questions must be tackled by compiling data from several independent sources and then devising a way to standardize those data. This approach requires collaboration, too, and it sometimes works okay, but rarely so well as having access to a single large-scale data set that was generated using standard field methods. Now that I am responsible for leading, managing, and funding my own field surveys, I have begun to appreciate how difficult it is to build something like EMAP. I can also attest to the incredible value of the process and endpoint. Having access to EMAP data has allowed me to answer many of the questions that I explicitly set out to address. But it has also provided a means of generating peripheral publications to help maintain my productivity during transitional periods—and continuous productivity is essential for anyone considering an academic research career. I am now teaching this philosophy and some of the necessary skills to Bianca as we try to answer some new questions regarding the factors that maintain such exceptional fish diversity in tropical rivers and how these factors may change in the future. And I am still pursuing opportunities to work with Bob; next up is a major grant proposal to build fish interaction network models from some of the data that he has worked so hard to collect.

Bianca: Teamwork and collaboration have been essential components of all my research projects. Both as an undergraduate and a graduate student, it was necessary to work as a coordinated group because research funding is more limited in Brazil than in the United States, and large, tropical rivers require many hands to adequately sample. All of the students in my major professor's lab (and in most Brazilian universities) worked together, regardless of individual thesis topic or level of experience. But this worked well because the more experienced students took it upon themselves to teach and mentor the newer students, and within this mix, new ideas were always evolving. For me, the more significant hurdle to large-scale collaboration, particularly with international collaborators, has been the need to master English. English is now the international language of science and it can be difficult for nonnative English speakers to integrate with the larger scientific community. Fortunately, the Brazilian government has created the Science without Borders program to immerse young scientists (me, for example) in English-speaking research environments. The program's goal is to send 101,000 Brazilian undergraduate, graduate, and postdoctoral researchers abroad by 2015. It is because of this program that I have been able to work with international researchers like Bob and Dan, focusing on large-scale questions that require new and exciting skills in geographic information systems, statistical analysis, and ecological theory. These skills are particularly important to me because tropical rivers are inherently large-scale systems! When I complete my postdoctoral research, I am excited about returning to Brazil to share what I have learned, to begin building new collaborative networks of my own, and to expand on those developed with Bob and Dan.

Reflections and Some Nonintuitive Advice for Those Starting on a Similar Path

Bob, Dan, and Bianca: Through independent experiences, we each came to understand that some of the most interesting and pressing research questions are, by nature, large-scale. And we learned that answering these questions requires large data sets that are beyond the capacity of any one of us to collect alone. Progress has therefore meant learning to work in teams and to manage and analyze large data sets. This last point is not trivial. Anyone working with very large, complex data sets will quickly learn that data management is an essential yet difficult skill. Basic spreadsheets will only take you so far before they corrupt the referential integrity of your data. (Just think about the last time you put the kibosh on your meticulously prepared data by accidentally sorting a spreadsheet without selecting all rows.) Eventually, you will need to invest some time in mastering a relational database system, such as Microsoft Access.

Finding creative ways to apply your past experiences and skills can also help to open new, collaborative doors. For instance, Bob has followed a uniquely nonlinear career path, transitioning among academic, agency, and consulting roles; working on four continents; and ultimately becoming president of the American Fisheries Society. At some level, this flexibility has been a product of his formal training and studies in psychology, environmental policy, and economics, which were all motivated by his personal interest in the human side of natural resource management. As an alternative example, Dan is now developing some precautionary decision-making tools for endangered species science. His past interest in environmental law and policy, which led to formal law school training (while working on his Ph.D.) and a law review article, helped him network with several leading scholars in environmental law. Together, they are searching for ways to better integrate the proactive intent of the Endangered Species Act with the evidence-based norms of scientific research. Some of your most productive and exciting collaborations will develop in ways that you did not originally intend, so be creative, remain open to new ideas, and remember that successful, large-scale collaborations begin with solid, small-scale collaborations.

Finally, we strongly recommend studying a second language. If you are a nonnative English speaker, this goes without saying. Becoming part of the international science network was and continues to be a significant challenge for Bianca. Basic English classes within the Brazilian school system did not fully prepare her to communicate at professional meetings or in English language science journals, and the opportunity to perfect her English is a big part of the reason she is now a postdoc in the United States. But even if you are "lucky" enough to have English as your first language, you should remain cognizant of the fact that other scientists, many of whom are brilliant and potentially fantastic collaborators, may not fully understand your work. Clearly, Bob's international research would be easier to conduct if he had mastered a second language. And in our anecdotal experience, others are more prone to like and respect you if you at least make a minimal effort to communicate in their native language.

In summary, we suggest the following points to young and aspiring professionals in fisheries science and management:

- Actively seek collaboration with other disciplines and interest groups, including nonscientists;
- Present your work at large scientific meetings, where you are likely to meet new professionals and future collaborators;

- Develop some large-scale research interests that can serve as a substrate for large-scale collaboration;
- Learn to manage and analyze large data sets collected over large spatial extents;
- Be creative in leveraging and applying skills that are not immediately relevant to your primary research;
- Study and use a second language to facilitate cross-cultural communication; and
- Seek out a capable and caring mentor, and when you find one, do everything you can to make sure he or she knows how much you appreciate him or her.

Our shared penchant for large-scale research has become a signature of both Bob's and Dan's research portfolios, and it is quickly becoming an asset that Bianca will be able to leverage in her own career. Of course none of us can say with certainty where this path will take us. But, we each trust that by walking it together, we will have more opportunities, have more fun, and go further than any of us would alone.

Biographies

Bob Hughes, the 2013–2014 president of the American Fisheries Society, is a courtesy associate professor at Oregon State University and senior scientist at the Amnis Opes Institute. He received his A.B. and M.S. from the University of Michigan and Ph.D. from Oregon State University. Prior to joining Amnis Opes, Bob worked as an on-site contractor for the U.S. Environmental Protection Agency. He is a member of Oregon's Independent Multidisciplinary Science Team and has been a guest professor at the Universiität für Bodenkultur, Vienna, Austria; Universidade Federal de Lavras, Brazil; and Universidade Federal de Minas Gerais, Brazil.

Dan McGarvey is an assistant professor at Virginia Commonwealth University. He received his B.A. from Wittenberg University, M.S. from Pennsylvania State University, and Ph.D. from the University of Alabama. Dan conducted postdoctoral research with the U.S. Environmental Protection Agency. He has served as an adjunct professor at the University of Georgia, a research assistant for Oregon's Independent Multidisciplinary Science Team, and a consultant for the National Council for Air and Stream Improvement.

Bianca Terra is a postdoctoral research fellow at Virginia Commonwealth University through Brazil's Science without Borders Program. She received her B.S., M.S., and Ph.D. from the Universidade Federal Rural do Rio de Janeiro. She previously served as a research intern at Oregon State University.

Evolution of a Fisheries Scientist: From Population Dynamics to Ecosystem Integration

PETER C. JACOBSON*
Minnesota Department of Natural Resources
27841 Forest Lane, Park Rapids, Minnesota 56470, USA

Key Points

- Managing fisheries in a world increasingly complicated by large-scale ecological stressors requires moving past single species models and towards incorporating systems-level concepts.
- Although challenging, students and young professionals should embrace systems-level thinking early in their careers, building on the ecological and mathematical concepts first pioneered with single-species models.

The trail of fishery science is strewn with opinions of those who, while partly right, were wholly wrong.
—Michael Graham, *The Fish Gate*

Fisheries science has a rich and proud history of studying the dynamics of important fish populations all around the world. Countless fisheries scientists have cut their career teeth modeling the population processes that drive many of these commercially and recreationally important fisheries. Quantifying those processes can be intoxicating for the mathematically inclined. Differential equations elegantly describe the fate of fish as they grow, reproduce, die, or are harvested. Parameterizing the equations is only limited by the amount of survey or fishery data available and the mathematical tools available to the analyst. Mathematically integrating them to calculate yield and population size is a wonderful and satisfying step. Imagine the feeling of exhilaration that Ray Beverton and Sidney Holt experienced when they first combined growth, recruitment, and natural and fishing mortality functions into a single yield-per-recruit equation (Beverton and Holt 1957). Envision the anticipation of Beverton eagerly awaiting another calculated yield as Holt diligently cranked out results for another step of fishing mortality from his hand-operated, World War II vintage, and now legendary Brunsviga calculating machine (Figure 1). Appreciate the sense of satisfaction when they published the remarkably complete *On the Dynamics of Exploited Fish Populations* and fully realized the impact that their work would have on fish population modeling around the world. The book is still the one of the most highly cited works in fisheries science—certainly satisfying at every step of their remarkable scientific journey and nourishing for the many fisheries scientists who have been developing and applying models of fish populations since.

* Corresponding author: peter.jacobson@state.mn.us

FIGURE 1. Ray Beverton (left) and Sidney Holt (right) working with a Brunsviga calculating machine in the fisheries laboratory in Lowestoft. Note the cardboard three-dimensional yield isopleth diagram in the background! Photo used with permission from Cefas (Center for Environment, Fisheries and Aquaculture Science, Lowestoft, UK).

Single-species population models are an important first step in the understanding and management of fish stocks. Unfortunately, many times they are insufficient and sometimes even wrong. Natural mortality rates vary. They vary across years, across stocks, at different stock sizes, under different climates, and under many different environmental conditions. So do growth rates. And recruitment does as well. Fishing effort and rates are notoriously unpredictable and, even worse, sometimes depensatory. Although single-species models can be successful for managing stable populations, when the underlying ecosystems that support these fisheries change, the models become less accurate. Sometimes devastatingly inaccurate and important fisheries collapse. Even Ray Beverton laments the "carefree days of the 1950s" compared to the "depressing series of stock collapses and depletions of the 1980s" (Beverton 1998). Although the inability of decision makers to act on the recommendations of scientists is commonly a factor in fishery collapses, Beverton specifically notes the "inadequacy of methods dealing with some multispecies fisheries." In addition, uncertainty of model predictions and system understanding is always present and cannot simply be solved by collecting additional data. Single-species, stationary yield models do not capture the varying dynamics required to adequately describe real world populations.

Beverton and Holt understood this. In fact, much of their book is devoted to extending the "theory of fishing" to incorporate the effects of density dependent mortality, variable recruitment, and changing fishery dynamics. However, as more varying dynamics are incorporated, the mathematical relationships become more complex. Easily solved integrations are replaced by systems requiring numerical methods, which, in and of itself, is not a problem, given modern computing power and techniques. But, it comes with loss of mathematical elegance. And as the complexity of the models increase, the number of underlying assumptions and the amount of data required to parameterize the models greatly increases. Uncertainty must be considered as well in the most quantitative fashion possible. All of which is much less satisfying for the savvy analyst who understands the house of cards complex models can be built on.

Models that recognize and account for complexity are required to manage complex systems. Fisheries scientists have long recognized the importance of understanding and incorporating ecosystem changes for effective fisheries management, but the resulting science is still young. Large-scale ecological stressors such as climate change, land use, and habitat change, and invasive species threaten fish populations around the world. The environments that fish live in have experienced extreme changes in recent years. Understanding how these environmental changes affect fish populations is now as important as understanding the basic population dynamics that traditionally has been the focus of fisheries science. And that makes understanding far more challenging. Ecosystem-based fisheries management encourages the formal incorporation of system productivity, trophic-level dynamics, community shifts, bycatch effects, habitat changes, and climate change—seriously complex stuff. Too complex to effectively manage fisheries on a large scale? Perhaps. Indeed, Sidney Holt questions that "humans are not yet ready to manage marine ecosystems" (Holt 2008), yet he also recognizes that if we can "ensure maintenance of their biodiversity and biological productivity, and their ability to continue to provide a variety of goods and services for ourselves and for future generations ... if that is ecosystem management, then I'm all for it"!

As fisheries scientists interested in modeling move through their career, the shortcomings of traditional and mathematically elegant approaches become painfully obvious. The nice, neat, satisfying mathematical equation solving that is so intoxicating gets replaced by the messy, sobering, and uncomfortable real world of ecosystem shifts that are difficult to measure and understand, let alone predict. Of course personal career satisfaction is not the most important reason why we model fish populations—benefits to society are. So, the transformation that comes from recognition of ecosystem complexity and uncertainty is necessary and common for fisheries scientists. Many have transformed from freshly minted population dynamicists to grizzled, old, systems scientists who understand that traditional population approaches are not sufficient to conserve fisheries in a world that is enduring extreme ecological stressors. Ecosystem complexity and uncertainty is then more thoroughly considered through this transformation. Model limitations are better understood, and there is explicit understanding that models can fail to provide sufficient predictive power for sustaining fisheries and ecosystems. And finally, late in their careers, many gray-bearded fisheries scientist come to realize that protecting ecosystems and habitats is one of the most important professional responsibilities that we have.

My own career trajectory followed that exact path, starting with a single-species commercial fisheries model and ending with supervising a habitat research group that integrates many important systems-level dynamics. Unbeknownst to me, my major professor Dr. William Taylor was preparing me for systems-level thinking. Dr. Taylor added a systems modeler to my graduate committee that greatly prepared me for that evolution. Dr. Erik Goodman was a modeler in the Electrical Engineering Department (home of many modelers in the day when electronic computing was just becoming available to the masses). Dr. Goodman had a specific interest in applying mathematical modeling to ecological questions and was known for encouraging students to think at a systems level. Dr. Taylor also had me take a course in systems modeling from Dr. Thomas Manetsch, also in the Electrical Engineering Department. Dr. Manetsch was known for modeling the economies of developing countries in the third world—hardly related to fish population modeling, or so I thought. Of course it was related, and the same concepts that make modeling economies at a national scale are entirely applicable for modeling fish populations at an ecosystem level. Although not fully appreciated at the time, Dr. Taylor was preparing me for a systems-level evolution.

Fortunately, universities have made great strides in formally incorporating ecosystem-level training and management that greatly accelerate the transition to systems-level thinking. A number of universities have created structural institutions that focus on integration of the sciences necessary for understanding ecosystem dynamics. For example, current students of my alma mater Michigan State University now have the Center for Systems Integration and Sustainability to formally guide incorporation of systems-level thinking into their academic development. The center recognizes that sustainability requires that multiple disciplines join forces to fully understand complex ecological systems. Interestingly, when human interactions are formally coupled with natural ecosystems, patterns emerge that would not have been evident when studying these systems individually. Unfortunately, the nonlinear, time-lagged nature of this coupling greatly increases the complexity of models required to understand both natural and human systems. Fortunately, exposing students to this complexity early makes the daunting transition to systems-level thinking much faster and with less pain. Students quickly understand the challenge that systems-level thinking entails and are much more capable of applying that thinking to solve real-world ecological problems when they start their professional careers. Current students that engage opportunities for incorporating systems-level thinking have a much greater head start than fisheries students from even a generation ago.

Resource agencies, which are forced to manage fish populations with shifting baselines and ecosystem dynamics, have also made significant advances towards systems-level management. Early career professionals are better prepared to engage in these emerging initia tives. Large habitat programs exist in a number of federal and state agencies. A renewed focus on coastal and marine habitats by the National Oceanic and Atmospheric Administration is predicated on the understanding that healthy fish populations capable of sustaining robust fisheries require high quality habitats. Similar programs for inland waters by the U.S. Fish and Wildlife Service, Bureau of Land Management, and U.S. Forest Service also focus on the importance of protecting and restoring fish habitats. Many states have a greatly increased emphasis on habitat programs as well. New federal and state collaborative ventures, such as the National Fish Habitat Partnership, highlight the need to manage fish habitats at a landscape level. Climate change adaptation strategies are being developed at state and federal levels. Invasive species now garner intense attention and management. Advanced spatial tools and data sets are allowing analysts to understand critical landscape–fish linkages. All of these efforts are designed to manage the ecosystems so necessary to sustain healthy and productive fish populations.

Are these profound changes and increased complexity less satisfying for fisheries scientists? Probably not. Incorporating these changes into the management of fisheries resources and working at an ecosystem level is certainly daunting. However, the seemingly formidable challenge of managing at a system level comes with opportunities. Young fisheries scientists can embrace these opportunities and do not need to be overwhelmed by the complexity. The same mathematical elegance that Beverton and Holt applied to fish populations in a commercial setting can now be applied to ecosystem effects on fish populations in an ecologically stressed world. Although these systems are inherently complex, there is a critical need to capture the important ecosystem dynamics in a mathematically coherent fashion. Complex processes can be distilled into integrated models using innovative and mathematically elegant approaches. Computing power is no longer a bottleneck. Processing power of hand-held smartphones in 2013 exceeds that of desktop computers in the 1980s. Students and young fisheries scientists now have computational tools that Beverton and Holt may never have dreamed possible.

How do they do this? One example is to start with the book by Walters and Martell (2004) that takes readers from the basics of population dynamics through the complexity of food webs and ecosystem contexts with emphasis on management alternatives. A second important reading (Hilborn 2012) takes readers through the social, economic, and political settings that shape fisheries management. Institutional structures now offer the multidisciplinary environment necessary to tackle systems-level problems. Resource agencies critically need systems-level thinkers to truly manage populations exposed to large-scale ecological stressors. No longer does a student and young fisheries professional need to trudge through a career-long transition from a single-species population dynamics focus to an ecosystem-level understanding. With current academic and resource agency support for systems-level thinking, that transition will happen much more quickly and proceed far more smoothly. Students and early career professionals can embrace these larger-scale concepts and build on the ecological and mathematical concepts first pioneered with single-species models. Young scientists are proving to be especially adept at modeling landscape–fish linkages. The future is certainly bright for the students and young fisheries scientists who immerse themselves into understanding ecosystems and landscapes–the management of our fisheries resources will greatly benefit from that.

Acknowledgments

George Spangler and William Taylor planted the seeds of systems-level thinking during my academic training, and Jack Wingate and Donald Pereira allowed those seeds to grow in a resource management setting. All are considered life-long friends and mentors.

Biography

Peter Jacobson received a bachelor of science in fisheries science from the University of Minnesota in 1980 and a master of science in fisheries science from Michigan State University in 1983. He has worked for the Minnesota Department of Natural Resources since 1988, most recently as the supervisor of the Habitat Group in the Fisheries Research Unit. His most recent area of research interest includes the effects of eutrophication and climate change on coldwater fish.

References

Beverton, R. J. H., and S. J. Holt. 1957. On the dynamics of exploited fish populations. Great Britain Ministry of Agriculture, Fisheries and Food, Fishery Investigations Series II Volume XIX, Lowestoft, UK.

Beverton, R. 1998. Fish, fact, and fantasy: a long view. Reviews of Fish Biology and Fisheries 8:229–249.

Graham, M. 1943. The fish gate. Faber and Faber, London.

Hilborn, R. 2012. Overfishing: what everyone needs to know. Oxford University Press, Oxford, UK.

Holt, S. 2008. Foreword. Pages ix–xx in A. Payne, J. Cotter, and T. Potter, editors. Advances in fisheries science: 50 years on from Beverton and Holt. Blackwell Scientific Publications, Oxford, UK.

Walters, C. J., and S. J. D. Martell. 2004. Fisheries ecology and management. Princeton University Press, Princeton, New Jersey.

Casting a Wide Net: Integrating Diverse Disciplines and Skillsets in Fisheries Policy Careers

KRISTINE D. LYNCH*
Washington, D.C., USA

KELLY M. PENNINGTON
Conservation Biology Graduate Program, University of Minnesota
135 B Skok Hall, 2003 Upper Buford Circle, St. Paul, Minnesota 55108, USA

Key Points

A career in fisheries policy is endlessly challenging, yet searching for solutions that balance the diverse aspects of fisheries can also be highly rewarding. We outline what you can learn as a fisheries policy professional and what characteristics might serve you well if you are considering such a career path:

- Fisheries are complex and composed of dynamic biological, ecological, economic, social, legal, and political systems.
- Finding durable policy solutions to fisheries issues requires an eagerness to learn, an open mind, and critical thinking skills.
- While a policy degree is not usually required to succeed in fisheries policy careers, interdisciplinary preparation will help policy practitioners understand and utilize diverse policy-relevant information.
- A career in fisheries policy provides many opportunities to continue learning, to be exposed to different worldviews, and to refine professional skillsets long beyond the classroom.

Introduction

Do you like to study many different fields, learn from people with diverse worldviews, find creative ways to adapt to changing conditions, and integrate everything you learn to solve complex social and scientific problems? If so, you may enjoy a career in fisheries policy. In this essay, we discuss how our backgrounds as interdisciplinary fisheries scientists served us well in the fisheries policy world. We hope that after reading this essay you will understand if and how to approach the policy field as one of many options available to you after completing your graduate degree.

* Corresponding author: kris.d.lynch@gmail.com

Who Are We?

Between us, we have 11 years of experience working on federal fisheries policies. We served as staff on the U.S. Senate committee that has jurisdiction over the management, conservation, and development of living marine resources in federal waters. As Congressional staff, we were responsible for informing Senators and their staff about fisheries policy issues by providing analysis and recommendations to inform their policy decision making.

In this essay, we are defining "policy" as society's response to problems at various scales of governance—local, state, federal, and international. For example, fisheries policy makers can help develop solutions, and the means to implement those solutions, to balance the ability of an ecosystem to sustainably regenerate fishery resources and the needs of fishing communities to sustain their livelihoods from those resources. In doing so, policies can take many forms, but we are not advocating herein for any specific policy solution. Any solution (and process for finding it) needs to take into account the unique characteristics of any given fisheries problem, and the analytical frameworks for finding such solutions are beyond the scope of this essay.

Instead, we aim to introduce you to what we have learned through our engagement in the policy process so you might assess whether to pursue a policy career yourself. We also speak from our experience working on policy from a federal legislative branch perspective. As you consider a policy career, remember that fisheries policy professionals work in the executive branch agencies, in nongovernmental organizations, and in the private sector at all scales of governance, and such work would likely yield similar challenges and rewards.

Who Can Become a Fisheries Policy Professional?

Fisheries policy professionals hail from diverse backgrounds. Nevertheless, we have noticed that many effective policy practitioners share the following characteristics: broad, interdisciplinary training; excellent communication and negotiation skills; keen critical thinking; open-mindedness; an ability to adapt to constantly changing issues; and an eagerness to continue learning and applying those lessons. As a graduate student in a fisheries-related discipline, you may be beginning to recognize yourself in this description. The same skills you have cultivated to write a concise abstract or to explain your research to conference attendees can be translated into writing succinct summaries of legislation or explaining the effect of a law's provision on a particular stakeholder group. The critical thinking skills that you need to pass examinations and to test hypotheses would serve you well as you look for creative policy solutions that work for multiple stakeholder interests and minimize unintended consequences. It is likely that you have become an expert on a single set of questions driving your thesis or dissertation work, but it is also likely that during your graduate work you have enhanced these other types of broader ways of thinking.

Policy professionals literate in many disciplines and skilled at weaving appropriate data and information together will likely be in the best position to identify, test, and advance robust solutions. There are very good reasons to be narrowly focused on your research during graduate school; however, if you have time to broaden your studies you will find interdisciplinary training will serve you well in a fisheries policy career. The term "fisheries" does not only refer to fish stocks; it also encompasses the social and economic systems—the people and communities—that have developed at small and large scales to harvest fish. Because fisheries require sustainable fish production, it is certainly useful to understand how modeling works, how assumptions inform outcomes, how data are collected and analyzed, and how uncertainty can be addressed. But these technical issues represent only a fraction of policy-relevant information. Many fields come together to inform fisheries policy, including ecology, oceanography, limnology, population dynamics, genetics, economics, business, law, political science, anthropology, psychology, communications, game theory, and many others. Because fisheries harvesting and other management problems are complex, any single discipline in the natural or social sciences is not likely to hold the silver bullet to solving these problems; therefore, it is important to know when and how to consult with experts in different fields. We urge you to consider taking courses and interacting with students and faculty in disciplines outside of your core focus.

Are you thinking "But what I love about graduate school is the opportunity to specialize. I don't have time to take courses in a variety of different disciplines"? Do not rule out policy work, because experts who can deeply investigate one aspect of a problem are vital to solving fisheries policy puzzles. For example, fisheries scientists use extensive data and complex ecosystem models to recommend changes to harvest levels. At the same time, social scientists are also needed to analyze the economic impacts on communities of harvest rate changes and propose systems of allocation or mitigation to make those changes workable. Attorneys are also indispensable when it comes to crafting and improving the legal frameworks that facilitate effective policies. If a policy does not take into account input from these experts and from many other stakeholders, it probably will not work as intended. You do not need to be an expert in any or all of these fields, but you will need to be eager to learn from others who are.

What Skills Will I Develop as a Fisheries Policy Professional?

You might be getting the idea that fisheries policy formation is far from an exact science, and the policy process will not always lead to an ideal, predictable solution that satisfies all parties (in fact, it rarely does). Still, the most interesting aspects of a career in fisheries policy are that one is constantly learning, adapting to changing social and ecological systems, and encountering diverse worldviews.

How to Communicate Effectively with Stakeholders

Working on fisheries policy will require you to learn about the network of stakeholders concerned with a fishery and to understand how they communicate. You will have opportunities to talk to, and to learn from, managers, scientists, fishermen, conservation advocates, economists, community leaders, law enforcement, lawyers, politicians, farmers, corporate executives, and many others. Through these exchanges you can better understand their attitudes and goals, as well as the different ways they may view the same problem. These conversations with diverse audiences will also help you find different ways to articulate how you see the problem, what approaches you are considering, and what actions you think would facilitate progress, which may in turn draw out new and relevant information from stakeholders. You will learn to be an effective listener, to incorporate new evidence, and not to make assumptions about the positions or motivations of concerned stakeholders.

How to Think Critically about Policy Solutions

For any given fishery, you will likely encounter some stakeholders saying harvest should be minimized (or eliminated) and others saying it should be maximized. These contrasting views will provide you the opportunity to think critically about the arguments on all sides and look for lines of reasoning and evidence rooted in the specific ecosystem or community in question. In different cases, you will likely apply different analytical frameworks and criteria for defining and evaluating success. At the same time, you will likely challenge your own assumptions about "how it works," acknowledge your implicit worldview, and be open to going where the new evidence and information take you.

How to negotiate, compromise, and build productive relationships

Most of the time, practical and effective policy agreements do not come from ideological extremes; instead, they come from compromises that land somewhere in the hard-won middle, after extensive give and take. You will hopefully learn to identify common ground, nurture productive dialogue, and cultivate patience about working toward durable solutions rooted in that common ground. You may become skilled at fostering creative approaches to help policies get adopted as intended, and you may come to value policies that incorporate the flexibility and adaptability that allow policy-based management measures to change as ecosystems, habitats, fish stocks, harvesting technologies, market prices, and fishing communities change.

How your work can affect people's lives

One of the most rewarding aspects about working on fisheries policy is the opportunity to experience the policy-making system firsthand. You will meet people who are eager to share their knowledge with you and to help you understand their experiences and points of view. You may even have opportunities to visit stakeholders where they live and work: perhaps you will go to the piers and coffee shops in fishing towns and maybe even get out on a fishing vessel or observe management councils at work. You will learn about how integral natural resources are to the socioeconomic context of a community. You will also gain a better appreciation of how difficult policy formation is, and even if you decide eventually that fisheries policy work is not for you, you will emerge a better-informed citizen for your experiences in the policy world.

In conclusion, fisheries policy—whether at the scale of local subsistence communities or globalized industrial trade or somewhere in between—can be a challenging and rewarding career. Almost every fisheries problem should be approached as a unique situation, since details of the ecological and social environments will vary considerably between and within regions. The fisheries policy field needs people who can understand and appreciate the diversity and complexity of fisheries issues. With solid academic preparation, genuine curiosity, willingness to think independently and critically, and dedication to facilitating the creativity and compromises it may take to advance sound policies, you can have great success in addressing (and hopefully solving) some of today's most challenging natural resource management issues.

Acknowledgments

We would like to thank a number of current and former Hill staff and other policy practitioners who were willing to discuss this topic with us and help improve drafts of this essay. We also appreciate the many fisheries stakeholders from all disciplines and walks of life for teaching us many of these lessons. Finally, we thank our mentors in academia who encouraged us to make our work policy-relevant and to undertake interdisciplinary courses of study.

Biographies

Kristine D. Lynch served as staff for the U.S. Senate Commerce Committee from 2002 to 2009. Before working on Capitol Hill as a 2002 Knauss Sea Grant fellow, she studied environmental policy and behavior at the University of Michigan School of Natural Resources and Environment and did graduate work at Michigan State University's Department of Fisheries and Wildlife. She earned her M.S. by conducting research on comanagement in small-scale artisanal fisheries in Malawi and worked with Dr. William W. Taylor for her Ph.D. doing a social network analysis of fisheries stakeholder organizations. While in school, she also provided research and administrative support for the Great Lakes Fishery Commission and the Michigan Department of Natural Resources.

Kelly M. Pennington came to the U.S. Senate Committee on Commerce, Science, and Transportation on a Knauss Sea Grant fellowship in 2010 and served as professional staff on the committee through 2013. Prior to her fellowship, she earned her Ph.D. in conservation biology from the University of Minnesota. Kelly's dissertation research, with her major advisor Dr. Anne Kapuscinski, investigated the risk of gene flow from genetically engineered fish to wild fish in different environments.

Managing Your Career

Jim Martin*
Berkley Conservation Institute
Post Office Box 1109, Mulino, Oregon 97042, USA

Abigail Schroeder
200 Jensen Road, Prattville, Alabama 36067, USA

Key Points

- Actively manage your career so that your job suits your personality.
- It is critical to choose your mentors, be an active learner, and know when to stay in a job and when you need to make a change.
- Do not depend on your place of work to worry about your happiness. That is your responsibility.

Introduction

Over the course of our careers, we have seen many young professionals get their first job and proceed to go on autopilot, assuming that one job will naturally follow another and that happiness and fulfillment will result. This is rarely the case. All too often, professionals find themselves stuck at mid-career with no clear idea of how to go forward. They are in a job that is okay but not where their real passion is. They find themselves on a career track that is not fulfilling for them. At that point, it is hard to back up and start over. The key for you as a professional is to know yourself and take responsibility for your career decisions early on. Mentors and colleagues can be key resources to help you make those critical career decisions, but ultimately it is up to you.

First and Foremost: Know Thyself

Before you can effectively manage your own career, you need to understand the person for whom you are managing it—yourself.

Think Carefully about the Big Decisions in Life

In our opinion, there are four decisions that are critical to your life's path: (1) Whom will you choose as a life partner? (2) What kind of work will you do? (3) Where will you live? (4) Do you intend to live an ethical life? These decisions will have an extraordinary impact on your happiness and the route your life takes. They also set the context for your career.

Once these big questions are settled, it is time to consider issues that will make a big difference in career effectiveness and happiness.

* Corresponding author: jtmartin@purefishing.com

Be Aware of the Long Game

When you are making career choices, remember that you are making them for the long term. You may not love the job you currently have, but it may be necessary in order to position yourself for your ultimate goal, whatever that may be. Be sure to spend time early in your career understanding exactly what you want out of it. Do you want to be making the big decisions at the top someday? Then plan your career to get you there. Maybe what you really love is fieldwork and you dread the thought of being a desk jockey. That is fine too; just make sure to plan your career so that you do not find yourself trapped behind a desk.

Ethics Are Critical

A tendency to lie, cheat, or steal will not be tolerated in our profession. No one can hide ethical problems for a whole career; they are career enders and eventually they will come to light. In science, and natural resource management in particular, there is zero tolerance for poor ethics. Bottom line: if you cannot justify a decision to your mother, do not do it!

Now that You Know Thyself, Manage Thyself!

Once you have a clear understanding of your personality, talents, and desires, use that understanding to make active career decisions that add value to both yourself and your organization. Actively learning and developing new skills is a critical part of any professional's career development. Based on our experience, here are some points to consider as you go about strengthening your value to your organization.

Make Yourself Indispensable

In order to survive the regular budget cuts that all organizations experience, you will need to develop excellence in the characteristics that are most important to an organization. Be a creative problem solver. Be an excellent communicator. Work effectively in a team. Creativity is critical, it makes you unique—and the more unique you are, the more difficult it is to replace you. Communication and teamwork skills are also vital—these are the skills that make you easy and even enjoyable to work with. When headcount is up for review, you want to be the one-of-a-kind, difficult-to-replace person whom everyone likes working with.

> While I was at Michigan State University working in the Department of Fisheries and Wildlife, I often had the occasion to work with Jim. As I was preparing to leave that position to go on to my next adventure, Jim told me something that has stayed with me throughout my subsequent career "You will be successful not only because you are smart and hardworking, but because you are pleasant to work with." That statement resonated with me and I have tried my best to live up to it ever since—I have found that being able to bond with and relate to your team is crucial to putting out high-quality work.
>
> —Abigail Schroeder

Raise Your Hand

When everyone else is stepping backwards or waiting to be called upon, you want to be the one to step up to the plate. Special assignments are terrific opportunities to learn new skills, work with new teams and mentors, and demonstrate a strong work ethic. They allow you to demonstrate abilities that are out of the normal work situation. These are chances to stand out from the crowd.

Work on What Matters

Most of us are competent enough to do a good job of whatever assignment we take; the key decision is which assignment we decide to go after and which we pass on. You want to be sure that you are spending your energy on the right things. There are only so many hours in the day and not every project or committee membership that comes along is going to move the needle in your organization. When you are looking at career development opportunities, ask yourself the following: Is this aligned with my organization's current objectives? Can I make a difference here? Will this put me out there and get me noticed by the right people, or will I be spinning my wheels? There will be times when you need to take on the things that matter less in order to gain the experience and respect to earn the right to work on more important things later. It is a balancing act you will need to be aware of and manage throughout your career.

Do Not Be Afraid to Create Your Own Opportunity

If you see potential for improvement or a problem within your organization that needs to be tackled, do not be afraid to put a proposal together and volunteer to take on the project. If it is aligned with your organization's strategic goals, then you may be surprised by how much support you garner from the powers that be.

> Alignment is everything here. One of my first large-scale projects with my current organization was developed specifically to tackle one of our highest profile issues on a micro-level. When I brought it before my then boss as a proposal, it ticked all the boxes and I was given leave to proceed with the project—a project that has since greatly expanded from its original scope. Timing, relevance, and a passionate individual to go after it—it is a potent combination.
>
> —Abigail Schroeder

Manage Your Relationships—People Count

In our experience, there are certain skills that help improve working relationships and thus enhance organizational effectiveness and professional achievement. Focusing on your relationships both inside and outside of the organization will go a long way towards increasing your effectiveness.

Give People Your Full Attention

Ever notice that some people are constantly distracted when you are visiting with them, while others give you their undivided attention? Remember how you felt about both kinds of people? If you are going to listen, then listen well. Give feedback and repeat what you have heard to be sure that you are actively engaged. This has a profound impact on people and will increase the effectiveness of any team. Governor John Kitzhaber of Oregon is a classic example of this quality. Even though extremely busy, the governor gives each person his whole attention for as long as they are visiting. People often cite this quality in surprise and delight. Listening carefully and without distraction is one of the best ways to show respect, and in relationships, respect is everything.

Practice Empathy

Most of the people you work with are people who have a genuine passion for the resource, whether it is fisheries, water, wildlife, or forestland. It is important to remember that no one is truly a villain, despite what differences in approach, opinion, or philosophy you may have with them. No matter where their experiences have taken them, they inevitably come back to a genuine love of and appreciation for the resource that hooked them in the first place. When working with constituents and the public, remember that people feel passionately about their points of view and can react angrily to setbacks or differences in opinion. Most angry people are afraid of losing control, of losing a way of life, or of a potential economic impact on family and community. Understanding that they are scared gives you an empathic view that helps in dealing with conflict and keeping your objectivity under fire. A good example of this situation is the often-controversial regulatory decisions that fish and wildlife agencies have to make. We have seen the use of empathy and understanding when reacting to criticism and conflict go a long way towards defusing an escalating conflict in cases such as deer management in Michigan and salmon management in Oregon.

> Once during a time of extreme controversy over salmon management in the Northwest (known as the "Salmon Wars" of the 1980s), Oregon Fish and Wildlife (ODFW) Director Jack Donaldson was burned in effigy by commercial fishermen during a demonstration staged for the media. The ODFW staff was furious and wanted to counterattack. Director Donaldson calmed the group by stating, "They are mad as hell because they are scared to death." His empathy calmed our group and gave us a different perspective of the conflict. None of us ever forgot this demonstration of empathy. This example has proven useful in subsequent conflicts.
>
> —Jim Martin

Never Lose Your Temper

As advisors on important political decisions, scientists are expected to remain objective in spite of the passion they feel for the resource. Losing your temper is a sign of lost objectivity and diminishes your professional credibility. In our experience, the vast majority of the time we lost our temper, we have regretted it later.

Communication Is Key

The most effective professionals are thorough and well prepared with their communications, skills that are highly valued in most organizations. Be brief and to the point when communicating. Why does your point matter? What are the alternatives, which are best, and why? How certain are you? If you cannot explain an issue to a neighbor, you are probably being overly technical. Agency leaders do not have time for extensive reports. You need to present an executive summary and attach the details for their review. The best way to give a good talk is to pick an interesting topic—if you do not get to pick your topic, then take the time to find an interesting angle to it. Once you have your topic and your angle, practice your delivery multiple times in front of an uninvolved audience and incorporate their feedback. Finally, respect your audience and the other speakers—always stick to your time.

Work with the Media

The media is the key to helping you tell a story to a wide audience and can be crucial in moving the needle of public opinion in a way that affects policy. Work with media you trust and be trustworthy in return. Spend time building that relationship. Resist the temptation to overstate your recommendation or distort your level of uncertainty. Remember your audience. Use real life analogies and plain language to explain what is important for people to know—ask yourself, why should anyone care about this? Always volunteer to be available for fact checking or last-minute checks for accuracy. Your goal is to be a trusted source, so act accordingly. A classic example of this is Dr. James Hansen of NASA on the topic of global climate change. His factual basis and common sense language makes him one of the most quoted authorities in the popular media.

History and Context Matter

At work, every problem appears in the now, but inevitably, the context and the history of the problem are crucial to finding a robust solution. Understanding the moving trajectory of a problem through history is critical. Key stakeholders view the problem differently, and therefore their criteria for a durable solution are different. These differences affect the perception of adequacy and fairness to people in a solution. Natural resource problem solvers need to always be aware of this and strive to balance biological adequacy and political acceptability within the context of the problem. The history of a problem is often an important insight. As an example, there is a lot to learn about problems in salmon management in the Pacific Northwest from the book *Salmon Without Rivers* by Jim Lichatowich, which explains the 150-year effort to restore salmon in the Pacific Northwest and provides a rich argument for the importance of taking history and context into account. The scope and history of a problem is almost always a key perspective in problem solving.

> In his book, Jim Lichatowich emphasizes the importance of testing assumptions that are often blindly accepted as cherished theories in the profession. The assumption that hatchery technology could make up for habitat loss in recovering wild salmon was a particularly relevant example.
>
> —Jim Martin

Finally, Manage Your Career

Managing your career implies that you are taking an active role in determining your destiny. As young professionals, we sometimes are so eager to get a job and do it well that we overlook the importance of reflecting on where our career is going. We mistakenly think it is the responsibility of our organization to be sure we are satisfied and happy in our work. We should be asking ourselves on a regular basis whether we feel good about the direction of our career. Is there alignment between our personality and our work? Are we continuing to learn and grow? Are we making a difference?

Think of Your Career in Three Phases

In the first third of your career, explore the profession as widely as possible. Try new assignments, take new jobs, learn from new mentors, and be active in your professional society. These things will help you find the job that best fits your passion. The key is to find the job that perfectly fits your personality and talents. In the second third, build your reputation by helping your teams within your organization meet their mission and goals. The final third of your career is your time to lead and teach. Thinking about your career in these phases early on will aid you in charting out where you want your career to go and will help you think both tactically, for the phase you are currently in, and strategically, for the long term.

Find the Right Mentor, but Do Not Stay Too Long

One of the keys to learning fast is to focus on the right mentor rather than the perfect job. There is a reason that certain units consistently produce excellence in their team members. Find out what that is. Find mentors, work with and for them, and learn from their example. You will be working in a team environment with others who are also attracted to that mentor. One of the downsides of doing this is that it can be easy to get too comfortable and stay longer then you should. Be conscious of when the learning curve is flattening out and seek out new mentors and challenges.

It Is Okay to Be a Little Scared

There is a tendency to coast once you have gained initial competence. We get comfortable in our surroundings and with our team. If you do not feel a little fear of having bitten off too much, that is a sign that you are leveling off in your learning. Never stop pushing yourself, and look for new challenges regularly. Most people want to work within their comfort zone. When we get outside of our comfort zone, we risk making mistakes and suffering criticism. But that risk is balanced against the opportunity to grow and learn at a faster rate. Knowing your own personality, your strengths and weaknesses, can help guide how much risk the learning experience is worth in the long term.

> When I was 6 years into my career after college, with a wife and two small children, I decided to go back to Oregon State University for a master's degree. That was scary for me, but I met my great mentor, Dr. Carl Schreck, there, who influenced my thinking on future directions. I never returned to research on the Rogue River upon completing my degree. Instead, I went to the ODFW headquarters to pursue policy work in the hot political crucible. This was a career-changing decision that I have never regretted. The change began with the scary decision to go back for more education at a critical time in my career. I was way out of my comfort zone!
>
> —Jim Martin

Pace Yourself

Be prepared to burn the midnight oil from time to time, but know that working hard is not the same as working smart. If you are always working too hard, you lose balance and perspective. You may think you are demonstrating commitment and dedication, but often it shows poor judgment in balancing work with life. Ultimately, an excessive workload is unsustainable in the long term and will curtail your health and your career.

Be Willing to Blaze Your Own Path

In any organization, there are a number of predefined career development paths. You start in an entry-level position and then climb the ladder until you stall or reach the pinnacle for that career path. Perhaps you spend several years in limbo waiting for the next wave of retirements to open the path back up. The established career path may be a perfect match for you and your career ambitions—it may take you exactly where you want to go, and that is fine. However, if you find that it is not making sense to you, do not be afraid to take the proverbial road less traveled. It may require more imagination, more sponsorship, and more input from your mentors, but it is entirely possible to blaze your own trail through an organization, or through multiple organizations, and to carve your own unique niche in the process. Keeping this in mind is important for three reasons. First, it strikes a balance between your own needs and the needs of your organization. Second, it prevents you from stalling out in limbo at a critical point in your career. Finally, following your own path gives you a unique and unusual skill set and perspective, one that is not easily replaced.

> My career path to date has hardly been what you could call linear, I started off with a degree in international relations, detoured through fisheries, and did a master's degree in supply chain management before finally ending up working in the forest industry—and my career's far from over. Work has taken me from academia to industry, from north to south, and to all points in between. Although it may be unconventional, the range of experiences has proven invaluable to me as I progress in my career and has given me a unique point of view, a precious commodity in any organization.
>
> —Abigail Schroeder

There Is No Substitute for Enthusiasm

Enthusiasm for your work is the key to your creativity and energy. If you are not feeling enthusiastic about your work or surrounded by enthusiastic people, it is time to rethink your choices. Finally, do not be ashamed to wholeheartedly commit to your organization. If we were to pinpoint a single factor that will make a difference in the success or failure of your career, it would have to be passion—or a lack thereof. Most of the greatest mentors in our careers have been universally enthusiastic…a contagious characteristic!

> My father has been an influence and mentor to me throughout my life. He has often told me that the key to his success is the fact that every day when he wakes up one of the first things he asks himself is "What can I do today to make my organization better?" That single question serves as both his focus and the key to his enthusiasm.
>
> —Abigail Schroeder

In Conclusion

Professionals who are the most effective and fulfilled in their work continue to develop and grow through their whole life. They recognize key decisions and take on the responsibility to make them thoughtfully. They use networks to bounce ideas off and learn from mentors and colleagues. They assertively pursue projects and work assignments that will benefit both their organization and their own professional development. They read constantly and learn from others to broaden their understanding and creativity. Above, we have suggested tips and approaches that, from our experience, have proven to be useful, but we are still learning and always will be. We challenge you to do the same.

Biographies

For eight years, Jim Martin taught natural resource problem solving at Oregon State University. During that time, he brought in experts from across the spectrum of natural resource management, as well as former students who had recently made the transition, to give advice to students on preparing for their careers.

Abigail Schroeder completed her graduate work at Michigan State University and is now working in the forest industry after taking an unconventional career route into natural resources.

Recommended Reading

Leopold, A. 1949. A sand county almanac: and sketches here and there. Oxford University Press, New York.

Lichiatowich, J. A. 2001. Salmon without rivers: a history of the Pacific salmon crisis. Island Press, Washington, D.C.

Covey, S. R. 1989. The Seven habits of highly effective people: powerful lesesons in personal change. Free Press, Washington, D.C.

Carnegie, D. 1936. How to win friends and influence people. Simon and Schuster, New York.

Martin, J. 2008. Leadership: a tale of six mentors. Fisheries 33(9):454–457.

What They Do Not Teach You in Graduate School: How to Be an Effective Listener, Be a Good Participant and Chair at Meetings, and Deal with Opposing Views

Russell Moll (Retired) and Shauna Oh*
California Sea Grant College Program
Scripps Institution of Oceanography, University of California, San Diego
9500 Gilman Drive, Department 0232, La Jolla, California 92093, USA

Key Points

- An important aspect of a professional career is to develop good listening skills by showing patience and respect when others are speaking.
- Those same listening skills can be put to good advantage in participating and chairing small group meetings.
- Good listening skills can be used with success and as a key step in resolving conflicts.

Introduction

With a newly minted graduate degree in hand, many of us depart school for a career with an impressive set of skills and knowledge. Often, recent masters of science, Ph.Ds., and post-doctorates are at the cutting edge of their fields, possessing the latest knowledge and research methods. They stand poised to make an outstanding contribution in the years ahead to their respective fields of science. There are, however, some skills that are not usually taught in a formal education setting and are best learned on the job. In the long run, these skills may be just as important to career success as those learned through formal education. This vignette addresses three key skills that are essential for a successful career: good listening, meeting management, and addressing opposing views.

Becoming a Good Listener

Perhaps the most challenging concept to grasp is why it is worthwhile to become a good listener. Our premise is that being a good listener is key to becoming a good scientist and advancing one's career. In departing graduate school with a wealth of new knowledge, there is an irresistible desire to share that knowledge. Yet, it is imperative to take the time to listen carefully to others to show respect and, in the case of colleagues, show support. Opinions formed or decisions made on partial information are unwise and can lead to misunderstandings or even hurt feelings and damaged relationships and outcomes. Showing respect and support to others comes in the form of taking the time to listen to what they have to say. This

* Corresponding author: shaunaoh@ucsd.edu

can be key in developing better working relationships with your colleagues and an important learning opportunity. We cannot learn from the wisdom of years of collective experience of others without taking the time to listen and process that information.

Although there are not hard and fast rules that make a good listener, some hard-won experience has served us well over the years in this regard. From our experiences, we have found a good first step is patience. In a first meeting with an individual, first minutes at a small group meeting, first moments of an interview, or similar first encounters, exercise patience in expressing your point of view. Be careful not to formulate a response while the other person speaks. As an example, we have found that during job interviews, candidates who are too quick to provide a response before listening to the complete question or all that an employer has to say can come across as impatient, presumptuous, or overly confident.

Your comments are more effective if they are shaped by what you learn as you listen. On many occasions we have disagreed with what is being said by another person, yet resisted the temptation to interrupt to correct or interject an opinion. Rather than offend the other speaker and risk changing the tone of the discourse to one that is defensive and or hostile, our patience in listening was rewarded with cordial discourse.

Another key step to becoming a good listener, yet one that is difficult to accept, is to suppress the desire to multitask. With today's technology, folks often grab devices such as laptops, smart phones, or tablets at the earliest opportunity. That is a sure sign that your attention is divided and you are not listening to what is being said or providing the respect due to the person speaking. Through numerous personal experiences, we realize that in this busy world, to be a good listener and an outstanding contributor to your field, one has to budget his or her time carefully and listen with undivided attention. Doing so will be foundational to succeeding in the related skills of meeting management and dealing with opposing views.

Meeting Participation and Management

Carrying the concept of being a good listener to the next level, we offer ideas on the topic of how to participate in small group meetings and make a lasting and impressive contribution. No matter how gifted your science might be, if you cannot work effectively with others in small groups, you will see your career stagnate. One of the best strategies for success in small group meetings is to come with an open mind and anticipate a large measure of listening and learning at the beginning. Often, at meetings, we use the strategy of letting others make their points at the outset, to let them play their hand first, thus giving us an advantage. We then formulate effective responses that range from agreeing wholeheartedly with prior viewpoints to sharing opposing views to countering with effective arguments. This strategy has been especially effective for us in strategic planning meetings.

Being a good meeting participant includes arriving with an open mind and a good attitude. Too often, we exit a meeting mumbling, slightly out loud, "That was a waste of time!" If you arrive at the meeting with the attitude of making your collective time productive, you will likely seek creative means to achieve that end. We often take quick notes at the beginning of the meeting as we listen to others. Use those notes to look for common threads among meeting participants, and seek to tie them together into a successful action plan. As mentioned above, do not multitask during meetings. The risk in doing so is to signal to others that you are disinterested in the meeting, that other matters are more important, and that listening to differing viewpoints is not a priority. Taken together, this can be viewed by others as an indication of disrespect. While at the meeting, your primary task is to be a meeting participant, and other things can often wait for either a break in the meeting action or meeting conclusion.

Another key aspect of success at small group meetings is to understand the meeting objectives and stay focused on those objectives. We recall many instances when the meeting conversation has wandered off point. Our approach, which is one that often garners respect as wise counsel, is to bring the meeting back to the task at hand. This can be done with a simple statement: "That is a very interesting point, yet our primary task here today is...." Hand in glove with keeping the meetings focused is listening carefully at the outset.

As your career progresses, you may find yourself appointed as the meeting chair. Although this can come as a natural progression with career advancement, a more common yet unexpected pathway is because of the absence of the appointed meeting chair. In some cases, during a meeting you may unofficially assume the role as meeting chair because an idea you have proposed takes hold and gains traction. In all cases, being prepared to serve as meeting chair is vital to the success of the meeting. Below are suggestions for chairing small group meetings based on our collective experience.

As emphasized above, you should listen as carefully as you can. Thinking back on the numerous instances where we have served as meeting chair, we found the role of listening becomes the principal task. As chair, you move from the realm of participant to facilitator and are tasked with hearing every point of view and making sure that each participant is able to express his or her points of view. If you see one person dominating the discussion and not listening and another or others not speaking, take the time to draw the nonspeakers out and thereby to some degree silencing the dominating speaker. It never hurts to speak directly to someone and ask him or her to share his or her view. Such an approach also stifles excessive multitasking. Again, make quick notes to keep track of the discussion and, if possible, assign a note taker so that, as meeting chair, you are free to facilitate the meeting. Those same notes can also help you keep track of the names of participants and where they sit at the table.

The success of the outcome of the meeting depends on the chair's ability to keep participants engaged to the objectives of the meeting. As such, make clear at the outset the objectives of the meeting and how they may be met. We have found that our successes as chair of small group meetings demonstrate leadership and propel a career forward in leaps and bounds. Likewise a good meeting outcome when one of us served as chair usually led to an invitation to chair future meetings.

Addressing Opposing Views

Inevitably, during encounters and meeting participation, you will find that you and/or colleagues disagree with what is being said by the others. We have found that when opposing views come during one-on-one conversations or small group meetings, there is an opportunity to strike a successful compromise and be viewed in a favorable light. Once again the success of approaching opposing views is to initially listen carefully. You may get the drift of the disagreement very quickly, but you will not understand its full depth and origins until you hear much more of what the other person has to say. Again, by listening to what others have to say, you will be more informed and better prepared to respond with constructive and effective ideas on how to resolve the conflict or change minds to another point of view. Do not be afraid to express your opposing views, but our suggestion is to do so after you have heard others first.

The concept that we have followed numerous times in suggesting a resolution or differing idea is to follow the adage "use soft words but hard arguments." Making your case in a firm and clear manner is effectively achieved with carefully chosen words and ideas. You also must accept the premise that not all disagreements will be resolved or that these may take a lot of time to resolve. However, with time and compromise, most positions will shift to a less strident disagreement.

In fashioning a resolution, while there are many ground rules to guide you, we offer these two common-sense suggestions that have worked well for us. First, begin with areas of common agreement and see how you can progress from there. Mercifully, we have learned that in science, most of us remain open minded and in the face of strong support for one position tend to come around to a common solution. Second, remain cool, calm, and professional in demeanor. We have observed this is difficult for some personalities. Someone who erupts in anger or frustration at a meeting does an enormous disservice to him or herself and his or her long-term career aspirations. Finally, if you have a firm belief of the validity of your position, especially in matters of new discoveries in science, you are always encouraged to stick with that belief with polite discourse that pushes the frontiers of knowledge forward.

Conclusion

Through this vignette, we have discussed three important skills to consider for career development. The wisdom of employing good listening skills will garner the respect and recognition from your colleagues. Perfecting those listening skills will help you better manage meetings and assume leadership roles in small group settings and more effectively resolve opposing views—the marks of a seasoned professional. We have discovered that much of what gets done in our profession happens in these small meetings, and with mastery of these skills, you will be poised to make impactful contributions throughout your career. Those same listening skills will also serve you well in dealing with opposing views either in individual discourse or larger group meetings.

Biographies

Russell Moll has conducted research in the nearshore marine environment, salt marshes, African mangrove systems, the Great Lakes, small lakes, and temperate and tropical rivers. His appointments include director of the Cooperative Institute for Limnology and Ecosystems Research, associate program director in biological oceanography at the National Science Foundation, and director of the Michigan Sea Grant Program and director of the California Sea Grant Program. He has a B.A. from the University of Vermont, M.S. from Long Island University, M.S. from the University of Michigan, and Ph.D. from Stony Brook University. He now serves as a private consultant and cruise ship lecturer.

Shauna Oh has studied the invasion ecology of green crabs *Carcinus maenas*, the implication of mating success of Dungeness crabs *Cancer magister* in their management, and the role of science in the decision-making process at a regional fishery body. She is the associate director at California Sea Grant, a university-based program that supports marine research, extension, and education. She strives to refine the skills discussed in the vignette while contributing to the scientific and policy aspects of research program management. She earned her B.S. at University of California-Los Angeles, an M.S. at Humboldt State University, and a Ph.D. at Michigan State University.

A Human Side of Fisheries

Shawn J. Riley* and Amber D. Goguen
Department of Fisheries and Wildlife, Michigan State University
480 Wilson Road, 13 Natural Resources Building, East Lansing, Michigan 48824, USA

Key Points

- Fisheries management requires engaging people.
- Understanding the importance of and encouraging public participation of all stakeholders will improve effectiveness of achieving fisheries objectives.
- Becoming a better people person is a skill all fisheries professionals should strive to achieve; we provide tips on how to go about being a better people person.
- There are other profound, more personal reasons for deepening relations with the people for whom you work for and work with that will enrich your career.

Introduction: "We need better people people!"

In the enthusiasm and excitement for fish and their watery habitats, it is easy to lose sight of the people in a fishery. As college educators, we continually pester resource agency directors and division chiefs about the traits they desire in professionals starting their careers. The nearly unanimous replies are "We need better people people" or "We want employees with better people skills." But what do they mean? When pressed further, administrators reveal that they are seeking employees who can communicate well with external partners and stakeholders and work effectively with others in their agency. These comments are reinforced by recent graduates in fisheries who often muse that although their education and training mainly focused on biology or ecology, they now find themselves working mostly with people.

Likely, this is not the first time you have read or have been told that you will spend much of your career in fisheries with people, not fish. You may have been attracted to fisheries because of a love for fish or aquatic ecosystems, or perhaps you imagined a career working alone in the great outdoors, or maybe you even hold a secret aversion to working with people. You are not alone! Most people are not drawn into fisheries because they want to work with people. However, the preponderance of evidence suggests that fisheries management is as much a social as a biologically scientific discipline.

We encourage professionals, young and old, to think more deeply about the human side of fisheries: the people who participate in fisheries, the people who manage fisheries, and the people who advocate for fisheries. As we will cover in this chapter, most of the world is moving toward increased participatory governance (fisheries being just one aspect of governance); the need for fisheries personnel who can work with diverse publics is acute. People skilled at the human side of fisheries will unlock rewarding career possibilities: the

* Corresponding author: rileysh2@msu.edu

knowledge, experiences, and friendships developed in close association with resource users and professionals.

Why People Matter in Fisheries: Fisheries Are a Public Trust Resource

So, how did people get so entwined with fisheries management in the United States? The Food and Agriculture Organization of the United Nations defines a fishery in terms of "people involved, species or type of fish, area of water or seabed, method of fishing, class of boats, purpose of the activities or a combination of the foregoing features" (FAO Fisheries and Aquaculture Department 2012). Just a little reflection on that widely accepted definition reminds us that although fish exist without people, a fishery does not. Fish are a public trust resource, which means they belong to everyone. This public trust arrangement necessitates quality communication between public trustees (government), public trust managers (fisheries professionals), and the beneficiaries of the public trust (current and future generations). The need to become a better people person thus becomes part of the job of being a fishery professional (Figure 1). Thinking about your role as a manager of valuable public trust resources—fish, the water in which they live, and the benefits derived from those resources—will increase your effectiveness as a professional.

More Reasons Why People Matter: Participatory Governance and the Concept of Stakeholders

Fisheries management began with expert-driven and passive-receptive models of public participation. However, as the array of people and interests in fish have grown and diversified, a

FIGURE 1. Children at the 4H Great Lakes Natural Resource (GLNR) camp are taught to clean a fish they caught charter boat fishing that morning. The GLNR camp engages youth with natural resource management professionals that emphasize the importance of future generations in fisheries conservation and management under the Public Trust Doctrine. Photo by Amber Goguen.

new era of public participation in environmental decision making was born. This movement away from representative forms of democracy toward participatory governance requires, by law, that public input be considered in environmental decisions. New management practices emerged that actively sought to incorporate diverse perspectives in decision making and coined the term "stakeholder." A stakeholder is defined as any person who is affected by or affects fisheries or fisheries management decisions and actions. Stakeholders are not only voices to include in decision-making processes, but may be a rich source of knowledge about ecosystems and changes taking place (Figure 2).

The concept of stakeholder recognizes that not everyone has the same stake in fisheries and not all stakes are equal. Some stakeholders are more interested or affected by decisions than others. Effective fisheries professionals work hard to improve how decisions are made; there is public demand for increased transparency and involvement. Disagreements frequently arise from decisions not about the technical attributes (e.g., stock assessment), but about the processes used in making decisions. In some of our recent studies, we found that the public's confidence and trust in resource agencies is affected more by the perceived quality of processes used to make decisions than the perceived technical competence of the biologists. These findings suggest that the job of being a fishery professional requires awareness and skillful practice of governance, as well as management.

FIGURE 2. A fisheries professional interviews an angler, one of the many stakeholders in fisheries management. The information gleaned from such surveys is used to help inform management decisions and improve communication between fisheries agencies and stakeholders. Photo courtesy of Mike Thomas.

The Importance of People in Politics: Internal and External Politics Affect Fisheries Management

An inevitable part of being a public trust manager of public fisheries is the politics that accompany people and public decisions about public resources. Fish biology can and should inform policy affecting fisheries. Nonetheless, if people are an integral component of fisheries, social science—the disciplines concerned with society and the relationships of individuals within a society—can and should inform policy as well. Policy decisions in a public trust arrangement are made and implemented by people for people, most of whom value and support fish and their habitats. The term "politics" originated from the Greek word *politikos*, meaning "of, for, or relating to the people, citizens." Despite the grumblings of many professionals, politics is by definition impossible to remove from fisheries.

Your skills as a people person—your desire to engage other people, your willingness to listen, your ability to communicate the knowledge you are privileged to possess, and your skills at bringing together people with differing values—will improve your chances of making lasting contributions to the conservation of resources about which you are so passionate. Knowing who the key players are and forming a positive relationship with them can go a long way in your ability to implement positive change in fisheries. Social sciences can help you anticipate issues by providing insights into the complexity of decisions that result from demographic and political shifts within society. Local, state, and federal politics affect fisheries management, but so can the politics internal to an organization and the connections that organization has to external players.

The Objective Role of Fisheries Professionals: "Park your Values at the Door"

As a word of caution, fisheries professionals in their role as public trust managers will more effectively fulfill that role if they "park their own values at the door." That is many, if not most, fisheries professionals also are anglers or at least passionate about fish. All professionals bring their own set of values with them in everyday life. It is all too easy to blur distinctions between whose values are being served in setting objectives and carrying out management. The difficult role of public trust managers can be best accomplished by focusing on providing the actual decision makers with the comprehensive information they need to make decisions, including insights about stakeholders. We are not advocating that fisheries professionals become valueless drones, methodically maneuvering through the motions of management–far from it! We are advocating, however, that professionals entrusted with public resources make more durable contributions by staying mindful of blurred distinctions.

Social Networks: It Is More than Just a Job; Fisheries Is a Vocation and a Community

As should be apparent, the evolving role of fisheries professionals necessitates communicating—formally and informally—with the array of people involved in fisheries. Among those people are a special set: your colleagues (Figure 3). You will amass a network of colleagues throughout a career. You would be smart to start early and never stop building that network. Those networks will support your continued intellectual and emotional well-being while providing valuable perspectives. Every professional we know has gone through low times in his or her career, times when programs are not working or, worse yet, not even given a chance to work. There will be times when some stakeholders, including those inside your organization, will work against your best intentions and efforts. It is at times like these that your social network functions like a safety net and prevents you from landing hard on the bottom. This same network helps propel you back up with new ideas and a renewed sense of commitment. It is not only your accomplishments within the field that matter, but how you overcome conflict, disappointment, and rejection to "fight another day." These are hard to do alone but happen more readily if you are a people person.

You may be amazed with how much can be accomplished in casual peer-to-peer interactions with stakeholders. Some of the best ideas are drawn out on the back of napkins over lunch or liquid refreshment after the workday is supposedly done. When you practice your profession as a vocation rather than a job, you never stop learning or adapting. Your intensive schooling likely prepared you for the technical aspects of the fisheries profession, yet to sustain that preparation you must become a lifelong learner and adaptor. The people with whom you surround yourself, including an array of stakeholders, will greatly influence your thinking. They will provide the needed honest critique of those ideas drawn out on a napkin. And, they will cheer you on to success when your ideas become reality.

Becoming a People Person: Learn to Connect with People about Fisheries

Scott Bonar, in his book *The Conservation Professional's Guide to Working with People*, states, "One of the best skills people can have is the ability to manage themselves." University degrees provide technical education, but most people skills typically are learned in the school of hard knocks. Of course, you can take courses, and likely have, in communications, group dynamics, and conflict resolution. Ultimately, it is those people who have the ability to

FIGURE 3. Researchers conducting an annual sturgeon population census survey in the Black River, Michigan. The people with whom you work and work for are important part of your career. Start early and never stop building your social network. Photo courtesy of the Molecular Ecology Laboratory at Michigan State University.

manage themselves and generate a strong enough desire to improve their people skills, and then persistently and creatively practice those skills, who notice significant changes in their abilities. The research on people skills is unanimous: people skills are not born to a person, skills are learned.

At its essence, being a people person is about communication. This trait of how well people communicate is one of the defining characteristics of humanity. Other animals—even fish!—can communicate, yet none do so with the sophistication and intent of humans. Pick appropriate situations in which your communication achieves small victories that lead to bigger successes. Jump at opportunities to share your privileged knowledge of fish whenever you can. Insights from social sciences–economics, sociology, social psychology–can provide the rigorous basis for developing communication. Yet, it is not just about being a polished communicator. It is about engaging people, stakeholders, in fisheries. Think about some of our great naturalists—Jane Goodall, Aldo Leopold, Teddy Roosevelt, and Carl Sagan, to name a few. The common characteristic among these people was their ability, but more importantly, all shared a deep desire to encourage people to think more, to value more, and to be more involved.

Conclusion: Profound Reasons to Engage People as a Fisheries Professional

There is a more personal and profound reason to consider the human side of fisheries. Your career in pursuit of fish will take you on a distinctive journey to unique places and along the way you will amass encounters—some good, some bad—with a remarkable array of people who enrich your life. All these people, many of whom are stakeholders you serve, make valuable deposits to your memory bank. You will make a little money along your journey as a fisheries professional, but it will be your memory account and not your individual retirement account from which you draw inspiration and satisfaction. Each of those deposits made by the people for whom and with whom you work accumulates interest. You can make withdrawals from that memory account anytime without diminishing the capital! Engage people about the wonders of fish and their watery habitats. It will be through people that objectives for fish are created and achieved, or not. Fish in today's world are sustained or lost through the actions of people.

Biographies

Shawn J. Riley is a professor in the Department of Fisheries and Wildlife at Michigan State University (MSU). He teaches a course at MSU titled "Human Dimensions of Fisheries and Wildlife" and is a frequent speaker on the subject in the United States and Europe. During 2009–2010, Shawn was a senior Fulbright fellow at the Swedish Agricultural University in Umeå. He earned a Ph.D. at Cornell University and B.S. and M.S. degrees at Montana State University. Between his studies at Montana State and Cornell, he was a research biologist, management biologist, and state-wide program biologist for Montana Fish, Wildlife, and Parks.

Amber D. Goguen is a National Science Foundation graduate fellow at Michigan State University, where she is pursuing a Ph.D. in the Department of Fisheries and Wildlife. Her research focuses on the human dimensions of wildlife management, with a particular interest in understanding the role consumption of fish and wildlife as food plays in coupling human and natural systems. Amber earned a B.S. in Wildlife and Fisheries Conservation and a B.A. in Studio Art from the University of Massachusetts-Amherst. After graduation she worked for the Ecology of Bird Loss Project and Buffalo Niagara Riverkeeper.

References

FAO (Food and Agriculture Organization of the United Nations) Fisheries and Aquaculture Department. 2012. Fisheries glossary. FAO, Rome. Available: www.fao.org/fi/glossary/. (February 2014).

Go Forth and Tell Them What You Have to Say!

Kelsey Schlee*
Center for Systems Integration and Sustainability
Department of Fisheries and Wildlife, Michigan State University
115 Manly Miles Building, East Lansing, Michigan 48823, USA

Stan Moberly
606 East Tenakee Trail, Box 599, Tenakee Springs, Alaska 99841, USA

Key Points

Good communication can help you earn the trust of others, translate knowledge into action, and help fisheries professionals maintain relevance and the support of society. To become an effective, communicator you must

1. Create a communication strategy:
   a. Be able to answer who you are communicating with, what you are communicating, how you are going to do it, and why.
      i. To help answer these questions, remember AMMO to define your **A**udience, **M**essage, **M**edia, and **O**bjectives.
2. Use your people skills:
   a. Be professional inside and out.
   b. Listen to your audience and others around you.

Introduction: The Eureka Moment

About a year or so ago, during one of our chats, Stan was giving me (Kelsey) a rundown of the status of fisheries science today, what we are doing right and what we are doing wrong, trying to give me some perspective on the strengths of our profession, as well as the challenges I might face throughout my career. After we had spent some time focusing on what was going well, marveling at the advancements that have been made since the beginning of his career, his tone suddenly changed. Throwing his hands up in dismay, he said, "You know what, Kels? Our new technologies, models, and theories are all well and good, but one thing people in our line of work still don't always realize is that you can't just chuck your science on the public's doorstep and run, expecting them to know exactly what you wanted to say and how to use the information! We really must make a stronger effort to communicate or risk being misunderstood and limiting the progress we can make toward our goals!"

After that initial proclamation, the necessity for effective communication between fisheries scientists and the rest of society became a theme in our interactions. He explained how, over time, he came to the realization that better communication helps create trust between

* Corresponding author: schleeke@gmail.com

fisheries scientists and other members of the community, translates knowledge into political and resource management action, and maintains fisheries scientists' relevance and support from the rest of the public. Realizing how important communicating science is, once he was promoted to director of the Fisheries Rehabilitation, Enhancement and Development Division at the Alaska Department of Fish and Game in 1982, Stan required all biological staff to make a concerted effort to engage with the public in some way, shape, or form at least twice each year.

Stan's dedication to promoting effective communication strengthened my belief that it is central to bringing out the best in fisheries science. The following is a synthesis of the best of what Stan and I discussed and learned about communication over the course of a year, information that we believe fisheries professionals should have to help them form an understanding of this valuable skill. The way we see it, effectively conveying information requires two major components: (1) a communication strategy, and (2) as Stan would put it, "people skills."

The Communication Strategy: Pick 'em, Convince 'em, Ask 'em

Before you can really throw yourself into trying to communicate your science to any audience, you need to seriously consider who you are trying to communicate with, what exactly you are trying to communicate, how you are going to do it, and why you have engaged in this undertaking in the first place. Along the way, Stan and I picked up a useful acronym to help remember the essential components of a good communication strategy, helping to answer those vital communication questions: AMMO (**A**udience, **M**essage, **M**edia, and **O**bjectives).

First, answer the "who" question by addressing the first letter of the acronym, A. The A stands for **A**udience and refers to the group of people with whom you are trying to communicate. When choosing your audience, think about who is most impacted by the information that you have, who might be in the best position to help you achieve your goals, or who might have competing interests that could benefit from your side of the story. There is a good chance that you will be speaking to very different audiences throughout your career. To quote Stan, "one day your audience could be the local kindergarten class learning about where fish come from and the next, a bigwig policy maker you are trying to persuade to ban the production of genetically modified fish." Once you have identified your audience, keep them in mind throughout the rest of the development of your communication strategy because the characteristics of your audience influence your answers to what you are trying to communicate, how, and why.

After you have identified your audience, you are ready for the second communication question: what is it that you are trying to communicate? For help, remember the second letter of AMMO, defining your **M**essage. The message is your main point; it is the information your audience should remember most. In the case of fisheries scientists, the message often boils down to the main finding of your research. This main message should be quickly evident and referred to often. We all know how unbelievably frustrating it can be to sit through a long presentation that takes darn near a millennium to get to the point; do not let this happen to you! Remember to keep in mind the attention span of your audience. Put yourself in their shoes and think of the extent of your own attention span, especially when learning new information from outside of your own area of expertise. In addition to making your message memorable and arriving at your main point quickly, be sure to tailor your message to your audience. Use language they are comfortable with, emphasize the importance of your message to them, and be sensitive to their ideas and concerns. Tailoring a message to your audience well means putting forth effort beforehand to get to know the group with whom you are trying to communicate. Read a bit about the people with whom you will be interacting, learn a little about their line of work, or perhaps meet with them ahead of time if possible. At the very least, practice delivering your message to someone outside your field and ask them to repeat your message back to you. If you do not hear what you were expecting from them, go back to the drawing board, rework your message, and try delivering it again!

With your audience defined and message in hand, it is time to consider how you will deliver it; you need to choose your **M**edia. Will you be delivering a PowerPoint presentation, writing a newspaper article, designing a pamphlet, or guest starring on a radio talk show? In order to decide which is best for you, again consider the characteristics of your audience to reach them as effectively and completely as possible. Consider the types of media with which they are comfortable. Is this a magazine-reading crowd, or would they be better served if you created an interactive Web site or blog? To better identify the characteristics of an audience with whom you are unfamiliar, use the power of observation; lean on the past experiences of others who have worked with this group before, or go ahead and ask them directly what they prefer if the situation is appropriate.

After choosing the best media for communicating your message, be sure you are prepared to effectively use your selected tool for engagement. If you are unsure how to create a visually appealing, easy-to-use, and effective Web site, for example, do your own research on Web design, take a class, hire help from someone more experienced, and always solicit constructive criticism to help you improve your product. Similar strategies can be used to help perfect your use of all types of different media. If you are relying on your personal performance (i.e., giving a presentation), make sure to throw practice into your toolkit of improvement techniques. Have others listen to what you have to say to make sure that it is engaging and direct. It can also help calm your nerves and allow you to develop your own method of coping with a racing heart and sweaty palms.

You are a persuasive communicator. You have convinced us of the message you delivered during your awe inspiring speech; we are on your side. Now what? What do you want us to do to help? You must answer the final communication question, why are you doing this?

Now, it is time for the last letter in AMMO, the O for **O**bjectives. To actually propel yourself closer to your goal through communication, you must know what results you are anticipating. While you may be simply trying to inform your audience in some cases, and this is a legitimate objective, oftentimes you are seeking some action from your audience. Maybe you want them to adjust a policy, provide you with financial support, or change their fishing practices. Make sure what you want your audience to do is made abundantly clear. Be sure to use your discretion to determine the tone in which you ask and that you have fully explained why you are asking. Do not forget to consider your objectives when designing your communication plan.

Tips for Every Professional: Say Your Please and Thank-Yous and Wash Behind Your Ears

Now that you have the skeletal structure of developing a communication strategy well in hand, it is time to address those so-called "people skills." People skills are important in communication because they can help to make you more approachable and believable to your audience, helping you to achieve your goals. Remember, the stereotype of a scientist, as some may see it, is a rather reclusive personality who lives in his or her head and has more than a little trouble relating to others, let alone explaining the value and application of his or her work. While this is indeed a stereotype and certainly does not cover the vast array of our undoubtedly wonderful personalities, many stereotypes contain a tiny grain of truth. In this case, that grain of truth is that we probably did not initially get involved in this field to talk with people about fish, what we do, and why we do it. So, we might need to beef up the people skills or at least reflect on them just in case. While this portion of the paper may come across a little more like a lecture from your mom than anything else, these tips are a worthy reminder and might help to reassure the rest of the world that we fisheries folks are not the scientist stereotype. We are a perfectly relatable, understandable, respectable group of professionals with something important to say.

Since we are professionals, it is important that we be professional. Maintaining the appropriate level of professionalism is a big part of improving the outcomes of your communication efforts. Without it, you may not command the respect you need; others may not take you seriously, and therefore, will ignore or overlook your message. First, always be polite and cool-headed no matter what situation you find yourself in. Undoubtedly, at some point, we all find ourselves communicating an unpopular message, and at those times, mastering this professionalism skill can be difficult. Remember that while you should stand up for yourself, rudeness and disrespect will not sway your challenger to your side. Instead, it might cause your message to be lost and certainly will not win you any friends. On the other hand, demonstrating composure and rationality might.

While respect for others is important to remember in times of conflict, it is equally as important to remember in your day-to-day interactions as well. Consider the following to up your game: address others by their proper title until otherwise specified, use correct grammar and punctuation when emailing, and remember to send thank-you notes when appropriate.

As superficial as it may sound, you also need to look the part. Of course, there is no need to wear a three-piece suit to work if that is not part of your workplace culture, but never ever show up for a meeting with the governor in your Birkenstocks. When you are trying to deliver a message that you want taken seriously, dress like you want to be taken seriously. Wash and groom as appropriate for the occasion. If you are not sure about how you should be presenting yourself, observe how others are dressed or simply ask.

Maintaining your network is also an essential component of professionalism and enhancing communication. Maintaining your network means continuously engaging the people you know inside and outside your field of expertise. You will want to do this because the people you know can help communicate your message, might be more receptive to a message you are trying to communicate to them, or might provide you with advice or another form of support. Make sure to acknowledge colleagues when you see them. If you have not seen or spoken with someone in a while, drop them an e-mail or hand-written note or be willing to provide assistance when requested, if you are able to do so. Do try, try, try your very best to remember names and important information about your contacts personally and professionally—like where they work, what do they do, or what they told you about their dog and their kids.

Along this same vein, always remember to listen. Many skills of professionalism can be easily and quickly adopted, but some other people skills take a little more finesse to master. Listening is one of these skills. If you have a message that you are trying to communicate, you must really listen to the feedback of others to understand why your message did or did not elicit the results you wanted. Sometimes, really deep listening requires a lot of patience and empathy, other people skills that will most certainly help to improve your communication skills as well.

In conclusion, communication is critically important to our profession. It can help us earn the trust of others, translate knowledge into action, and maintain our relevance and the support of society. To become better communicators, we should be able to answer what we are communicating, why we are communicating it, who we are communicating with, and how we are going to do it. Creating a communication strategy using our acronym, AMMO, encourages us to critically consider these questions and answer them thoroughly. In addition to our communication strategy, we must also remember employ our people skills to be the relatable, understandable, respectable group of professionals that we are. Now, go forth and tell them what you have to say!

Acknowledgments

We would like to extend our thanks and appreciation to all the natural resource professionals who have kindly contributed their thoughts about communication with us over the past year. Additionally, we would like to thank the creative minds behind this book, Bill Taylor, Nancy Léonard, Abby Lynch, and Marissa Hammond, for helping us to share the knowledge "we wish we would have had" with our peers for the betterment of our profession.

Biographies

Kelsey Schlee is a native of mid-Michigan and a Great Lakes enthusiast. While writing this vignette, Kelsey was a master's student working under Bill Taylor at Michigan State University. There, she studied the effects of global climate change on thermal habitat availability for Brook Trout *Salvelinus fontinalis* and took a special interest in learning to communicate science to policy makers, with the hope of helping to "move the needle."

Stan Moberly is a dedicated and experienced member of the fisheries community, specializing in fish habitat. He began his journey as a biologist when he graduated with his B.S. from Kansas State University and went to work for the Nebraska Game and Parks Commission. From there, he moved on to Alaska to work for the Alaska Department of Fish and Game, where he held a variety of positions, including director of the Fisheries Rehabilitation, Enhancement, and Development Division. Stan finished his career working with Northwest Marine Technology in Washington State. In addition, Stan has been a celebrated and influential member of the American Fisheries Society.

Learning Is an Ongoing Experience: What Fishers Have Taught Us during Our Careers

Vahdet Ünal* and Huriye Göncüoğlu
Ege University, Faculty of Fisheries, 35100, Bornova, Izmir, Turkey
and
Mediterranean Conservation Society, 370 Sk. No: 13, 35680, Foca, Izmir, Turkey

Key Points

- If you want doors to open, do not give up on your readings and fieldwork, even if you feel like a fish out of water in the beginning.
- Your science should serve not only you, but also your interested society.
- The big picture is fisheries management itself. Try to grab it.
- Developing true and lasting communication with fishers and other stakeholders is essential.
- Political will and commitment are as important as the quality of the science. Actually, they are essential for successful fisheries, and without them there is no success.

Introduction

It was 1994 when I (VÜ) first started my career in fisheries. I had been studying agricultural economics, but love of the sea caused me to change my discipline. At first, I was like a fish out of water, confused and helpless, and spent days and nights on board different types of Mediterranean fishing vessels. My impressions of the fishers were that they thought they knew all about the sea and fish. Those simple solutions to the problems they talked endlessly about, they knew them better than anyone else.

There were two sentences I heard constantly: "Those who manage the fisheries don't know anything about it," and "University is no use to us." It was actually the fishers who helped me choose my field of study: fisheries management. Since the fishers thought those managing fish did not know anything about it, I was interested in learning how fisheries should be managed. How were fisheries managed in developed countries?

During the time that I was doing my master of science in fisheries, I had the opportunity to go to Canada, one of the most developed coastal countries, for three months, where I learned that the depletion of northwest Atlantic Cod *Gadus morhua* stocks in the early 1990s caused 40,000 fishers to lose their jobs. I remember an ironic headline in the *Washington Post* entitled "Fishermen Stranded" and was intimidated by the field of fisheries management since I saw that even such a developed country had failed to manage its own fisheries. With

* Corresponding author: vahdetunal@gmail.com

multiple stakeholders, including biologists, economists, ecologists, engineers, politicians, sociologists, lawyers, and even anthropologists, along with the intricate nature of marine species, the big picture of fisheries management was very complex. Fisheries management was a dynamic, complex, uncertain, multidisciplinary web of problems, and establishing a successful solution seemed to be impossible.

The undeniable concepts of fisheries management are renewable resources, tragedy of the commons, marine ecosystems, fishing communities, scientists, decision makers, coast guards, administrations, managers, governance, targets, strategies, policies, priorities, dynamism, complexity, uncertainty, multiple disciplines, professions, and many other factors. Although I was frightened by all of the above factors included in fisheries management, I took on the challenge in order to both understand how management was done and to study further so that I could be the first in my country to both teach fisheries management and implement it.

Here, we reflect on years of experience collected mostly from our own studies and the supervision of master's and Ph.D. theses, as well as on our successes and failures in the same geographic area (Gökova Bay-Southern Aegean Sea-Eastern Mediterranean, Turkey), dealing especially with small-scale fisheries. We have, of course, been influenced by many, including students, fishers, and managers, as well as recognized scientists such as Harald Rosenthal, editor in chief of the *Journal of Applied Ichthyology*, who stated,

> Education does not only provide you with an ample opportunity to develop your own career and comfortable life-style. Having received a high quality education burdens you also with the responsibility to serve your society.

This quote suggests that when selecting a subject or formulating a project or a thesis, we should consider how it will serve our fisher society, as well as where, and how, our research will fit in the above-mentioned puzzle of fisheries management. For us, this philosophy is both meaningful and satisfactory. We felt responsible for the marine ecosystem and the fisher community, thinking substantially about developing true and lasting communication with fishers and other stakeholders.

Developing a true and lasting communication with fishers and other stakeholders is essential. If you do not take sufficient time and effort to understand fishers, you may be disillusioned to find that your research is not welcomed by them. So, what is wrong? There may be nothing wrong in terms of the science, but your work may have no practical use. Fishers and other stakeholders would like to know about the results of your study, explicitly if it has yielded results and whether you have provided them with any solutions. Your focus should be on concrete, applicable results and be able to meet current requirements.

Be Persistent and Doors Will Open

I (HG) have been pursuing my academic career for seven years with growing enthusiasm and faith, but I have yet to get a permanent paid position. An important factor that helps me continue, aside from my eagerness to learn and discover, is persistence, inspired by a fisherwoman I met during one of my field studies. Memduha Dinc (in her late 40s at the time of this writing) lives in a log cabin and works alone on her boat to provide for her sick parents, struggling against the harsh conditions of the sea and life. She said in one interview, "I never thought of a job other than fishing."

The fisherwoman's resolution struck me. As a young scientist studying fisheries, I was moved by this daring woman and her stand. I then started questioning what I could do for fisherwomen, how to empower them and improve their life conditions. After finishing my master's thesis on fisherwomen in 2008, I coordinated two projects involving fisherwomen, with budgets of US$60,000 and $57,000, with two of the most active nongovernmental organizations (Underwater Research Society, Mediterranean Conservation Society) in Turkey. These projects were widely covered by the national media as well as many international Web sites. I am the first person to study fisherwomen in my country, working as a consultant for nongovernmental organizations to empower women in Turkey. I owe this accomplishment to my persistence, to choosing a unique subject and placing my belief in it following the principles mentioned in this article.

Try to Be an Ordinary Person, Not Superhuman!

Try to be humble. Traditional fishers of the Mediterranean are generally hot-blooded; it is easy to communicate with them but difficult to sustain this communication. Being able to receive continuous dependable data from fishers, keep them involved in the project, and have them attend your meetings can be an art form. Your schooling is never enough. You either do this with an experienced master to guide you or work your own way through. But do not forget that you only have one chance with a fisher. The keys to success are your sincerity, integrity, power of adaptation, and, most importantly, knowledge. In terms of establishing a long-term relationship, it helped us a great deal to give feedback to the fishers by providing updates on the projects and research findings.

Choose the Right Instruments and Attitude

Within the framework of a large-scale project (approximately $2 million budget) on integrated coastal zone management in Gökova Bay, a fishermen's pledge was agreed upon. Fishers and project coordinators sat down to draft the "I shalls" and "I shall nots" of the pledge, which supported sustainable fisheries. A small no-take zone (NTZ) was also proposed in the inner bay for the protection of an ecologically important sea grass species in the Mediterranean marine ecosystem. This NTZ aimed to provide a nursery from which the fishers would benefit. These facts were explained in the meeting, where both parties agreed to the content. When the project was concluded, however, there was neither a pledge nor a NTZ. The failure of the project was caused by the fishers finding the consultation process disingenuous and that the project was already a fait accompli, even without the fishers' involvement. Leaving a bad impression is an obstruction that not only you will face, but will affect others after you.

The above NTZ had failed in the $2 million project, but a couple of years later, six different NTZs were proposed through a mere $20,000 project (UNDP GEF Project TUR/SGP/OP4/RAF/), this time with a different team and a different attitude. Shortly after, these zones were officially announced and remained intact. So, how did this approach succeed?

This project relied on the results of scientific studies. Most of the project budget was spent on meetings that served to delegate to, and consult with, all stakeholders. We (VÜ and project team) introduced fishers and other stakeholders to comanagement, which is the sharing of power and responsibility between the state and resource user groups in the management of natural resources. We informed the fishers and other stakeholders of the benefits of NTZs. For meetings, we chose lecturers from independent, experienced, born-leader fishers who could influence the fishers. We gave examples of success stories from different parts of the world, especially Mediterranean countries with similar characteristics. We evaluated earlier scientific studies and went through current records of local fishery cooperatives. These findings detected a significant decline in the cooperative incomes generated by fishers, showing that the fishery needed assistance to remain economically viable and that shrimp and grouper fishing, which constituted one-third of the total income, was in deep trouble. Shrimp had already been depleted, and grouper were being threatened by illegal spearfishing.

We were able to convince the fishers that these threats to the fishery needed to be handled in a new way rather than through traditional management approaches and measures. We never gave false hope or overpromised. Eventually, a final meeting was held in the capital, Ankara, with the attendance of all stakeholders and an official from the government. The fishers were happy with the results but had one single condition for going forward with the implementation of the project: very strict protection of the closed area. Today, there are six sensitive areas designated NTZs in the Gökova Bay reserved only for juvenile growth. But now those areas are threatened due to illegal fishing. The reluctance and inefficiency of legal authorities was the exasperating reason leading us to establish a nongovernmental organization, the Mediterranean Conservation Society, which helps fight illegal fishing. With fishers among the cofounders, we made sure the fishers continued to be involved in the entire process and immediately took upon ourselves the task of stopping illegal fishing by buying a fast and powerful boat to start a ranger system, employing trained locals and fishers, in the protected areas. At the time of this writing, the project is still ongoing.

Big budget projects may not yield big results, but sometimes, with a very small budget, you can achieve big goals. The magic lies in the right attitude and instruments.

In the end, all stakeholders agreed that when there is insufficient political will and commitment, there is no hope for success. Therefore, we started comanagement movements in the area. On September 6, 2013, we organized another meeting on NTZs, which all stakeholders attended, where we informed the fishers and other stakeholders of the results of the three-year monitoring studies, which included videos taken by submarines, cooperative records and numbers submitted by representatives of nongovernmental organizations, scientists who worked in the field, and cooperative managers. Following the meeting, fishers were asked whether they wanted these zones to be open to fishing. All representatives, especially leaders of the fishing community, greatly appreciated the studies and the positive results except for a number of fishers who demanded that the coordinates of two of the six NTZs be reorganized, protesting that winter-time fishing zones would be too limited. This demand, which meant that portions of the zones would be open to fishing, was approved by the majority of the attendees, as well as decision makers from Ankara. We are happy that the NTZs are still largely supported by the fishers three years after their creation. Strong enforcement and proper consultation with stakeholders are essential components of successful fisheries management.

Concluding Remarks

I (VÜ) find the FAO (1995) definition of fisheries management quite appropriate; I spend many hours on it in my classes. I could even dedicate an entire term to this definition:

> The integrated process of information gathering, analysis, planning, decision making, allocation of resources, and formulation and enforcement of fishery regulations by which the fishery management authority controls the present and future behavior of interested parties in the fisheries, in order to ensure the continued productivity of the living resources.

Don't the majority of the problems we are facing today stem from the fact that fisheries managers are unable to predict, and thus control, the actions of the stakeholders, especially fishers? The answer is obviously yes, but there is even more to it. We came across further problems on almost every concept present in the definition of fisheries management, including lack of data and management plans, weak enforcement, and so forth. Establishing successful fisheries management in a given location requires much more than what is written in the definition of successful fisheries management.

In conclusion, we believe

- Comanagement could be a good approach, but it requires the will and ability of stakeholders to work together.
- Although fisheries management requires high quality scientific studies, studies alone are not sufficient, and there is no certain recipe or guarantee for success.
- Political will and commitment are as important as the quality of the science for successful fisheries management.

Acknowledgments

We would like to thank T. Bodur and S.-J. Youn for their assistance in English editing and all the referees who helped during the reviewing process of this vignette.

Biographies

Vahdet Ünal studied agricultural economics at Ege University-Turkey until 1989 and continued his education at the same university as well as Central Institute of Fisheries Education in India and Barcelona University in Spain on fisheries economics and management. Besides his research studies and lectures on fisheries economics and management, Vahdet Ünal is also the current chair of International Relations & European Union Commission at Ege University, Faculty of Fisheries. He coordinated the Sub-Committee on Economic and Social Sciences under General Fisheries Commission for the Mediterranean of FAO from 2009 to 2013. He is a cofounder of the Mediterranean Conservation Society. Ünal speaks English and Spanish.

Huriye Göncüoğlu is a Ph.D. student at Ege University, with a master's degree focusing on fisherwomen. Her Ph.D. dissertation is on the efficiency of small-scale fisheries in southern Aegean, Turkey. She is also tutoring and mentoring on leadership and social entrepreneurship at the Turkish Educational Association and the Association of Businesswomen in Izmir on behalf of the Change Leaders Association. She is a cofounder of the Mediterranean Conservation Society. She speaks English and Spanish.

References

FAO (Food and Agriculture Organization of the United Nations). 1995. Guidelines for responsible management of fisheries. In Report of the expert consultation on guidelines for responsible fisheries management, Wellington, New Zealand, 23-27 January 1995. FAO Fisheries Reports 519.

The Odyssey of a Fisheries Scientist in Greece in the 21st Century: When the Journey to Ithaca Is Still Harsh but Probably More Interesting than Ever!

VASSILIKI VASSILOPOULOU* AND PARASKEVI K. KARACHLE
Hellenic Centre for Marine Research
46.7 km Athens Sounio Avenue, Post Office Box 712, Anavyssos Attiki 19013, Greece

ANDREAS PALIALEXIS
European Commission, DG Joint Research Centre
Institute for Environment and Sustainability (IES) Water Resources Unit (Unit H.01)
Via E. Fermi 2749, Building 46 (TP 460), Ispra (VA) I-21027, Italy

Key Points

- Build trust between scientists and stakeholders through high quality science and proper communication.
- Put theory into practice through a meaningful participatory process.
- Define scientific priorities according to policy needs.

The Ithaca Analogy

In his poem "Ithaka," the Greek poet Constantine Cavafy uses the familiar story of Homer's Odyssey as a metaphor for the journey of life. Odysseus's 10-year voyage home from the Trojan War, with its many twists and adventures, is a metaphor for a fulfilling life. The most important symbol in "Ithaka" is Ithaca itself. Ithaca, Odysseus's island kingdom, represents both the starting and the ending place. Cavafy suggests that the meaning of life is the journey and the experiences along the way, not the destination itself; it is the path in between that makes life worth living:

Ithaca

As you set out for Ithaka
hope the voyage is a long one,
full of adventure, full of discovery.
Laistrygonians and Cyclops,
angry Poseidon—don't be afraid of them:
you'll never find things like that on your way
as long as you keep your thoughts raised high,
as long as a rare excitement

* Corresponding author: celia@hcmr.gr

stirs your spirit and your body.
Laistrygonians and Cyclops,
wild Poseidon—you won't encounter them
unless you bring them along inside your soul,
unless your soul sets them up in front of you.

Hope the voyage is a long one.
May there be many a summer morning when,
with what pleasure, what joy,
you come into harbors seen for the first time;
may you stop at Phoenician trading stations
to buy fine things,
mother of pearl and coral, amber and ebony,
sensual perfume of every kind—
as many sensual perfumes as you can;
and may you visit many Egyptian cities
to gather stores of knowledge from their scholars.

Keep Ithaka always in your mind.
Arriving there is what you are destined for.
But do not hurry the journey at all.
Better if it lasts for years,
so you are old by the time you reach the island,
wealthy with all you have gained on the way,
not expecting Ithaka to make you rich.

Ithaka gave you the marvelous journey.
Without her you would not have set out.
She has nothing left to give you now.

And if you find her poor, Ithaka won't have fooled you.
Wise as you will have become, so full of experience,
you will have understood by then what these Ithakas mean.

—C. P. Cavafy

Introduction

Being a fisheries scientist in Greece is a personal challenge; neither fisheries nor research have ever been a priority in the country, and hence, once you have picked that path, you know it is going to be an adventurous journey. We present experiences of three fishery professionals at different stages of their careers: the first one has traveled a rather long, interesting journey while the other two, not far from their starting points, have set off for different destinations (i.e., different Ithacas). It is not the destination that counts but the journey itself, sailing full of dreams on a long voyage "full of adventure, full of discovery" to be part of the world of fisheries science. As told in these three stories, the packing list for this journey should include high-quality science aimed at answering policy and societal needs through effective collaboration with experts from other scientific domains and close interaction with end users.

High-Quality Science Is the Most Crucial Item on a Fisheries Professional's Career "Packing List"

My (VV) career started more than 25 years ago, with a B.S. in biology, an M.S. in biological oceanography, and a Ph.D. in fish biology and ecology. There were few fisheries scientists at the time and the main task was to acquire basic knowledge of the life history strategies of key, commercially important species through a few targeted national research projects. Then, the first European Union (EU) funded projects began to flow in. They had a rather short duration (2–3 years) and dealt with specific issues of concern. Hence, the data and knowledge, although important, had a rather sporadic and fragmented nature and did not provide an overview of the state of the stocks and sustainability of the fisheries.

Meanwhile, through EU subsidies, the commercial fleet renovated and became more efficient, with landings increasing until the mid-1990s. Then, the situation changed dramatically as demersal stocks showed severe signs of overfishing! The lack of management plans based on sound, scientific data and knowledge became evident. Fishers began complaining, putting the blame on either pollution or overfishing, the latter resulting from competing fishery sectors, never from their own practices. In addition to the medium-scale fisheries of about 700 trawlers and purse seiners in Greece, there was a huge fleet of some 17,000 artisanal vessels scattered all around the country, and each sector blamed the other for the decline in catches. At that point, fishery administrators seemed incapable of dealing with the problem, and the time for fishery scientists to take a more prominent role came with the emerging need for data and knowledge with which to conduct robust assessments that would shed light on the picture. Being a young fishery scientist at that time, I realized that knowledge provided through our research could aid in decision making. Communication platforms were gradually built, and scientists were invited by authorities to present results on issues of concern. Since then, we have focused on keeping science relevant in order to answer evolving needs. There are many examples of policies that have failed dramatically because they have not taken into account scientific findings, so it became obvious that sound decision making can only be based on solid scientific advice.

Deterioration of fish stocks, however, was a fact in almost all EU countries. The Common Fisheries Policy[1] proved ineffective for sustainable fisheries management, even in spite of numerous reforms. Basic pillars contributing to the evolution of policy were the development of policy-oriented science and the establishment of suitable monitoring projects, while visualization of outcomes through appropriate tools enabled interaction interfaces with managers and fishers. However, communication with fishers was another major challenge! Until a few years ago, communication was rare and fragmented, but things have improved recently through targeted communication activities outlined in specific EU research projects. A good example of an effective interaction with fishers was at a meeting with representatives from trawler associations, where results from 10 years of onboard and scientific survey data were shown; it was obvious that mean lengths of an important target species of Mediterranean demersal fisheries were decreasing, and the mean size of landed fish was below the established minimum landing size. A constructive debate took place on the need to increase awareness of those results, also raising issues of poor enforcement of existing measures by

[1] General information on the Common Fisheries Policy (CFP) is provided in Box 1. The official Web site of the European Commission contains details on the CFP (http://ec.europa.eu/fisheries/cfp/) and CFP reform (http://ec.europa.eu/fisheries/reform/index_en.htm).

Box 1. Common Fisheries Policy

Due to the uniformity of the seas in terms of nonexistence of borders and boundaries, the European Union (EU) launched the Common Fisheries Policy (CFP) in order to manage its fisheries resources more effectively and under a collaborating umbrella among the Member States. Through the CFP, which was formally established in 1983, the EU set the rules for sustainable fisheries and provided Member States with the tools to enforce these rules. One of the key points of the CFP was monitoring the size of the EU fleet, preventing its expansion and moving towards size reduction. Negotiations with third-party countries, fisheries product prices, and development of aquaculture were also included in the CFP. Funding scientific research and data collection were also important in order to ensure a sound basis for policy and decision making (European Council 2013). Yet, the 2008 CFP has proven inadequate to achieve its goals, and for that reason the EU has moved towards its reform, with the new CFP, effective January 2014.

authorities and weak compliance on the fishers' side. They finally agreed to adopt responsible fishing practices if they were based on solid scientific advice and, hence, fair measures. On the research side, we committed to doing our best to provide high-quality science and establish effective mechanisms of communication and interaction with the fishers. Since then, there have been many steps towards building and maintaining trust between scientists and stakeholders from different action arenas. Although this is a continuous participatory process that takes effort and time, it is both necessary and rewarding.

Collaboration between Stakeholders and Scientists Is a Not-to-Forget Item on the Packing List

I (PKK) got all my degrees in Greece: B.S. in biology, M.S. in hydrobiology, and Ph.D. in ichthyology. When the time came to search for a job, I wanted a real job with long-term prospects, a job that I could build upon and move forward. I found one as a researcher in my field of expertise and in the best research center in my country! But before that, a variety of part-time jobs made up my curriculum vitae. Yet I was fortunate that these jobs were closely related to my desired profession and from those jobs I gained valuable experience!

Probably one of the best professional experiences I ever had was serving as a scientific advisor for fishermen's unions. I had the unique opportunity to work closely with fishers on issues important to them, mainly legislation (the making of, amending, and enforcement) and the economic problems they faced (from fuel costs to fish prices), and represent the fishers in domestic and European arenas. I also had the opportunity to participate in meetings of the Regional Advisory Council for the Mediterranean[2], which is an EU body exclusive to stakeholders. Through that work experience, I came to realize that what I was taught at the university was the theory of fisheries, but putting theory into practice in everyday life was definitely a different story and a great challenge for me!

[2] Regional Advisory Councils, http://europa.eu/legislation_summaries/maritime_affairs_and_fisheries/fisheries_sector_organisation_and_financing/c11128_en.htm.

Through my efforts to combine the best possible theory with the most applicable and widely accepted practice, it became evident that partnerships, not only with scientists from other fields, but also with fishers, should be established; for example, when creating management plans of fishery resources, the socioeconomic consequences of possible measures should be considered through the development of alternative scenarios. I remember that during a Fisheries Congress there was a psychoanalyst working with fishers who faced huge economic problems due to the collapse of the fish stocks. The psychoanalyst was literally crying for help from us fishery scientists to support fishers by proposing management measures that took into consideration the socioeconomic impacts on local communities. A well-esteemed ichthyologist replied by saying that "we are merely ichthyologists and can only look at the fish." Yes, that is true, we are ichthyologists, and we need to admit that we do not know everything! However, putting theory into practice is like putting the pieces of a big jigsaw puzzle together with scientists from different disciplines collaborating with end users. From my experience so far, I think that this is a major role for fisheries professionals: to be the connective link between their science and other fields, scientific or not. We should aim to provide the best possible science, but we need to keep in mind that it is only by putting all those pieces of information from different sources together that we get a holistic perspective on issues of concern.

"Packing Smart" for Sound Decision Making: Linking Science with Policy Requirements

I (AP) am a biologist, with my M.S. and Ph.D. in environmental biology and management of marine resources. I initiated my research career as a research assistant in the Hellenic Centre for Marine Research. At that time, I considered research as a means to study marine ecosystems following my personal interests and ambition to publish my work in scientific journals. Getting grants for competitive projects forced me to expand my field of research and, consequently, to acquire a wider perspective of marine science and its role. I realized that research is not a personal ambition or a way to advance your career, but a public service. Fisheries science is an applied science and thus has direct linkages to and impacts on local societies, ecosystems, and policy. Then the question is "Who decides and orients the priorities of research?" We should expect a variety of answers to this question, depending on the perspective: politicians, decision makers, managers, and scientists. This process may start with the scientists, but it is implemented with the contribution of all the other relevant players. This is the twofold social role of research: (1) prioritize the issues that should be incorporated into policy (e.g., manage overfished stocks) as related to environmental, social, and economic needs, and (2) provide robust techniques, data, and tools to support policy implementation (e.g., management plans for sustainable fisheries).

After more than a decade in a purely scientific organization, I moved to a position where my main responsibility was scientific support to policy. I have acquired a new perspective of marine science, and I have appreciated the research I have done more than ever. I have come to recognize and value the impact of marine research on decision and policy making, as well as scientists' responsibility and duty towards society and the environment. High-quality science is the best evidence for policy, and its importance is best highlighted by the ecological and economic failures of former decisions. Such failures resulted in the recent reform of the Common Fisheries Policy in the EU, aimed at management plans for all fisheries based on the best possible scientific advice.

Hence, young professionals in fisheries should consider bridging between their science and policy requirements. They should bear in mind that science-driven policy is based on concrete evidence and clear actions that ensure the implementation of any fisheries-related political vision (e.g., sustainable management of specific stocks and economical sustainability of the related sectors). Emerging professionals in fisheries science should challenge themselves with different perspectives in fisheries management, experience varying scientific approaches, and communicate with stakeholders who usually have different thoughts and priorities. Communication creates more open-minded scientists, contrary to the introverted nature of scientific work. It constitutes the means to transfer scientific knowledge to nonscientific stakeholders, and this process is quite demanding due to the conceptual heterogeneity against similar terms (e.g., sustainability) and functions (e.g., stock assessment).

Conclusions

Our recommendation for young professionals is that they should try to produce high-quality science, keeping in mind that the science should be applicable, relevant, and meaningful for solving everyday fisheries management problems. However, fishery scientists are not miracle workers; they are part of a chain where managers should ask the right questions and scientists should provide relevant and meaningful answers. The latter should feed into a transparent and participatory process in which fishers play an important role. This process will ensure wide acceptance of measures and will contribute to sound decision making aimed at the sustainability of resources for generations to come.

Fisheries science is Ithaca, the starting and ending point of the fisheries scientist's professional journey ("without her you would not have set out"), and as important as she may be, she is the means for a beautiful journey "wise as you will have become, so full of experience," with all the challenges and controversies you have to face and overcome!

Biographies

Vassiliki Vassilopoulou is a research director at the Institute of Marine Biological Resources and Inland Waters of the Hellenic Centre for Marine Research in Athens, Greece.

Paraskevi Karachle is an associate researcher at the Institute of Marine Biological Resources and Inland Waters, of the Hellenic Centre for Marine Research in Athens, Greece.

Andreas Palialexis is a scientific/technical officer at the Joint Research Center in Ispra, Italy.

References

European Council. 2013. Common Fisheries Policy regulation 1380/2013. Official Journal of the European Union L 354/22. Available: http://eur-lex.europa.eu/LexUriServ/LexUriServ.do?uri=OJ:L:2013:354:0022:0061:EN:PDF. (January 2014).

Leadership in Practice

Leadership Starts at the Beginning

Douglas Austen*
American Fisheries Society
5410 Grosvenor Lane, Bethesda, Maryland 20814, USA

Key Points

- Assuming a leadership role requires preparation.
- Prepare by thoughtfully developing and exercising your leadership skills.
- Explore formal and informal opportunities to develop your leadership skills both inside and outside your profession.

Introduction

> Leadership is accomplishing things that reach beyond solitary abilities by acting—and getting others to act—with a maturity that surpasses limited self-interest.
> —John Baker, president of READY Thinking (an organizational and leadership development firm)

Stepping into a leadership role can be a daunting challenge but can also be one of the most rewarding experiences of your career. For most of us, making a difference as a natural resource professional is a fundamental objective that defines our career and is a reason why many of us chose this field. We want to improve the resources that we love and leave them better, not only for us to enjoy, but also for future generations. We do this through many avenues—research, management, teaching, administration, advocacy, and much more. Through all this, one consistent challenge that is a component of nearly every one of these pathways is the need for leadership.

What is leadership and how does it differ from management? One way to differentiate leadership from management is that leadership is a vision of where to go while management is the process of getting there. Peter Drucker (2003) states, "Management is doing things right; leadership is doing the right things." And, sometimes, leadership is doing hard things. Marty Linsky's (2012) preferred definition of leadership is "disappointing your own people at a rate they can absorb." Providing good leadership (think of Lincoln, Eisenhower, Martin Luther King, Jr., and Mandela) comes in many forms and at every level, but without it we struggle to advance in the right direction. Leadership takes on very difficult challenges that need to be addressed but often go heavily against the grain of people's comfort zone. Unfortunately, examples of poor leadership, often resulting from a misguided or failed vision of the future or based on inappropriate or corrupt values, are many, and stories abound of programs and projects that faltered due to inadequate leader-

* Corresponding author: dausten@fisheries.org

ship. Yet, developing leadership skills and maturing as a person who can lead, as well as a good person to be led, is something that all of us can accomplish with a little planning, some homework, practice, and simple tenacity.

Take just a moment to consider some of your experiences with leadership, either as one being led or as a person in the role of leading. Why was that person (or you) placed in that role? Were you prepared for the situation? What did you think about that person's ability to be successful in that role? How did you feel about the consequences?

From our early days as a youth in sports, in music, or with friends, we have dealt with leadership challenges and opportunities. Sports teams need leaders (captains) and teammates to be led, girl and boy scouting was designed, in part, to help build leadership skills, and your garage band needed to have someone willing to pick the music. As an undergraduate, the leadership opportunities expand through classroom-related activities, campus organization involvement, and job or internship positions. In graduate school, the introduction to professional society committee involvement and elected positions and leading and managing a research project opens a new door to leadership opportunities and challenges.

My experience with this was probably very typical. I was not a team captain or in school government and the only time in school that I ran for an elected club office, I lost. Yet, when an opportunity to do something that I was interested in was suffering due to lack of leadership, I only had minor trepidation about stepping into that role. Was I prepared? Generally, not. Did I help the situation? Not always. What I did learn was that we can all become better leaders and that the skills and emotional maturity needed to be an effective leader can be developed. From those early and probably awkward attempts at leadership to being in a position to lead organizations and conservation programs, I have learned a few lessons that may be helpful to others. You will have to find your own pathway, but possibly one or more of these lessons will give you some ideas to advance your own leadership capacity.

Learning and Practice Is as Important with Leadership as It Is with Succeeding in any Activity

Clearly, a person must practice in order to develop a reasonable level of skill at any activity. How many of you know of friends, or maybe yourself, who play a sport but rarely practice or practice poorly? Can you really expect to substantially improve your game if you never practice the skills that you regularly have difficulty performing?

We suffer the same lack of preparation for leadership. We put people in the position of leading a project, work team, or organization without ensuring that they have the skills to be successful. The unfortunate consequences of this common situation even have a name, the Peter Principle (Peter and Hull 1969), which describes people being advanced to positions beyond their competency. While it may appear that some people seemingly have greater natural abilities to lead than others, all of us can improve our skills with careful practice and learning. The real lesson here is to not wait until you are thrown unprepared into leading a major project to develop your leadership skills. Learning to be a good leader in little activities will set you up well for success in larger ones. Take on that committee leadership role or volunteer to lead a fundraising effort. Do something to get yourself in the game. In fact, you may never get that chance on the large projects unless you can show that you have developed at least a moderate level of leadership competence in the small ones.

Opportunities to Practice Are Everywhere

Identify leadership opportunities in all walks of life, not just as a professional. One of my passions has been running, and my first real leadership opportunity came through a running club that needed a president and also a race director. Both provided early opportunities to develop new skills and learn leadership in a friendly setting. We must be open to these calls. In most of our lives, we are constantly balancing many other demands on our time, such as family, school, personal interests, sports, church, or other activities, that add value to our lives and that of others. In many of these, we simply want to participate; we want to enjoy or be enriched by the activity or just be a part of the occasion or event. In others, we sense the need, or are actually called on, to provide leadership. These can be immensely valuable opportunities not only to enrich our own lives and those of people we care about, but also to learn and practice skills.

One of my most valuable leadership opportunities came not in work, but in our church, a place where politics, personalities, and beliefs can conspire and mix in ways that can challenge even the best leaders. In this case, consolidating two church locations, selling a facility and expanding another, and creating a new vision were wrapped together. Leading a group that was charged with creating a sense of common vision for this type of change was both daunting and fraught with landmines while simultaneously affording an opportunity to work with skilled leaders from other professions. The experience changed me and helped to change a church. In other situations, I have had the wonderful opportunity to lead youth groups, be a running race director, and lead mission trips. None of these have had anything to do with conservation of our natural resources. All of these were situations where I simply responded to a need that was important to me. The resulting leadership opportunity provided unique challenges that led to growth as a leader, contributed to a larger cause that was important to me an others and, as I have looked back, were situations that built the various skills needed to be a more effective leader. Professionally, this has also played out in leading American Fisheries Society (AFS) committees and holding AFS offices at the state and regional level, as well as chairing or co-chairing state and national conservation projects. The leadership challenges are similar across the spectrum, and it was clear to me that skills developed in one forum are easily translated to others.

Leadership Skills and Attributes

There is no end to the litany of skills that leadership professionals have attributed to good leaders: communication, motivation, honesty, creativity, ability to delegate, listening, humor, confidence, positive attitude, intuition, and many more. No leader has all of these skills and having all of these skills does not make you a leader, but they certainly do help. Developing these skills will help you, regardless of the activity. Possibly a better way to look at leadership is through three key attributes defined by Roselinde Torres (2013). She expressed these attributes through three questions:

1. Where are you looking to anticipate change? Are you reading widely, experiencing different cultures, and expanding your input channels of information? A leader needs to have awareness of the factors leading to change. They need to be able to look around corners to anticipate change rather than at walls where change is an obstacle.
2. What is the diversity measure of your network? Do you interact with people of different cultural, political, racial, and socioeconomic character? Understanding and relating to people different from you will help you understand their needs, perspectives, and

challenges and make you better able to lead in a direction that embraces them rather than creating obstacles.

3. Are you willing to abandon the past? Breaking out of comfort zones is the hardest activity that many of us will deal with, but it is essential. Leadership is about change and change does not happen without breaking from the past.

All of these skills and attributes require commitment and hard work in order to become fluent. But like developing any new habit or skill, some of the work can and must be done on your own while others may require or benefit from assistance. Finding help along the way is critical and one solid approach is to find a coach or mentor.

Find a Mentor or a Leader You Admire

There are many important discussions that each of us have with people who influence our lives. It could be with parents about some of the basic rules of behavior, companions or spouses about relationships, or bosses about work expectations. One of the conversations that you should have with someone who has an interest in your professional development is about the opportunity to develop as a leader. Frequently, this can be a mentor in your agency or company, a colleague who has developed good leadership skills, or a formal leadership coach. A mentor, leadership coach, or just a good friend can help you assess your skills, work on your weaknesses, and play to your strengths. They can also help to identify and place you into situations that can increasingly allow you to practice leadership, as well as identify and incorporate your personal strengths and weaknesses to be better at that role.

Seek Out Situations Where There Is Good Leadership

Do your part and willingly be led, but also observe and learn from those who have shown the ability to be effective leaders. Growing as a leader is not only dependent upon being afforded or finding a place to lead, but is also built upon being willing to be led and being an active and attentive member of a group with good leadership. What are the traits and characteristics that you observe in good leaders? How do they interact with the people that are supporting them or those that oppose? How do they listen, delegate, and decide? We learn from others in all other parts of our lives, so why not from good leaders? Unfortunately, we all experience bad leaders as well. But, they can also provide learning experiences. The warning here is that we must be truthful in recognizing that we all have weaknesses. Acknowledging and addressing our weaknesses is the only difference between our ability to practice good leadership and simply repeating the mistakes of the ineffective leaders we are observing.

Formal Programs Are a Great Source but Not the Only Pathway

In a direct response to the challenge of leadership, a number of agencies or professional societies have responded by developing programs or classes on leadership. The National Conservation Leadership Institute (www.conservationleadership.org) and the Wildlife Society's Leadership Institute (http://wildlife.org/professional-development/leadership-institute) for emerging professionals are excellent examples. Many of the federal agencies have their own internal leadership programs, such as the U.S. Fish and Wildlife Service's Advanced Leadership Development Program, geared toward staff who have been identified as showing some of the potential for future leadership roles. States offer a smattering of leadership training that vary considerably in focus and quality. When looking beyond the relatively narrow confines of our profession, the opportunities to engage in leadership training are vast and include such widely renowned programs as the Harvard Kennedy School Executive Education Program (http://ksgexecprogram.harvard.edu) and others found in many top universities. While these are tremendous programs that provide opportunities for some, they are often out of reach to many of us due to limited class sizes, location, timing, or expense, and they clearly provide a limited amount of training or practice in the development of a good leader. Similarly, a quick search will find hundreds of books, Web-based classes, and journals that provide a wealth of information on leadership. Clearly, the importance and timeliness of the challenge of leadership has spawned its own industry. When one of my previous employers (Pennsylvania Fish and Boat Commission) recognized a need to develop leadership and determined that the current offerings in state government were insufficient, we worked with consultants to develop our own leadership program. Be innovative and even employ leadership to develop more leadership!

Motivation and Commitment

At the same time, consider your motivation. There is a substantial and important difference between seeking a leadership role solely to be in charge and taking on a leadership role because the health and success of the program requires a change in direction or lacks direction altogether. In the latter of these cases, you may feel compelled to provide that leadership role because it will advance a cause that is important to you and your profession; you are, in a sense, a reluctant but appropriately motivated leader. You want to help the activity be successful and someone needs to provide the leadership necessary for that success. The difference between leading to satiate an ego and leadership to enrich lives is immense. Leadership motivated by the wrong reason leads to failed leadership, which is a product of leading but to an inappropriate endpoint. Consider why you feel called to step into a leadership role and check your ego at the door. Consider also the time and your emotional investment in the activity. Common sense dictates that you can only invest so much time and energy in any set of activities before your ability to effectively contribute is compromised. Leading takes time and energy; underestimating your ability to invest sufficient time to result in a good outcome is a serious lack of leadership maturity.

Conclusion

Leadership is a skill that all of us can work on and improve. It will not only prepare us for the opportunities to lead, but will also make us more effective as members of teams led by others. In working on leadership, you need to acknowledge that there is no single path, no simple set of courses, and no easy solution. But there is an abundance of opportunities out there just waiting for the right person, with the right motivation, to step into the void and make a difference. Before taking on a leadership role, however, be thoughtful about your reasons. Be sure that your motivation is sincere because if it is not, it will be the first attribute that will be seen by those you intend to lead. Do what you can to also put yourself in the best position to succeed as a leader. Develop a larger vision of your community, your profession, and the world around you. Expand your network of interactions with people of all walks of life. Finally, prepare yourself to lead change because that is what leadership is all about. We need to do this because the success of our conservation mission increasingly depends upon effective leadership to ensure that the natural resources we all cherish will be there for future generations to experience and enjoy.

Biography

Douglas Austen is executive director of the American Fisheries Society. Most recently, he served as national coordinator for the Landscape Conservation Cooperatives for the U.S. Fish and Wildlife Service. In that capacity, he worked with a broad and diverse array of federal and state agencies, nongovernmental organizations, tribes, universities, and others to implement a landscape conservation approach to management of our natural and cultural resources. Doug has been working in the conservation field for more than 30 years with time spent as executive director of the Pennsylvania Fish and Boat Commission, in various positions with the Illinois Department of Natural Resources, and as a research biologist with the Illinois Natural History Survey where he also was an adjunct faculty with the University of Illinois. Doug received his Ph.D. from Iowa State University, M.S. from Virginia Tech, and B.S. degree from South Dakota State University. He is also a graduate of the National Conservation Leadership Institute.

References

Drucker, P. F. 2003. The essential Drucker: The best of sixty years of Peter Drucker's essential writings on management. HarperCollins Publishers, New York.

Linsky, M. 2012. One-party control doesn't always eliminate political turmoil. Linsky on Leadership 2012. Available: www.ncsl.org/legislators-staff/legislators/legislative-leaders/linsky-on-leadership-march-2012.aspx. (April 2014).

Peter, L. J., and R. Hull. 1969. The Peter Principle: why things always go wrong. William Morrow and Company, New York.

Torres, R. 2013. What it takes to be a great leader. TED Talks. Available: www.ted.com/talks/roselinde_torres_what_it_takes_to_be_a_great_leader. (April 2014).

Leading for Conservation Success

Hannibal Bolton*
U.S. Fish and Wildlife Service
1849 C Street NW, MS 3344, Washington, D.C. 20240, USA

Cecilia Lewis*
U.S. Fish and Wildlife Service
4401 North Fairfax Drive, MS 770, Arlington, Virginia 22203, USA

Key Points

Successful conservation outcomes in the 21st century will require an understanding that

- The conservation landscape is no longer dominated by federal and state agencies.
- The role of resource management agencies will depend upon their ability to expand beyond the traditional role of the lead agency into the role of exercising leadership.

Introduction

When I (H. Bolton) started my career as a fisheries biologist, the world of fisheries management was straightforward—one either worked for a federal agency or one worked for a state agency. I chose to work for the U.S. Fish and Wildlife Service, and my job was to conserve fish, wildlife, and their habitats. Forty years later, I am still doing what I love. At the start of my career, success seemed simple. All I needed to succeed was my passion, technical skills, and hard work. I have since learned that successfully doing my job requires more than just my desire and commitment to save the world. Over the course of 40 years, I have learned a few lessons that have helped me grow and successfully lead others through various resource management challenges, lessons that, I believe, federal and state agency leadership must adopt and embody to remain relevant and successfully navigate the conservation challenges society will face in the 21st century.

My introduction to leadership came early in my career when I was leading fish surveys and stream inventories on the Menominee Indian Reservation in Minnesota. I was determined to save the world. One day, I was called into the tribal chairman's office. Surrounded by members of the tribal council, the chairman asked me why I was on the reservation and what I thought I was doing. I told the chairman I worked for the U.S. Fish and Wildlife Service (Service), that the Service had tribal trust responsibilities under the auspices of several treaties, and that I was on the Menominee Reservation to assist and support the tribe with

* Corresponding authors: hannibal_bolton@fws.gov; cecilia_lewis@fws.gov

fish and wildlife conservation. I will never forget the chairman's curt response to my righteous certainty: "Did it ever occur to you [that] we don't want you here?"

I was humbled by my first lesson in leadership. I thought I was leading, but no one was following, in part because I did not understand the meaning of partnership and, more to the point, because I did not see the Menominee as my partner. What should have been an exercise in collaborative conservation had instead been an exercise in hubris by a conservation neophyte oblivious to the importance of people, the people closest to the land, in achieving conservation success. My misstep was perhaps understandable by the standards of the time, when conservation was seen by many to be the exclusive purview of federal and state agencies. We all knew, or thought we knew, who was best qualified to make conservation decisions. It was the experts, people like me—trained in college, agency employed, and armed with a lot of passion and a solid mission to conserve natural resources for the greater good.

Today, I know better. I know that conservation is not just for experts and that successful conservation takes more than a technical education and the desire to do good things. Now, I see the complexity of multiple (and sometimes) overlapping decision-making jurisdictions, a myriad of different ownership and land-use types, mounting regulations, and changing societal attitudes and needs. I am also acutely aware that the challenges we face, now and in the future, exceed the fiscal and technical resources and capabilities of our federal and state resource agencies. Making matters more complex is the fact that the number of organizations involved in conservation has grown by leaps and bounds. The expanding universe of nonagency players (i.e. nongovernmental organizations, citizens groups, advocacy groups, etc.) in conservation challenges the traditional role of agencies like mine. We can no longer do the business of conservation as I once tried to do on the Menominee Reservation. Undoubtedly, if the agencies that led conservation in the past century are to be among its leaders in the next, they must adapt to a world in which collaborative and networked governance of natural resources is the rule and no longer the exception.

My experience points to six leadership lessons that were instrumental in achieving successful management outcomes. Incorporating the principles from these lessons is, I believe, essential to achieving conservation success in the 21st century.

Leadership Lesson #1: Exercise Humility in Agency Mission and Vision

Public servants work on behalf of the public and stand for the belief that conserving natural resources is in the best interest of the nation. We should never step down from that commitment; however, we also should not let our zeal for process and procedure or our commitment to tradition and institutional culture stand in the way of achieving the public good. Too often, resource agencies don blinders or construct barriers that impede problem solving, sometimes conceiving of solutions in the narrowest context allowed by the agency mission or only implementing methods that were used in the past.

For example, years ago I was asked to break an impasse at the Upper Mississippi River National Wildlife and Fish Refuge. In an effort to protect habitat for canvasback ducks, the refuge designated its waters off-limits to fisherman during the ducks' migratory stopover on the refuge. Moreover, the Wisconsin Department of Transportation needed to dredge excess sediment from the refuge waters, which would destroy Bluegill *Lepomis macrochirus* habitat. This sparked uproar in the local fishing community because the arrival of the ducks corresponded with the peak of the Bluegill fishing season. Unfortunately, the refuge's decision to restrict access to its waters fell short of reconciling conservation with sustainable community harvest.

I hosted multiple public meetings and met with all sides in the conflict, and together we came up with a win-win solution. As it turned out, the ducks primarily relied on areas of the refuge with stands of wild celery, which was concentrated in just 12,000 of the refuge's 195,000 acres. The fisherman agreed to stay out of the areas of the refuge used by the ducks, and the refuge agreed to open the remaining waters to Bluegill fishing. No less important, many of the fishermen were also duck hunters who understood the need for duck conservation. They simply wanted to have their ducks and eat their Bluegills, too. Likewise, the Wisconsin Department of Transportation agreed to restore Bluegill habitat as part of the deal to dredge excess sand from refuge waters.

Collaboration prevented us from completely overlooking and dismissing the interests of stakeholders in an effort to deliver the agency mission and vision on their behalf. In retrospect, the solution was obvious, yet putting it into practice required the humility to recognize that the agency solution is not always the best or only solution. Better solutions can sometimes be realized by working collaboratively with interested parties rather than constructing rigid solutions that ultimately isolate us from citizens and potential partners.

Leadership Lesson #2: Traditional Leadership Has Limits

Traditional resource agency leadership functions within a bureaucratic environment characterized by well-defined lines of authority for decision making. This may work for agencies, but it does not work for partnerships that, by their very nature, transcend public and private jurisdictional boundaries. Regardless of noble and well-intentioned missions or grand visions of future conditions, we are ultimately limited in what we can do alone. We should always remember Theodore Roosevelt's words: "Do the best you can with what you have where you are." Yet, to achieve more—which we must in light of the conservation challenges before us—we need more. We need collaborative thinking and tools to supplement traditional leadership. We must, in other words, seek ways to make meaningful conservation decisions and actions among parties that are joined only by voluntary consent.

During my Senior Executive Service training, I spent several months at the Golden Gate National Recreation Area under the tutelage of an amazing leader, Brian O'Neill, the park's superintendent. Golden Gate is an amalgam of park units, including Golden Gate Bridge, the Presidio of San Francisco, and Muir Woods. Superintendent O'Neil understood Roosevelt's maxim and the use of collaboration to reach conservation goals that neither he nor his staff could achieve alone. In particular, he understood that the success of the park, and his vision for the park, depended on others owning what he believed in. He could not demand that the San Francisco Bay community support what he was trying to do. And he had far too few internal resources to manage the park in isolation.

What he did was remarkable. He gave the Golden Gate community a stake in the park's future by sharing the responsibility for managing and funding the park. More than 10,000 volunteers became caretakers of the park with real responsibilities, including hands-on supervision and day-to-day stewardship of designated areas and resources within the park. Additionally, small and large businesses in the Bay Area bought into his vision and helped finance the park's operations. Traditional leaders may shy away from this approach due to a failure to comprehend the importance of public-private partnerships or fear that this type approach will erode their authority. Superintendent O'Neill understood that leadership was not always a function of command and control; in a pluralistic setting, it is a function of sharing, collaboration, and joint ownership. Applied more broadly to natural resource conservation, the lesson is clear: partnership means more than just sharing a vision; it entails sharing the day-to-day work needed to realize that vision. For this to work, a leader must be willing to step outside his or her bureaucratic comfort zone, explore innovative ways to partner with others, and be prepared to share decision-making responsibilities with the parties who make the critical difference between conservation success or failure.

Leadership Lesson #3: When Possible, Lead from Behind

Superintendent O'Neill's collaborative leadership approach underscores the third lesson, which, I believe, is critical for natural resource management in the 21st century—leading from behind. He never put his ego ahead of the target conservation outcome. He also understood the implications of full ownership in order to achieve conservation success. Others must be given opportunities to lead if they are to embrace, in a meaningful way, the vision you hold and the mission you are striving to achieve. This can be done in many ways, but principally, it is done by creating—not exempting—opportunities for others to lead. Encourage new leaders to step forward, defer to others while offering a helping hand, and champion your ideas by making them the ideas of others. Leading from behind can be an effective catalyst for encouraging and developing new and emerging leaders.

In the early 2000s, I was one of several leaders that helped launch the National Fish Habitat Action Plan, now known as the National Fish Habitat Partnership (NFHP). I also played a role in fostering one of its partnerships–the Reservoir Fisheries Habitat Partnership (RFHP). My leadership consisted of helping keep the idea of a reservoir partnership alive and facilitating the logistics needed to make it happen. This included bringing people together, pushing for regular meetings, maintaining communication between the most interested states, providing staff support, and encouraging the most committed individuals to assume leadership roles, with a pledge of my full support if they did. The RFHP thrived, and I proudly maintained my position in the background, facilitating, supporting, and encouraging a state-led initiative. I take great pride in knowing that much of my leadership was invisible and my greatest reward was watching others take the lead and establish a state-led, nationwide fish habitat partnership.

Leadership Lesson #4: Promote Scientific Transparency

Supporters of the National Fish Habitat Action Plan understood that science serves leadership best when it is accessible, understandable, and shared among all conservation partners. We should never use science as a shield to hide behind or as a justification to ignore our partners. Yet, for many years, the U.S. Fish and Wildlife Service's fisheries program operated as if the state fish and wildlife agencies did not exist. We pursued our own science agenda—selecting priorities and studies that served our needs alone—and thought we were making a difference when in actuality our science had few discernible outcomes. The state fish and wildlife agencies were angered by the lack of collaboration; they saw us going off on our own, drawing our own conclusions, and widening the growing gulf in scientific understanding of fisheries management. The lack of collaboration led to a lack of scientific transparency, and the fisheries program and its science were becoming irrelevant to its most vital partners, the state fish and wildlife management agencies.

To reverse that growing scientific chasm and promote transparency, we sat down with the state fish and wildlife agencies and other fisheries management partners and came to a common understanding of the threat we all faced—declining fish habitat—and the need for a unified approach to fisheries management. In other words, apply a common approach to the common problem that transcends our individual authorities and operational boundaries. The 18 individual fish habitat partnerships that make up the NFHP are made up of thousands of agency and citizen conservationists working in tandem with hundreds of other conservation partners to make science and science-based tools readily available to all. One of the hallmarks of the NFHP is the commitment to making fisheries science and fisheries management practices a shared resource and a cooperative experience, respectively. Collaborative research, management efforts, and scientific transparency will help facilitate successful conservation outcomes in the 21st century and beyond.

Leadership Lesson #5: Commitment to Accountability

Our legitimacy as leaders is no greater than our ability to produce and demonstrate results. We are accountable to the American people. Ours is a public trust for which there must be the highest standards of accountability.

Shortly after I became assistant director for Wildlife and Sport Fish Restoration (WSFR), I learned that the Office of Management and Budget (OMB) gave my program a failing grade for its failure to demonstrate results. Congress followed on OMB's heels with its own set of expectations for administering State Wildlife Grants, which are part of the WSFR grant portfolio. WSFR grants are funded through excise taxes on the sale of hunting, shooting, fishing, and boating equipment. The manufacturers that make up the wildlife and sport fishing industry also agreed that the program was not producing desirable results. They wanted to see the actual impact of the levies on hunting, fishing, and boating activities across the country. I took these challenges seriously: the WSFR program funds approximately one-third of all state fish and wildlife agency budgets. Anything that threatens the credibility of the WSFR program also threatens the viability of fish and wildlife conservation in America.

I came to the WSFR program a firm believer in accountability. One of my first actions was to initiate development of a tracking and reporting database to provide the level of accountability and transparency demanded by OMB, Congress, and industry manufacturers. Wildlife TRACS (Tracking and Reporting Actions for the Conservation of Species) is the outcome of that action, and it is now the official performance reporting tool for the WSFR program. TRACS provides the means to track and report the outputs of WSFR-funded projects and the interim effectiveness and long-term outcomes of those projects. We can now demonstrate what we have long known; the WSFR program makes a difference for species, habitat, and people.

Developing TRACS is just one example of how to be accountable. The willingness of others to support our conservation efforts depends on our ability to show how public funds are spent and demonstrate the conservation benefits those funds help create. There is a natural tendency to resist accountability, and natural resource managers are no exception. We just want to be left alone so we can do our good work; after all, we are serving the public interest, right? But here is where leadership is so essential. I had to overcome skepticism and fear to make TRACS the vehicle of accountability for WSFR. I did it by paying heed to the advice and constructive criticism of supporters and the fears and concerns of the skeptics and using the information from those sources to establish the level of transparency and accountability expected from OMB, Congress, manufacturers, and the American people.

Throughout your career, you will face the same burden of proof and persuasion. However, the necessity for accountability should never make us averse to new ideas and new approaches. Remember, our legitimacy as leaders and the conservation legacy we leave for the next generation depends on our ability to produce and demonstrate results.

Leadership Lesson #6: Be Receptive to Innovation and Entrepreneurship

The sixth and final leadership lesson I have learned is to always be open to innovation and entrepreneurship. Federal and state agencies should not be satisfied with their past successes or seek refuge behind walls of process and procedure. Conservation works best when leadership is responsive, flexible, adaptive, and inventive in the face of constantly evolving environmental challenges. In a certain sense, this lesson brings us full circle to humility–the realization that we do not have all of the answers and that the best answers may be those that we could never have imagined on our own

Typically, good leaders support scientific and technological innovations that enhance our capabilities and entrepreneurial advancements to make our agencies more effective and efficient. These are important; they are not, however, most vital to future leadership. The greatest innovations and entrepreneurial contributions will be those that determine how we govern natural resources in the future across jurisdictional lines. Earlier in this discussion, I pointed to the success of Superintendent O'Neill in pioneering new ways to engage citizens in conservation and new ways to expand the decision-making arena. We must also expand that partnership model to include new modes of natural resource governance that are adapted to 21st century resource management complexities.

It is happening. We are on the cusp of unprecedented and creative systems of natural resource governance that will build on landscape-scale networks of jurisdiction and ownership. New collaborative infrastructures fashioned by these systems will redefine the roles of traditional resource agencies and partners and will make collaborative conservation more effective than ever. One can see evidence of it at the Detroit River International Wildlife Refuge, where private and public players from two nations are working together voluntarily to restore and protect species and habitat across multiple jurisdictions and property ownerships. It is in this crucible of challenge and complexity that new agency leaders must emerge and garner the skills to lead us to conservation success in a world that is anything but straightforward.

These leadership lessons are not a panacea; they are a distillation of my experiences and things I have learned over my career. They are intended to help current and emerging leaders become better leaders by encouraging them to view conservation from a broader perspective and, therefore, improve the odds of conservation success. Moreover, they reflect what I believe is a fundamental shift in conservation, away from centralized agencies to shifting alliances of public and private entities formed to address conservation issues that have historically resisted resolution in the face of jurisdictional and governance challenges. Conservation will always reinvent itself anew to meet the exigencies of changing times. For that to happen with clear design and purpose, current and emerging leadership must be attuned to its evolving environment, and it must always be aware of its roots, but never bound by them.

Biographies

Hannibal Bolton is a 40-year veteran of the U.S. Fish and Wildlife Service. He is currently the assistant director for the Wildlife and Sport Fish Restoration Program in Washington, D.C. He is charged with the administration and oversight of federal grant programs, including the Wildlife and Sport Fish Restoration Programs, State and Tribal Wildlife Grants, and the Coastal Impact Assistance Program. These grant programs provide hundreds of millions of dollars annually to help fish and wildlife agencies in the states, U.S. territories, and District of Columbia conserve, protect, and enhance fish, wildlife, their habitats, and the hunting, sport fishing, and recreational boating opportunities they provide.

Cecilia Lewis is an employee of the U.S. Fish and Wildlife Service in the Fish and Aquatic Conservation program, headquarters office. Her duties include working with headquarters staff to coordinate and execute National Fish Habitat Partnership activities, and she is the liaison to the Reservoir Fish Habitat Partnership.

Lessons on the Road to Leadership Effectiveness

WILLIAM A. DEMMER*
Demmer Corporation
1600 North Larch, Lansing, Michigan 48906, USA

Key Points

- Break through emotional paradigms that inhibit leadership growth.
- Focus on effectiveness.
- Empower organizational leadership.

Introduction

Many skills for leadership are developed while on the job. I realized as my career was evolving that I had certain emotional issues that were inhibiting my ability to lead effectively. I knew that I needed to isolate those issues and find tools that would allow me to work through them in order to more fully develop my leadership potential. This chapter will describe certain inhibitors that challenged me. To challenge my inhibitors, I became proactive with self-help books, management seminars, personal development coaches, and eventually psychological counseling. Developing my self-confidence and a successful leadership style took time. I learned first that if I faced my inhibitors honestly and courageously and then executed a plan to work through them, I became better prepared to lead more effectively.

Your challenge will be to find your inhibitors, face them early in your career, and work through them courageously with tools that you will find or learn about. Once you do, confidence, joy, and effectiveness will then flow more quickly into your professional and personal life.

Building Blocks of Communication

Communicating with diverse personalities can be most challenging. When I began my career, I was dealing with employees and customers who had lived a lifetime within their job responsibilities. They were not about to let a young pup influence their output! I needed to find a way to bridge over their anxieties about me and mine about them. In your career, you will encounter arrogant types, reticent types, insecure types, and people who view the world as a glass half full. Your challenge will be to develop skills that will enable you to communicate with them accurately and inspirationally. Developing a working relationship with older employees or difficult associates can be a challenge. I learned through the teachings in Dale Carnegie's (1936) seminal book, *How to Win Friends and Influence People*, that by becoming interested in other people you make more friends than by trying to get them interested

* Corresponding author: bdemmer@demmercorp.com

in you. Talk to people about themselves and they will talk for hours. Through this process of communication and relationship development, a bond of trust can evolve. Think about how you can help the people around you evolve. Focus your thoughts outward not inward. Think in terms of your employer's and your team members' needs. See if you can find that something that will make their glass half full. How can you help your team members and employers be successful, too?

Do not forget humor, as it can be a great stress reliever. Oftentimes during a meeting when I sense that the audience is tense, that is the time that I begin to look for that right opportunity to inject some levity into the environment! Other times, the audience might be day dreaming, which is also a good time to introduce a little humor. Just remember that effective communication is the first step in building relationships and trust.

Building Inner Peace

The more at peace you are with the outside world, the more emotional energy you will have to deal with the challenges of leadership. If you approach leadership from a happy and confident perspective, your potential for success will be greater.

I discovered a formula in my mid-thirties that put me in charge of my own happiness. After years of trying to make my friends, family, employer, employees, and customers happy, I said, "Who is trying to make me happy?" When I thought about it, it seemed that only I was worried about other people's happiness! Next, I began to think about the messages that I was receiving from other people: "what can you do for me"; "you are not quite good enough"; "if only you had done it this way"; "why are you wearing that tie"; "you did it okay, but look at what I accomplished"; and so forth. Having reflected on how I was receiving and responding to others' needs, I had an epiphany: I was not taking care of myself or my needs. In Dr. Wayne Dyer's (2001) infinite wisdom, I discovered that you have the opportunity to choose how you react to outside influences and the imposed needs of others. If someone says something to you or about you, you have the power to accept, reject, or ignore his or her implications. In effect, if the shoe fits, wear it; if not, ignore it.

I discovered that we have power to decide how we react to almost any situation. We consequently have the power to control our moods and ultimately our happiness. We are in charge of our own happiness. Happiness is generated from within ourselves and should not be dependent on another person, place, or thing. If something unexpected happens to negatively impact your research project, your team's goals, or individual projects, I suggest that you take a deep breath, decide what you own, and then develop an action plan to right the situation. My experience has taught me that if you are not part of the solution, you are part of the problem! These suggestions will help you develop a calmer demeanor under pressure and move you closer to inner peace.

I was surprised to understand just how powerful the control that I had created over my own joy and happiness was. When I stopped requiring or looking for other people to make me happy, I truly took emotional destiny into my own hands.

Keep the Focus on Effectiveness

When I had earned the right to move up to a leadership role in my career, I began to lead like my predecessors. I was now to be the authority in my area of responsibility. I was the person who was expected to be right, to have the right answers, to be the oracle of knowledge. Trying to manage and lead like my predecessors was causing me inner turmoil that

I was unable to reconcile. When I finally came to terms with my turmoil, I realized that I was living and managing a lie; I did not have all of the answers, and I was not an oracle of profound knowledge! I was, in fact, trying to live the incorrect equation of life:

To be right is to win
To win is to have power
To have power is to control
And to control is to be safe

I finally matured as a manager when I realized that truth is constantly being revealed. What was revealed as truth yesterday can be modified by what was revealed or learned today. The lesson for me from these situations was that there are often many sides to being right. Some of these sides can be an even more effective way of being right. I changed my paradigm of presentations and arguments to become more like Socrates who used discussions and soft debates as a way to peel back those onion layers as a search for the most effective rightness and truths.

I grew to understand that people who need to be right, look right, and or act right are defending a level of insecurity. I suggest living and leading with a different equation:

With humility you listen
When you listen you can hear
When you hear you can learn
And when you learn you can live more effectively

I also like to think that there is a difference between listening and hearing. To me, most people who listen just go through the motions of attentiveness. But people who listen and hear are really trying to understand what is being communicated. I have tried to be a good listener, really hearing and absorbing the cogent points of a conversation or presentation. I am always looking for opportunities to differentiate myself and our business offerings from the competitions. Really hearing what our customers' needs are has been a big help in that effort.

A fisheries professional needs to get that same deep level of feedback, whether it comes from outside organizations, decision makers, or the public. The ability to really hear and understand is critical in any career. My suggested equation beginning with humility is a great place to start.

Process in Leadership

Developing communication skills, creating a level of inner peace, and focusing on effectiveness were all critical building blocks in my evolving leadership potential. But even with those tools evolving, managing people was still an emotional challenge for me. When I discovered the beauty of management by process and the tools of process improvement and began incorporating the tools of process improvement in my approach to organizational situations, my leadership capability soared!

Being biologists and scientists, you do not need me to explain what a process-based research project is or what a process-based environment looks like. What I am suggesting at this point is that you apply your process-based thinking to your leadership and management style. Every department or team project has some kind of output expectation. That output expectation could be research, data generation, or project-based completion, all of which have a value. That value can be measured by timing, cost, and quality. The value of those characteristics can then become your team's metrics. All of your team's input can then be designed as subprocesses to support your metrics. Your employees or team members should all have defined roles to play to ensure that those metrics are met. When job descriptions include clearly defined process metrics, employees have a much clearer vision of how to be successful. This system of management is often referred to as "performance management"; much has been written about this systematic approach (e.g., Daniels 2004).

When metrics are well understood, you, as the leader, have the opportunity to be a more of a coach and mentor than the heavy-handed boss. Even the thought of being heavy-handed does not sit well with me, emotionally. Include your team members in the development of your metrics, get their metric buy in, and then coach them to success. You replace potential management anxiety with the potential of feeling the satisfaction of nurturing. Also, when you incorporate the tools and facilitate process improvement, you have the ability to create leadership within your team and staff. Dr. W. Edwards Deming has written extensively on the subject of process and process improvement. Watching people flourish under my mentorship has been one of the greatest contributors to my sense of joy on the job.

Emotional Maturity in Leadership

Thirty years into my career, I looked back at my evolution into a more confident leader and reflected on the emotional paradigms that I had faced, challenged, and changed. Most of those challenges occurred unexpectedly. Had I the courage to face many of my fears aggressively prior to ignoring them and their potential to inhibit my effectiveness as a leader, my career path would have been smoother earlier. Twenty years into my career, when I discovered the world of process and process improvement that I described, any pretense of my being an oracle of profound knowledge was shattered. As the processes that we were trying to improve got closer to touching my responsibilities, I had to make a big decision: was I going to defend the nonvalue-added nonessential parts of how I defined my work day, or was I going to engage aggressively in our company's process improvement program? Much of my nonvalue-added time spent was, as it turned out, designed (by myself) to keep myself away from the more difficult and emotionally challenging work tasks, like negotiating for engineering changes or collecting customer payables.

Using a process-based approach to my business challenges and by incorporating my lessons learned in communication, effectiveness, and confidence, I shifted my paradigm of leadership from the oracle of knowledge to one that leads by example. I emerged into the world of management by process and engaged aggressively in our company's process improvement program. I was evolving into an inspirational leader. The emotional tools that I had assembled had equipped me with the confidence to challenge the issues of today and create a more effective and satisfying tomorrow.

Your challenge will be to look at your leadership inhibitors honestly and then, with courage, develop a plan to work through them. A positive attitude will help overcome any associated fear of change and will create excitement in the process.

Biography

William (Bill) A. Demmer is a graduate of Michigan State University (MSU) College of Engineering and a recipient of the college's Claude E. Erickson Distinguished Alumni Award. He is also the recipient of MSU's Distinguished Alumni Award. Bill Demmer's work with the MSU Department of Fisheries and Wildlife associated with the Boone and Crockett

Club earned him an Honorary Alumnus Award from the MSU College of Agriculture and Natural Resources. Demmer is president and CEO of his manufacturing business, the Demmer Corporation, headquartered in Lansing, Michigan. He is also president of the Boone and Crockett Club, the oldest conservation organization in North America. He and his wife Linda are the parents of five adult children, three of whom are married, all of whom were raised hunting, fishing, and celebrating the wonders of American wildlife.

References

Carnegie, D. 1936. How to win friends and influence people. Simon & Schuster, New York.

Daniels, A. 2004. Performance management: changing behavior that drives organizational effectiveness, 4th edition. Performance Management Publications, Atlanta.

Dyer, W. 2001. Pulling your own strings: dynamic techniques for dealing with other people and living your life as you choose. William Morrow and Company, New York.

Recommended Readings

Dyer, W. 1993. Your erroneous zones. Avon Books, New York.

Dyer, W. 1980. The sky's the limit. Simon & Schuster, New York.

Fellers, G. 1994. Why things go wrong: Deming philosophy in a dozen ten-minute sessions. Pelican Publishing, Gretna, Louisiana

Moore, T. 1994. Care of the soul: a guide for cultivating depth and sacredness in everyday life. Harper Perennial, New York.

Nillsson Orsini, J., editor. 2012. The essential Deming: leadership principles from the father of quality. McGraw-Hill, New York.

Peck, S. 1980. The road less traveled. Simon & Schuster, New York.

Quinn, R. 1996. Deep change: discovering the leader within. Jossey-Bass, San Francisco.

Walton, M. 1988. The Deming management method. Perigee Books, New York.

Leading with Vision

Kenneth Haddad*
Post Office Box 35, Lloyd, Florida 32337, USA

Jessica McCawley
Florida Fish and Wildlife Conservation Commission
2590 Executive Center Circle East, Suite 201, Tallahassee, Florida 32301, USA

Key Points

- Fisheries management in the 21st century is challenging and complex. Inspirational visions of the future of our fisheries are critical; otherwise, fisheries management will be rife with conflict.
- A vision must set the direction to the future, and leadership should provide the inspiration to those people who will help accomplish that vision.
- A young professional with skills in leadership and strategic thinking and an understanding of human dimensions who can lead with vision will be more desirable in the fisheries workforce of the future.

Strategic thinking for any individual, business, or industry must begin with a vision. Think of a vision as an outcome you would like to see sometime in the future. Strategically, a vision should be well defined and clear enough to create a path for accomplishment. President John Kennedy's 1961 vision of seeing a man on the moon within the decade is an excellent example. Simple and clear, it was a visionary challenge at a time when man was just entering space and the thought of putting a man on the moon in such a short time period did not appear in most people's wildest dreams. President Kennedy created a vision; he charted a path for technology development and daring risk, which led to tremendous success, a man on the moon in 1969. Just about everything the space program did in that eight-year period contributed to achieving that well-articulated vision.

Where are those visions in fisheries? The management, use, and conservation of our freshwater and saltwater fisheries have not had meaningful visions to create a path to the future. Strong and inspirational visions are an absolute requirement in the 21st century, and a vision should always recognize that fisheries resources are a public trust resource to manage for future generations. We believe that fisheries professionals need to be better equipped to think strategically and have the skills to develop clear visions. Curricula for college and graduate level fisheries professionals should prepare students to become visionary leaders.

Ken Haddad: Early in the 1980s, as a young biologist, I was challenged by the governor's office to give my vision of Florida's marine fisheries relative to regulatory needs in, I think,

* Corresponding author: kenhaddad50@gmail.com

15 years. Well, I made a somewhat bold statement that "every fishery species in Florida will be managed by rules and regulations." Although this vision became reality, I was really just predicting an outcome based on my state of knowledge at the time and not what would have been best for Florida fisheries. It was not a meaningful vision. A real vision in the early 1980s should have gone something like this: In 15 years, fishery rules and regulations in Florida will be minimal due to strong conservation ethics, excellent scientific knowledge, and trust and cooperation between user groups. Visions must be simple, set a clear direction to the future, and include how scientists, managers and stakeholders will all work together towards the future. A vision must emerge from the chaos and provide the inspiration and direction that all can embrace. A vision must also recognize the science-based disciplines of human dimensions that bridge the gap between the physical and social sciences and provide a huge and emerging set of tools and techniques for those of us concerned with managing our public trust fisheries.

Creating a meaningful vision is not easy, and looking at Florida fisheries provides some insight. I have been involved in marine science and fisheries in Florida for more than 35 years. It was not until I moved from the research perspective to the management perspective of fisheries and wildlife late in my career that it became obvious that there was little strategic thinking involved in fisheries management at local or national levels. It seemed that management was being conducted on a day-to-day basis with no look to the future or vision to set a path to the future. When I had the opportunity, as executive director of the Florida Fish and Wildlife Conservation Commission (FWC), I asked that leaders of our Divisions of Marine Fisheries Management and Freshwater Fisheries Management work with stakeholders to develop a vision and a path to the future for both of these fisheries in Florida. Both divisions went through an extensive process that included stakeholder surveys, facilitated meetings with diverse groups, and email feedback to better gauge expectations for the future. One of the outcomes of each of the processes was vision statements that can provide a short case study about vision.

> Freshwater Fisheries Vision: To encourage partnerships between stakeholders and FWC so that freshwater fisheries strategic management plans can be jointly developed.

I cannot say that it jumps off the page as an eloquent statement, but it does provide a very succinct direction. The vision recognizes that building the partnerships with stakeholders is the future of freshwater fisheries.

> Marine Fisheries Vision: In 10 years, Florida will have sustainable and productive fisheries that fulfill the broad range of resource user interests and provide substantial economic benefits for Florida. To protect what we have and maintain sustainable fisheries, Florida must have excellent water quality; sufficient, high-quality habitat; and effective control of resource harvest levels.

Other than a bold timeline, this is a vision that is safe, aims to please all, and does not really incorporate a human dimension focus. It has become more of a vision on a shelf rather than providing explicit directions into the future.

I would say neither vision qualifies as inspirational. I favor the freshwater fisheries vision because it clearly focuses on the human dimension and recognizes that it will be people, through partnerships, who make the future for the fish and habitat. The vision is also simple and concise. This vision has resulted in an entirely new direction in Florida's freshwater fisheries that is driven through fisheries management plans built through the merging of science, stakeholder partnerships, and a common view of the future. New programs such as Trophy Catch and new partnerships with anglers and business leaders are leading to a bright future for freshwater fishing in Florida because the vision sets the direction. Marine fisheries management is often complex and prone to conflict, and creating a meaningful marine fisheries vision for Florida has been challenging. Missing in the vision is a reference to partnership. User interest is included but really implies status quo fisheries management as opposed to giving solid new direction. We say this because many marine resource stakeholder groups are well organized and actively participate in the decision-making process, but definitely not as trusted partners. While this is great for day-to-day management, it does not easily lend itself to looking at the future and instead reinforces the status quo. Without an in-depth understanding and application of leadership and strategic thinking and information and insight from the social sciences, it will be virtually impossible to move from conflict to a common future and create meaningful and direction-setting visions. The marine fisheries vision really needed to focus on the human dimension because the people are both the problem and the solution to the future of marine fisheries. In fact, creating the right visions at local, national, and international levels will not be easy due to the fact that we have let conflict, and not vision, drive management for so many years. This will be a great challenge for all upcoming and young professionals.

So, what do you need to do to prepare yourself for meeting the challenge? I do not think we have the complete answer, but we do have some observations and thoughts:

> Most fisheries professionals are trained as biologically focused scientists, and that training does not include leadership, vision, or dealing with people. Early in my career, I was fortunate to recognize that I needed these additional skills. Taking occasional training courses and reading volumes of books on business planning, leadership development, visioning, and dealing with people and difficult situations put me in a position to be effective in fisheries and wildlife management. As students and young professionals, you have the opportunity to equip yourself early in life with these skills, and they will be valuable in all aspects of your career and life.
>
> In all fisheries, but particularly in marine fisheries, human conflict due to competing interests will become the absolute number-one fisheries issue. We will need more adaptive human-dimension solutions based on vision and fewer technical solutions based on fisheries science alone. Certainly, all fisheries must still remain biologically driven, and your core education should be in the natural sciences, but as students and aspiring young professionals you need almost equal focus on the science of human dimensions, along with leadership and strategic thinking.
>
> You will note that we have not given you a textbook prescription for creating a vision. You can find many approaches to creating a vision online, and it is important that you take a look. Meaningful visions do not necessarily come from a textbook approach. However, if you focus on leadership, strategic thinking, and an understanding of the full suite of human dimensions disciplines, you will be able to lead the development of meaningful visions and, better yet, ensure that those visions are reached.

Jessica McCawley: I have been the director of the FWC Division of Marine Fisheries Management for just two years and inherited the vision statement for this division that Ken

Haddad described as inadequate. Early in my career with the FWC, I worked with members of Florida's marine life (aquarium trade) industry to help create a long-term vision for this fishery. The vision was created by working hand in hand with the industry and relevant stakeholders over a multiyear process. This process was highly successful as trust was developed and people were honest about their opinions and listened to the the opinions of others to form a set of regulations that would sustain the resource and fishery well into the future. So I know that meaningful visions can be created for marine fisheries.

When I attended graduate school in the mid-1990s, formal course work in leadership, visioning, strategic thinking, and human dimensions was not part of the required curriculum for a marine science degree. Although I had an idea that fisheries management was not a field that came with a cookbook, I did not fully grasp that managing marine fisheries is often about managing the resource by working side by side with the people that use, interact with, and are impacted by the resource. Even without formal coursework, my thesis research and the work I completed in graduate school necessitated interactions with commercial and recreational fishermen, which gave me hands-on experience with the people side of fisheries. However, I wish I was more prepared to define a vision and work with stakeholders to see that vision to completion. As Ken recommends above, I have since acquired many of these skills through my work with the FWC, including intensive individual study, the National Conservation Leadership Institute and other training programs, and real-life stakeholder interactions. I agree that the aforementioned vision for marine fisheries in Florida does not really set a direction, but instead defines what is already being done. In my role as a division director, I plan to set a more challenging marine fisheries vision that will set a clear direction for Florida's marine fisheries management. One of the keys to setting this vision is crafting it in partnership with Florida's marine fisheries stakeholders since they will play a significant role in achieving that vision. The real challenge to creating a vision that emerges from the chaos of marine fisheries, and actually carrying out that vision, is having leadership and strategic thinking skills and seeking out social science information and insight to effectively lead, work with, and provide inspiration to staff and stakeholders. If I had understood this early in my career, I believe my effectiveness would have been greater. That is why training in these seemingly out of profession areas is so important to students and young professionals.

We have some advice for upcoming students and young professionals:

> As you select your elective coursework for your degree, seek out courses that deal with strategic thinking, leadership, human dimensions, and effective communication. The already graduated young professional can find university courses, shorter professional courses, books, and private-sector and agency-based training that can better equip you with the tools needed to excel in your profession. Keep in mind that creating visions can sometimes be overlooked or glossed over, even though it is one of the most important elements for the future of our fisheries.
>
> Consider a thesis involving research that is applicable to science and management decisions and includes work with marine fisheries stakeholders. This type of experience will help you gain additional skills in the human dimensions arena. Skills gained while completing this type of thesis project or independent research are also helpful in learning how to set a vision that is both challenging and achievable.
>
> Symposia, conferences, and new journals and books that focus on human dimensions of natural resources management and those that focus on leadership and strategic thinking are also helpful in gaining additional experience and skills that will

lead to the ability to develop meaningful visions. Also, you should be aware that the American Fisheries Society (AFS) now requires course work in communications and human dimensions as part of the AFS professional certification program.

I can assure you that you will be more desirable in the job market if you develop the skills we recommend and if you have the ability to lead with vision.

In conclusion, we believe that it is important that the next generation of fisheries scientists, managers, and stakeholders create challenging visions together for the future in order for our fisheries to prosper. Worldwide, our fisheries have been managed for the moment and not through a clear vision of the future. Global changes in areas such as economics, technology, demography, and climate all create a state of urgency for those entering the fisheries profession. You need to be able to develop a vision of the future and have the skills and ability to help create and carry out the visions necessary for successful future fisheries science and management at the local, national, and international levels. We are counting on you.

Acknowledgments

We would like to thank the reviewers for their excellent input. This is was not an easy topic to review, and they truly helped create a better manuscript.

Biographies

Ken Haddad retired as executive director of the Florida FWC and currently works with the American Sportfishing Association. He has served as the director of the Florida Marine Research Institute and the FWC Division of Marine Fisheries Management. Ken has a B.S. in biology from Presbyterian College and an M.S. in marine science from the University of South Florida.

Jessica McCawley is the director of the Division of Marine Fisheries Management of the Florida Fish and Wildlife Conservation Commission. She has been with the agency for 10 years. Jessica received a B.S. in marine biology from Spring Hill College and an M.S. in marine science from the University of South Alabama, where she did research on Red Snapper *Lutjanus campechanus* and artificial reefs. Jessica is also a graduate of the National Conservation Leadership Institute.

Considering Habitat in the Interdisciplinary Fisheries Management Profession

TERRA LEDERHOUSE*
National Oceanic and Atmospheric Administration, National Marine Fisheries Service
Office of Habitat Conservation
1315 East-West Highway, Silver Spring, Maryland 20910, USA

THOMAS E. BIGFORD
American Fisheries Society
5410 Grosvenor Lane Suite 110, Bethesda, Maryland 20814, USA

Key Points

- Consider habitat conservation and ecosystem interactions as a higher priority in your slice of the evolving fishery field. This will help you tackle the growing opportunities for fisheries professionals.
- Step outside your comfort zone. Seek greater depth and breadth as you sharpen your skills. Prepare to make decisions with less than complete knowledge. Learn how to communicate with traditional and new audiences. The habitat thread connects all disciplines and peoples.
- Think outside your technical box, beyond your routine network of experts and issues.

The Changing World of Fishery Management

Aquatic habitat conservation has always been an important component of effective fishery management. Rebuilding and maintaining sustainable fisheries requires an approach that goes beyond controlling fishing effort and includes strong conservation of the habitats that provide the essential nursery, feeding, and breeding grounds that result in productive fisheries. Despite this link, habitat conservation and fisheries management are all too often treated as distinctly separate management approaches by natural resource organizations. Fisheries management has struggled to move beyond traditional approaches dominated by stock assessment models and actions that emphasize fishing mortality. Habitat conservation often proceeds with little consideration of how changes in habitat function influence the productivity of a fishery. The growing number of threats facing fisheries, from climate change and sea level rise to increasingly developed coastlines, hastens the need to bridge the gap between these disciplines.

Within the past few decades, our habitat conservation efforts have not kept pace with rapidly changing environmental conditions and disappearing coastal habitats, perhaps hindering the ability of some stocks to rebuild and decreasing the resiliency of stocks to respond

* Corresponding author: terra.lederhouse@noaa.gov

to a changing climate. Winter Flounder *Pseudopleuronectes americanus* may be an example of the former, with the continued loss of mid-Atlantic wetlands appearing to complicate efforts to rebuild populations declining from years of overfishing. These realities suggest that we must adopt a more complex yet promising ecosystem approach to fisheries management that considers habitat conservation as a core component of an effective management strategy. From our perspective, expertise in habitat conservation will become a critical skill for a successful fishery manager, not a separate career path.

Emerging professionals who are able to bridge this gap will tap into an abundance of resources that can guide them toward a successful career in the increasingly interdisciplinary field of fishery management. A habitat conservation specialist often interacts with a broad range of partners outside of the fishing industry who can provide new expertise to inform management decisions. For example, a habitat specialist may interact with other industries and sectors to plan for and develop new sources of energy or will advise local government or community organizations on construction projects affecting aquatic and coastal environments. More traditional fishery managers may interact more directly with the recreational and commercial fishing industries to set fishing limits or to adjust fishing gear configurations. Rarely, especially in marine settings, have habitat and fishery managers worked together to consider habitat contributions to fish stock productivity or to establish stock objectives or set harvest levels. Despite the disparate sectors and approaches, that dichotomy must be bridged.

As they progress, emerging professionals will note a continual mix of new state and federal mandates and economic imperatives that contain new goals for aquatic ecosystems and the habitats within them. These changes are good, indeed overdue, and fishery professionals would do well to embrace them. Each change equates to new career opportunities. In this scenario, we expect an increased need for fishery professionals with expertise in habitat and ecosystem science to provide critical input to priority issues in fish stock assessment, fishery management, and natural resource policy. Fisheries professionals need to be ready to seize this moment by engaging outside the usual industries or disciplines, communicating knowledge and ideas clearly and succinctly, and making crucial decisions for our fish, commercial and recreational fisheries, and the habitat that sustains them.

Adapting to a Changing Environment

As the approach for managing fisheries becomes more complex, so will the skillsets necessary for fishery professionals to successfully address the challenges facing our natural resources. The new fisheries career must embrace shifting environmental and socioeconomic realities and will require skills and abilities not usually taught in traditional academic programs for fishery science and fishery management. Each aspiring professional must embrace those challenges while in school and then while emerging into their chosen field.

The fisheries profession is blessed with many brilliant scientists and managers who have been trained, for example, to develop and run ecosystem models and to design smaller and more reliable fish tags. Brilliance in one or a few arenas is laudable but no guarantee of success. Broad aptitudes not traditionally associated with a fisheries science program might be the most important assets you can bring to your new role as a fisheries professional, whether in the habitat world or with another focus. Fisheries scientists and managers will need to broaden their training beyond traditional disciplines to include topics like landscape conservation, conflict management, and structured decision making. They will also need to develop key skills and hone traits that may not be taught in a traditional classroom setting. In our experience, these skills and traits include the following:

1. *Willingness to Make Decisions*

Managers need to be willing to make decisions based on the best available data and information. With the addition of habitat conservation, fisheries management decisions are becoming increasingly complex. As the experts, our challenge remains to make the best decision with available science. We must not shy from difficult decisions or postpone crucial actions. In the past, and not just in the fisheries professions, public and private sector professionals have used the lack of perfect knowledge as the basis for delay while arguing for increased funding for research on topics such as the status of stocks or vulnerability of fish habitats to natural and manmade threats. While these investments are still essential to successful fisheries management, it has become abundantly clear that managers will need to make decisions based on the information available to them. We may never have the data we need to quantify the contributions of habitat quantity or quality to stock productivity, but as professionals we need to make difficult decisions about our fishery resources. We must accept that decisions must be made with the best available data, usually with the option of adaptive management offering the opportunity to revisit actions based on improving information.

One example from the hydropower world offers a glimpse at the difficulties of working with incomplete information. When the Edwards Dam was proposed for removal from the Kennebec River in Maine in 1993, the best models left public and private sector players wondering how the river, several populations of imperiled fish, and affected businesses would respond when the dam was breached. Those concerns eased following dam removal and when it became evident that benefits outweighed costs. Now, two decades later that rare decision not to relicense an existing hydropower facility reminds us that careful analysis with imperfect knowledge can open rare opportunities and then yield success. This willingness cascades through the ranks to include everyone from hydrologists to adjacent land owners, and certainly includes the fishery professionals who contributed their research, analysis, synthesis, and recommendations. As we look forward, be ready to apply your expertise to future decisions.

2. *Ability to Effectively Communicate with Your Colleagues, Your Bosses, and the Public*

Verbal and written communication skills are vital to every successful career. This essential skill is crucial for communicating your ideas, keeping the public informed and gaining their support, working with your peers, and ultimately advancing your own career. Most of our time in any discipline is spent communicating through writing, so learning to write succinctly and with a mission is critical to conveying your point. That is true with email or a journal article or talking points for a PowerPoint presentation. Communication skills are particularly important when data are complex, incomplete, and even nonexistent. That is certainly the situation for much of the habitat arena, where specific connections to fishery management remains to be deciphered.

Kris Gamble, a communications specialist in the NOAA (National Oceanic and Atmospheric Administration) Fisheries Office of Habitat Conservation, exemplifies this principle, showing that communication skills are an important component of a successful career in fisheries and habitat conservation. Here, Kris shares her experience using her expertise in communications to advocate for habitat conservation:

> After working in radio and television for 20 years, I had returned to school to study science and was looking for a way to combine my communications skills with my new field of study. My biggest challenge was convincing the people interviewing me for a job as a communications specialist at NOAA that sometimes a communicator

interested in science can be equally or even more effective than a scientist interested in communications. By always maintaining a distance from the subject matter, I am better able to determine what will and will not need translation to a public audience, for example, communicating the concept of a habitat's ecosystem services.

The general public does not understand what ecosystem services are, and therefore they do not see a direct connection between fish habitat and their own lives. However, when we communicators step in to explain that without healthy habitat, people could not enjoy seafood, have fresh water to drink, or even protect their life and property from storm surge, suddenly the services provided by the ecosystem become clear and the public cares about our science.

A key point is that a skilled communicator can help with our messaging. The lack of training in a fisheries field proved to be a short-term hurdle while improved communications continue to yield long-term benefits.

3. Willingness to Step Outside Your Comfort Zone

Most fisheries professionals leave school with a particular specialty or expertise. That specialty may be a certain species, geographic area, or technical tool. We have observed that those who are willing to step outside their comfort zones, to tackle new opportunities, often gained perspectives that translate into even more successful careers. A wider comfort zone reflects burgeoning pressures on aquatic habitats from traditional and new threats. The implications of new threats cannot be addressed with a narrow set of technical skills, character traits, or work experiences. For example, experts in marine fisheries can learn from their freshwater colleagues. Researchers can benefit from participating in the policy process and vice versa. Volunteering for a review panel to select projects for state or federal funding gives a great perspective on the financial side of our fields. Expanding your horizons will open doors throughout your career. Make it a point to say "yes!" Do not let fear block your career path.

Advice for New Recruits

All of this is to say that the profession is changing, and you must be ready to embrace those changes. Below, we offer a collection of tips for new recruits to the fisheries professions, with anecdotes from our personal experiences as emerging professionals and even as seasoned veterans. (It is never too late to seize that special opportunity!)

- Make the most of opportunities provided by internships, fellowships, and professional societies. These are often the best way to gain the experience necessary to make you a competitive candidate for your dream job. Internships and fellowships can also be the best way to test drive a job or employer while also getting your foot in the proverbial door. Just remember that potential employers are also testing you, so make the most of your opportunity.

  I was fortunate to receive two fellowships that helped me pursue a fisheries career. The National Science Foundation's Research Experience for Undergraduates program gave me the opportunity to participate in summer field studies at a marine science lab. This research experience was critical to my acceptance to and success in a graduate research program, as I had little opportunity in my small liberal arts college to participate in marine biology research. Having the research experience certainly gave me an advantage as a prospective graduate student and taught me the skills necessary to design and execute my own graduate research project. Following graduate school, through the Dean John A. Knauss Marine Policy Fellowship, funded by the NOAA National Sea Grant College Program, I worked for one year on international fisheries management issues (in my case, at NOAA Fisheries). This was an eye-opening experience. Coming from a technical background in habitat restoration and fisheries management, this experience with international and federal policies and legislation that govern commercial and recreational fisheries was certainly outside of my comfort zone, but it also allowed me to view firsthand how government officials make critical decisions with often incomplete data. This experience made me much more competitive when I started applying for jobs within the federal government. (TL)

In my career, I have gained immeasurably through relationships with professional societies. American Fisheries Society (AFS) President Ken Beal gave me the opportunity to chair a committee in 2001, serve on the executive board of its Estuaries Section for more than a decade, and chair a symposium at an annual AFS meeting. Those opportunities have proved to be great ways to grow personally and expand my network. I met people outside my federal job, developed personal skills that benefited me back in my day job with NOAA Fisheries, and gained the confidence to do more. I encourage you to use AFS to your advantage while also enhancing your career satisfaction. And, in keeping with a recurring theme in this chapter, do not stop with AFS. Many of our issues connect to the interests of other professional associations. Become active in other societies, too. And encourage them to invest in AFS. (TB)

- Look for leadership opportunities, whether pursued yourself or imposed by your supervisor. Opportunities include an offer to brief your leadership on a trendy topic, shadow ing a leader at their next meeting, representing your leaders at an important meeting, offering to lead a group to develop a pilot project, or pursuing an officer position with an AFS unit. Just say "yes!" to new opportunities.

When more-senior staff moved on to other opportunities, I was charged with taking on leadership roles in two habitat programs at NOAA. This was the first time I was tasked as the lead for a major agency effort. Where I once could count on lots of advice and direction from someone more senior to me, I was now the one making decisions and advising senior government officials. Stepping outside of my comfort zone forced me to become confident in my decisions and assertive with my advice. These skills are essential to my career and will be useful as I move towards more-senior leadership positions. (TL)

I remember being a new federal employee working on fishery management policy and protected areas. Out of nowhere, I was told to lead a team developing an environmental assessment of deep seabed mining in the eastern tropical Pacific. Another time, I was told to lead an appeal process for a limited entry program in the West Coast groundfish industry. In both instances, I had great support from my supervisors but absolutely zero technical knowledge about the subject matter. I quickly learned that the experience outweighed my anxiety. Say "yes!" Test your limits. The confidence you build will extend to other instances, and the friends you make will be with you for years. (TB)

- Be creative! Many opportunities to add habitat knowledge are within reach in current jobs (e.g., filling that habitat void in a stock assessment task). Our point is that you might not need to change jobs to gain more satisfaction or increase your success; maybe you just need to adjust your role or your program's goals or add another key piece of the complex fish puzzle. If you would prefer to test your interests, perhaps you could learn about projects being implemented in another part of your work place or help managers in different disciplines. Look for opportunities and make strategic decisions, and the happiness and success you seek may suddenly be within reach.

  My background and expertise, from graduate school to my professional life, has been in habitat protection and restoration, yet I have always had an interest in working directly with fishermen and the fishing industry. I have found ways to connect my expertise in habitat issues with my interests in the fishing industry in various aspects of my professional life. I have worked directly with watermen in the Chesapeake Bay to restore oyster reefs and to comanage the harvest of oysters in special reserves. At NOAA, I have sought opportunities to work with fishing industry representatives to incorporate habitat information into their management decisions. Seeking these kinds of opportunities has allowed me to work on more traditional fishery management issues from my current job in habitat protection. (TL)

  I have found great rewards in expanding my professional networks. Traditional partners in my field are environmental groups like Clean Ocean Action and professional societies like the Society of Wetland Scientists. I learned my success with known partners increased when I expanded my network to include industry groups like the National Hydropower Association or more-focused groups such as dredgers or home builders. I learned much, and I would like to think that those groups gained perspective as well. A second example occurred when I hired interns and contractors. Several times a sharp candidate prompted me to think well beyond my initial intentions. Instead of hiring another biologist, I hired a political scientist or a business major; both of those hires taught me plenty, and most continue to excel in NOAA years after their initial internship. (TB)

Conclusions

The expanding habitat arena translates directly into new career opportunities for fish professionals of all stripes and colors. Whether you are new to the field or a grizzled veteran, view every task in school or your job as a personal challenge to do better for fish and your own career. Resolve to take chances, to meet new people, to extend your confidence and expertise in different arenas, and to take advantage of all that AFS has to offer through its many units and events. Most of all, say "yes!" much more than "no!" We think you will be surprised with the rewards.

Acknowledgments

The authors thank their colleagues in the NOAA National Marine Fisheries Service Office of Habitat Conservation for their contributions to the points articulated in this chapter. Their personal experiences coupled with our own careers helped to shape our primary points. Special thanks to Ms. Kristian Gamble for sharing her personal experiences.

Biographies

Terra Lederhouse is a marine habitat resource specialist in the Habitat Protection Division of NOAA Fisheries. She spends her time protecting habitats essential for rebuilding and maintaining sustainable marine fisheries. Before coming to NOAA, Terra waded in the mud of the Chesapeake Bay to restore oyster reefs and fish habitat.

Tom Bigford is policy director at the American Fisheries Society. He retired in early 2014 from NOAA Fisheries, where he spent the past 30 years working on fish habitat issues at the regional and national levels. In his new career at AFS, he will rely on his Michigan roots (and Michigan State University undergraduate coursework) to add freshwater perspective to his marine experience. Together they will help him tackle new policy issues and his strong interest in mentoring young professionals.

Lessons in Leadership

William F. Porter*
Michigan State University
480 Wilson Road, 13 Natural Resources Building, East Lansing, Michigan 48824, USA

Key Points

- The pivotal lessons in leadership often originate in trial-by-fire experiences.
- Mentors can help separate the lessons from the hot embers.
- The important lessons of leadership boil down to vision, communication, positioning and self-management.

Introduction

My introduction to leadership was not auspicious: "The only problem with this new guy is that he's from Iowa, is a still-wet-behind-the-ears Ph.D., and, as best we can tell, doesn't know the first thing about fish or wildlife or forestry in the northern forests." That was the essence of the opinions, sans profanities, expressed by the staff of the Adirondack Ecological Center where I was the newly appointed director. When you consider that the heir-apparent to the director position was a long-serving staff member, and the previous director's son-in-law, you begin to see the larger picture.

The staff critique was largely correct but missed the point. Regardless of whether the focus is fisheries, wildlife or forestry, or rocket science, every organization is built on a substantial body specific knowledge, and new leaders are often technically ill-prepared. That technical knowledge is certainly important, but it has only a little to do with being a leader. The core of leadership fits into a structure offered by Warren Bennis and Burt Nanus (e.g., Bennis and Nanus 1997). The simplicity of their structure enabled me to continually focus on improving my leadership skills. This structure has just four elements:

Vision
Communication
Positioning
Self-management

Learning to live this structure and understand it in depth is a process of trial and error. Perhaps most importantly, learning leadership skills is a process of self-reflection with the help of mentors along the way. This chapter is about these elements and the good fortune I have had to find mentors who could help me realize the lessons.

* Corresponding author: porterw@anr.msu.edu

Vision Is the First Criteria for a Successful Leader

Leadership is first about vision. In my experience, possessing a sense of vision is a rare human quality. Few people really think about big possibilities. Those people who do are good at synthesis. They read widely, talk with many kinds of people, and combine the information in creative ways. Their creative thoughts allow them to see opportunities before others do. They are good at strategic planning because they are constantly listening to the needs expressed by people around them. They are particularly sensitive to the unstated values of groups of people they encounter, and they see how the needs and values and new ideas can come together. If they are in leadership positions, they are able to see how to reshape an organization to take advantage of an opportunity. Importantly, their thinking is not limited by the resources currently available. Indeed, they recognize that it is actually easier to attract resources to bold ideas.

I learned the power of vision and bold ideas when a new college president commissioned each academic unit to prepare a strategic plan. When our unit presented the ideas that we were exploring for our plan, he asked whether we had considered an even bolder set of ideas. Our prior experience was that getting the institution to invest even US$10,000 in a field station 150 miles away from campus took an enormous effort. We thought we *were being bold* when we presented ideas estimated to require $50,000. When we laid out our ideas and our reading of the financial realities, he just said, "Don't worry about the money; think big!" We went back to him three times and each time he asked if this was the boldest idea we could conceive.

The last plan we presented called for a $13 million investment: more student housing, expansion of the dining center, fiber-optic and wireless computer networks, biogeochemistry monitoring systems, a conference center, and new staff positions in research, teaching, and operations. We developed a contextual framework based on *place*, the Adirondack Park, and cast a compelling case for it as the world's foremost experiment in building sustainable human economies while preserving the integrity of a wilderness ecosystem. We said we wanted to be the center of science for this grand experiment. Hyperbole? Perhaps. But only a little, in our minds. At that point he said, "Okay, now I'm on board."

The reason for our success was that this college president was a visionary thinker and a superb judge of ideas that would excite the imagination of others. He came from a background in business and understood that attracting attention is the most important step to getting financial investment. Exciting ideas attract attention. We finally presented a program plan that would be exciting to a broad array of the college's constituencies, and he gave us priority with the college's development office (the fund-raising arm of the college). When they identified a prospective investor, he stood with us to close the deal.

Where other administrators expressed skepticism that we could raise even $100,000, this visionary leader encouraged us to have faith in the power of big ideas. Over the next five years, we attracted nearly $11 million to implement our plans. All of that led to more students and faculty gaining exposure to the Adirondack ecosystem. It also nurtured confidence in our abilities among our friends in the area. That confidence led to new opportunities for research, instruction, and especially educational outreach. Success begets success.

Reflecting on this experience, the lesson I learned about vision and leadership was pretty simple on one level: do not be afraid of thinking big, and do not listen to the pessimists. On another level, mentors helped me realize that our success was associated with finding a combination of individuals at different levels in an organization who could conceive of bold ideas and then persuade still others with vision to buy into the ideas.

It Is All about Communicating

Selling a bold idea to a visionary thinker is one thing. Persuading others to help you garner the resources and implement a bold idea successfully is quite another. That persuasive ability is what we often think of when we talk of leadership. The root of persuasion in leadership is communication.

Most people believe they are good at communication and persuasion because they are doing it every day. Leaders, like professional athletes, take common, everyday abilities to a much higher level. When most of us have an idea, we express it without a lot of forethought. Leaders realize that the competition for the attention of those who can help achieve bold plans is intense. An idea needs to be refined and polished so that it is tailored to each person or group. One of the approaches to refining and polishing an idea is to cast it as a concise vision document. The standard is one page, and leaders slave over each of the approximately 500 words that will fit on that one page. From that arises the *elevator speech*: the one-minute description of the idea. What goes into that? An associate of mine once crystalized the content of the one-pager in terms of four questions: What is the issue? Why is it important? Where is the specific problem? How are we going solve the problem?

The one-pager and the elevator speech are just the beginning. Style is as important as content. We think of style in terms of clichés like dress for success. Indeed, the same principles apply because we are seeking to make each audience comfortable with us and excited about our ideas. There are many audiences, and each has a little different perspective on the issue, its importance, the problem, and the solution. Leaders are constantly tweaking the same idea in different ways to appeal to different audiences. Communication is about knowing the audience and knowing the specific cues that will get their attention. Some audiences like colorful brochures and view-booklets that intersperse photos with text that conveys a rich message. Others are more comfortable with a conversation over coffee. Either way, we need to frame the idea in terms that are central to the values, and the vocabulary, of that specific audience.

The importance of communication became clear to me once we set our $13 million plans in motion. As we worked with the college's Development Office, it became apparent that their forte was communicating ideas in a way that will excite others to invest. One grant for $1 million came from an agency that was completely unknown to us when we began writing the plan. To acquire their financial support, we framed the issue, the problem, and our solution in the language they used on their Web site. When we approached the local town board, we spoke of the importance of our plans for fisheries and wildlife research to boosting local tourism. When we approached the state economic development agency, we spoke of how our plans would create jobs in a rural community and education. When we pitched our ideas to the state conservation agency, we spoke of the research that would provide the science to address issues like acid rain and impacts of forestry on wildlife. With each audience, we looked for ways to assimilate their goals into our work. We were successful because they not only understood our ideas, but sensed that if we were successful, they would be too.

Here, my mentors moved my thinking away from the cynical attitude of considering all of this to be akin to hucksterism and political spin. Dissecting a sales pitch or a political speech shows that the approach follows a formula that has worked throughout human history: identify a problem, persuade people that it is in their interest to help you solve the problem, and then offer a solution.

Positioning Is What Others Call Luck

Most people consider luck to be just happy coincidence. To leaders, the definition of luck is preparation meeting opportunity. Leaders have the vision to see soon-to-emerge opportunities before others do and the communication skills to persuade others to join then in pursuing the opportunity. Well before the opportunity becomes evident to others, leaders begin positioning their group: developing plans, identifying milestones, investing in infrastructure, and attracting talent. When the opportunity arises, they are more ready than their competitors to take advantage.

Timing is the crucial determinant of successful positioning. Too early and few others will buy in, or if they do, the investments risk being squandered. Too late and time will be inadequate to prepare, and the investments will be again be at risk. It is important to recognize, though, that some lead times can be long. Decades are not unrealistic. For instance, 20 years ago, leaders in fisheries and wildlife programs foresaw the explosion of computing and statistics. When they had a faculty vacancy to fill, they included quantitative skills as one of the important criteria. They were ready when the explosion occurred in satellite imagery and Global Positioning System technology and brought the enormous volumes of data of a quality never before seen. These groups of faculty had the expertise in massive computing systems, spatial statistics, and Bayesian analysis to take advantage of the data. The fisheries and wildlife programs that are flourishing today are those that began hiring and growing the skills of faculty well in advance of that explosion. Was it luck … or leadership?

Whether the goal is adding faculty with particular strengths or lobbying for a change in fishing regulations, the process is the same: anticipating a future opportunity and preparing to take advantage. Among my most successful efforts at positioning while in a leadership position was preparing for a soon-to-emerge research opportunity. The reading I had been doing, and conversations I was having at professional conferences, suggested that diseases in fish and wildlife soon would take center stage as a hot-button issue. Wisconsin had just experienced an outbreak of chronic wasting disease in its deer herd, a disease related to mad cow disease. Like recent outbreaks of viral hemorrhagic septicemia in fish, chronic wasting disease brought a host of challenging implications for natural resource managers. My conclusion was that the question was not *if* these diseases would occur but *when.*

With a major disease program at a competing university in the state, I had little obvious leverage. I began positioning for the future opportunity to expand into the arena of disease by inviting the chief of the state wildlife agency to lunch. I shared with him the advances we were making in applying new quantitative tools to behavioral ecology and how this might relate to disease management. I then made the time to get to know the head of the Fish and Wildlife Division and talk about the role of science in controversial decisions pertaining to fish and wildlife management. Finally, I finessed an opportunity to spend an afternoon in the field with the commissioner, the leader of the state natural resources agency. I knew personal rapport could go a long way in helping them trust that we could do the work when a disease outbreak occurred.

At the same time, I began watching for a potential graduate student to do the work. On a snowy day in February, a prospective student arrived to discuss doing a doctoral degree in my laboratory, and I suggested she might explore how diseases spread for her dissertation. She crafted a proposal, and we drove to the state capital to pitch it to the natural resource agency. We did not succeed. Indeed, she remembers it as a spectacular failure. The problem was that there was not wide agreement that disease was the top priority for research, and even if it was, there was no money. Still the effort was not a complete failure because our idea was now on the table. Within six months, a disease outbreak occurred. Federal funds were suddenly available to support research and management action. We were invited back to present our proposal again. The efforts in building rapport with key individuals paid off because this time we walked out with a research contract for $1.1 million.

The mentoring lesson I drew about positioning was twofold. First, the experience taught me to trust intuition. Looking over the horizon is a process of synthesizing trends from across broad sources of information. Leaders are those who trust their judgment enough to push forward, even when those around them disparage the ideas. Second, what mentors helped me understand was that leaders are continually working to build rapport with those who they anticipate will be in a key position when opportunities arise.

Self-Management Is Critical to Achieving the Best

A central principle to the management of successful ventures is that good is the enemy of great. If you settle for good, you'll never strive for great (Collins 2001). The challenge to living up to that principle is that as we move into leadership positions, we accumulate such a large suite of responsibilities, that our day-to-day agendas become overwhelming. Leaders separate themselves from others because they recognize that they need to focus on a limited set of priorities. On these, they do not settle for good. They strive for great.

Sometimes striving for great means dispensing with myriad important responsibilities. As the director of a research station in the middle of the Adirondack Park, I was surrounded by one of the world's great experiments in conservation. Could society really achieve the goals of building a vibrant economy while ensuring the ecological integrity of a wilderness ecosystem? The past 30 years had been a particularly formative period in the history of this experiment. While I had a lot of my plate, I decided the most important problem was to capture the voices of the many visionaries and managers who were instrumental in creating this great experience before these people passed from the scene.

To focus on this effort, I collaborated with two close colleagues to plan a book. Individual chapters were solicited from people who we identified as having played the most important roles in the evolution of the Adirondack Park over the past 30 years. My two colleagues and I edited the chapters and create a coherent structure to the book so that the storyline would be evident to readers. We successfully pitched the book to a major university publisher, and I arranged to take a six-month sabbatical leave from the day-to-day responsibilities of administration and teaching. Those six months were crucial because they enabled me to become immersed in the issues and find the common threads among the various chapters. The entire project took three years because we refused to compromise on quality. We kept seeking the best from each author. Some would say we pushed them awfully hard, but the final product gained a five-star rating and won a literary award.

One of the two colleagues with whom I partnered in writing the book was an early mentor. The lesson that he helped me see is that achieving the best is not just a product of remarkable talent. We had extraordinary talent. Bringing the book to fruition was a product of discipline. Attention to timelines and to details and helping others stay focused was absolutely essential.

Mentoring: Putting Muscle on the Skeleton of Leadership

Leadership is learned amid trial by fire. As we know from most subjects, the lessons are most efficiently learned with the help of teachers, coaches, or mentors. Good teachers help us see the skeletal structure with which to organize important ideas, and they help us to build the muscle to put those ideas in motion. If we view the four elements of leadership—vision, communication, positioning, and self-management—as the skeleton, then the challenge is finding good mentors to help us build those elements into a successful leadership style. While sometimes mentors are formally assigned to us, our best mentors are more often colleagues or supervisors who are successful leaders and with whom we connect on a personal level. Mentors challenge and critique our growth. Often more important, they just listen and help us to find insight and perspective on the leadership challenges we face day to day. The best mentors become our cheer leaders, bolstering our confidence as we grow.

Acknowledgments

I owe much of my leadership skill to five mentors. John R. Tester at the University of Minnesota helped me learn the importance of clarity in communication. Robert L. Burgess, a department chair, helped refine my communication skills and taught me the essence of self-management. Three individuals showed me the power of strategic planning as a means to visioning and positioning: Ross S. Whaley and Cornelius B. Murphy, both presidents of the State University of New York College of Environmental Science and Forestry, and Donald F. Behrend, chancellor of the University of Alaska system.

Biography

Dr. William Porter is Boone and Crockett professor of wildlife conservation at Michigan State University. He earned a B.A. in biology and teaching at the University of Northern Iowa and M.S. and Ph.D. degrees from the University of Minnesota. He spent 32 years on the faculty at the State University of New York College of Environmental Science and Forestry in Syracuse where he taught courses in wildlife ecology and management and ran a large research lab of graduate students focused on the application of quantitative tools to wildlife research. His formal leadership roles there included serving as chair of the College Faculty, and director of both the Adirondack Ecological Center and the Roosevelt Wild Life Station.

References

Bennis, W., and B. Nanus. 1997. Leaders: strategies for taking charge, 2nd edition. HarperCollins Publishers, New York.

Collins, J. 2001. Good to great: why some companies make the leap and others don't. HarperCollins Publishers, New York.

Fisheries Decision Making: Advice from the Introduction of Pacific Salmonids into the Great Lakes

Howard A. Tanner
Michigan Department of Natural Resources
5755 Green Road, Haslett, Michigan 48840, USA

Abigail J. Lynch*
Center for Systems Integration and Sustainability,
Department of Fisheries and Wildlife, Michigan State University,
115 Manly Miles Building, East Lansing, Michigan 48823, USA

Key Points

Fisheries decision making is a difficult process but can be supported by addressing the following questions before a decision is made:

- Why make a decision?
- Who makes the decision?
- How should the decision be made?
- When should the decision be made?
- What if there is a mistake in decision making?

Isn't it strange, that princes and kings,
and clowns that caper in sawdust rings,
and common-folk like you and me,
are builders for eternity?

To each is given a bag of tools,
a shapeless mass and a Book of Rules;
and each must make 'ere time has flown,
a stumbling block or a stepping stone.
—R. L. Sharpe, "A Bag of Tools"

Introduction

Dead Alewives *Alosa pseudoharengus* strewn across beaches, a stinking, rotting mess. Empty fishing ports with the collapse of native fish communities and traditional commercial fisheries. Great Lakes state residents more willing to make donations to save the whales and dolphins than the Great Lakes in their own backyards. In the 1960s, Great Lakes fisheries

* Corresponding author: ajlync@gmail.com

managers were faced with a mandate of conservation with little resource to conserve. What decisions needed to be made?

At the time, Alewives were estimated to make up more than 95% of the fish biomass in Lakes Michigan and Huron. Floating masses of dead Alewives closed hundreds of miles of beaches. Victims of the "tragedy of the commons," once plentiful Lake Trout *Salvelinus namaycush* and the native coregonids, were depleted by individuals acting in personal, rather than the common, interest. Lake Trout were, in effect, commercially extinct. With little connection to these native fishes and no desire to connect with invasive Alewives, many Great Lakes residents had lost their connection with the lakes and sense of place. They felt little attachment to the region and little pride in the Great Lakes resources. While some managers may have viewed this situation as a problem, problems are often opportunities in disguise. The bursting population of Alewives represented a huge biomass, a huge food supply upon which a new fishery could be built. How should the opportunity be pursued?

Alewives, and the potential to establish a new recreational fishery in the Great Lakes, provide an example of why fisheries decision making should be a careful, deliberate process. Remember that decisions seek to incite change and many people generally do not like institutional change. A structured decision-making process can make evaluation of options more transparent, especially in complex situations. Being able to communicate the answers to the following questions will lead to a more informed and potentially supportive public, in addition to more effective fisheries decisions: (1) why make a decision, (2) who makes the decision, (3) how should the decision be made, (4) when should the decision be made, and (5) what if there is a mistake in decision making?

Why Make a Decision?

> In any moment of decision, the best thing you can do is the right thing, the next best thing is the wrong thing, and the worst thing you can do is nothing.
> —Theodore Roosevelt

When faced with the need to address a fisheries issue by making a decision, one of the most important questions to ask is if this decision being contemplated is best for the resource. Fisheries are a public resource, so fisheries decisions have both immediate and long-range impacts, for the public now and in the future. These decisions are not easy and require discerning judgment. Perhaps the best decision is to maintain the status quo, but it should not be made by default because it is easier than making another choice. Decisions should be made with an eye to the long-term future, in anticipation that future impacts of the choices are likely more important than the immediate impacts on the current situation. Long-term consequences are more permanent, are often larger-scale, and have a broader reach.

With no natural predators and little or no commercial value, invasive Alewives were a nuisance. Measures were in place to restore Lake Trout populations but Sea Lamprey *Petromyzon marinus* parasitism was hindering recovery. Machines were used to clean dead Alewives off the beaches. Attempts were made to develop profitable markets for Alewife—fish oil and fish meal. It was clear, however, that these small-scale efforts were not making much headway in addressing the full-scale of the Alewife problem—they were not sustainable solutions. A decision needed to be made outside of the box to manage the excess Alewife biomass because it was clear that without intervention, the situation would not improve, and the commercial fishery approach would not solve the problem.

Who Makes the Decision?

> A lot of people don't want to make their own decisions. They're too scared. It's much easier to be told what to do.
> —Marilyn Manson

In fisheries management, decisions are often difficult. There are often many stakeholders with conflicting interests. As a general rule, decisions should be made as far down the chain of command as possible. When there are many decisions that need significant attention, efforts to compile the necessary information should be delegated among a highly capable team. A supervisor should recognize his or her own strengths and weaknesses and select team members to complement and balance. More complex decisions may need to be moved up the chain of command, but no higher than necessary. Beware of the staff person who avoids making decisions that he or she should make and, instead, passes them up the chain, taking the easy way out. This is reverse delegation—detect and avoid it.

For various reasons, over the first half of the 20th century, the Michigan Department of Conservation basically deferred Great Lakes fisheries management decisions to the federal government, the Department of the Interior's U.S. Bureau of Commercial Fisheries, which focused on encouraging commercial rather than recreational fisheries. But, the Alewife issue posed an opportunity ripe for change. A blue ribbon committee had recently declared the Fisheries Division nonfunctional. Dr. Ralph McMullen was newly appointed director of the Michigan Department of Conservation with orders to affect change. He hired me (H. Tanner) as fisheries division chief, the first professionally trained fishery biologist and first brought in from outside the department to fill the position. Addressing the Alewife problem was my responsibility and decision to make, but it helped to know I had the support of my director to take innovative approaches to affect significant change.

How Should the Decision Be Made?

> Examine each question in terms of what is ethically and aesthetically right, as well as what is economically expedient."
> —Aldo Leopold

Defining the scope of a decision is particularly important. If the dimensions are too narrow, the sources of and scope of the solution(s) will be limited. Conventional thinking will only result in conventional and status quo decisions. Good decisions often shift standards and understanding. So, remember that the only constant is change and embrace that change with enthusiasm. There can be strong pressure to conform to the status quo and be content with "good enough." Risk inherently has a negative connotation. But as the old adage goes, with the greatest risk comes the greatest reward. It is important to always remain open and supportive of constructive and innovative change that goes beyond good enough.

Dogma within biological sciences has been againsts introduction of an exotic species. When the Fisheries Division was faced with the need to address the Alewife problem, conventional solutions, such as rehabilitating native species, had not been not effective. But introduction of exotics faced some formidable foes: federal agencies, commercial fishing interests, and their allies. Did introduction of Pacific salmon fit into the definition of conservation? Gifford Pinochet, the first director of the U.S. Forest Service, defined conservation as management of public resources for "the greatest good for the greatest number for the longest period of time." In the case of the Alewife problem (opportunity), the Fisheries

Division and I rationalized that it was inappropriate for most areas of the Great Lakes to continue to be allocated to a few commercial fishermen rather than realign the management of lakes to support a recreational fishery serving tens of millions of area residents and visitors. This realignment was still a very risky choice. Recreational fishing in Michigan had been declining, and there was little to no tradition of sportfishing on the Great Lakes. I was going to be glorified or vilified for my decision, but it was a risk I was willing to take to restore biological balance to the Great Lakes and create a highly lucrative sport fishery. It was also a risk that could be rationalized to the public. The Fisheries Division spent many hours and speeches selling the program and the funds required to implement it. This time and effort demonstrated the Fisheries Division's commitment to the decision and won the support of the stakeholders.

When Should the Decision Be Made?

> If you think you can do a thing or think you can't do a thing, you're probably right.
> —Henry Ford

Is now the time to make the decision? Though the automatic response may be "yes," there may be reasons, such as pending additional information, that might substantially improve your ability to make the best decision. Support is also very valuable for any decision. Making a decision before knowing that stakeholders and those implementing the decision are in favor of it is not advisable. If a decision is overturned, it destroys confidence. But, if all the appropriate information and support is available, make a decision decisively and make a decision promptly; otherwise, waiting is just procrastination. Again, having a highly dependable and supportive team can assist in this process. While the decision is yours, input from others can often provide a helpful perspective.

While the Fisheries Division had identified a need to address the Alewife issue and create a recreational fishery based on introduced fish, there was still no course of action and no infrastructure to start this new fishery in the Great Lakes. Then, late one evening, I got a call informing me that, for the first time in decades, there was a surplus of Coho Salmon *Oncorhynchus kisutch* eggs from Columbia River hatcheries. The window of availability of eggs was very short. By the next morning, I knew that I would try my best to make the introduction. I discussed the opportunity with all available staff that day and no serious dissent emerged. Though I was painfully aware of the huge responsibility I was about to accept, I knew it was time to act. So, I made the call, arranged for the shipment of Coho Salmon eggs to Michigan, and initiated the first successful large-scale introduction of Pacific salmonids into the Great Lakes.

What if There Is a Mistake in Decision Making?

> If you are going to grow something, go where it will grow.
> —John A. Hannah, Michigan State University President (1941–1969)

Everyone makes mistakes sometimes. If everything was known about fish, there would be no need for fisheries management. If a mistake is made, it is important to acknowledge it; a cover-up would be deceitful and devastate credibility. "I don't know" is also a perfectly acceptable answer if there is no evidence suitable for answering the addressed question.

The first shipment of Coho Salmon eggs from the Columbia River provided opportunity for the introduction of a recreationally viable species into the Great Lakes. After that first Coho Salmon introduction, the Fisheries Division attempted to introduce a number of other species, including Chinook Salmon *O. tshawytscha* and Striped Bass *Morone saxatilis.* While Chinook Salmon introductions have had an overwhelmingly positive impact on the Great Lakes fisheries, Striped Bass were seriously considered, but rejected. After a lengthy examination, including commitment of staff time and resources to raise Striped Bass in a Michigan hatchery, the species was never introduced into the Great Lakes. The objective of the trial was redundancy, to ensure the success of the introductions into the Great Lakes. Striped Bass ended up not being the appropriate fit, biologically, socially, or economically. They were not anticipated to reach their full growth potential because of cold temperatures in the Great Lakes, they appeared less desirable to Great Lakes recreational anglers, and they were more expensive to produce. By being transparent in why I chose not to pursue introduction of Striped Bass, the Fisheries Division was able to retain credibility with the public and its support for subsequent trials for Atlantic Salmon *Salmo salar*, lake-dwelling Brown Trout *Salmo trutta*, steelhead *O. mykiss*, and other Pacific salmonids.

The introduction of Pacific salmonids to the Great Lakes, restoring biological balance to the largest system of surface freshwater on earth and generating one of the world's most economically valuable freshwater recreational fisheries, was an unconventional decision made in response to a need to address invasive Alewives. The Fisheries Division recognized the need to reestablish the predator–prey dynamic, as well as to meet the huge potential and unsatisfied demand for recreational fishing. I asserted the Fisheries Division's authority to create a new sport fishery to fill both needs. When the opportunity to introduce Coho Salmon presented itself, I took it, along with the full responsibility of the decision. This decision was basin-scale—and not just any basin, the largest system of surface freshwater on the planet. What I decided not only affected Michigan, but also the other Great Lakes states and Canada. No other fishery biologist ever had such a decision to make. Depending on the outcome, I would be a hero or a scapegoat. Ultimately, the decision resulted in restoration of ecological equilibrium, creation of new fisheries, new sources of income for the region, greater opportunities to utilize the public resource, and control of Alewives. While the introduction of exotics was surely controversial, and there have been many issues with the subsequent management of Pacific salmonids in the Great Lakes since this initial introduction, the decision was made with the public good in mind.

Introducing Pacific salmonids into the Great Lakes is one such example of an approach that can be applied in any fisheries decision-making context. This approach includes answering the following questions: (1) why make a decision, (2) who makes the decision, (3) how should the decision be made, (4) when should the decision be made, and (5) what if there is a mistake in decision making? Structuring a decision around these five questions streamlines the process and communicating these steps makes the process more transparent to those outside of decision making. Remember that conservation is management of the public resources to produce the greatest good for the greatest number for the longest period of time. This is the ultimate rationale for any fisheries decision.

Acknowledgments

We would like to thank Bill Taylor for introducing us, Nancy Léonard and three anonymous peer reviewers for strengthening this vignette, and the Howard A. Tanner Fisheries Excellence Fellowship for supporting graduate fisheries research within the Great Lakes, connecting waterway, and tributary streams.

Biographies

Howard Tanner, known as the father of the Great Lakes salmon program, served as the Michigan Department of Natural Resources (DNR) chief of the Fisheries Division (1964–1966) and director (1975–1983). The introduction of Pacific salmonids, considered by some to be the most successful bio-manipulation ever attempted, occurred under his watch. Before serving as Michigan's chief of Fisheries Division, he taught at Colorado State University and led the Colorado Department of Game, Fish, and Parks research program. Before accepting the position of DNR director, Howard served eight years as professor and director of natural resources at Michigan State University. In retirement, Howard is an adjunct professor at Michigan State University and also serves on the Lake Huron Advisory Committee. He has been inducted into the National Fisheries Management Hall of Excellence, the National Freshwater Fishing Hall of Fame, and the Michigan United Conservation Clubs' Hall of Conservation.

Abigail Lynch is a research fisheries biologist at the U.S. Geological Survey's National Climate Change and Wildlife Science Center. She is a recent graduate from Michigan State University with a Ph.D. in fisheries and wildlife; a secondary degree in ecology, evolutionary biology, and behavior; a doctoral specialization in environmental science and policy; and a College of Agriculture and Natural Resources certificate in college teaching. Her dissertation was designing a decision-support tool to assist Great Lakes fisheries professionals and fishermen prepare for the impacts of climate change on Lake Whitefish *Coregonus clupeaformis* production. Her research interests include marine and freshwater fish conservation with a management focus on fisheries systems.

Saving Fish? Saving Fisherfolk? Reflections on Designing Governance Policies for Fisheries

Anastasia Telesetsky*
Natural Resource and Environmental Law Program, College of Law, University of Idaho
875 Perimeter Drive, MS 2321, Moscow, Idaho 83844, USA

Rebecca Bratspies
CUNY School of Law
2 Court Square , Long Island City, New York 11102, USA

Key Points

In a changing world, be flexible and expect to change. Designing effective governance policies for fisheries requires a genuine interest in problem solving coupled with a long-term investment in human relations and an optimistic outlook on human behavior.

- Be bold in your thinking about socioecological problem solving.
- Actively seek constructive feedback on your ideas from colleagues across disciplines.
- Understand where old policy ideas come from, and, where appropriate, revive old policy ideas.
- Find a mentor or be a mentor.

In the field of fisheries regulation, there are few clearly delineated rules regarding how to draft effective policies and laws that will protect the lives of fish as well as the long-term livelihoods of fisherfolk, including all of those men and women who either directly participate in marine or freshwater fishing or indirectly benefit from fishing by providing fishing-related goods and services. This makes the field a challenging one for lawyers since we tend to like clear rules. Sustainable fishery management cannot follow ironclad rules because fisheries science requires adaptive management.

Fish move. They sometimes do not reproduce on schedule or according to human plans. They ignore our efforts to create open and closed seasons. Just as fish refuse to cooperate with our rules and regulations, environmental conditions are also changing as the ocean warms and acidifies or freshwater systems undergo eutrophication. We are faced with constantly evolving fishery research, which challenges scientific findings that have become accepted as immutable facts by the law.

As fisheries governance scholars, we work at the intersection of biology, marine ecology, economics, and law. The science is fraught with uncertainty, the lines of legal authority are

* Corresponding author: atelesetsky@uidaho.edu

ambiguous, and the humans involved can be frustratingly unpredictable. While as lawyers we prefer clearly delineated rules that can be easily interpreted, as academics dedicated to sustainable fisheries, we need to embrace uncertainty. Fishery laws and regulation cannot be static and rely on outdated management concepts such as maximum sustainable yield. Instead, these laws and regulations must be designed to be flexible—otherwise, they will swiftly become obsolete in this rapidly changing world where ocean temperatures are warming, fish stocks are depleted to the point of collapse, and policies of national food security are boosting unprecedented levels of industrial aquaculture.

Our great challenge is to develop policies capable of protecting the livelihood needs of a wide range of different kinds of fishing communities today while looking out for the needs of future human generations whose livelihood and well-being will also depend on healthy natural ecosystems. Fishery policymakers responding to political pressures have spent far too long pretending they could ignore the physical limits of the earth's biogeophysical systems by, for example, creating fishing quotas that exceed scientific recommendations. We cannot continue to prioritize the needs of today at the expense of tomorrow.

Creating policies capable of simultaneously saving both fish and fisherfolk requires patience, inspiration, and perseverance. It requires a working knowledge of social, legal, and ecological factors. If we knew the secret socioecological formula for effective fishery management that would protect both fish and fisherfolk, we would be zealously promoting it. We know it has something to do with integrated ecosystem management, marine and freshwater protected areas, use rights that mirror property rights, meaningful participation by fishing communities, transparent decision making that expressly considers that participation, comanagement, and credible enforcement, but we have no magic formula for balancing these proposed policy interventions that is guaranteed to ensure both long-term jobs and viable fish populations. Finding a balance among the various competing stakeholders is our struggle—and it may wind up being your struggle as well.

We would like to offer a few suggestions of how to begin your career as a fisheries professional with an interest in creating effective fishery management laws and policies. Each of these suggestions is intended to help you stay engaged for a lifetime in this rewarding field of helping fish and fisherfolk.

1. Dare to imagine a different sociopolitical world (even when your ideas may seem unpragmatic), be open to constructive criticism, and seek interdisciplinary opportunities.

New ideas are essential for innovation. While, there are many old problems, like overfishing, habitat destruction, pollution, and illegal fishing, there are also new problems, such as exploiting genetic material on the high seas, genetically modified fish farming, impacts of increasing temperatures on freshwater species, ocean acidification, or disputes over seabed mining. New problems demand innovative thinking.

Always remember that the pragmatic solution to addressing the needs of fish and fisherfolk will differ in part on who your client is and what they view as their core interests. Perception is everything! If you are working for a traditional profit-oriented industrial trawler, the status quo seems pragmatic because it takes into consideration both conservation measures and stock capture. By contrast, if you are working with a restoration ecologist, the status quo is unacceptable because many stocks under the current legal regime are overfished or are in danger of being overfished. If you are working for a government regulator, you may find yourself uncertain about how to react depending on what policy directions you are receiving from high-level, political decision makers.

Promoting fishery policies for long-term equity and sustainability frequently challenges entrenched economic interests. This can be an uncomfortable place as an emerging professional. You may not want the less-than-friendly attention from older colleagues that articulating innovative policy ideas might bring. It is certainly professionally safer to stay inside the mainstream of old policies because you will not expose yourself to academic, industry, or civil society criticism. But, sometimes it is worth taking an intellectual risk and imagining new possibilities. Your unpragmatic policy proposal, which requires political will, supplementary financing, or cooperation that may be absent today, may still open up new governance possibilities by creating a dynamic intellectual vision in an area that may be trapped by political inertia. For example, few policymakers were talking about natural capital accounting or ecosystem services a decade ago. Now, these terms create a common language for both national and global policymakers.

In order to avoid undue criticism of wishful thinking when you propose a policy intervention that has been historically stymied by a lack of political will, you may want to categorize your new policy proposals as think pieces in order to elicit more helpful reactions from your senior colleagues. This may help to avoid the frustration of having your ideas peremptorily dismissed as impossible due to lack of political will, financing, or necessary motivation on the part of nonstate actors. Several possible outlets for your creative thinking include journals that allow for opinion pieces, such as *Marine Policy*; conference panels for works in progress, such as the mid-year meeting sponsored by the American Society of International Law; and academic-oriented blogs, including Yale Environment 360.

Good ideas will gain traction over time. For example, the now well-accepted proposal to restrict commercial fishing activity through the property mechanism of individual fishing quotas was initially resisted as an intrusion of privatization into the commons. While some concerns remain, individual fishing quotas have been successfully embraced by certain fishery managers as a rational economic and ecological response to the "tragedy of the commons." Be patient when proposing something new and open to constructive criticism. The value of the peer review process is in helping to refine and build intellectual momentum for good ideas.

As the ocean warms and as fish stocks continue to be overexploited, fisheries governance challenges are only going to increase. Legal thinkers willing and able to engage in lateral thinking outside of the mainstream as well as in interdisciplinary collaboration will become increasingly important. Fishery management is neither exclusively a legal problem nor an ecological problem. It is a socioecological problem requiring creative synergistic thinking to deal with novel challenges. New professionals can seek interdisciplinary opportunities by joining groups that might not be their obvious professional home in order to learn from others. For example, fishery biologists interested in learning more about natural resources law might join the American Bar Association Section of Environment, Energy, and Resources or the International Environmental Law section. A lawyer who is interested in fisheries science might join the American Fisheries Society or might collaborate with fishery biologists, climate scientists, and/or public health officials.

2. Show humility.

As an emerging professional, you may find that your colleagues have designated you as the institution's expert on all things related to fisheries. You may feel that you are expected to design effective policies without adequate time or resource support. Resist the temptation to pretend an omniscience you do not have. Do not be afraid to consult with colleagues outside of your area of expertise and build a social network of problem solvers. If you have a specialty in law or policy, take the initiative to ask fishery biologists what human activities or laws are having the most immediate negative impact on fishery stocks. If you are a fishery biologist, go out of your professional comfort zone and ask lawyers about the institutional barriers in creating regulatory policy. Academic and practitioner silos are created by individuals seeking echo chambers where their ideas are already well understood. Effective natural resource policy is rarely created within a silo, but instead requires a cross-pollination of ideas across disciplines.

In the world of fishery policy, there are some excellent examples of cross-pollinating, multidisciplinary, problem-solving projects. The University of British Columbia Fisheries Centre "Sea Around Us" project involves economists, biologists, ecologists, and legal specialists. While this sort of institutional collaboration may not be available to you where you have been hired for your first job, be willing to reach outside your own institution to build a social network. If you work in a government agency, you can reach out to academics at a local university. If you work in a university, you might call someone in a civil society conservation group.

If you are new to the profession, where do you find colleagues that may be interested in collaborating in developing new or revising existing policies and regulations? One possibility is to join professional groups and volunteer to be on their planning committees or in their work groups. Another possibility is to attend conferences and seek out panelists that might have an overlapping interest area and make a point, no matter how awkward it may feel, to introduce yourself. If you present yourself as genuinely concerned about a specific problem that others are also thinking about, most professional colleagues will find time for you to share your ideas. They may even possibly make an offer to collaborate on a research or policy project. If you are short on time and travel funds, you might still be able to start a professional problem-solving network with a once-a-month multidisciplinary reading group where you might meet at a cafe and discuss new journal articles with both senior and junior colleagues.

Of all the suggestions that we offer, this one may be the hardest to act upon; from our experience in legal circles, incomplete knowledge of a subject matter often seems like a character flaw. There is an implicit threat to one's self-esteem in asking questions that arguably might have been mastered earlier through classwork, fieldwork, or career experiences. Yet, being ready to ask those questions, and then to seek out their answers, is a sign of intellectual maturity. Experts too often overvalue their own area of knowledge and discount the importance of areas they do not know. This results in an intentional self-"blindering," where some experts see everything in the light of their expertise. For example, some lawyers see every problem as a legal problem that simply needs an amendment of the law or an investment in litigation. Such problem-solving approaches fail to understand that law is really only a tool that reflects preexisting or emerging political, economic, psychological, and social values. The self-knowledge and self-confidence to say "I do not know this area and I need help" will serve you well. Not only will the quality of your ideas deepen and improve, but you will also obtain a competitive advantage in the institutional marketplace of ideas by understanding that complex problems require multidisciplinary solutions. Any meaningful career is a lifelong learning process, and there will always be opportunities for learning from both your junior and senior colleagues.

3. Sometimes old problems need old policy solutions. Be ready to reconceive and repropose old solutions.

In our first suggestion, we recommended that you think imaginatively and not be afraid to ask "what if" questions regarding the institutional boundaries of fishery management policymaking. We also believe that many great policy suggestions have already been made but not yet acted upon by decision makers. Maybe the initial policy idea was good, but the monitoring arrangements were too difficult, the financial incentives were insufficient, or the legal penalties were too lenient. A difficulty with the implementation of a given policy does not mean that policymakers should simply dispense with such policy recommendations. It may still be appropriate to resuscitate old policy solutions when your research leads you to believe that earlier policy proposals should still be acted upon.

For example, one of the chronic problems with marine overfishing is the overcapacity of fishing vessels plying the oceans with no regulation of their activities because the flag state either does not have the political will or the legal capacity to monitor vessels flagged to the nation state. It is not a new policy suggestion to suggest that the international community should regulate these so-called "flags of convenience." Indeed, the debate over what constitutes a genuine link between a vessel and its flag state still rages decades after the question was initially raised. Here, it is perfectly reasonable to add your voice and to reiterate, once more, the need for a treaty or some other binding document to define "genuine link" for purposes of the management of fishing vessels flying flags of convenience.

A historical lack of political will does not necessarily mean that you should abandon old policy solutions in search of new approaches. If old policy solutions seem to be the best approach, then use your expertise to help build the political will to make it happen, even if the changes were proposed decades earlier. As a young professional, you will have exposure to new ideas and new fields that may suggest new arguments for old policy solutions.

This advice of recycling old policies may at first seem counterintuitive. After all, you will not build an academic, agency, or industry reputation for being a creative thinker by calling for policies that have already been proposed. Yet, your new voice and energy may be just what is needed to renew an old proposal or to bring needed momentum to otherwise inert ideas. There is no way to predict the tipping point when a seemingly unpopular idea accelerates into policy. The recent international restrictions on shark catches under the Convention on the Trade in Endangered Species illustrate this well. Opponents to certain practices in shark harvesting have been calling for bans on shark finning for years, but it was not until March 2013 that there was finally critical mass to implement an old policy solution for an old problem. Understanding the value of persistence and patience is a must for successful long-term policymaking.

4. Find a mentor. In fact, find more than one.

The word "mentor" comes from the Latin noun mentos meaning "intent, purpose, spirit, passion." A mentor is not necessarily someone who is at the end of his or her career and now is giving back to the profession through training. If that were the only definition of "mentor," most of us would never experience mentoring. Instead, a mentor is someone who shares with you his or her spirit and passion. A mentor can be a peer who graduated at the same time you did but decided to take their career in a different direction or someone who may be just a few years ahead of you in their career and is generously willing to share their spirit and passion with you by recognizing your potential and helping you to realize it. Even though you are an emerging professional, you may find yourself in the position to mentor a colleague. Do not be afraid to take this opportunity because it may help you to learn not just more about yourself, but also about your field of interest.

There are many paths to building a successful career in fishery policymaking. Common to all of these paths is a genuine interest in problem solving plus an investment in human relations. In trying to solve complex socioecological problems, use your active imagination to propose innovative solutions but also be willing to revive old but solid policy solutions. Problem-solving is not merely an intellectual exercise in theory. Good problem-solving is a group effort over time, so we also recommend acknowledging to yourself what you do not know and seeking help from others, so that you can build both a professional network for problem-solving and, hopefully, some life-long friendships.

Acknowledgments

We are deeply grateful to the many colleagues who have supported our work and have been willing to share their creative energy and enthusiasm with us. We are also grateful to each other—our successful comentoring experiment has enriched our work and created a valued friendship.

Biographies

Anastasia Telesetsky is an associate professor at the University of Idaho College of Law in the Natural Resources and Environmental Law Program. Her research areas are ecological restoration and the law of the sea.

Rebecca Bratspies is a professor of law at the CUNY School of Law and the founding director of the CUNY Center for Urban Environmental Reform. Her research focuses on transparency and participation in public decision-making processes.

Fly-Fishing for the Future: How the Michigan State University Fly Gals Are Mentoring Future Conservation Leaders

Kerryann Weaver*
U.S. Environmental Protection Agency
Region 5, Water Division, Watersheds and Wetlands Branch WW16J
77 West Jackson Boulevard, Chicago, Illinois 60614, USA

Tom Sadler
Outdoor Writers Association of America
615 Oak Street, Suite 201, Missoula, Montana, 59801, USA

Key Points

Mentoring workshops, such as MSU Fly Gals, strengthen the connection between recreation and conservation by creating a unique opportunity for professional networking. This model works best by

- Attracting participants with an outdoor recreation skill,
- Using that skill to reinforce a larger goal, and
- Engaging previous participants and a broader professional network.

Introduction: The MSU Fly Gals Model

When considering the future of fisheries, science often dominates the conversation. While science is clearly essential to the future, the ability to encourage others to follow the professional path is often more difficult to accomplish. In this vignette, the authors provide a dialogue about an innovative approach used by Michigan State University (MSU). This nontraditional form of mentoring provides participants with an opportunity to see real-life examples and impacts of fisheries science, recreation, economics, and outdoor activity in a nonacademic setting. Most importantly, the participants build a network of colleagues that share a common experience and who can offer support and advice in the future.

The value and importance of engaging in this nontraditional mentoring experience with emerging professionals is that it brings participants back to the outdoors and nature, a place where their existing interests in fisheries likely began. We feel that, in this atmosphere, the younger generation of fisheries and conservation-minded professionals can be empowered and their sense of direction, professional goals, and interests in fisheries rekindled.

MSU Fly Gals is a program taking an innovative mentoring approach to cultivate a new generation of conservation leaders. Using a back-and-forth dialogue approach, the au-

* Corresponding author: kewaco@gmail.com

thors share their views of the history, evolution, and importance of MSU Fly Gals; why the program has become such a success; and why it serves as a good model for others to use to mentor future professionals. Tom Sadler, in plain text, details his experience as a fly-fishing instructor and conservation lobbyist, while Kerry Weaver, in italics, shares her experience as a participant and fishing enthusiast turned fly-fishing instructor.

MSU Fly Gals is an excellent model for mentoring because participants are immersed in an outdoor recreational activity, which better enables linkages with conservation issues. It fosters connections between participants and establishes a network with conservation specialists on the front lines of current issues in natural resource management. A key aspect of the program is that it attempts to strengthen the connection between recreation and conservation. This model can easily be applied in other contexts as the model is built on people's interests in outdoor recreational activities, which link directly to conservation issues.

From Barroom to the North Branch: History of the Program

Typical of many good ideas, the genesis of MSU Fly Gals was a barroom conversation between Dr. Bill Taylor and me. Taylor, an MSU University distinguished professor, was Chairman of the Sport Fishing and Boating Partnership Council at the time. He and I were talking about the future of conservation, where the leaders would come from, and how many state and federal folks were not familiar with some of the traditional outdoor activities like hunting and fishing. As both a conservation lobbyist and fly-fishing guide, I used fly-fishing as an example: "if you teach people to fly-fish, they will usually become conservationists, whether they intended to or not."

Dr. Taylor, being quick to seize an opportunity when he sees one, asked if I would be willing to teach his female graduate students as well as some female natural resource-based career professionals how to fly-fish and maybe share some of my experience as a longtime conservation advocate. We saw an opportunity to create a program that would take an innovative mentoring approach to cultivation of new conservation leaders. The fact that I would be teaching on the renowned North Branch of the Au Sable was an added incentive. So began an eight-year run that has become more than a weekend retreat to learn about fly-fishing and evolved into a coveted invitation to not only learn to fly-fish, but also share ideas about the important work of a conservation professional. In the eight years since the program began, more than 40 women have become MSU Fly Gals.

At the time Dr. Taylor and Tom were solidifying their idea, I was a student of Dr. Taylor's, in my first year of graduate school, pursing a Master of Science from MSU's Fisheries and Wildlife Department. I and eight other women were invited by Bill Demmer, the executive vice president for Conservation and Education of the Boone and Crockett Club, and Dr. Taylor to Demmer's Big Creek Lodge in north central Michigan for three days in the summer of 2007. It was an invitation I was glad to accept, as I enjoy fishing and my background, experience, and education have been focused on fish, fish habitat, and natural resource management. However, I had not taken up fly-fishing on my own and the task seemed rather daunting to me at the time.

More Than a Weekend Retreat: Using an Outdoor Recreation Skill to Attract Participants

The use of outdoor recreation, in this case fly-fishing, was the primary inducement for participants to attend. What transpires, however, is a lot more than just learning a new skill.

The first cohort of women to participate was of varying disciplinary and recreational backgrounds from the MSU Department of Fisheries and Wildlife. Tom and Dr. Taylor's vision for this program was for it to be self-perpetuating, first targeting graduate students who would then go on to recruit and teach new cohorts of female fly-fishers. The women who participated were selected because they had a curiosity and passion for the natural world that united them with a shared goal—learning to fish—in a beautiful setting, away from the traditional academic environment. Further, their educational and career pursuits had a natural resource management and conservation focus. As such, they were perfect candidates to foster the program's goal of further cultivating conservation in participants as they evolved and grew professionally.

MSU Fly Gals are well-educated, accomplished women in their own right. They give as good as they get. It can be pretty entertaining to listen to the conversations at meals or on the river and discover just how amazing the MSU Fly Gals are. They are not all students either; a variety of professionals outside of MSU have been invited to participate. As each year has passed, there have been participants from the wider community from graduate students and faculty at MSU to women from the state and federal agencies like the Michigan Department of Natural Resources and U.S. Fish and Wildlife Service.

The purpose of getting participants from the wider community is to bring together individuals with diverse backgrounds and experiences who share the same concern and passion about the environment. As such, we have a group of individuals who can begin to get a better understanding of conservation issues and can bring unique perspectives about and solutions to conservation problems and issues. This specific program strives to get women more actively involved in fly-fishing and conservation issues related to fish and fish habitat.

According to a 2010 Recreational Boating and Fishing Foundation report, women currently make up 25.2% of all fly-fishing participants and represent the fastest growing segment of the fly-fishing industry. This statistic makes programs like MSU Fly Gals all the more valuable. It empowers women with the confidence to pursue an activity historically dominated by men. The program presents an opportunity to build a network with relationships that are much stronger because everyone has this shared extraordinary experience that goes well beyond typical academic and professional associations.

However, this model is not exclusively applicable to women. It can be used with a variety of participants as long as the model incorporates the main components: using a recreational activity to empower emerging professionals and future leaders toward a larger goal, thus enabling them to gather a unique understanding and appreciation for the goal.

Getting Started: Using the Skill to Reinforce a Larger Goal

The attraction of learning a new skill is exciting, but when that new skill also leads to a broader learning experience, in this case fisheries science and its impacts in society, then the skill, if chosen properly, has the ability to reinforce larger lessons.

We began the experience with a meet and greet, if you will, to ring in the opening of the fishing season and form friendships and professional networks. The second day, we were taught by "Dr." Sadler about fly rod anatomy, basic casting, tying knots, and essential fly-fishing gear. We spent the final day on the river practicing our new skills. These seemingly simple interactions and learning sessions were important in building strong social ties to facilitate reaching out to people easily and connecting and learning from them. It was an opportunity to build communication skills, put learned skills to practice, and get through it with the help of others, important attributes for any emerging professional and leader.

The MSU Fly Gals example is successful because it has gone beyond a fun weekend retreat to creating a network of professionals who share a common experience that is grounded in both learning a recreational skill and how that skill can lead to a better understanding of why the conservation of our natural resources is so important. While book learning and scientific study are important, adding a hands-on outdoor recreational skill to help explain the importance of conservation incorporates elements of fun and networking that might not otherwise be readily available. MSU Fly Gals uses fly-fishing for that purpose, but other outdoor recreation skills would serve the same purpose. Each element has built on the foundation of the initial purpose of the program. For example, having a guest speaker who offers a view on fish and wildlife conservation issues and the importance of our outdoor heritage to conservation has added depth and breadth to the program. Those participants from outside of the MSU graduate program increase the intellectual horsepower of the program by bringing their real world experiences to the table and offering diversity to the group that raises the level of shared knowledge.

The opportunity to engage with the guest speakers was a key component to the program. We were learning fly-fishing and doing so in a manner that developed a heightened awareness of fish and wildlife resources and conservation of these natural resources. To me, it was a great way to experience fly-fishing, support the research I was conducting for my master's at the time, and, unknowingly, make connections with individuals who were able to provide a greater link to the conservation world as a whole and help instill in me a greater sense of advocacy for resources. More importantly, participation fostered a greater appreciation for the natural resources surrounding us all. As such, the MSU Fly Gals model has a broader applicability to natural resource conservation and sustainability. The model's simplicity justifies its utility: using an outdoor recreational activity to attract like-minded individuals, making linkages between the activity and a larger conservation goal, and fostering dialogue and interactions among its participants as well as through interactions with specialists in the field of conservation. We feel the interactions and discussions that take place through such a model will help guide emerging professionals through their career pursuits.

The program has broader applicability because its initial vision has been carried out and we feel it is a success. As such, it is suitable for replication. The initial goal of Dr. Taylor and Tom of the program being self-perpetuating is becoming a reality. The program has matured. New elements have been added to the program since I first attended/experienced it: float trips on the north branch; guest speakers like Jim Martin, conservation director of the Berkley Conservation Institute (BCI); tours of Fuller's historic North Branch Outing Club; and, most recently, some of the women from the first cohort and I had an instructor class so that we could help teach the skills that we had learned.

At the last MSU Fly Gals outing in which I participated, Jim Martin and I discussed how current legislation regulating aquatic resources affects the conservation movement and the roles and abilities of state and federal agencies tasked with ensuring regulatory compliance. Jim, and BCI, have been advocating for the protection of wetlands and ensuring the quality of fishable and swimmable standards of the Clean Water Act. At the time, it was valuable to learn the role of his organization's effort in issues related to my career pursuits in the federal government. This interaction with Jim fostered my engagement in conservation policy and advocacy and broadened my professional network.

Better Conservation Professionals: Engage and Network with Previous Participants and Outside Influencers

Learning a new recreational skill is interesting in its own right, but it is only a tool to accomplish the goal of creating more well-rounded professionals. Equally important is working with previous participants and other influencers to build a network of relationships for the participants.

During the MSU Fly Gals weekend, the give and take around the dinner table and the fireplace is where the magic happens. What you have in MSU Fly Gals after they have been through the program are women who push themselves out of their comfort zone, learn a new skill that connects them to the natural world outside of the scientific arena, and provide experiences that connect them with previous participants who can both mentor them and network for them. But they can accomplish this with that finesse and nuance that creates a more positive mentoring relationship. I firmly believe this form of mentoring leads to a better professional. They have a common bond of shared experience that reinforces the tie to the work they do and the larger outcome they seek.

Returning alumnae serve as unofficial guides and instructors to the program. They routinely return for some part of the weekend to welcome the new recruits and share their experiences. This mentoring has been the key to the success and longevity of the program. This is an important element to making the mentoring aspect durable. Keep the connection from early participants to new recruits helps reinforce the core purpose and share lessons learned.

More women are coming into the conservation and natural resources fields, fields traditionally populated by men. Outdoor recreational opportunities with coworkers or colleagues therefore were also male-dominated. What the MSU Fly Gals model does is provide an opportunity for women to enjoy an outdoor activity with other women and to enjoy the same kind of bonding and learning experiences the men do.

The first cohort of MSU Fly Gals have all found jobs in various parts of the country, but we are all working and living everyday with conservation issues at the heart of all we do and will continue on that way in perpetuity. It was an excellent, yet humbling experience to help teach the new class of cohorts with Tom. I learned twice as much as I had initially, formed an even tighter bond with Tom as our instructor, and reconnected with the women in my cohort. I also bonded with the new cohort. I felt that I had become a mentor, connecting with them now as an instructor and conservation professional, like those that attended the inaugural MSU Fly Gals program. Where I had once had an experience that helped develop my passion and stewardship for our fisheries resources, I now felt that I was helping develop the same passion and stewardship in others. This is why this model is so important for other fisheries professionals to use.

MSU Fly Gals alumnae are landing in increasingly important positions in the fish and wildlife community. Those who follow in their footsteps will benefit from those shared experiences. That network is far different from what would be found in normal academic and professional associations. The program is building a cadre of fly-fishers who know a lot more than just how to cast a fly rod. MSU Fly Gals will be a great asset wherever they wind up. It is fun to think of where MSU Fly Gals alumnae will be 10 and 20 years from now, and I know they will be kicking butt and running the world.

Overall, the experience has allowed participants to engage and network with previous MSU Fly Gals as well as other fly-fishing enthusiasts. MSU Fly Gals now know the joy of hooking a fish with a fly, which allows them to connect with fly-fishermen in their new networks. MSU Fly Gals has allowed participants to develop a better rapport and tie with individuals they now work and interact with in state and federal agencies, nonprofit organizations, and the private sector. Participants Dr. Nancy Léonard (Northwest Power and Conservation Council) and Dr. Dana Infante (Michigan State University) reflect positively on their experiences. MSU Fly Gals improved and strengthened existing relationships not only with other MSU Fly Gals, but also with professional peers and coworkers. The program demonstrates the importance of participation, social interaction, and engagement. The experience has provided a new sense of confidence in developing and enhancing networks and provides a chance to develop better relationships with peers and leaders in fisheries conservation, as well as attaining the larger goal of teaching the conservation ethic.

MSU Fly Gals has made me a different fisherwoman and, most importantly, a better conservationist. This is the premise of MSU Fly Gals. This model builds new conservation professionals by conducting recreational activities in the environment, whether they are fly-fishing, kayaking, on nature hikes, hiking, or boating, and so forth. These conservation activities and discussions can be transformational, particularly for emerging professionals. I continue to stay in touch with gals from the original cohort and also with the gals I helped teach in following years. This model holds participants accountable for staying true to a conservation agenda and lays a foundation or network of individuals to continue to collaborate with and work with as emerging professionals to strengthen the connection between recreation and conservation. Overall, the model can instill a confidence about a recreational activity, like fly-fishing, and bond a group of like-minded conservation professionals as participants.

Closing: Collective Recommendations

We believe a program similar to MSU Fly Gals can be easily replicated, but there are key elements to making it successful:

- Keep previous participants engaged.
- Have a clear goal for the program, like mentoring conservation professionals, as in the case of MSU Fly Gals.
- Bring in outside thinkers as participants, instructors, and speakers.
- Add elements of fun and reward, such as receptions at different locales and equipment donations.
- Choose your venues wisely so a legacy can be created and perpetuated.
- Remember the goal is to use the recreational skill as a social glue and not an end to itself.

What has been created in MSU Fly Gals is a conservation legacy. By bringing together women who share a common interest in natural resource conservation, it establishes a professional network of conservation-minded people on a personal level. Each cohort is able to share ideas and grow together, learning as much from each other about fisheries and conservation issues as they do about fly-fishing.

After eight years, the program is not only going strong, but has become a sought-after invitation. This is a good indication that the program has had a positive impact and has been successful in its purpose.

The MSU Fly Gals program happens to use fly-fishing as the recreational catalyst for both education and mentoring. The specific activity is less important than the need to foster an experience-sharing opportunity for the participants at a venue that enhances that experience. The key is finding an activity and location that creates those elements noted in the recommendations above. This, coupled with bringing in fisheries and conservation professionals who can share their experiences, knowledge, and wisdom, makes the program unique, engaging, and successful.

We believe similar mentoring programs can provide cadres of professionals that have shared recreational experiences and shared appreciation for the work they and others do. Using this model, participants will instinctively look to mentor others in a similar manner. We encourage others to give this place-based, recreational activity approach to mentoring a try.

Biographies

Tom Sadler is the executive director of the Outdoor Writers Association of America and a former conservation consultant and advocate. Tom, a lifelong fly fisherman, guide, and instructor, and his wife Beth live with their black Labrador Lily in the Shenandoah Valley town of Verona, Virginia.

Kerryann Weaver is an environmental scientist with the U.S. Environmental Protection Agency's Region 5 Office in Chicago. Her career exploits range from working as a marine biologist on commercial fishing vessels in Alaska, to teaching small-scale aquaculture to farmers in rural Zambia as a Peace Corps volunteer, and working as a research assistant at Michigan State University. An avid outdoorswoman, Kerry and her husband Ben, daughter Rose, and dog Libby live just outside of Chicago, Illinois.

Disclaimer

This work is not a product of the U.S. Government or the U.S. Environmental Protection Agency (U.S. EPA), and the author is not doing this work in any government capacity. The views expressed are those of the author only and do not necessarily represent those of the United States or the U.S. EPA.

Emerging Topics

Riding with the Drivers of Change in Fisheries Science: A Holistic Approach for the Future

DEVIN M. BARTLEY*, NICOLE FRANZ, CARLOS FUENTEVILLA, AND KOJI YAMAMOTO
Fisheries and Aquaculture Department
Food and Agriculture Organization of the United Nations
Viale delle Terme di Caracalla, Rome 00153, Italy

Key Points

- The next generation of fisheries scientists will have a greater need to become multidisciplinary in their expertise so that they can integrate social, political, and economic disciplines into a changing field of fishery science to address the future challenges facing the world's aquatic ecosystems.
- Fisheries scientists should be prepared and willing to embrace the complexities and challenges of a new era of fisheries science.

Introduction

The fast pace of technological development (e.g. computers, nanotechnology, Global Positioning System, and smart phones) will provide fisheries professionals with unprecedented tools to apply to the field of fisheries science. A planet with nine billion people will put profound pressure on the world's aquatic resources. Fishery professionals will have to deal with a wide variety of challenges and will require a variety of skills to deal with this pressure. The field of fisheries science is becoming more multidisciplinary with greater need to integrate social, political, and economic disciplines within the sector.

We provide some insight and perspective on how emerging professionals came to fisheries science and how they see their role as the sector and the world evolve. We briefly examine the present situation in the world's fisheries and look at some drivers that will influence their future. In a world more connected than ever by technology, trade, travel, and flow of information, fisheries professionals will need to address major global issues. Perhaps the most significant issue from our point of view is ensuring that the world's food supply from fish is produced and distributed in a sustainable and responsible manner.

The Present

To see into the future, it is necessary to look at the present. In 2011, the world produced about 153 million metric tons of fish.[1] Improved technology, better distribution channels,

* Corresponding author: devin.bartley@fao.org

[1] "Fish" refers to fin fish, aquatic mammals, aquatic invertebrates, and aquatic plants; "fisheries professional" includes aquaculture professionals; and "fisheries science" includes aquaculture science.

and increased production from aquaculture have allowed world fish production to grow at a rate of more than 3% per year over the past five decades, outpacing the 1.7% per year growth in human population. This rapid growth is depicted by the increase in per capita food fish supply from an average of 9.9 kilograms (liveweight equivalent) in the 1960s to 18.4 kilograms in 2009. However, global averages mask local variation: per capita fish supply ranged from 8.5 kilograms per year in Africa to 24 kilograms per year in North America, and in parts of Sub-Saharan Africa fish consumption has declined (FAO 2012).

Along with these increases in fish production came concerns over the sustainability of the sector:

- Capture fisheries production from most of the oceans' major fish stocks has plateaued, and several stocks have been depleted;
- Finfish from inland waters are the most threatened group of vertebrates used by humans;
- Aquaculture relies heavily on limited supplies of forage fish to provide fish oil and fish meal to many commercially important farmed species; and
- Aquaculture has been in some areas a source of pollution (e.g., uneaten feed from salmon pens), habitat loss (e.g., when mangroves are cleared for shrimp ponds), and introduction of invasive species (e.g., the golden apple snail introduction to the Philippines).

The fisheries sector is also being negatively impacted by other uses of the world's waters, wetlands, and coastal areas. Land-based sources of pollution and unsustainable forestry, agriculture, and land use degrade aquatic environments and threaten the livelihood of fishery-dependent communities; hydroelectric development further degrades inland water ecosystems; and agriculture currently uses about 80% of the world's freshwater withdrawals, making them largely unavailable for fisheries. Impacts of climate change, such as sea-level rise, changes in precipitation, ocean acidification, and increased frequency and intensity of extreme weather, will further impact the sector.

To address these issues, fisheries scientists today are looking at broad-based or holistic approaches to deal with often conflicting objectives of economic growth, environmental conservation, and basic human rights to resource access and use. This trend will surely continue. Traditional, biology-focused single-species approaches to fisheries management have given way to more holistic approaches that support governance systems that integrate human concerns, such as right to food and livelihoods, and environmental concerns. These approaches include

- Ecosystem approaches that acknowledge the knowledge of fisher workers and other stakeholders and address their concerns to put fisheries and aquaculture into an ecosystem context;
- Livelihood approaches that include three major capitals–those of natural resources, of human and social value, and of economic and financial capital; and
- Human rights-based approaches that recognize the universality, inalienability, indivisibility, interdependence, and interrelatedness of human rights. They also include the principles of nondiscrimination, equality, participation, inclusion, accountability, and the rule of law.

Emerging Professionals in the Present

Because of the need to address the numerous factors and disciplines affecting the fisheries and aquaculture sector, many opportunities for education, research, employment, and career choices are available to the fisheries professional today. As a result, the present job for many emerging professionals may be somewhat different from their initial expectations in relation to their career path and professional focus.

Some emerging professionals did not even aim to be fisheries professionals. KY: *My original intention was to become a farmer and to have my own aquaculture business. I worked in a number of aquaculture farms and in an aquaculture research station. Eventually, I enrolled in a master's program for aquaculture. However, an opportunity to work with the Network of Aquaculture Centres in Asia-Pacific (NACA) inspired me to get into international development work contributing towards food security through aquaculture. This job was an important step to broaden my interest and provide me with additional opportunities.*

For other emerging professionals, initial plans for a career in one discipline changed toward a different discipline. NF: *When I started my studies in economics, my objective was to work in marketing. But then, development economics and food security issues attracted my interest. When I started in the Fisheries and Aquaculture Department of the Food and Agriculture Organization of the United Nations (FAO), instead of contributing to the noble cause of increasing food security I found myself, to my disappointment, counting fish, extracting and reorganizing production data from huge FAO data sets, and not fully grasping the broader picture within which I carried out this activity.*

But my disappointment was short-lived; I quickly realized that fisheries are actually ALL about food security and development. With the support of experienced supervisors, I learned about development challenges and opportunities in relation to capture fisheries and aquaculture. I learned about the impact of globalized value chains on small-scale fisheries in developing countries in terms of domestic food and nutrition security and income distribution. These lessons are now being applied to developing international guidelines[2] *for improved governance and sustainability of small-scale fisheries.*

Responsible fisheries management must address these broader issues of livelihoods, food security, and governance. NF: *We realize that we cannot expect a poor fisher to engage in sustainable fisheries management if his main concern is how to feed his family, how to enable his children to attend school, or how to access medical care for his sick mother. How can the fisheries sector develop its full potential to contribute to food and nutrition security and poverty alleviation if it is excluded from national development strategies and is geographically, socially, politically, and economically marginalized?*

The multidisciplinary nature of fisheries and aquaculture development and management will certainly not diminish in the future. With the increased complexity of fishery issues, partnerships will be necessary to address them adequately; no one organization will have the expertise to solve biological, social, and economic fisheries issues. Emerging professionals will need to be able to communicate effectively to ensure these partnerships are mutually beneficial and efficient. DMB: *In most every project, proposal, or meeting in which I have participated over the past 10 years, issues of equity, human rights, economics, and social acceptability have been as important as the biological management and conservation of fish stocks. Fisheries professionals that have this perspective are especially effective at engaging stakeholders and forming partnerships.*

The Future

Fisheries and aquaculture are crucial to livelihoods and food security for millions of people. However, significant increases in production will be required to feed nine billion people in the future. The nonconsumptive aspects of aquatic ecosystems (e.g., cultural, spiritual, rec-

[2] For more information, please see www.fao.org/fishery/ssf/guidelines/en.

reational, and tourist) are expected to play important roles in human well-being in a future with more wealth. There will be advanced fishing and communication technologies as well as expanded application of fishery products to other sector (e.g., algae fuel).

Six main drivers of change affecting the food system have been identified that will also shape the future that the next generation of fishery professionals will need to face: (1) a global population increase to nine billion people by 2050 and increased urbanization; (2) increased consumer wealth and gross domestic product, with greater demand for fish and fish products; (3) the ability to achieve good governance of food production and conservation; (4) climate change and effectiveness of adaptation and mitigation strategies; (5) competition for resources where demand for global energy and for fresh water could double between now and 2050; and (6) changes in values of consumers that influence policies and food and lifestyle choices.

The drivers that we deal with most directly in this chapter are good governance, competition for resources, and consumer values. These drivers are not mutually exclusive and may act together to influence fisheries and aquaculture.

The Ability to Achieve Good Governance of Food Production

NF: *We, as emerging professionals, will need to recognize the interrelatedness of development and human rights, looking beyond the boundaries of traditional fisheries management issues and engaging with broader relevant processes. In October 2012, the United Nations (UN) special rapporteur on the right to food submitted a report on fisheries to the UN General Assembly. He stressed that linking guidance on sustainable fisheries and aquaculture to international human rights law, including the right to food, is essential. He also called on countries to fulfill their obligations with regard to the right to food, ensuring active involvement of fishing communities to meet these obligations. I (NF) have seen evidence that active participation of fishers and their communities through different forms of self-governance and management can provide the right incentives for sustainable behavior. Factors for success and failure of collective action need to be further explored to establish an enabling framework for fisheries communities to improve their livelihoods.*

A broader approach to sustainable development is also called for in the Rio+20 outcome document, "The Future We Want" (United Nations Conference on Sustainable Development Rio+20 2012), which has food security, nutrition, and sustainable agriculture, including fisheries and aquaculture, among its priority areas. The Rio+20 document is one of the elements that inform the currently ongoing definition of the post-2015 Sustainable Development Goals under which the next generation of fishery professionals will work. NF and CF: W*e had the chance to feed into the global Rio+20 process. We saw the importance of small-scale fisheries for achieving sustainable development. Healthy ecosystems, sustainable fisheries, and aquaculture were appreciated for delivering food and nutritional security and in providing for the livelihoods of millions of people. Seeing all this recognized at the highest political level reinforced our professional motivation and commitment.* Many of the fisheries- and aquaculture-related topics that fisheries professionals will have to address in the future are likely to stay the same (e.g., fishing overcapacity, IUU (illegal, unreported, and unregulated) fishing, bycatch reduction, and environmental concerns). The way to address them, however, is likely to change—by taking into account the involvement of all stakeholders in processes to achieve common goals, by adopting truly holistic and integrated approaches, and by being action and results oriented.

A goal of future fishery professionals is to maintain biodiversity and ecosystem services while feeding the world and ensuring supply meets demand. However, it is unclear how to do this practically. CF: *Following presentations on fish marketing and sustainable fisheries manage-*

ment in a fishing area with major fishery problems, an exasperated fisheries officer said that he could not find a reasonable way to conserve fish stocks and lower fishing capacity while increasing national per capita consumption, protecting national fishers from cheap imports, and providing a source of employment for thousands of fishers, fish workers, and their families.

The holistic approaches to fishery and aquaculture management where stakeholders' wishes and rights-based approaches are included will help assure that these multiple functions are addressed. CF: *As an emerging fishery professional in the first few years of a career, I expected to face complex problems daily. The training of fisheries professionals continues to evolve and prepare professionals for a sector that demands multidisciplinary, integrated, holistic, and comprehensive solutions, leaving aside the mentality of isolation that has hampered management in the past. Managers will need to appropriately assess the environmental, social, and economic aspects of fisheries and decide on actions based on conservation, sustainable use, and fair and equitable benefit sharing. We must become effective communicators and engage all stakeholders in the process of protecting our aquatic resources and improving poverty alleviation and food security, through facts and science, including traditional knowledge, and through empathy and ethics. We must improve our understanding of the opportunity costs involved in fisheries and aquaculture related decisions, and feel confident making difficult but substantiated choices.*

Competition for Resources

Competition over resources—namely fish, forests, land, and water—is likely to further increase in importance for the fisheries professional. At country and local levels, this competition can originate within the sector (e.g., among small-scale fisheries or between small-scale and large-scale fisheries). Fishing communities have often developed their own mechanisms for dealing with conflict. NF: *During a three-month stay in Aceh, Indonesia, I learned about a 400-year-old customary system dealing with fisheries-related conflict. The Panglima Laot (Commander of the Sea) is an elected local member of the fishing community invested with certain institutional powers, who calls for weekly meetings to discuss and solve potential and existing disputes. For example, the Panglima Laot prohibited the clearing of trees and mangroves from coastal areas by local communities in order to protect breeding areas for important fishes. Emerging professionals will need to be aware that such traditional practices are extremely important and must be incorporated into a holistic approach to resource management.*

Competition over resources can also originate from other sectors. For example, development of tourism in coastal areas in the Caribbean is encroaching on traditional fishing settlements, threatening traditional livelihoods, which are often based on customary rights rather than on legally recognized use and access rights. To secure fisheries and aquaculture tenure systems, which determine who can use which resources, for how long, and under what conditions, fisheries managers will have to engage with broader spatial planning and development processes. At the global level, fisheries managers will have to deal with competition over resources. This competition can result in fisheries agreements between countries, with important potential impacts on domestic fisheries of the host countries.

Fisheries professionals will need to understand and communicate the risks and benefits of new opportunities and technologies and also know how to develop them in a practical manner. With increasing demand for seafood and a leveling of production from capture fisheries, aquaculture will clearly be an opportunity in the future of many fisheries scientists working to help reduce hunger and poverty. However, there will also be competition for resources and markets between aquaculture and capture fisheries. KY: *Aquaculture as the fastest-growing food sector comes with a great responsibility. I believe the continued development of the sector, particularly related to application of new technologies that increase efficiency for resource usages, and knowledge transfer can enable aquaculture to meet the growing demand for fish in the future. For example, recirculating aquaculture systems, alternative fish feeds, genetic improvements, and offshore mariculture are technologies that can open a range of opportunities for higher and sustained productivity. On the other side, looking at the realities of many developing countries, improving the basic farming practices of small-scale farmers could significantly improve the overall efficiency of the aquaculture sector and improve millions of livelihoods.*

Changes in Values of Consumers

More and more, consumers are demanding socially and environmentally friendly products. In effort to address sustainability and social responsibility issues of the sector, certification and ecolabelling play and will continue to play an important role. Fishery professionals will have the opportunity to help these schemes achieve their multiple goals of protecting fishery resources and providing economic opportunity for the sector. KY: *I am currently involved in the work related to aquaculture certification, such as supporting an implementation of FAO aquaculture certification guidelines. I spent over a year with shrimp-farming communities in Thailand, where I witnessed the significant difficulties (e.g., cost of certification), as well as a few success stories, related to certification schemes. Special consideration is needed for small-scale farming communities and engagement with producer organizations (e.g., farmer cooperatives) to move forward. Measuring sustainability is a truly challenging task, and knowing what level of tradeoff between environmental and human well-being is acceptable to consumers will be a challenge.*

Conclusion

The next generation of emerging fishery professionals will need to be alert to new opportunities, adaptable, and capable of multisectoral/cross-disciplinary thinking and communication. They will need to address complicated and important ecological, social, and economic issues in order to (1) balance future demand and supply sustainably, (2) ensure adequate stability in supplies of fish and fish products, (3) help achieve access to food and end hunger, (4) help the sector adapt to and mitigate the impacts of climate change, and (5) maintain biodiversity and ecosystem services while feeding the world. Although the fishery professional of today may face these same challenges, the challenges for the next generation of fishery professionals will be of greater magnitude, due to the certainty of more people and more competition for resources and the uncertainties of future governance systems and climate change.

For some of the more established professionals, the move toward holistic multidisciplinary thinking and cross-sectoral collaboration has been interesting and challenging. To modify an old axiom, "Teach a person to fish or to farm fish and you feed them for a lifetime, but only if the social, political, economic, and environmental climates encourage fishing or farming." Many of the past failures of resource management, development, and conservation have come from not understanding the broad mix of issues and priorities of the system under consideration. In this broad mix, there will be no substitute for scientific excellence, good interpersonal skills, and professionalism from fisheries scientists. Looking from the present into the future, we strongly believe that emerging professionals that embrace the complexities of fisheries science and are willing to take on challenging tasks will be well equipped to make a significant shift toward responsible and sustainable fisheries and aquaculture. We believe the next generation will be an exciting period for fishery professionals of all ages.

Acknowledgments

We gratefully acknowledge the following people who supported and encouraged us and provided us with many opportunities: Eddie Allison, Pedro Bueno, Rudolph Cleveringa, Brian Davy, Helga Josupeit, John Kurien, Audun Lem, C. V. Mohan, Michael Phillips, Antonio Rota, Carl-Christian Schmidt, Rohana Subasinghe, Jessica Thomas, Robin Welcomme, and Rolf Willmann.

Biographies

Devin M. Bartley is a fourth-generation Californian now working for FAO as senior fishery resources officer. After receiving a doctorate in ecology from San Diego State University and the University of California, he spent a short period of time applying genetics to the management of native fishes in California. In 1991, Devin left California to become the first aquatic geneticist at FAO. After a brief return to California to serve as state aquaculture coordinator in 2008, Devin is again working for FAO on inland fisheries, genetic resources, stock enhancement, ecolabelling, and introduced species.

Nicole Franz, a German national, holds a bachelor's degree in economics and a master's degree in international cooperation and project design from the Sapienza University of Rome. She joined the Policy, Economics, and Institutions Service of the FAO Fisheries and Aquaculture Department as a fishery planning analyst in January 2011, working primarily on small-scale fisheries policies. She has previously worked for the OECD (Organisation for Economic Co-operation and Development) Fisheries Policies Division in Paris and as fisheries and aquaculture consultant for FAO and for the International Fund for Agricultural Development. As a consultant for FAO, she was involved in extensive field work on fish marketing in Namibia and in Indonesia.

Carlos Fuenteville, born in Mexico, is a junior professional officer at FAO in the Fishery Policy, Economics, and Institutions Division. His current work involves small-scale fisheries policy at international and national levels, including the human rights-based approach, capacity building, and right to food in fisheries. He holds a master's degree in marine affairs and policy from the University of Miami's Rosenstiel School of Marine and Atmospheric Science, where he was a research assistant at the experimental fish hatchery specializing in the development of larval rearing protocols and information dissemination. He also had brief stints at the National Fisheries and Aquaculture Commission in Mexico and the Mexican Center for Environmental Law.

Koji Yamamoto, a Japanese national, joined the Aquaculture Service of FAO in 2010 as an associate professional officer. His current work involves small-scale aquaculture management, cluster farming, aquaculture certification, and aquatic biosecurity governance in different regions, especially in Asia. Prior to joining FAO, he was a research associate at the intergovernmental Network of Aquaculture Centres in Asia-Pacific, based in Thailand. He holds a Bachelor of Science in marine biology from Bangor University (Bangor, UK) and a Master of Applied Science in aquaculture from James Cook University, (Townsville City, Australia).

References

FAO (Food and Agriculture Organization of the United Nations). 2012. State of world fisheries and aquaculture. FAO, Rome. Available: www.fao.org/docrep/016/i2727e/i2727e00.htm. (January 2014).

United Nations Conference on Sustainable Development Rio+20. 2012. The future we want: final document of the Rio+20 Conference. Available: http://rio20.net/en/iniciativas/the-future-we-want-final-document-of-the-rio20-conference. (January 2014).

Recommended Reading

Bruton, M. N. 1995. Have fishes had their chips? The dilemma of threatened fishes. Environmental Biology of Fishes 41:1–27.

Comprehensive Assessment of Water Management in Agriculture. 2007. Water for food; water for life. A comprehensive assessment of water management in agriculture. Earthscan, London.

FAO (Food and Agriculture Organization of the United Nations). 2011. Technical guidelines on aquaculture certification. FAO, Rome.

Foresight. 2011. The future of food and farming: challenges and choices for global sustainability. Executive Summary. The Government Office for Science, London. Available: www.bis.gov.uk/assets/foresight/docs/food-and-farming/11-547-future-of-food-and-farming-summary.pdf. (January 2014).

Naylor, R. L., R. W. Hardy, D. P. Bureau, A. Chiu, M. Elliott, A. P. Farrell, I. Forster, D. M. Gatlin, R. J. Goldburg, K. Hua, and P. D. Nichols. 2009. Feeding aquaculture in an era of finite resources. Proceedings of the National Academy of Sciences of the United States of America 106:15103–15110.

Office of the High Commission on Human Rights. 2004. Human rights and poverty reduction: a conceptual framework. United Nations, Office of the High Commission on Human Rights, New York.

Sinderman, C. J. 2001. Winning the games scientists play. Perseus Publishing, Cambridge, Massachusetts.

How Will Invasive Species Impact the Future of Fisheries?

D. ANDREW R. DRAKE* AND NICHOLAS E. MANDRAK
Department of Biological Sciences, University of Toronto Scarborough
1265 Military Trail, Toronto, Ontario M1C 1A4, Canada

Key Points

- Invasions are characterized by uncertain outcomes to ecosystems and the fisheries they support.
- Fisheries professionals must go beyond traditional fish population and angler management. Invasive species risk assessment is a core element of successful fisheries science and management.
- The impacts of invasions to fisheries depend on the willingness of society to support, and governments to adopt, long-term adaptive strategies to prevent future arrivals of key invasive species.

Introduction

A good way to determine how invasive species will impact the future of fisheries is to learn from the past. The Laurentian Great Lakes ecosystem has one of the best documented histories of fisheries and biological invasions. The earliest known exotic fish, Common Carp *Cyprinus carpio*, was intentionally introduced into the Great Lakes in the mid-1800s to establish a fishery for European settlers who desired a familiar species for food and sport. Subsequently, similar intentional introductions of a variety of species, such as Brown Trout *Salmo trutta* and Pacific salmons *Oncorhynchus* spp., have repeatedly occurred and continue to this day. Unintentional introductions also played a role in the introduction of nonnative species. Alewife *Alosa pseudoharengus*, a small forage fish, invaded the Great Lakes during the 1930s through newly constructed canals that provided connections between the Great Lakes and to watersheds beyond the basin. Ruffe *Gymnocephalus cernua*, another small species, invaded the Great Lakes years later through ballast water in ships that used the canals.

Species have been introduced for a variety of reasons and through various vectors, and these introductions have had a range of ecological and socioeconomic consequences. Many of the early stocked species now support recreational fisheries but compete with native species that were once the basis for commercial and recreational fisheries. Other invasive species, such as Rainbow Smelt *Osmerus mordax*, foraged on, or competed with, native forage species and became key forage themselves for the stocked fishes. Yet other invasive species indiscriminately impacted native and introduced species alike through direct predation, such as by the Sea Lamprey *Petromyzon marinus*, or insidious food web

* Corresponding author: andrew.drake@utoronto.ca

disruption, such as by the Round Goby *Neogobius melanostomus.* Some of these effects were predictable, but most were not. Surprisingly, in rare instances, invasive fishes have even been marginally beneficial to native species. In short, invasive species have had every possible conceivable (and inconceivable) effect on fisheries, both negative and positive, so what can we learn from the past?

Uncertain Outcomes: Winners and Losers

To highlight some ecological and socioeconomic factors about fish invasions and illustrate the difficulty (and possibly even harm) of drawing uniform conclusions about their outcomes, we describe a relatively recent fish invasion whose uncertain outcomes, winners, and losers are common to many species invasions across North America.

Round Goby, a small, benthic fish, invaded the Great Lakes in the early 1990s by hitching a ride in the ballast water of commercial ships arriving from eastern Europe. Many ecological changes have occurred following its rapid establishment and spread. Most lakes experienced declines in native invertebrates and small benthic fishes due to competition and predation by Round Goby. Lake-wide benthic fish biomass increased because Round Goby preyed heavily on zebra and quagga mussels *Dreissena* spp., themselves bottom-dwelling invasive species, leading to fundamentally altered food webs. These were somewhat expected outcomes with mechanisms (competition, predation, food-web changes) common to invasions, but the direction, key players, and magnitude of changes were far less predictable.

Most game fishes have now incorporated Round Goby as a prey source. In some cases (e.g., native Walleye *Sander vitreus*), consumption of Round Goby has probably reduced fish condition because capturing Round Goby requires increased energy while providing lower food value compared to typical prey. In other cases (e.g., native Yellow Perch *Perca flavescens*, Lake Whitefish *Coregonus clupeaformis*), it has probably increased condition by offering a new food source in greater abundance than typical prey. For Smallmouth Bass *Micropterus dolomieu*, this new food source has strongly increased condition and growth rates. In short, one of the many surprising outcomes of this species invasion is that Round Goby has simultaneously led to decreases and increases in fish condition.

Changes to fisheries in the Great Lakes have been widespread in response to these new ecological winners and losers. Some anglers perceive Round Goby as beneficial, due to positive effects on Smallmouth Bass growth and new foraging behavior, as evidenced by the proliferation of fishing lures in Round Goby shapes, sizes, and colors. Perceptions by Yellow Perch anglers are mixed, with heavier catches of perch but increased landings of gobies during perch fishing and, as yet, unknown contaminant impacts given high loads in gobies. Most recreational anglers dislike nearshore catches dominated by gobies, which are considered to have low sporting and food value, but may provide opportunities for catches by young or occasional anglers. Commercial fisheries for Lake Whitefish have probably benefited from increased fish condition, but with unknown stock abundance and contaminant effects. Commercial baitfish fisheries involving wild harvest from the Great Lakes have suffered due to the increased potential for Round Goby bycatch, which requires greater sorting of commercial catches to remove the invasive fish, thereby reducing harvest efficiency. Outcomes for other recreational and commercial fisheries are as yet unknown, but continued changes to fisheries are inevitable as ecosystems conform to new baselines involving invasive species (see Figure 1).

The Round Goby story is one of many examples across North America of uncertain outcomes, winners, and losers to ecosystems and the fisheries they support. Similar stories

THE STAKEHOLDERS

1. SMALLMOUTH BASS ANGLER

2. BAITFISH HARVESTER

3. WALLEYE ANGLER

4. NATURALIST CLUB MEMBER

5. PARENTS FISHING WITH KIDS

6. WHITEFISH TRAPNETTER

7. INNOCENT BYSTANDER

8. FISHERY PROFESSIONAL

FIGURE 1. Post-invasion perspectives: conversations overheard from fisheries stakeholders that illustrate uncertain outcomes for ecosystems and the fisheries they support

could be written for Lake Trout *Salvelinus namaycush* in Yellowstone and Largemouth Bass *M. salmoides* in California, to name just a few examples. In an increasingly complex world, the uncertain outcomes and clear losers for ecosystems and the fisheries they support justify invasive species prevention efforts. We should recognize that invasions cause uncertain changes to complex systems and be transparent when documenting and communicating their outcomes: losers are common, winners are possible, but we should avoid characterizing them as universally harmful. Species invasions transform fisheries with changes that are rarely unidirectional due to varied resource use by stakeholders. Viewpoints surrounding invasive species, and their consequences for fisheries, vary strongly. Given the responsibility of fishery professionals to manage aquatic resources for the public good, rates of ecological and socioeconomic change imposed by invasions pose one of the greatest management challenges of our time.

Managing Invasive Species in the Face of Uncertainty

Managers are faced with difficult decisions about how to manage invasive species given varying stakeholder viewpoints and the strong potential for ecological losers. Given the uncertain outcomes of invasive fishes, simply allowing the ongoing invasion of North American ecosystems is a risky game of chance given billion-dollar commercial and recreational fisheries at stake. Beyond the obvious ecological changes, failing to prevent invasions by adopting a do-nothing approach is made even more risky by fisheries management frameworks that are unable, due to logistical constraints, to respond dynamically to increased uncertainty and unpredictable consequences expected with increased invasions. But what other opportunities for management exist? Controlling established invasive species is costly (e.g., control of Sea Lamprey, an invasive species in the Great Lakes, costs US$16 million per year). For many small species, such as Round Goby, control technologies simply do not exist or have undesirable effects on native species. Preventing the arrival of invaders is less costly, but it is more difficult to provide tangible evidence of success and effective use of resources. Nonetheless, a risk-based approach to preventing the arrival of key invasive species is the only realistic opportunity for reducing the uncertain outcomes involving invasive species. Risk assessments provide guidance to managers on the ecological and socioeconomic risks and associated uncertainties for species and their pathways of introduction (e.g., ballast water, bait, and stocking). From these risk assessments, prohibition and watch lists of key species can be developed and management of their pathways of introduction can occur (Box 1). However, these actions are only effective if prohibitions are enforced, watch-list species are monitored, and managers are prepared to take action, such as rapid response, if listed species are detected in the wild. Results of risk assessments can be used to identify potential mechanisms of ecological change (e.g., food web alteration), which will allow for proactive dialogue with stakeholders about uncertain outcomes (i.e., games of chance with fishery resources) and the potential for clear losers. If we had only known how widespread and dramatic the ecological changes would be following the arrival of Common Carp, Alewife, or Round Goby, we probably could have done a better job of communicating and preventing their arrival in the first place. In an increasingly complex world, fisheries professionals are forced to think beyond traditional fish population and angler management. Invasive species will continue to threaten ecosystems and the fisheries they support with uncertain outcomes, so fishery professionals should embrace invasive species risk assessment as a core element of successful fisheries science and management. As resource managers move towards risk-based decision frameworks

Box 1. Risk Assessment

Risk assessment identifies how likely an event, and associated consequences, are to occur. Risk management reduces the risk associated with the event through management actions. Both risk assessment and risk management play important roles in identifying and reducing the risk of species invasions.

For example, an ecological risk assessment may be conducted to identify how likely a species is become established, based on factors such as introduction effort and habitat suitability, and the magnitude of ecological consequences if established (e.g., predation, food web changes). Risk assessments can also be used to determine the relative risk of establishment and impact by pathway, such as canals, aquaria, or bait. Risk assessments may be based on qualitative data such as expert opinion or quantitative data derived from scientific studies, such as environmental tolerances, movement models, and food web models. Identifying uncertainty, both in input data sources and risk outcome, is an important component of risk assessment, regardless of the type of data available.

Results of risk assessments are used to inform approaches for risk management. For example, the "Binational Ecological Risk Assessment of the Bigheaded Carps (*Hypophthalmichthys* spp.) for the Great Lakes Basin" (DFO 2012) indicated that bigheaded carps *Hypothalmichthys* spp. pose strong ecological risk to the Great Lakes. The assessment indicated that preventing entry into the Great Lakes basin is the best approach to reduce the risk of invasion (i.e., a do-nothing approach, or control following establishment, would likely be ineffective). Therefore, reducing ecological risk would involve reducing the likelihood of introduction through known pathways, such as canals, by restricting movement through use of barriers, and the live food trade by prohibiting the possession of live bigheaded carps in the Great Lakes basin.

for many fisheries issues (including invasions), one of the best ways for fishery professionals to prepare for the challenges of invasive species is to gain a working knowledge of risk assessment, which will foster effective communication about risk and uncertainty among peers and stakeholders.

Despite our best risk assessments, intentions, and policies, some invasions will occur. How will we handle future invasive species, and what should (or can) be done with existing invaders? Where possible, these species should be managed to preserve the core ecological and social values of fisheries, as is currently done through the Sea Lamprey control program in the Great Lakes. This program requires strong cooperation and commitment of fiscal resources among agencies, such as the Great Lakes Fishery Commission, American and Canadian federal governments, tribal and First Nation governments, basin states, province of Ontario, and stakeholders across Great Lakes jurisdictions. The ecological and socioeconomic effects of Sea Lamprey are extensive, so a well-coordinated control program is the only opportunity to reduce the impacts associated with this parasitic species. An individual Sea Lamprey typically consumes 40 pounds of fish per year, and the Sea Lamprey population as a whole consumes 20 million pounds of Lake Trout and Lake

Whitefish annually basinwide. Without active control of Sea Lamprey, these and other coldwater species would crash; that, in turn, would limit the productivity of commercial and recreational fisheries for native and introduced species alike. For other invasive species without active control programs, control of species may occur through emerging capture fisheries, such as White Perch *Morone americana* in the lower Great Lakes and Asian carps *Hypophthalmichthys* spp. and Grass Carp *Ctenopharyngodon idella* in the Mississippi basin. However, capture fisheries may transform undesirable species into desirable ones, potentially leading to unintended consequences such as creating future public demand for invaders. Therefore, one of the biggest challenges facing fishery professionals is knowing when to forego species control measures and adapt to a new ecological baseline that includes established invasive species.

Unfortunately, due to logistical and fiscal constraints across many jurisdictions, many professionals are given no choice but to manage fisheries in the face of invasion, with little opportunity for assessment, prevention, or control. When in this situation, fishery professionals must clearly document ecological winners and losers and changes to fisheries as new baselines unfold. Communicating with stakeholders is an extremely important goal given the many misbeliefs, rumors, and hearsay held in public opinion surrounding species invasions and their effects to fisheries. Proper communication helps stakeholders make sound choices in daily life, such as by draining live wells to prevent the movement of species among water bodies while also creating a more scientifically literate public that makes better voting decisions and encourages science-based lobbying. Educating stakeholders and generating social consensus about the concerns associated with invasive species is the only opportunity for future financial and logistical support from government once these management priorities hold widespread public opinion.

How Will Fishery Professionals Impact the Future of Invasive Species?

So, what have we learned from the past? Invasive species behave in unpredictable ways with uncertain outcomes. Simply allowing ongoing invasions is a game of chance with fisheries resources. Control is very expensive, but inaction is even more expensive with the potential for clear losers and billion-dollar fisheries at stake. As invasive species policies, with greater emphasis on prevention and pathways, are fully implemented in the future, the number of new invasions should decline and the total number of invaders should plateau, thereby reducing overall biological, social, and economic impacts (Figure 2). The impact of invasions to fisheries depends on the willingness of society to support, and governments to adopt, sound science and adaptive policies to prevent future arrivals of key invasive species, which will succeed only through long-term fiscal and operational commitments. Sound legislation, regulations, and policies, such as the Great Lakes Water Quality Agreement which includes specific legislation about the risks of invasive species, are positive first steps towards the development of coordinated invasive species management programs. Moving forward, the success of these programs requires fishery professionals to advance the science of invasions by developing risk assessment tools that identify harmful species from those likely to be inconsequential if released to the wild. Implementing these tools within prevention management frameworks will minimize games of chance with fisheries resources and reduce the uncertain outcomes for invasive species and the fisheries they threaten.

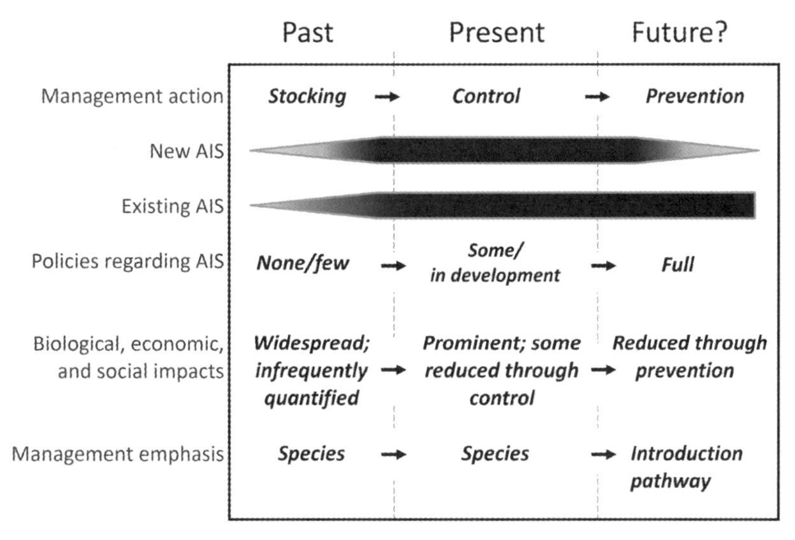

FIGURE 2. Trends in aquatic invasive species (AIS) prevalence, impact, and management.

Biographies

Andrew Drake is a postdoctoral fellow in biological sciences at University of Toronto Scarborough. His interests are fisheries and aquatic science, modeling, risk assessment, and environmental management.

Nicholas E. Mandrak is an associate professor in biological sciences at University of Toronto Scarborough. His interests are the biodiversity, biogeography, and conservation of freshwater fishes.

References

DFO (Fisheries and Oceans Canada). 2012. Binational ecological risk assessment of the bigheaded carps (*Hypophthalmichthys* spp.) for the Great Lakes basin. DFO, Canadian Science Advisory Secretariat, Science Advisory Report 2011/071, Winnipeg, Manitoba.

Three Sides of the Same Coin: Aquatic Animals–One Health–Ecosystem Health

Mohamed Faisal*
Departments of Pathobiology and Diagnostic Investigation and Fisheries and Wildlife
Colleges of Veterinary Medicine and Agriculture and Natural Resources
Michigan State University, 784 Wilson Road, East Lansing, Michigan 48824, USA

Key Points

- The health of aquatic animals is integral to the health of terrestrial animals, including humans, and the ecosystem's health.
- The economic and ecological devastation caused by several aquatic emerging diseases could have been avoided had a comprehensive aquatic animal health plan been in place.
- A comprehensive aquatic animal health plan requires making fish health part of veterinary curricula and making fish health classes readily available to fish and wildlife management undergraduates.
- We must apply a collaborative, interdisciplinary approach to rehabilitate aquatic ecosystems for aquatic and terrestrial animal health, human health, and ecosystem health.

Health for One, Health for All

In the wake of several animal-borne infections of humans, health professionals, managers, and politicians tried to explain the factors that may have led to the increased emergence of zoonotic diseases. Undoubtedly, human health is linked to ecosystem health, terrestrial animal health, and aquatic animal health. However, there is no universally accepted benchmark for a healthy ecosystem; frequently, the gauge of an ecosystem's apparent health status depends on which societal aspirations are driving the assessment. For example, physicians claim that human health is the best metric to assess ecosystem health. If this were true, then habitat degradation of forests, for example, would lead to a decline in wild canines, which will reduce rabies prevalence in humans, a sign of a healthy ecosystem according to the definition.

The term "one health" more properly conveys the connection between terrestrial animals and ecosystem health. For example, when mammalian wildlife begins to suffer from new infectious or toxin-related diseases, more likely than not humans in the area will be also affected. In many presentations, I have heard one health and ecosystem health referred to as two sides of the same coin because of the linkage between human and terrestrial animal

* Corresponding author: faisal@cvm.msu.edu

health and the health of their ecosystems. But, not once have I heard mention of aquatic animal health. Where is that side of the coin?

The aquatic biosphere, water, covers more than two-thirds of earth. While aquatic animal health is heavily influenced by the water quality, it also dictates how healthy aquatic ecosystems are and, in turn, how healthy terrestrial ecosystems are and the terrestrial species depending on them. For example, with the advances of diagnostic assays, scientists reported that aquatic animals can transmit human and animal pathogens, such as avian flu transmitted by shrimps and the Ebola virus transmitted by cichlids to humans.

The linkage between terrestrial and aquatic systems is not limited to negative influences. We also know that aquatic organisms have high nutrient value and are important food sources for many terrestrial organisms, including humans, bears, and piscivorous birds. We are also discovering that aquatic organisms provide great promise for the development of new pharmaceuticals that can cure terrestrial diseases. For example, scientists have already isolated more than 20,000 biochemical compounds from aquatic animals, leading to several new classes of human drugs, such as zidovudine (or AZT [azidothymidine]) for AIDS treatment, cytosine arabinoside for leukemia treatment, and Prialt for human chronic pain.

So, while my colleagues recognize *terrestrial* one health and ecosystem health as two sides of the same coin, they fail to note that this coin is, in fact, a three-sided coin. The aquatic environment is the indispensable caregiver for all life on Earth and *aquatic animals* are the foundation of the terrestrial animal health and ecosystem health. Consequently, I recommend that we revise the two-sided coin to include aquatic animals and their ecosystem: aquatic animals, terrestrial animals, humans, and ecosystem health, which I will refer to as "aquatic animals–one health–ecosystem health" (see Box 1 for more information). My awareness of this three-sided coin comes with a lifetime of experience in the aquatic animal health profession. But for others to appreciate this three-sided coin, aquatic animal health needs to be better integrated into veterinary science and fish and wildlife management science curricula. There is a disconnect within the education system that leads to a failure to perceive the importance of healthy fish to healthy humans. For the benefit of aquatic and terrestrial animal health, ecosystem health, and human health, students must learn the importance of rehabilitating aquatic ecosystems.

Veterinary Medicine to Aquatic Animal Medicine

I am a veterinarian born and raised in Cairo, Egypt. As a young student, I loved the veterinary discipline and was very eager to know how pathogens and hosts fight each other, resulting in a balance between both communities. I also wanted to learn more about disease control and therapy.

Unfortunately, I did not enjoy the clinical veterinary classes as much as I enjoyed the preclinical ones. I thought that animals were not treated as they deserved in terms of providing the right medical approach, even if that cost of the treatment is worth more than the market value of the animal. Following graduation, I toured many facilities where animals were kept for fattening before slaughter. I remember seeing a broiler battery in southern France where laying hens were not permitted to turn around in tiny steel cages, fearing that their eggs would be broken or destroyed. I looked the chickens in their eyes and could feel their misery.

That was the turning point in my career when I decided not to practice traditional veterinary medicine. Rather, I decided to study viruses, structures that most scientists did not even consider to be organisms, yet can cause serious diseases in all living organisms. The

Box 1. One-Health Terms

Aquatic animals: They live in water for most or all of their lives. They can be vertebrates or invertebrates. They include both freshwater and marine animals such as cnidarians, annelids, fish, amphibians, aquatic reptiles, aquatic birds, and aquatic mammals. Aquatic animals have special anatomical and physiological adaptations to help them survive in the aquatic environment. Current challenges facing aquatic animals include emerging infections, overfishing, chemical, biological, and physical pollutions, and severe climatic changes.

Aquatic animal medicine: A new discipline in veterinary medicine that deals with health and diseases of aquatic animals. This discipline also includes **conservation medicine**, which seeks to rehabilitate aquatic populations of concern. Aquatic veterinarians inspect and certify health status of aquatic organisms and assess the risk of their trafficking from one waterbody to another.

Ecosystem health: This metaphor describes the condition of an ecosystem. There are no metrics to assess what a healthy ecosystem should look like since ecosystem conditions are subjected to continuously fluctuating environmental factors. Fire, flooding, drought, extinctions, invasive species, climate change, mining, fishing, farming, logging, and oil spills can all affect an ecosystem.

One health: This worldwide concept expands collaborations and communications in an interdisciplinary fashion, including all aspects of health care for humans, animals, and the environment. By fostering this concept, the aim is to accelerate biomedical research discoveries, enhancing public health efficacy, expeditiously expanding the scientific knowledge base, and improving medical education and clinical care.

virus I chose to study for my master's degree is called the Quranfil virus, a tiny virus that is transmitted between pigeons and humans via ticks. I was fascinated by this virus and its ability to choose the proper tick organs to infect in order to ensure its ease of transmission to both hosts. Because of this study, my research interests focus on the ecology of diseases as much as on their pathogenesis. From this research, I learned that knowing the ecology and pathogenesis of each disease is important for its control. Otherwise, we risk the potential for outbreaks that can have severe repercussion on animals, humans, and the ecosystem.

I then stumbled on an old ichthyology book written by Lagler et al. (1962), where I found a statement that intrigued me: "Without piscine ancestry, man might never have evolved." If fish are that important to the history of mankind, why were they not a part of the veterinary curriculum? Do fish get sick? What clinical signs will they show? Can they sneeze or cough? Do they get a fever? And how do we treat them? Are there treatments available? Not knowing an answer to these questions, I was curious to learn more about the host–pathogen relationship in fish. This is when I dove into the fish world, seeking knowledge in aquatic animal medicine. The deeper I dove, the more I fell in love with the underwater world. This love has never faded, still to this minute.

Throughout my career in aquatic animal medicine, which has extended over 37 years, I have witnessed many fish diseases, in fact too many to count. I have observed fish as they struggle to survive and fish populations as they struggle to persist. I have closely watched fishery managers as they try to manage declining fisheries, and in some instances, I have been fascinated by what they did.

However, deep inside, I feel that the unprecedented challenges facing the aquatic environment and its resident organisms are far above the scope of regional, and even national, levels of fishery managers. Overcoming these challenges requires others to have a broad understanding of factors that affect fish health and their habitat. Fisheries professionals must advocate for the three-sided coin approach to aquatic and terrestrial animal health, human health, and ecosystem health. Fish health is a globally important issue and must be approached from a global and interdisciplinary perspective. Below are some examples of lessons I have learned over my career and a few recommendations to offer the emerging professional.

Unintentional Repercussions of Segregating Fish Health from One Health–Ecosystem Health

Reflecting back on several incidences that I have witnessed from practicing aquatic animal medicine in three continents, Europe, Africa, and North America, I have come across multiple emerging diseases that are now impossible to control. If the aquatic animals–one health–ecosystem health three-sided coin approach had been implemented back when I began my career, many health issues related to how aquatic resources are managed could have been avoided or minimized.

Nonnative Fish Result in an Increase in Malaria

I was called to examine a mortality event of a broodstock of cyprinids that occurred in Egypt during 1983. The fish had just been imported from Hungary into a state-of-the-art hatchery that had been built for more than US$20 million. I wondered why in a country like Egypt that is rich in its diverse native fish species would nonnative fish, particularly those that have had a history of being invasive, be imported. The fish were dying from a heavy infection of anchor worm *Lernea cyprinacea* that damaged the skin and musculature of these giant fish. From these imported fish, I isolated an extremely unique strain of bacteria called *Pseudomonas fluorescens.* While I was able to treat the infected broodstock, both the parasite and bacterium spread to nearby waters and aquaculture facilities, killing millions of native fish.

This incidence left me wondering how efficient local fishery managers were in controlling nonnative fish and the pathogens they might be carrying and how such an outbreak was permitted to occur by international authorities. I took the matter to Egyptian veterinary authorities and my effort ended successfully with a regulation that prohibited the import of live fish or their products unless accompanied by a health certification and a quarantine of no less than 3 months. Unfortunately, the introduced Hungarian cyprinids spread in Egypt and out-competed some native fish that feed on mosquito larvae. As a result, human malaria cases increased sharply.

Toxic Chemicals and Fish Disease

I have witnessed first-hand the negative impact of releasing low levels of toxic chemicals over a long period of time on the health of aquatic animals. One incident I witnessed occurred in the southern branch of the Elizabeth River in Virginia, USA. Over the past five centuries, several wood treatment facilities were built along the banks of the river. The current treatment involved immersing tree trunks in creosote to prevent fungal growth. Creosote consists of multiple compounds, 85% of which are polycyclic aromatic hydrocarbons (PAHs). Treated trunks were then loaded onto ferries, with the creosote still dripping into the water as the ferries cruised on to their destination. Knowing that some PAHs are calcitrant compounds whose average half-life time is approximately 500 years, one can estimate the magnitude of accumulated PAHs in the environment of these waterways.

When a group of Virginia Institute of Marine Science researchers (including myself) investigated the health of the resident fish, cancers, cataracts, and severe skin erosions were seen in almost all fish examined. In a population of resident Mummichog *Fundulus heteroclitus*, more than 80% developed one form or another of liver cancer. The effects of toxic chemicals on fish health are far-reaching, as they have the ability to cause changes in the genetic makeup of exposed fish. The federal regulatory authorities and state fishery managers have tirelessly attempted to minimize further influx of toxic chemicals in the Chesapeake Bay and to remediate the contaminated sediments. Human health professionals benefitted immensely from the findings of these studies because they unraveled previously unknown links between toxic chemicals, immunosuppression, hepatic biotransformation of PAHs, and liver cancers.

In the first of these examples, fisheries professionals made decisions based on a new food resource for people but without considering the ecosystem impacts. In the second example, fisheries professionals sounded the alarm on the ecosystem impacts caused by another water resource user. In both examples, the economic and ecological devastation caused by the aquatic diseases could have been avoided had a comprehensive aquatic animal health plan been in place. As an emerging professional, I encourage you to take classes that will expand your understanding of the aquatic animals–one health–ecosystem health three-sided coin approach. Engage with professionals from outside your immediate area of expertise; they may provide you with insight that will prevent a catastrophe similar to those faced by me in the above examples. Remember that the challenges we are facing require a comprehensive, integrative, and multidisciplinary approach.

A Need for Action

The geographic ranges of pathogens are changing quickly; they are invading new waters and affecting more species of ecological and commercial importance. The frequency of epidemics and number of new diseases in corals, sea turtles, and marine mammals is also increasing, with many linked to anthropogenic activities. Decreased availability of domestic seafood has led to increased importation from countries with lower public health and food handling standards. The shortage in seafood has caused an exponential expansion in aquaculture. While aquaculture's contribution to annual fish consumption is increasing at relatively high levels, diseases continue to be detrimental to their success. We can no longer afford to be unaware and ignore that third side of the coin: aquatic animals and their habitat. Indeed, never before has there been a greater need for integrating aquatic and terrestrial animal health, human health, and ecosystem health together as one health.

The need to be proactive in addressing health of aquatic animals is recognized by the comprehensive fish health plan being developed by the U.S. Department of Agriculture. This plan aims to promote health and well-being of aquatic animals, thereby rehabilitating aquatic ecosystems. Along with the advances in managing aquatic animal health, a systematic prioritization of hazards and risks are being evaluated. In specific, this plan identifies five components where progress needs to be made to improve the aquatic animal health contribution to the aquatic animals–one health–ecosystem health three-sided coin:

1. Surveillance systems to recognize emerging diseases, pathogens, contaminants, and toxins that affect aquatic animal health.
2. Diagnostic laboratories to provide real-time processing of samples during mortality or morbidity events.
3. Emergency response plans to address animal die-offs, reproductive failure, and disease outbreaks.
4. Bioassessment, which is effective for evaluation of ecosystem health. Bioassessment indices usually incorporate multiple metrics, such as percentages of pollution-tolerant species and total numbers of species (species richness). This is important since aquatic biota respond to a combination of physical and chemical stressors.
5. Integration of fish health in curricula of the veterinary and fish and wildlife disciplines. While most of fish health professionals' time is devoted to research and diagnostics, several of them started producing fish health teaching materials to supplement the background of veterinarians and fishery biologists. These lectures are filling a missing niche. It will be even more effective if fishery managers learn more about fish health management.

Developing the above plan is a critical first step. Implementation of the plan, however, requires that professionals be trained in all three sides of the aquatic and terrestrial animal health, human health, and ecosystem health coin. As emerging fisheries professionals, I encourage you to focus on fish health. We can improve our understanding of fish pathogens and inform best practices of fish and fisheries management professionals to decrease accidental impacts like I have seen throughout my career. Take opportunities to work with state, regional, and national agencies and implement the three-sided coin approach to aquatic and terrestrial animal health, human health, and ecosystem health. Only through a joint effort can protection, preservation, and enhancement of aquatic animal health be achieved.

Acknowledgments

I would like to thank Nancy Léonard and So-Jung Youn for their guidance, support, and assistance in the production of this vignette.

Biography

Mohamed Faisal is the S.F. endowed Snieszko scholar and professor of aquatic animal medicine (2001–present) in the Departments of Pathobiology and Diagnostic Investigation and Fisheries and Wildlife, Colleges of Veterinary Medicine and Agriculture and Natural Resources, Michigan State University (MSU). He joined MSU in 2001 where he developed an aquatic animal health program that has attracted many graduate students from veterinary medicine and fish and wildlife disciplines. His program oversees all health issues associated with fishery conservation and rehabilitation programs of the Michigan Department of Natural Resources. In 2006, he cofounded the World Aquatic Veterinary Medical Association and was the 2013 president. He got his doctoral degrees from the University of Ludwig Maximilian, Munich, Germany in 1982. He authored and coauthored ~400 publications on research findings and practical applications regarding diseases of fish, mollusks, and amphipods.

References

Lagler, K. F., J. E. Bardach, and R. R. Miller. 1962. Ichthyology. John Willey & Sons, New York.

Recommended Reading

Aguirre, A. A., R. S. Ostfeld, and P. Daszak, editors. 2012. New directions in conservation medicine: applied cases of ecological health. Oxford University Press, New York.

Centers for Disease Control and Prevention. 2013. History of one health. Centers for Disease Available: www.cdc.gov/onehealth/people-events.html. (May 2014).

Faisal, M. 2007. Health challenges to aquatic animals in the globalization era. Pages 120–155 *in* W. W. Taylor, M. G. Schechter, and L. G. Wolfson, editors. Globalization: effects on fisheries resources. Cambridge University Press, New York.

Faisal, M., M. Shavalier, R. K. Kim, E. Millard, M. R. Gunn, A. D. Winters, C. A. Schulz, A. Eissa, M. V. Thomas, M. Wolgamood, G. E. Whelan, and J. Winton. 2012. Spread of the emerging viral hemorrhagic septicemia virus strain, genotype IVb, in Michigan, USA. Viruses 4:734–760.

Johnson, P. T. J., R. B. Hartson, D. J. Larson, and D. R. Sutherland. 2008. Linking biodiversity loss and disease emergence: amphibian community structure determines parasite transmission and pathology. Ecology Letters 11:1017–1026.

Lindenmayer, D. B., and G. E. Likens. 2009. Adaptive monitoring: a new paradigm for long term research and monitoring. Trends in Ecology and Evolution 24:482–486.

Web site: One Health Initiative. Available: www.onehealthinitiative.com/.

Enforcing Fishery Laws: The Key to Protecting the Commons

MARC GADEN*, JILL WINGFIELD, AND CHRIS GODDARD
Great Lakes Fishery Commission
2100 Commonwealth Boulevard, Suite 100, Ann Arbor, Michigan 48105, USA

Key Points

- The effectiveness of fishery management policies depends on successful implementation.
- Policy implementation often demands that people act sustainably, which sometimes requires people to restrain their behavior.
- When a strong link exists between management and enforcement, natural resources benefit greatly.

Individual Behavior and the Commons

Most people likely have had experience with law enforcement and, we suspect, have not found the occasion to be pleasant. The infraction was probably low-level: a speeding ticket, a parking violation, leaving trash cans at the curb for weeks; in other words, routine citations that only the most pious manage to avoid. Some even may have had a more serious brush with the law. Offenders are held to account for their actions and are punished: fines, demerits on their driver's license, probation, and perhaps prison. Those sanctions are in place to punish an offender sufficiently to deter bad behavior and to set an example for others.

Based on personal experience, we also note that few offenders believe, let alone admit, that the punishment is fair. At some point, they likely have thought, "Really? A $125 fine for five over the speed limit? C'mon!" Yet, that is the point: law enforcement exists to punish violators and prevent behavior that, as rational or minor it may be at the individual level, could have detrimental effects on others, on society, or on a natural resource. There's the rub. Is it possible a few individuals could affect a system as large as, say, the Great Lakes? Is it possible a couple of extra fish in the box would have any bearing on a fishery? Certainly not if you are the only fisher out there. But you are not, just as you are not the only car on the road, the only person who needs that last parking space, the only citizen who grimaces at your abandoned trash cans. You could rationalize (and might even feel badly about) keeping an undersized fish or importing a cool snakehead for your aquarium, but what about that guy on the other side of the lake with an illegal trap net? Throw the book at him!

Garrett Hardin, in his 1968 piece "The Tragedy of the Commons," articulated reasons why some actions might be perfectly rational at the individual level, yet are detrimental to a

* Corresponding author: marc@glfc.org

resource as a whole when everyone does it. He warned the cumulative effect of small actions (in his example, adding just a few more cows to an alpine meadow) leads to an unsustainable outcome (overgrazing). He also noted a person might be motivated to discharge his pollution into a common resource because doing so spreads the cost of the pollution around, sparing the person from shouldering the entire cost. Hardin's lessons certainly apply to fisheries. People generally believe that their own actions—whether they be harvesting a few extra fish or dumping the last few minnows from a bait bucket that might contain invasive species—are not significant enough to harm a resource as large as the ocean, a lake, or even a pond in Hardin's meadow. While that might be true at the individual level, cumulatively humans can have an enormous impact on a fishery, even one as large as in the Great Lakes. Hardin saw enforcement mechanisms, such as laws or fines, as a solution to this propensity to engage in unsustainable behavior.

The Management–Enforcement Nexus

Fisheries throughout the planet are stressed, and with a strong demand for fish comes continued economic incentive to overharvest. This propensity toward unsustainable behavior increases the fishery manager's reliance on effective enforcement mechanisms to support fishery management programs. Yet we have to wonder, is effective, equitable law enforcement of an activity as popular as fishing, especially in an area as large as the Great Lakes basin, possible? In our experience, the answer is yes, particularly if law enforcement officers are energized, educated, given the tools to do the job, and encouraged to work directly with fishery managers.

Fishery managers and law enforcement officials need to have a mutual appreciation for how each complements the other and then must work vigorously to support each other's efforts. Law enforcement is essential to fishery management because natural resources need protection from small, individual actions that have a negative, cumulative effect on the sustainability of the fishery. Just as a driver might resent being pulled over while others speed past him, or a fisher might think it unfair to be fined for the "inconsequential" infraction of keeping a few fish over the limit, a rational person, deep down, nevertheless knows that "if everybody drove or fished like me, we would all be in trouble." Moreover, by being in the field with an eye on human behavior, law enforcement officers have the added benefit of speaking authoritatively on the effectiveness of existing, or the need for new, policies and regulations.

This linkage between management and behavior ensures that policies and regulations designed to promote sustainable harvest and long-term ecosystem health, at both the personal and societal levels, are enforceable and implemented. Put another way, regulations must not only be related to a conservation goal, but they also must be enforced with the principle that even small, individual actions need to be considered in the context of large-scale, ecosystem-wide sustainability.

Policy and Enforcement: The Case of Asian Carp

Consider a situation where law enforcement officers have made major contributions to fishery management: addressing the looming Asian carp invasion of the Great Lakes. Asian carp escaped from southern aquaculture facilities in the 1970s and have been expanding their range ever since. The carp are abundant in the Mississippi River basin, which, of course, connects with many other major river systems, including the Ohio River, the Missouri River, and the Illinois River. Asian carp are most likely to enter the Great Lakes through two vectors: by swimming into Lake Michigan through manmade canals that connect the Great Lakes to the Illinois River, or through either the accidental or purposeful release of live Asian carp via the trade (sale) of live organisms.

The sale of live fish is of particular concern to law enforcement, as Asian carp are still being raised in the South. Until recently, it was legal to ship these fish live to fish markets, including markets in Great Lakes cities like Chicago and Toronto. During transport, and certainly once at the fish market, the threat of escape was ever present. While the U.S. Federal Government was working to prevent the carp from swimming into Lake Michigan, astonishingly it prohibited neither interstate transportation nor sale of these fish. The mindset was to simply erect a barrier on the canal connecting the Illinois River to Lake Michigan and assume that the other vector—the trade of live fish—was just not much of a threat. This approach ignored a simple fact that law enforcement officials and fishery managers knew all too well: the actions of a few individuals (e.g., a fishmonger selling live Asian carp legally at a Great Lakes fish market) could be enough to establish a population. The approach to keeping Asian carp out of the Great Lakes, in other words, had to address the risks at each point in the supply chain and needed to be far more comprehensive than it was.

Starting in 2002, members of the Great Lakes law enforcement community, working together through a joint, basinwide committee, began to press their judgment that the trade of live Asian carp presented a considerable threat to the Great Lakes and that law enforcement officials, given the authority, could halt such trade. At that time, and working with fishery managers, the officers went into high gear to establish as tight a net as possible around the Great Lakes basin. Absent a national (or binational) approach, the officers first worked at the provincial, tribal, and state levels to successfully prohibit the importation, possession, transportation, purchase, sale, release, and exportation of live Asian carp—a daunting task, as policies and regulations concerning Asian carp were either nonexistent or widely inconsistent. However, given that local regulations cannot interfere with interstate commerce, law enforcement officials also argued strongly and persistently that the U.S. Fish and Wildlife Service should declare all species of Asian carp as "injurious" under the U.S. Lacey Act, a federal law that would prohibit the interstate (and, thus, the international) transportation of live Asian carp. After more than a decade of pressing the need to stop the live shipments, officers and many others convinced the government to add Silver Carp *Hypophthalmichthys molitrix* and Bighead Carp *H. nobilis* to the injurious list in 2007 and 2011, respectively. In other words, these species of carp are now banned from entering any part of the Great Lakes basin alive. This policy would likely not be in place had the law enforcement and management communities not insisted upon it.

Despite the ban, the demand for live Asian carp (particularly at ethnic fish markets) remains high. As a result, a strong incentive still exists to transport and sell the live fish inside and outside the basin. Today, law enforcement officials trained to spot Asian carp confirm the presence of an illicit trade of these species. Vigilant officers continue to apprehend carp-filled tankers at the Canadian border, even though cross-border transportation should not, by law, occur. Inspectors also visit live fish markets, bait shops, and pet stores routinely to ensure that live Asian carp are not present. Even with several successful prosecutions and the imposition of heavy fines for the illegal transportation and sale of live Asian carp, some individuals still knowingly and repeatedly test law enforcement's determination.

So what is wrong with just a few Asian carp coming into the Great Lakes? A lot. For one, the recently published binational risk assessment for bigheaded carps (DFO 2012) concluded that as few as 10 male and 10 female Asian carp in a localized system could be enough to create a population. Thus, even one release of a shipment of live Asian carps could be catastrophic to the Great Lakes basin. Yet, as Hardin cautioned, individuals are prone not to know or care about the consequences of their seemingly small actions unless enforcement coerces them to behave sustainably. The driver of the shipping truck or the chef who buys live Asian carp at the market may not think his actions alone are detrimental to the region, but that ignores the power of an individual to cause serious harm to the resource, not to mention the potential snowballing effect of multiple shipments to multiple markets that each add to the problem.

Law enforcement officers know better; they are the best line of defense against unsustainable action, and without knowledge, coordination, and determination among the basin's officers, a few scofflaws could cause an ecological disaster. Diligent law enforcement ensures that both sides of the management-enforcement equation are addressed: officers are critical to the management process by putting heft behind Asian carp and other fishery policies, by leveraging resources among the disparate governments, by informing jurisdictions on the policies needed to minimize risk, and by deterring infractions and facilitating compliance among a sometimes reluctant public.

Advice for Natural Resource Professionals

What advice, then, do we have for natural resource professionals? First, they should always remember just how much of a difference law enforcement makes in affecting behavior. Individual actions might not seem like a big deal, but cumulative actions or, in the case of Asian carp, even the actions of a few individuals, could cause serious, irreversible harm to the resource. Laws are in place for a reason—to stop wanton harvest, to promote sustainability, and to prevent the movement of invasive species. Laws without enforcement are just words on a page. The recent, high-profile busts at the Canadian border have thwarted the movement of carp; they happened because the law enforcement community worked together to share strategy, to educate themselves on how to identify Asian carp, and to impress upon each other the importance of being vigilant. Law enforcement is, and has for more than a decade been, the linchpin to successful Asian carp policy.

Second, law enforcement officers should never underestimate their ability to influence policy, and fishery managers should never forget how much they depend on law enforcement for their fishery management policies to be successful. The manager–enforcer relationship is a two-way street: enforcement officers are not relegated to simply carrying out what policymakers want; they help identify policies that need to be changed and recommend new policies and regulations that should be created. Law enforcement officials are in an excellent position to witness human behavior firsthand and to reflect on how policies should be improved. A legal traffic of live Asian carp would likely still be commonplace today had the law enforcement community not been relentless in pressing for an interstate trade ban. Elected officials and bureaucrats listened and acted when law enforcement pleaded for stronger laws.

Law enforcement officials and managers must forge links to support each other's efforts. Through such cooperative action, natural resource users gain a sense of overall fairness and, at the same time, fishery policies are implemented in ways where small, individual actions do not as a whole lead to ecosystem loss.

Biographies

Marc Gaden is the communications director and legislative liaison for the Great Lakes Fishery Commission. He is also an adjunct assistant professor at Michigan State University and at the University of Michigan, where he teaches courses in environmental policy. He received his Ph.D. from the University of Michigan.

Jill Wingfield is the senior communications associate at the Great Lakes Fishery Commission. She earned a first-of-its-kind master's degree from Michigan State University in law enforcement and fishery management.

Chris Goddard is the former executive secretary of the Great Lakes Fishery Commission. In a previous position, he was a law enforcement officer for the Province of Ontario. He is an adjunct professor at Michigan State University and the University of Michigan. He received his Ph.D. from York University in Ontario.

References

DFO (Fisheries and Oceans Canada). 2012. Binational ecological risk assessment of the bigheaded carps (*Hypophthalmichthys* spp.) for the Great Lakes basin. DFO, Canadian Science Advisory Secretariat, Science Advisory Report 2011/071, Winnipeg, Manitoba

The Growing Importance of Communication for the Future of Interjurisdictional Fisheries Management, Using Bluefin Tuna as a Case Study

Fábio H. V. Hazin*
Department of Fisheries and Aquaculture, Universidade Federal Rural de Pernambuco
Rua Dom Manoel de Medeiros, s/n, Dois Irmãos, Recife-PE 52.171-900, Brazil

Felipe Carvalho
School of Forest Resources and Conservation, University of Florida
7922 NW 71 Street, Gainesville, Florida 32653, USA

Key Points

- Invest in the development of strong communication skills; they will increase your chances of professional success.
- Do not underestimate the importance of properly communicating management actions and the results achieved by those management actions.

A Bit of History

The Bluefin Tuna *Thunnus thynnus* is an iconic fish that may grow up to 700 kilograms and sell for shocking prices. In January 5, 2013, a 222-kilogram (489 pound) Bluefin Tuna fetched a record US$1.7 million (about $7,650 per kilogram or $3,500 per pound) in the year's first auction at Tokyo's Tsukiji fish market. This record-breaking fish was probably the most expensive fish specimen ever sold for food purposes. Although these figures are more than 100 times higher than the usual prices achieved by the Bluefin Tuna, this price surely highlights the great value and cultural importance of the species.

Besides being the most expensive fish species ever sold, the Bluefin Tuna has also been exploited by some of the oldest fisheries on the planet, being regularly fished for thousands of years, for instance, in the Aegean Sea and in the Strait of Gibraltar. The fishery for Bluefin Tuna in the eastern Atlantic Ocean and in the Mediterranean Sea intensified in the early 1950s, resulting in a significant increase in catches from long-liners and purse seiners, with a total production peaking close to 40,000 metric tons (mt) in 1955. The landings then declined gradually, fluctuating from about 10,000 to 20,000 mt in 30 years, from the early 1960s to the late 1980s.

Since the establishment of the Convention of International Commission for the Conservation of Atlantic Tunas (ICCAT) in 1969, the Bluefin Tuna fishery in the Atlantic Ocean and the Mediterranean Sea has been managed by ICCAT, which presently consists of 47

* Corresponding author: fhvhazin@terra.com.br

contracting parties and 5 cooperating noncontracting parties. ICCAT is thus the regional fisheries management organization with the highest number of participants. Throughout the 1990s, Bluefin Tuna catches in the eastern Atlantic and the Mediterranean Sea started to increase strongly, reaching an all-time peak of 50,807 mt in 1996.

The introduction of fattening and farming activities into the Mediterranean Sea in 1997, coupled with good market conditions, associated with a growing demand for sushi and sashimi worldwide, resulted in a significant change in the structure of the fishery for the species, with a sharp increase in purse-seine catches (ICCAT 2012). From 2000 on, declared catches declined substantially, fluctuating around the total allowable catch (TAC) level imposed by ICCAT (32,000 mt in 2003–2006, 29,500 mt in 2007, and 28,500 mt in 2008), but information available from other sources has demonstrated that these declared catches have been seriously underreported. For reported catches in 2006 (30,647 mt) and 2007 (32,398 mt), the ICCAT Standing Committee on Research and Statistics (SCRS) estimated actual catches of 50,000 and 61,000 mt, respectively.

In the western Atlantic, catches were very low until the early 1960s, peaking in 1964 at 18,671 mt, due to large catches of adult fish from Japanese long-liners operating off Brazil and U.S. boats fishing for juvenile fish. After the collapse of the Bluefin Tuna longline fishery in 1967, catches remained between 3,000 and 5,000 mt until the early 1980s, dropping to less than 1,500 mt in 1982. Since then, the total catch in the western Atlantic has been relatively stable at about 2,500 mt until 2002 when it reached 3,319 mt, declining in the following years to values around 2,000 mt.

The majestic beauty and grandeur of this unique fish species, the very long history of its fishery in the Atlantic Ocean and Mediterranean Sea, and the sharp deterioration of the condition of its stocks in more recent years draw a considerable degree of attention from the world community. The Bluefin Tuna was turned into a symbol on how wrong fisheries management could go, highlighting, at the same time, the crucial importance of communication in both assessing the actual situation and helping to fix it.

The Bluefin Tuna Management Crisis and the Lessons Learned

In November 2007, during its 20th regular meeting, ICCAT was facing a very gloomy scenario, including heavy underreporting of catches. Actual landings that year were estimated to be about twice the TAC, which in turn was about twice the catch limit advised by the SCRS. At that point, the SCRS warned that if the fishing continued at those levels, there would be a high risk of stock collapse. Such a situation put the commission in the spotlight of the international community, turning ICCAT into a preferred target for the criticism and bashing of several environmental nongovernmental organizations (NGOs). This crisis also resulted in a proposal, which was not successful: to include the Bluefin Tuna in Appendix I of the Convention of the Parties to the Convention on International Trade in Endangered Species of Wild Fauna and Flora (www.cites.org/eng/disc/text.php), which would have resulted in a ban on international trade of Bluefin Tuna.

Due to several measures adopted by the commission, explained in more detail further down, five years later during the last ICCAT meeting, held in November 2012, the TAC of eastern Bluefin Tuna was at 12,900 mt (for 2011 and 2012). The reported catches, in turn, were close to 9,800 mt (in 2011) and considered accurate by the SCRS. Finally, the stock was already showing clear signs of recovery, with all catch per unit effort indices exhibiting an increasing trend in most recent years. Consequently, the projections that the stock would be fully recovered by 2022 were advanced by six years to 2016. The previous headlines from

NGO press releases then changed, for example, from *End of the Line for Tuna Commission: ICCAT Fails to Provide Protections for Atlantic Bluefin* to *Decisions on Eastern Atlantic and Mediterranean Bluefin Tuna Follows Scientific Advice: WWF Applauds.*

The management of the Bluefin Tuna fishery by ICCAT, in 5 years, turned from a shameful disgrace into a success. Learning what drove such a change and prevented the imminent collapse of a valuable stock might shed light on what is required to ensure the successful future of interjurisdictional fisheries management. The success of the management strategy adopted by ICCAT included (1) strict adherence to scientific advice in setting the TAC; (2) successful negotiation in setting quotas for contracting parties and cooperating noncontracting parties; (3) reducing fishing capacity and fishing effort; and (4) ensuring compliance through several tools, such as the vessel monitoring system, having observers on board, and catch documentation schemes.

The Main Challenges

The two main challenges to ensuring effective management of interjurisdictional stocks are compliance/enforcement and agreement on allocation of fishing rights among nations, with both challenges being closely linked. The linkage between compliance/enforcement and allocation resides in the need for all members to perceive the allocations and rules adopted by the regional fisheries management organizations (RFMOs) to be fair and equitable or else they will not be fully committed to effectively implementing the measures adopted. Usually, in an RFMO, getting agreement for a management measure required for the conservation of a given stock is much easier than effectively ensuring compliance with the measure adopted. This difficulty stems from the fact that, on the one hand, member states are sovereign and have exclusive jurisdiction over the vessels flying their flag, but on the other hand, states have dramatically discrepant monitoring, control, and surveillance (MCS) capacities. RFMOs, therefore, are coming to the inescapable realization that simply adopting conservation and management measures is not enough. The effectiveness of the measures adopted will be greatly compromised unless they are backed up by significant capacity-building efforts. RFMOs are beginning to understand that, in order to ensure the translation of adopted measures into reality, a lot of capacity building is required to enable all members to have the same minimum MCS standard. This problem becomes particularly difficult with regard to new entrants into the fishery, especially if they are developing coastal states. With most of the highly migratory fish stocks already at or often below the level required to ensure the maximum sustainable yield, the growing aspiration of developing coastal states to participate more actively in the fisheries for those stocks is increasingly clashing with the desire of developed nations already fishing for those stocks to keep the status quo or even to increase their share. This scenario led ICCAT to develop, from 1998 to 2001, 27 criteria for the allocation of fishing possibilities (fishing quotas or limits). Although these criteria have never been enacted literally, their creation signifies the recognition that all ICCAT members do have the right to participate in the fishery for the stocks present in the convention area. Being able to achieve agreement on the individual quotas for the Bluefin Tuna fishery in the eastern Atlantic and the Mediterranean Sea, for instance, was surely pivotal for the success of the measures adopted, but that does not mean that everybody was happy.

Presently, the measures are not mandatory for two member states that have objected to the quota allocation, which they perceive as being unfair. In spite of that, they have agreed to apply the measures anyway, so the objection was more a political gesture without further consequences. The overall adherence to the catch limits adopted, however, happened, to a great extent, only because the market for Bluefin Tuna products is largely restricted to a couple of ICCAT members who have been able to exert pressure to ensure the observance of ICCAT limits, even by states who have not agreed to them.

The challenges of compliance/enforcement and allocation are well recognized and tackled by RFMOs worldwide to varying degrees. A third challenge, however, which is most of the time grossly overlooked or bluntly ignored, is the need for proper communication–an issue that requires urgent attention by RFMOs and will, therefore, become increasingly important for new professionals.

Doing It Right Is No Longer Enough...Communicating What Has Been Done Has Become Almost as Important!

In the process to rectify the mismanagement of the Bluefin Tuna in the eastern Atlantic and the Mediterranean Sea, two meetings have been particularly important: the 21st regular meeting of ICCAT in 2009, in Recife, Brazil, where the basis for the new management regime was laid out, and the 17th special meeting of ICCAT 2010, in Paris, where the new management regime was consolidated. At the 2010 meeting, a TAC was set at 12,900 mt, which gave a 70% probability that the eastern Bluefin Tuna stock would be restored to maximum sustainable yield level by 2022, meaning that there was about 30% probability that the recovery could take longer. The probability of the condition of the stock improving in this period, however, was very close to 100%. The headlines that came out in the international media, however, conveyed a pretty different picture. One of them, for example (followed by an excerpt of the published article), stated,

> Environmentalists: Fishing Quota Could Be Death Sentence for Bluefin Tuna
>
> ICCAT's own scientists say that the current quota gives the species a 70% chance of recovery. "The word 'conservation' should be removed from ICCAT's name. Governments here have just agreed to a Bluefin fishing plan that scientists conclude has a shocking one-third chance of failing to protect the species. Would you get in an airplane if you were told that it had a 30 percent chance of crashing?" said one NGO, adding that "this is a monumental failure of the way governments are supposed to protect our oceans." [From Hance 2010.]

It is clear that the future of interjurisdictional fisheries management will also depend heavily on proper communication, an issue which has been heavily underestimated and which presents several challenges:

1. Timing: Old news is no news! RFMOs are usually very slow and incompetent in communicating with the public. Press releases, when they are done, come out a few days or even weeks after meetings are over. NGOs, on the other hand, understand public relation to be a vital element of their existence, so their press releases are broadly distributed even before the end of a meeting. Consequently, NGO press releases are usually the only source of information for the international media; therefore, their views are the ones that make the headlines.
2. Professionalism: Public relation is not for amateurs! The difficulty RFMOs face in communicating properly with the public is aggravated by the fact that neither fisheries scientists nor fisheries managers are prepared or trained to communicate with the public. Public relations and communication is not for amateurs! It is a profession, just as fisher-

ies science and fisheries management are professions. It cannot and should not be improvised. I (F. H. V. Hazin) used to say that a fisheries manager or a fisheries scientist is probably as good in communicating to the public as a public relations professional is in assessing and managing fish stocks.

3. An intrinsic handicap: Good news, no news! Bad news simply sells better than good news! Therefore, any media portraying the collapse of fish stocks will always receive much more public attention than media informing the public that the stocks are healthy. It is human nature. And it cannot be helped. RFMOs should, therefore, stop complaining about bad media coverage and start to factor this intrinsic reality into their public relations strategy!

What Is Necessary to Become a Fisheries Manager in Interjurisdictional Fisheries?

Interjurisdictional fisheries management requires cooperation and collaboration between countries that often have different management aims, regulations, and capacities. Fisheries managers, therefore, need the ability to efficiently and effectively deal with a diverse range of affected parties. It is an interesting but complicated process that demands sound scientific knowledge, some legal background, and lots of patience. More and more, however, emerging professionals need to realize that these skills alone are no longer enough. In a globalized world with increasingly volatile information, properly communicating what has been done and the results achieved by those actions is becoming almost as important as the management actions themselves. Emerging fisheries professionals, therefore, will significantly increase their chances of professional success if they understand this trend, invest in developing strong communication skills, and abandon the illusion that communication is so simple that it can be improvised. As highlighted by George Bernard Shaw, "The single biggest problem with communication is the illusion that it has taken place."

Biographies

Profesor Fábio Hazin was born in Recife, Brazil in 1964. He graduated as a fisheries engineer from Universidade Federal Rural de Pernambuco and got his doctorate in fisheries oceanography from the Tokyo University of Marine Science and Technology in Japan in 1994. He also completed a course on the law of the sea at The Rhodes Academy of Oceans Law and Policy. He has an academic background in fisheries biology of highly migratory fish species and fisheries management. He chaired a Food and Agriculture Organization of the United Nations (FAO) technical consultation to draft a legally binding instrument on port state measures in 2008–2009 and served as the chair of the International Commission for the Conservation of Atlantic Tunas from 2007 to 2011. Presently, he is the vice-chair of FAO Committee on Fisheries.

Felipe Carvalho, a native of Recife, Brazil, received a bachelor's degree in fisheries engineering from Federal Rural University of Pernambuco. In 2007, he moved to the United States to earn a master's degree in the program of fisheries and aquatic sciences at the University of Florida and is now a doctoral student there, continuing his research on Blue Shark *Prionace glauca* stock assessment. After he finishes his Ph.D., he will seek postdoctoral appointments and a subsequent position as a stock assessment scientist to help fisheries management agencies plan the future of our marine resources.

References

Hance, J. 2010. Environmentalists: fishing quota could be death sentence for Bluefin Tuna. Mongabay, Menlo Park, California. Available: http://news.mongabay.com/2010/1128-hance_bluefin_quota.html. (February 2014).

ICCAT (International Commission for the Conservation of Atlantic Tunas). 2012. Report of the Standing Committee on Research and Statistics (SCRS). ICCAT, PLE-104/2012, Madrid, Spain.

The Urban Future of Fisheries

SARA HUGHES*
National Center for Atmospheric Research
Post Office Box 3000, Boulder, Colorado 80307, USA

Key Points

- Municipal government officials and urban residents are important players in fisheries management, and young professionals should consider working with and for these groups.
- The better you know your stakeholders, including urban ones, the more successful you will be in reaching your management goals.
- Urban water conservation can be a useful entry point for linking municipal policies to fisheries management goals.

Why Cities?

Cities are probably not the first thing that comes to mind when thinking about fisheries management. We often think instead of wild streams, big ocean vessels, or maybe colorful coral reefs. But for the future fisheries professional, cities are important sites. In my own research, I have come to appreciate, and be fascinated by, the dynamic role cities play in where and how people live, the way land and natural resources are used, and how we structure our economies. In this vignette, I hope to impart the importance of cities for fisheries management and managers, and to offer advice based on my own experiences in navigating these urban dimensions.

Urban areas are growing rapidly, with consequences for land use, water demand, and water quality. Between 1970 and 2000, an estimated 58,000 square kilometers of land became urbanized (Seto et al. 2011)—this is equivalent to an area nearly twice the size of Belgium. While the highest urban growth rates are occurring in Asia and Africa, North America is the most heavily urbanized region of the world, with nearly 86% of the population living in urban areas (UNDP 2010). The patterns and processes of urban growth mean that land once acting as infiltration basins, floodplains, or forests is no longer serving these ecological functions. This urban transformation of our landscapes can have permanent consequences for aquatic systems and their fisheries. The hydrologic cycle can become altered as runoff, evaporation, and even precipitation patterns change with rapid urbanization. The sustainability of fisheries—both freshwater and ocean—will be determined in part by these changes.

The growth of cities also means they are now home for the majority of the world's population. More than half of all people on the planet live in urban areas; most of the population of industrialized countries is urban and many developing countries are highly urbanized.

* Corresponding author: shughes@ucar.edu

Developing countries are expected to cross the urban majority threshold in the next 20–30 years. This concentration of population means that cities are important sites of consumption, citizenship, and conservation. Cities are where people will decide what fish to eat, what candidates to vote for, and what lifestyle changes to adopt. As people continue to move to cities, and incomes subsequently rise, the demand for fish is predicted to increase dramatically (FAO 2006). Cities are also sites of fish production and fisheries management. Small-scale and artisanal fisheries, including urban fish farms and open-water fishing, are increasingly found in urban and peri-urban settings; as a result, we can no longer assume that a fishing community is a rural entity. With population growth, cities face increasing pressure on their use of waterfront space for fisheries and fish markets, and urban water sources are often polluted. Future fisheries managers will need to contend with these urban dynamics.

Urban residents and decision makers have significant potential to act as valuable partners and should be brought into the fisheries fold. However, these groups have not historically been partners in fisheries management and conservation efforts. Instead, fisheries professionals have most often partnered with outdoors enthusiasts, farmers, and recreationists in their efforts to forward local, regional, and national management goals. Cities have been leaders in addressing other environmental challenges, such as the drivers and consequences of climate change. Cities are incentivizing green buildings, funding energy efficient transportation, and reducing residents' exposure to hazards. Indeed, in 2010, mayors from more than 200 cities from around the world signed the Mexico City Pact, pledging to take action to address climate change; in addition, more than 700 cities and communities are part of the International Council for Local Environmental Initiatives' (ICLEI) Cities for Climate Protection Program. These actions, innovations, and programs can have major implications for fisheries as they will lead to cities changing their land and water use patterns, infrastructure systems, and decision-making processes; whether they help or hurt fisheries management goals will be in part the result of our efforts to partner with urban residents and decision makers.

Lessons from the Field

One effect of growing urban populations is the concomitant increasing pressure on scarce water resources, with consequences for fisheries upstream and downstream. The politics and economics of watershed and water resources planning mean that fisheries often lose out as cities grow. Water is often used at a rate that diminishes a water body's ability to support fisheries, and nonpoint source pollution from urbanized areas is often left untreated. I have spent time in Los Angeles, Delhi, and Mexico City, three of the largest cities in the world, to undertake research that will help us better understand the complex decision-making processes in cities and their implications for water resources. These experiences may provide lessons for future fisheries professionals working in an urban world.

Cities Can Conserve Water to Further Fisheries Goals in the United States

Water conservation can be a useful entry point for aligning municipal decision making with broader fisheries goals. Water that cities save has the potential to be used for instream purposes that help meet fisheries management goals. Domestic water use (water used in people's homes) has been declining on a per capita basis in the United States for the past 20 years—from 185 gallons per capita per day (gpcd) in 1990 to around 145 gpcd in 2010, a decrease of more than 20% (Kenny et al. 2009; Kenny and Juracek 2012). This is due, in part, to water saving technologies, public awareness, and water conservation policies. Despite these gains, there is the potential for significantly more domestic water conservation in the United States (Christian-Smith et al. 2012).

I spent a summer interviewing decision makers, elected officials, nongovernmental organizations (NGOs), and academics about water management in Los Angeles and its important links to the state's fisheries. Los Angeles is a city that has become synonymous with irresponsible water use, as depicted in the movie *Chinatown*. Los Angeles was notorious for using large quantities of water and for transporting it into the city over very long distances. But today the city is undergoing some major changes in the way it approaches water (Hughes et al. 2013). While Los Angeles used to have very high per capita water use, it has implemented such successful water conservation programs that while the population has increased by 1,000,000 people in the past 25 years, its water demands have remained constant.

Water conservation efforts in Los Angeles have been driven in part by a need to restore the Mono Lake ecosystem, which had been devastated by water diversions for the city. In addition, climate changes have made water scarcer in the region, local politics have demanded a green agenda from city politicians, and regulatory requirements—particularly those protecting threatened and endangered fish species—have limited the accessibility of some of the city's traditional water sources. In this case, policies designed to protect fish have actually helped the city of Los Angeles transition toward more sustainable patterns of water resource use. The fisheries needs in the Bay–Delta area of northern California and in the Owens Valley prompted regulators to limit the amount of water that Los Angeles could take from these areas (Hughes et al. 2013). Local demands for sustainable approaches to water management have led to the election of environmentally friendly city council members and the appointment of conservation minded managers within city agencies.

These types of changes can happen in other cities as well and will be components of sustainable fisheries in the future. One winning strategy is to work collaboratively with urban decision makers. For example, the Mono Lake Committee (www.monolake.org/mlc/) worked with the Los Angeles Department of Water and Power (LADWP) to identify ways the city could reduce water demand, and therefore dependence on Mono Lake, and helped LADWP implement a low-flow toilet retrofit program in the city. These types of partnerships will be critical for fisheries professionals navigating an urban world.

Know the Players: Municipal Governments

Working effectively with municipal governments requires that you get to know them. City governments are structured differently from one another and are likely to approach environmental decision making differently as well. One important difference for fisheries is the way that cities—and, more specifically, decision makers—use and interpret scientific information. My experiences in Delhi and Mexico City provided some key insights into how different cities find, use, and interpret scientific information. Understanding these differences between cities could make the difference between fisheries science being used or discarded in decision-making processes.

If a municipal government is contracting for much of its work, it will be useful to position yourself as someone they would like to work with, or get in touch with those who are already in that position. This might mean staying up to date on the most recent regulatory requirements for cities and contacting a department head to let them know how your areas of expertise could help the city meet these requirements. For example, in Delhi, science for climate change policy making was primarily contracted out to academics and research institutes; very little of the actual analyses or modeling was done in house. Different research groups, NGOs, and consultants were hired or partnered with to undertake different tasks. Bureaucrats in the city's government agencies would then use that information for writing reports and making decisions.

If a municipal government has a formal system for finding the best science, then it may be less effective to try for a direct contracting role. Mexico City had a very different approach to acquiring and using scientific information in decision making. In the case of climate change planning, for example, the city government worked with the National University of Mexico to establish a specialized organization, the Mexico City Virtual Center for Climate Change, with the explicit purpose of bringing the best science to policy makers and communicating decision makers' needs to scientists.

Understanding these different processes can mean the difference between good fisheries science being a part of decision making or not. Find ways to be involved with decision-making processes so that your science is used by decision makers and you have a better understanding of the city's needs. This can also mean interacting with decision makers beyond government, such as NGOs, community groups, and local businesses.

Global Processes Matter Locally

It will be critical for future professionals to understand the global drivers behind urban dynamics. Cities are more directly linked to the global political economy than ever before.

Cities are increasingly taking on the role of global actors. Mayors are negotiating directly with international leaders for investment opportunities, and international networks of cities are helping to establish best practices (Unexpected co-operation, and investment, beside the Maumee river 2012). The decisions cities make about natural resources, therefore, will be an outcome not only of local politics and processes, but also global dynamics, networks, and economies. Cities often compete for international exposure and want to be seen as global leaders. Environmental issues are one area where cities are taking the lead; mayors can gain significant international exposure by being green and innovative. For example, Mexico City has gained global attention for the work its leaders have done in the area of the environment. The former mayor of Mexico City, Marcelo Ebrard, was awarded the World Mayor prize in 2010 from the City Mayors Foundation in recognition for his environmental efforts in the city (vom Hove 2010). The city's Minister of Environment, Martha Delgado Peralta, is the First Vice President of the International Council for Local Environmental Initiatives (ICLEI), an international network of cities aimed at improving local sustainability.

Thoughts for the Future

Cities are where future anglers, citizens, conservationists, businesses, and decision makers will be and where future fisheries professionals can make major contributions to sustainability. Many decisions about sustainability will be made by municipal governments and urban residents, and fisheries will increasingly be centered in urban areas. Innovations are accelerating: New York City is quickly becoming a leader in experimenting with indoor fish farms (Anonymous 2012). My experiences have shown that there are particularly important links between urban water use and fisheries that will help future fisheries professionals be more effective when working in an urban world.

A fisheries professional working for (or with) a municipal government (departments dedicated to the environment or natural resources), urban NGO, or urban university, rather than being disconnected from fisheries, may actually be ideally positioned to have significant influence on the future sustainability of fisheries. An urban fisheries professional may also have greater access to legislative and judicial bodies, fishery management councils, and state and federal regulatory agencies. The Mono Lake case in Los Angeles is a great example of how one urban NGO has capitalized on this possibility—rather than only demanding that the city reduce its water use, the Mono Lake Committee worked successfully with the city of Los Angeles to identify and implement strategies for achieving water conservation. Living and working in cities allows fisheries professionals to interact with important fisheries stakeholders and decision makers and motivate large numbers of people to support sustainable fisheries management. Engagement, cooperation, and communication will be the keys for fostering urban partnerships for fisheries sustainability.

Acknowledgments

I would like to acknowledge support from the Institute of Environment and Sustainability at UCLA and the Integrated Science Program and Advanced Study Program at the National Center for Atmospheric Research.

Biography

Sara Hughes is a postdoctoral fellow at the National Center for Atmospheric Research (NCAR) with a fellowship from the Integrated Science Program and the Advanced Study Program. Prior to joining NCAR, Sara was a postdoctoral associate at the University of California, Los Angeles in the Institute for Environment and Sustainability with the Center for Sustainable Urban Systems. Sara received her Ph.D. from the Bren School of Environmental Science and Management at the University of California, Santa Barbara in June 2011. Her specialization areas are political institutions, science in policy, and water resources management. Sara's dissertation was supported by fellowships from the Australian–American Fulbright Commission and the CALFED predoctoral science program. Sara received an M.S. in 2006 and a B.S. in 2004 from the Department of Fisheries and Wildlife at Michigan State University. She has a master's degree specialization in gender, justice, and environmental change, and her master's thesis examined a new model of collaboration in policy making for groundwater resources in Michigan.

References

Anonymous. 2012. New York: urban fish farms. Dac & Cities. Available: www.dac.dk/en/dac-cities/sustainable-cities-2/all-cases/food/new-york-urban-fish-farms/. (June 2013).

Christian-Smith, J., P. H. Gleick, H. Cooley, L. Allen, A. Vanderwarker, and K. A. Berry. 2012. A twenty-first century U.S. water policy. Oxford University Press, New York.

FAO (Food and Agriculture Organization of the United Nations). 2006. Fish trade and food security. FAO, Santiago de Compostela, Spain.

Hughes, S., S. Pincetl, and C. G. Boone. 2013. Understanding major urban transitions: drivers and dynamics in the city of Los Angeles. Cities: The International Journal of Urban Policy and Planning 32:51–59.

Kenny, J. F., N. L. Barber, S. S. Hutson, K. S. Linsey, J. K. Lovelace, and M. A. Maupin. 2009. Estimated use of water in the United States in 2005. U.S. Geological Survey Circular 1344, Washington, D.C.

Kenny, J. F., and K. E. Juracek. 2012. Description of 2005–2010 domestic water use for selected U.S. cities and guidance for estimating water use. U.S. Geological Survey Scientific Investigations Report 2012–5163.

Seto, K. C., M. Fragkias, B. Gunerlap, and M. Reilly. 2011. A meta-analysis of global urban land expansion. PLOS ONE 6(8):1–9.

UNDP (United Nations Development Programme). 2010. World urbanization prospects: the 2009 revision. United Nations, Department of Economic and Social Affairs, Population Division, Nairobi, Kenya.
Unexpected co-operation, and investment, beside the Maumee river. 2012. The Economist (August 25).
vom Hove, T. 2010. Marcel Ebrard, Mayor of Mexico City awarded the 2010 World Mayor prize. World Mayor (December 7).

Climate Change and the Future of Freshwater Fisheries

Daniel J. Isaak*
U.S. Forest Service, Rocky Mountain Research Station
322 East Front Street, Suite 401, Boise, Idaho 83702, USA

Key Points

- Climate change will cause profound environmental and biological changes during the course of your career.
- Prioritize and be efficient with your time and allocation of limited conservation resources to have the biggest long-term effect on fisheries resources.
- Strive to collect long-term monitoring data and archive it in digital databases so that trend assessments for freshwater ecosystems are improved.
- Realize that the decisions you make in the next few decades may have important consequences for fisheries resources a century from now.

My first awareness of the importance that climate has for fish came during my summer field seasons as a Ph.D. student at the University of Wyoming. While conducting electrofishing surveys in the climatically diverse Salt River basin along the mountainous border between Wyoming and Idaho, I observed spatial patterns in species distributions and abundance that strongly paralleled gradients in stream temperature. At the same time, through the research that Dr. Frank Rahel and his students were conducting at the same university, I became aware that climate could change in ways that might have significant repercussions for those fish distributions. A few years later, climate change moved into the American consciousness with the sensation around Al Gore's movie *An Inconvenient Truth*, and peer-reviewed papers began appearing in the literature that documented environmental trends consistent with global warming. My interest was piqued, and I began seriously digging into the topic of climate change to better understand how freshwater ecosystems could be affected. What I found transformed me and, because of the profound implications that climate change has for everything aquatic, caused me to transform my research career.

As background for this vignette, it is worth noting that air temperatures measured at weather stations across the contiguous United States were warmer in 2012 than any year since direct measurements began 117 years ago (Figure 1). Rather than being a random event, those record temperatures are symptomatic of the long-term warming trend in the Earth's climate during the past century. The majority of warming is caused by human economies and population growth, which have altered the chemical composition of the atmosphere through the release of carbon dioxide and other heat-trapping greenhouse gases. The Earth's temperature is now warmer than any time in the previous 10,000 years, which

* Corresponding author: disaak@fs.fed.us

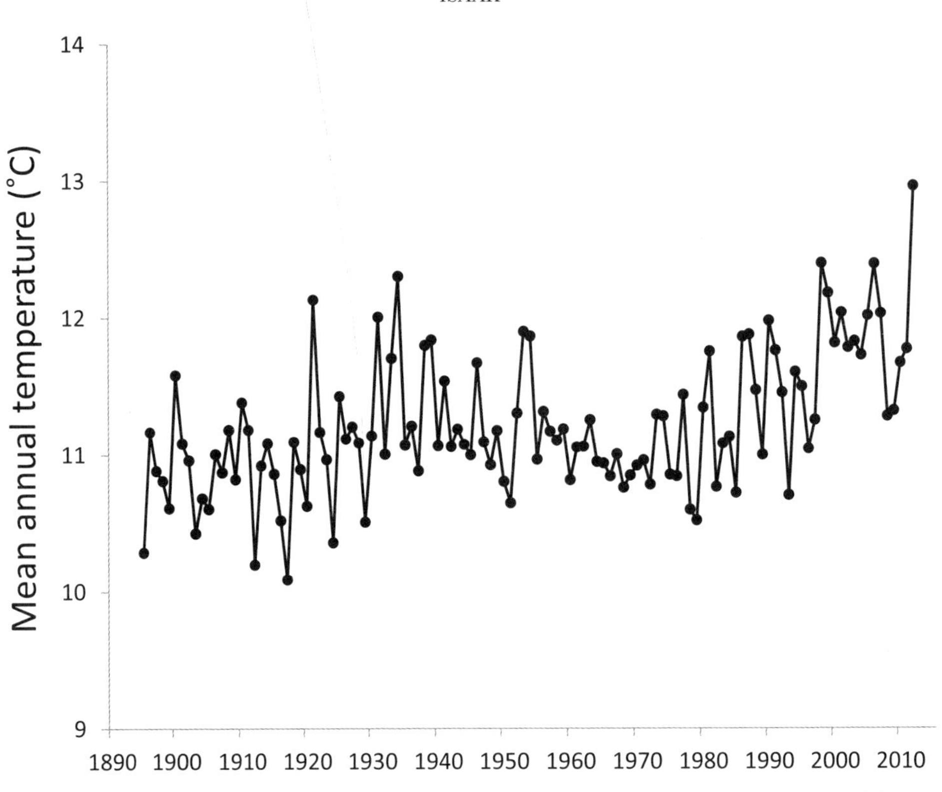

FIGURE 1. Mean annual air temperatures from 1895 to 2012 for the contiguous United States. Data are from the National Oceanic and Atmospheric Administration's National Climatic Data Center.

roughly coincides with the development of human civilization. Barring a series of major volcanic eruptions in future decades, or unforeseen breakthroughs in clean energy technology, it is highly likely that temperatures, and record warm temperatures, will continue to increase through at least the middle of the 21st century. Global warming, and the series of environmental, habitat, and fish population trends that accompany it, will be a defining theme in your career and the careers of fisheries professions that follow you.

Considerable evidence already exists for changes in freshwater environments that are associated with the historical increases in Earth's temperatures (see Recommended Readings). Lake and stream temperatures have been increasing while precipitation patterns and stream hydrologies show a variety of climate-related trends. Because fish and other aquatic organisms are ectothermic and confined to well-defined water bodies, their population dynamics and distributions must adjust to those changes. Adjustments consist of shifts in the timing of migrations, spawning, and other life history events, as well as poleward and upstream distribution shifts to cooler habitats. During those shifts, some species and populations are encountering barriers to migration, running out of room in ephemeral headwater streams, or being overtaken by shifting thermal conditions that cause local extirpations. Within the core areas of species' distributions, many populations are at no immediate risk, but even there, changes are occurring slowly that will cause habitat suitability to drift relative to the tolerances of individual species and communities.

Trends associated with climate change run somewhat counter to the "balance of nature" and "dynamic equilibrium" concepts that were central to natural resource education and stewardship in the 20th century. A dynamic disequilibrium is a better conceptual model for the way the world works now. The new model holds that aquatic environments are becoming more variable and gradually shifting through time as the Earth warms. Those trends are exacerbating the desire of growing human populations to have reliable water supplies. That means additional water development projects to store or divert water, and greater alteration of its movement across landscapes, which will have repercussions for freshwater organisms. Into this shifting milieu, nonnative species are being more frequently introduced or spreading from areas where they had previously been confined. A disequilibrial world is one that will systematically create winners and losers because species are predisposed either favorably or unfavorably toward the changes that occur. We have entered uncharted waters in many regards, and your generation is confronted with rapid changes at massive scales. You will have to learn, adapt, and evolve more rapidly to a greater set of challenges than your predecessors to be an effective resource steward in today's world. It will be important to seek out mentors and learn from those with experience in your new career field, but also realize that their experiences may be less of a guide than was previously the case.

More than ever, success will require clear goal setting, frequent reassessments as novel situations are presented, and communicating with diverse publics. In addition to traditional concerns about maintaining, improving, or understanding fisheries, you may find yourself confronted with questions about whether some long-term fisheries can continue to support harvest in the future. In other cases, broad geographical declines (or increases) in species that might once have been short-term phenomena may instead become the new normal as environmental trends related to climate lead to dissolution of contemporary biological communities and development of new ones. There will be questions and significant concerns from a public that is sometimes poorly informed and antagonistic regarding the issue of climate change. Some may question your credibility for even mentioning climate change and its effects on fish populations. Yet others will clamor for preserving biodiversity and native species within historical ranges that are no longer suitable for the species in question. Goals and management programs that were logical to your predecessors and seemingly achievable at the beginning of your career may become unrealistic in future years. At the same time, currently unforeseen challenges and opportunities will present themselves.

Navigating the complex environmental and social dynamics triggered by climate change requires broad understanding of many issues and superb communication and people skills. Push yourself to read and think broadly while continuing to hone your communication skills. Take extra writing and speaking classes while in school and look for opportunities to apply those skills after graduation. I was a terribly shy and nervous public speaker well into my 30s, but I took extra speaking classes during college and have always forced myself to prepare and give talks during my professional career. Though by no means a brilliant orator, I now give passable talks and am invited to do so frequently. One mentor early in my professional career, Bruce Rieman, was well known for his dynamic and engaging speaking style. Nearing retirement, he confided in me that he thought giving good talks was the most useful thing he ever did in terms of influencing other people. And how you influence and work with other people will be critical to building collaborative relationships and your effectiveness over time. No one person or organization can do it all. Teams of professionals with complementary skills are required to address climate change, so developing partnerships within your organization, between your organization and others, and across disciplinary boundaries will be key to leveraging resources and team building. Learn what your biggest strengths are as a researcher, manager, database designer, geospatial analyst, or team leader, and contribute your skills accordingly. People want to be engaged in meaningful work, and issues surrounding climate change and fisheries present young professionals a wealth of such opportunities.

Developing better information, and systems for creating information, is a powerful teambuilding endeavor that may also reduce many of the uncertainties about how climate change affects fisheries. Information ultimately flows from data and models that allow us to accurately describe the status and trends of aquatic resources. At this point, however, most of what we know is based on studies and monitoring efforts that are of short duration and small spatial extents, which limits our ability to think and manage at the broader spatial and temporal scales that are required by climate change. In some instances, developing the necessary information for key parameters like species distributions and abundance, stream and lake temperatures, or stream discharge may be accomplished simply through better coordination of sampling efforts among organizations or by aggregating existing data. One project I participated in, for example, developed an interagency stream temperature database for dozens of agencies in the northwest United States. The project was prompted by the fact that there were literally millions of temperature measurements in existence but the data were not usable because they were scattered among hundreds of biologists. Once a centralized database was created and made broadly available through a Web site, it provided the regional aquatics community with a resource worth millions of dollars. Far more valuable, however, is the information that the database continues to yield for better decision making and improving the efficiency of new temperature monitoring efforts among multiple agencies.

Because climate change is ultimately about gradual trends in the environment, long-term monitoring programs are especially important for describing the dynamics of freshwater systems. I can still hear Dr. David Willis in my undergraduate fisheries management class at South Dakota State University intoning, "Collect the same data, in the same way, at the same time and place each year" with regards to developing monitoring records. It is not rocket science but it gets the job done and, over time (e.g. several decades), may provide direct evidence regarding how climate is affecting species composition or habitat quality at specific sites. The longer any trend monitoring program can be maintained, the more valuable it becomes, so early in your career, seek out opportunities to establish, or participate in, monitoring programs. Modern temperature and flow sensors are relatively inexpensive and can record continuous measurements for many years, so in some instances, monitoring networks can be established even with limited budgets.

Better data and information will provide an important foundation for decision making during your career, but are not panaceas. There will be difficult decisions ahead, sometimes with only unappealing options from which to choose. Being effective may mean picking the least bad option and prioritizing resources where they serve important needs in the short term while also accomplishing the most good over the long term (decades [and maybe centuries?] rather than years). You may not see, or be credited with, the ultimate benefits and consequences of some of these decisions, but they will be important determinants of fisheries resources available to future generations. As a simple guide to effective action, ask yourself where desired resources, values, and fisheries may be most resilient to future climate trends, where they are at risk of disappearing, and whether there are intermediate areas where intelligent and proactive interventions could figure decisively in the outcomes? That final case is where resources will be most profitably spent, and in some cases, the threat posed by climate change could be used to marshal resources for long-needed habitat improvements and restoration activities. Effective restoration in some instances could offset some of the detrimental effects of climate change because many aquatic environments are already significantly degraded. In cases where desired conditions are impossible to maintain, keep an open mind because doors to new possibilities will be opening. For example, it might be possible to facilitate the establishment of important new fisheries or biodiversity reserves for more southerly species that were previously limited by cold temperatures. In many rivers across the western United States, Brown Trout *Salmo trutta* and Smallmouth Bass *Micropterus dolomieu* appear to be rapidly expanding upstream and could provide significant angling opportunities if properly managed. In all cases, think about designing and implementing systems and management strategies that are resilient, are flexible, and have the capacity for adaptation and evolution in the future.

Conclusion

It has been said that it is impossible to ever step in the same river twice. That statement remains true, but now the river is flowing ever faster. Your generation of fisheries professionals faces challenges unlike any preceding generation. Climate model projections suggest that air temperatures will warm another 1–3°C by mid-century. Even at the low end of that range, more warming would occur in the first half of the 21st century than occurred during all of the 20th century. Human populations will also continue to grow, and their per capita consumption rates will be higher than today as living standards continue to rise. Given those changes, this century will be a monumental and transitional one for the Earth, aquatic resources, and the fisheries profession. The stresses placed on the natural world will be unprecedented, and significant changes are inevitable. The choices that human society makes, and that you make as a fisheries professional during your career, will play a large role in shaping the world a century from now. I do not envy you for the challenges that lay ahead, but exceptional times create exceptional opportunities and people. Many among you will rise to the occasion and do great things that ensure important fisheries legacies for future generations. I wish you the best in those endeavors.

Biography

Dan Isaak is a research fisheries scientist with the U.S. Forest Service Rocky Mountain Research Station Aquatic Sciences Laboratory in Boise, Idaho. He holds a B.S. degree from South Dakota State University, an M.S. degree from the University of Idaho, and a Ph.D. from the University of Wyoming. His research focuses on understanding the effects of climate change and natural disturbance on stream habitats and fish populations, stream temperature monitoring and modeling, development and application of spatial statistical models for stream networks, and use of digital and social media to connect people, information, and landscapes. He also writes the Climate-Aquatics Blog (www.fs.fed.us/rm/boise/AWAE/projects/stream_temp/stream_temperature_climate_aquatics_blog.html).

Recommended Reading

Burkhead, N. M. 2012. Extinction rates in North American freshwater fishes, 1900–2010. BioScience 62:798–808.

Burt, T. P., N. J. K. Howden, and F. Worrall. 2014. On the importance of very long-term water quality records. WIREs Water 1:41–48.

Comte L., and G. Grenouillet. 2013. Do stream fish track climate change? Assessing distribution shifts in recent decades. Ecography 36:1236–1246.

Comte L., L. Buisson, M. Daufresne, and G. Grenouillet. 2013. Climate-induced changes in the distribution of freshwater fish: observed and predicted trends. Freshwater Biology 58:625–639.

Crozier, L. G., and J. A. Hutchings. 2014. Plastic and evolutionary responses to climate change in fish. Evolutionary applications 7:68–87.

Dudgeon, D., A. H. Arthington, M. O. Gessner, Z.-I. Kawabata, D. J. Knowler, C. Lévêque, R. J. Naiman, A.-H. Prieur-Richard, D. Soto, M. L . J. Stiassny, and C. A. Sullivan. 2006. Freshwater biodiversity: importance, threats, status and conservation challenges. Biological Reviews 81:163–182.

Hansen, J. E. 2010. Storms of my grandchildren: the truth about the coming climate catastrophe and our last chance to save humanity. Bloomsbury Press, New York.

Intergovernmental Panel on Climate Change. 2013. Climate change: the physical science basis. Cambridge University Press, New York. Available: www.ipcc.ch/. (February 2014).

Isaak, D. J., and B. E. Rieman. 2013. Stream isotherm shifts from climate change and implications for distributions of ectothermic organisms. Global Change Biology 19:742–751.

Isaak, D. J., C. C. Muhlfeld, A. S. Todd, R. Al-Chokhachy, J. Roberts, J. L. Kershner, K. D. Fausch, and S. W. Hostetler. 2012. The past as prelude to the future for understanding 21st century climate effects on Rocky Mountain trout. Fisheries 37:542–556.

Isaak, D. J., D. Horan, and S. Wollrab. 2013. A simple protocol using underwater epoxy to install annual temperature monitoring sites in rivers and streams. U.S. Forest Service, Rocky Mountain Research Station, GTR-RMRS-314, Fort Collins, Colorado.

Isaak D. J., S. Wenger, E. Peterson, C. Luce, J. Ver Hoef, D. Nagel, S. Hostetler, J. Dunham, J. Kershner, B. Roper, D. Horan, G. Chandler, S. Wollrab, and S. Parkes. 2011. NorWeST stream temp regional database and model. Available: www.fs.fed.us/rm/boise/AWAE/projects/NorWeST.html. (February 2014).

Isaak, D. J., S. Wollrab, D. Horan, and G. Chandler. 2012. Climate change effects on stream and river temperatures across the northwest U.S. from 1980–2009 and implications for salmonid fishes. Climatic Change 113:499–524.

Katz, J., P. B. Moyle, R. M. Quiñones, J. Israel, and S. Purdy. 2013. Impending extinction of salmon, steelhead, and trout (Salmonidae) in California. Environmental Biology of Fishes 96:1169–1186.

Kaushal, S. S., G. E. Likens, N. A. Jaworski, M. L. Pace, A. M. Sides, D. Seekell, K. T. Belt, D. H. Secor, and R. L. Wingate. 2010. Rising stream and river temperatures in the United States. Frontiers in Ecology and the Environment 8:461–466.

Keleher, C. J., and F. J. Rahel. 1996. Thermal limits to salmonid distributions in the Rocky Mountain region and potential habitat loss due to global warming: a geographic information system (GIS) approach. Transactions of the American Fisheries Society 125:1–13.

Loarie, S. R., P. B. Duffy, H. H. Hamilton, G. P. Asner, C. B. Field, and D. D. Ackerly 2009. The velocity of climate change. Nature 462:1052–1055.

Milly, P., K. A. Dunne, and A. V. Vecchia. 2005. Global pattern of trends in streamflow and water availability in a changing climate. Nature 438:347–350.

Mote P. W., A. F. Hamlet, M. P. Clark, and D. P. Lettenmaier. 2005. Declining mountain snowpack in western North America. Bulletin of the American Meteorological Society 86:39–49.

Rahel, F. J., and J. D. Olden. 2008. Assessing the effects of climate change on aquatic invasive species. Conservation Biology 22:521–533.

Rieman, B. E, and D. J. Isaak. 2010. Climate change, aquatic ecosystems and fishes in the Rocky Mountain West: implications and alternatives for management. U.S. Forest Service, Rocky Mountain Research Station, GTR-RMRS-250, Fort Collins, Colorado.

Schneider, P., and S. J. Hook. 2010. Space observations of inland water bodies show rapid surface warming since 1985. Geophysical Research Letters [online serial] 37:L22405.

Steffen, W., P. J. Crutzen, and J. R. McNeill. 2007. The anthropocene: are humans now overwhelming the great forces of nature? Ambio 36:614–621.

Stewart I. T., D. R. Cayan, and M. D. Dettinger. 2005. Changes toward earlier streamflow timing across western North America. Journal of Climatology 18:1136–1155.

U.S. Environmental Protection Agency. 2014. Guidelines for continuous monitoring of temperature and flow in wadeable streams. U.S. Environmental Protection Agency, Washington, D.C.

Vorosmarty, C. J., P. B. McIntyre, M. O. Gessner, D. Dudgeon, A. Prusevich, P. Green, S. Glidden, S. E. Bunn, C. A. Sullivan, C. R. Liermann, and P. M. Davies. 2010. Global threats to human water scarcity and river biodiversity. Nature 467:555–561.

Wenger, S. J., D. J. Isaak, C. H. Luce, H. M. Neville, K. D. Fausch, J. D. Dunham, D. C. Dauwalter, M. K. Young, M. M. Elsner, B. E. Rieman, A. F. Hamlet, and J. E. Williams. 2011. Flow regime, temperature, and biotic interactions drive differential declines of trout species under climate change. Proceedings of the National Academy of Sciences of the United States of America 108:1475–14180.

Making Adaptive Management Work: Lessons from the Past and Opportunities for the Future

MICHAEL L. JONES*
Department of Fisheries and Wildlife, Michigan State University
480 Wilson Road, Room 13, East Lansing, Michigan 48824, USA

GRETCHEN J. A. HANSEN
Wisconsin Department of Natural Resources, DNR Science Operations Center
2801 Progress Road, Madison, Wisconsin 53716, USA

Key Points

- Adaptive management is a tool for managing fisheries that is designed to reduce uncertainty; every time a management action is implemented, opportunities for learning are there for the taking.
- Key ingredients of good adaptive management include specifying objectives and options, engaging stakeholders, identifying critical uncertainties, and a commitment to monitoring and evaluation.
- Fisheries management abounds with opportunities for adaptive management. We are optimistic that future fisheries managers can, and will, implement this tool successfully.

Introduction

One of us (MJ) recently gave a talk at a national fisheries conference in Canada with the title "Maybe We Should Just Admit We Have No Idea What We're Doing." This title was of course chosen with tongue firmly planted in cheek, although it did have the effect of drawing a crowd! But why would this be an appropriate title for a fisheries talk? Because decision makers are nearly always very uncertain about the consequences of the management actions they contemplate taking. And being uncertain is not just annoying; it matters. It matters because the choices we make need to account for risk, which results from uncertainty. And it also matters because it means that reducing uncertainty is a good objective for management—so that we are less uncertain in the future. Establishing the reduction of critical (i.e., policy-relevant) uncertainty as an objective of management is the essence of adaptive management (AM). Both of us have had a lot of experience with AM during our careers: one of us was a new member of the group that first presented AM to environmental scientists and managers, both of us have taught about it, and one of us made it the central theme of her graduate research project. In this essay, we will share our perspective on AM gained from these experiences and offer an optimistic outlook for the future.

* Corresponding author: jonesm30@anr.msu.edu

Adaptive management is learning by doing (Box 1). However, AM is more than just trial and error; the "doing" component is structured to ensure that the "learning" component occurs in spite of the complexity and variability of fisheries systems. Adaptive management includes two general approaches: active and passive. Both approaches involve making predictions about the consequences of potential management actions, implementing an action or a set of contrasting actions, and then monitoring the outcome to evaluate the accuracy of the predictions (that is, test the hypothesis). Adaptive management is viewed as passive when the management action taken is selected because it is believed most likely to achieve management objectives. For example, a manager might decide to modestly decrease stocking rates in a lake because this is predicted to result in the highest growth and survival rates of the stocked fish, but would subsequently monitor the food web in the lake to learn more about interactions between the stocked species and its prey that were the basis of the original prediction. Adaptive management is viewed as active when the chosen management actions are deliberately experimental and not necessarily the choice that is predicted to be most likely to achieve objectives. Active AM often involves implementing contrasting actions in different ecosystem, just as one would do in a controlled experiment with contrasting treatments. For example, a manager might stock sport fishes varying densities across a set of lakes to more directly test hypotheses about food web interactions and determine the

Box 1. Defining Active Management

Adaptive management (AM) promotes learning about complex systems by designing regulations with specific objectives and questions in mind, implementing regulations strategically, monitoring the response, evaluating management outcomes, adjusting (i.e., adapting) the plan based on these evaluations, and designing an alternative plan to restart the cycle. Management actions are treated as experiments and can reduce uncertainty about system response to interventions. Regulations can be designed explicitly to test system boundaries and maximize learning (active AM), or for other objectives (passive AM).

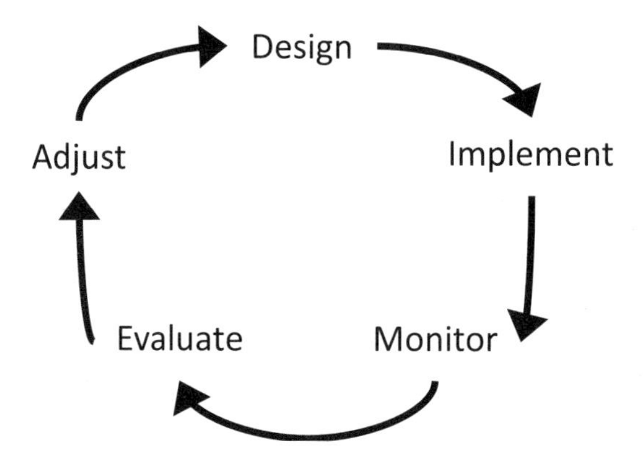

A simplified adaptive management cycle (modified from Murray and Marmorek 2003)

stocking density that results in the best outcome. In both cases, learning more about these interactions should aid future decisions about suitable stocking levels. Both approaches are valuable; the optimal management alternative is likely to be identified more quickly when active AM is used, but passive AM is often less costly and less controversial to implement (see Murray and Marmorek 2003).

Despite its conceptual appeal, AM has not been used in resource management nearly as much as we would have expected. Even so, we remain very optimistic about its promise for the future. Why? First, because AM simply makes too much sense to be a bad idea. Second, we have experience with a successful AM effort that resulted in major changes in the management of a notorious invasive species, so we know it can be done. Third, we think fishery management offers numerous opportunities for implementing AM—even active AM—that should be comparatively easy to implement. In the following sections, we briefly describe our success story with AM and use this and other examples to illustrate what we believe are the four requirements of successful AM. We conclude with a discussion of what we see as great opportunities for implementing AM in the future.

An Adaptive Management Success Story: Sea Lamprey Management in the Laurentian Great Lakes

Our success story involves the management of invasive Sea Lamprey *Petromyzon marinus* in the Laurentian Great Lakes and is based on GH's master's research. Sea Lampreys parasitize economically important Great Lakes fishes such as Lake Trout *Salvelinus namaycush* and naturalized Pacific salmon *Oncorhynchus* spp. Sea Lampreys are controlled primarily through the application of chemical lampricides that target stream-dwelling life stages before they enter the Great Lakes; in recent years, the annual budget for Sea Lamprey management has been roughly US$20 million. A substantial portion of this budget is spent on assessment to prioritize streams for treatment each year. However, given the fixed budget available for both assessments and application of lampricides (i.e., control), greater investments in assessment result in fewer resources available to treat streams and control Sea Lampreys by reducing larval populations. To explore the trade-off between assessment and control in this system, we developed an assessment method that allocated fewer resources to assessment and more to control, and used AM to learn whether this strategy would result in more cost-effective Sea Lamprey control. After 3 years of AM, we concluded that the alternative assessment method would result in more cost-effective control, and soon thereafter (2008) the Great Lakes Fishery Commission decided to switch to the alternative assessment method (see Hansen and Jones 2008 for more details).

Ingredients of Successful Adaptive Management

Our experience with AM of Great Lakes Sea Lampreys taught us that to be a successful adaptive manager, you need to pay attention to at least four things. First, as with all good management practices, you have to think carefully about management objectives and options. What objective(s) do you want management to achieve? What are the available alternative management actions? While these two ingredients of management seem obvious to nearly everyone, in our experience, it is not hard to find examples of management where it is not clear what the objectives were, or where a surprisingly narrow range of options has been considered. In our Sea Lamprey example, the objective was to design a management strategy that would kill at least as many Sea Lampreys as the status quo method while allocating fewer dollars to assessment. We considered many management options to achieve this objective, such as developing new assessment technologies and using different criteria for ranking streams. After much discussion with our stakeholders (see below), we ended up evaluating a management option that focused on rapid assessment. This was an aspect of the assessment program that managers were interested in changing prior to the initiation of our project (but without much justification), so conducting a management experiment was an idea that they found easy to support. This brings us to the second ingredient of successful AM: stakeholder involvement.

To engage in successful AM, you must be prepared to meaningfully involve stakeholders in the entire process of conceiving, designing, implementing, and monitoring an AM process. This is no easy task; involving stakeholders requires a significant investment of time and resources. However, in today's world, stakeholder involvement is becoming the rule rather than the exception. Adaptive management requires a commitment to learning over the long haul, and since fisheries are a public resource, public support for long-term actions is essential. Support from people who interact with fisheries on a daily basis can translate to support from higher-up administrators if they hear that their constituents approve of the approach. Getting stakeholders to buy in to the AM strategy is vital to its success. In our experience with Sea Lamprey management, the stakeholders were primarily fisheries managers and biologists from the U.S. Fish and Wildlife Service and Canadian Department of Fisheries and Oceans, who perform the on-the-ground assessment and control of Sea Lampreys. We collaborated closely with these managers throughout the AM process, and their engagement was essential to our success. They identified our management options, helped develop the alternative assessment protocol, and played a large role in the monitoring and evaluation efforts. Because they were engaged from the outset, the stakeholders—and in turn the decision makers to whom they reported—agreed with our findings, despite the fact that accepting them would result in a major restructuring of the Sea Lamprey management program.

Third, to manage adaptively you have to explicitly recognize critical uncertainties. Acknowledging uncertainty can make both managers and stakeholders uncomfortable. We do not know any managers that like to admit they do not know how to manage the fisheries for which they are responsible, and few stakeholders want experimentation occurring on their fishery. Still, things have really changed in the past decade or two, with spectacular collapses of commercial fisheries leading stakeholders and managers alike to be more candid about uncertainty and the associated risks. It is easy to think of uncertainties for any management situation, but that is not enough. Critical uncertainties are ones that affect the best management choice: that is, the truth about a critical uncertainty affects what you should do. Identifying which uncertainties are critical in management decisions is difficult, and most of the time this involves building a model. To apply AM, critical uncertainties are represented as hypotheses whose evidence can be assessed through monitoring the effect of management actions. In our Sea Lamprey example, management was used to test the hypothesis that the status quo assessment procedure was the most cost-effective approach, and the result was that the hypothesis was rejected.

The final essential ingredient of AM is a commitment to monitoring and evaluation to test the hypotheses. Monitoring is obviously essential for learning, even under the most passive of AM strategies. Without monitoring, there is no means of assessing the effect of management interventions. We believe the lack of rigorous monitoring and evaluation is by far the most significant reason why we do not have a vastly better record of successful

AM. Anyone who has worked in fishery management for even a few years will find it easy to think of instances in which, but for the lack of follow-up monitoring and evaluation, we could have learned so much about a system by observing its response to a management intervention. Good AM requires managers to meet all of the demands listed above, but without monitoring, AM is not just bad, it is impossible. Monitoring is expensive, so it is usually impractical for fisheries managers to monitor and evaluate every variable of interest. We suggest that monitoring and evaluation be tied explicitly to management decisions, such that future decisions cannot be made in the absence of monitoring data. In our Sea Lamprey example, monitoring (assessment) comprises a large portion of the annual Sea Lamprey management activities because determination of where to apply lampricides cannot occur in the absence of monitoring data. As a result, there is solid support for monitoring in Sea Lamprey management. For many exploited fisheries, where the goal is to determine future allowable harvests, monitoring and evaluation is equally important. harvest policy decisions are often explicitly linked to assessment (monitoring) information.

Future Opportunities for Adaptive Management

Although the challenges are considerable, we believe that fisheries management offers innumerable opportunities where AM is relatively easy to implement, if enough fisheries professionals (you!) became advocates for this approach. Much of fisheries management involves the implementation of actions for conserving and enhancing fisheries that are replicated across large spatial scales (i.e., inland lakes and streams). Replication creates great opportunities for good AM experiments, just as it does in the research laboratory. Sportfishing harvest regulations, stocking programs, and habitat enhancement activities are good examples of replicated management actions. It is not difficult to imagine what the objectives might be for these actions, even if they are not explicitly stated. For example, a harvest bag limit might be reduced for a particular sport fish with the objective of increasing the adult population size of the target species. It is also quite easy to imagine a set of alternative hypotheses associated with this management action, even if the action is not formally framed as a hypothesis. In our example, a bag limit reduction could occur because it is believed (hypothesized) that the current harvest rates are not sustainable. Despite a lot of wishful thinking, rarely do managers actually know that this is the case. Of course, what is nearly always missing from these situations is a monitoring and evaluation effort to determine whether the hypothesis was in fact supported based on the outcome of the implementation of the action.

These situations strike us as an incredible opportunity for AM. Imagine that a management agency decides to implement a habitat enhancement program for its coldwater streams. Numerous streams or stream reaches could be candidate sites for the proposed habitat management actions. To implement the program in an AM framework, the agency would first need to engage relevant stakeholders (e.g., local fishers, the Trout Unlimited chapter for the area, and landowners) in a discussion of objectives, options, and uncertainties. Once the problem had been framed, an experimental program of habitat manipulations could be designed to efficiently test critical hypotheses (e.g., about trout demographic responses to certain types of structures), with controls and replication and with a commitment to monitoring and evaluating outcomes. Specific criteria would be agreed upon a priori for determining if a manipulation was successful; for example, an increase in juvenile trout density of at least 20% in at least five years. Ultimately, the habitat manipulations might be similar to those planned in the absence of an AM approach but with the crucial addition of the explicit opportunity to learn. Engagement of stakeholders from the outset is very likely to create opportunities for citizen involvement in the monitoring efforts, which both saves money and fosters program buy-in.

Although AM might sound relatively easy, decades of failed AM experiments have proven that the challenges are often insurmountable. One challenge is designing an experimental approach that is acceptable to stakeholders and will allow learning in the face of high levels of environmental variation that are outside of managers' control. Another is identifying appropriate replicates and reference systems among myriad lakes or streams that each has a unique ecology and management history. Nevertheless, we believe that even a compromised experimental design can provide useful information for informing future decisions and is an improvement over a blind-faith approach to management implemented with no monitoring or evaluation and thus no possibility of learning (Hilborn 1992). Of course, not all management situations are suited to an adaptive management approach. For example, certain situations may have sufficiently little uncertainty such that an adaptive approach is unwarranted. Conversely, high natural variability or situations where key outcomes cannot be measured with much precision may make testing hypotheses with management experiments essentially impossible. Similarly, if insufficient resources and commitment exist for monitoring and evaluation, it is difficult to see the value in implementing AM. There are also circumstances where the risk of certain, plausible outcomes is too great to warrant management experiments. This may be arise in the case of management of an endangered species, for example. Finally, if stakeholders cannot agree on a set of objectives to guide management, a key ingredient of adaptive management (a clear set of objectives) will be missing.

As you become more familiar with the practice of fishery management, we are nevertheless certain that you will discover innumerable instances where adaptive management could be applied. Making fishery management adaptive requires more effort than not doing it. Engagement of stakeholders; being publicly honest about uncertainty; thinking carefully about objectives, options, hypotheses, and experiments; and building support for effective monitoring and evaluation are all challenges. This is almost certainly why AM has not happened more. But, we believe the unrecorded costs of not doing AM are huge, both in terms of lost opportunities for better knowledge and in terms of building a well-informed stakeholder community that understands the impacts of uncertainty on management outcomes and values learning as a management objective. We are very hopeful that the next generation of fishery scientists and managers will usher in an exciting partnership between research and management that will come to be called the "age of adaptive management" in fisheries.

Biographies

Michael Jones is a professor and co-director of the Quantitative Fisheries Center at Michigan State University. He studied with the early developers of the adaptive management concept at the University of British Columbia in the late 1970s and has worked subsequently as an environmental consultant, government research scientist, and university professor. When he does believe he has some idea what he's doing, he works with managers and stakeholders to improve decision making for Great Lakes and other fisheries.

Gretchen Hansen is a research scientist at the Wisconsin Department of Natural Resources. She worked as a postdoc and received her Ph.D. from the Center for Limnology at the University of Wisconsin-Madison. She received her master's degree in 2006 from Michigan State University, where she learned from her advisor (MJ) that we should not let the fact that we have no idea what we are doing prevent us from taking action. She is trying to promote the idea of adaptive management to anyone who will listen while researching historical trends in Wisconsin's fish communities and working with managers to understand the implications of these trends for current management systems.

References

Hansen, G. J. A., and M. L. Jones. 2008. A rapid assessment approach to prioritizing streams for control of Great Lakes sea lampreys (*Petromyzon marinus*): a case study in adaptive management. Canadian Journal of Fisheries and Aquatic Sciences 65:2471–2484.

Hilborn, R. 1992. Can fisheries agencies learn from experience? Fisheries 17:6–14.

Murray, C., and D. R. Marmorek. 2003. Adaptive management and ecological restoration. Pages 417–428 *in* P. Friederici, editor. Ecological restoration of southwestern ponderosa pine forests. Ecological Restoration Institute, Flagstaff, Arizona.

Recommended Readings

Clemen, R. T., and T. Reilly. 2001. Making hard decisions. Duxbury Press, Pacific Grove, California.

Hammond, J. S., R. L. Keeney, and H. Raiffa. 1999. Smart choices. Broadway Books, New York.

Holling, C. S., editor. 1978. Adaptive environmental assessment and management. Wiley, Chichester, England.

Irwin, B. J., M. J. Wilberg, M. L. Jones, and J. R. Bence. 2011. Applying structured decision making to recreational fisheries. Fisheries 36:113–122.

Lee, K. N. 1993. Compass and gyroscope: integrating science and politics for the environment. Island Press, Washington, D.C.

Nichols, J. D., M. D. Koneff, P. J. Heglund, M. G. Knutson, M. E. Seamans, J. E. Lyons, J. M. Morton, M. T. Jones, G. S. Boomer, and B. K. Williams. 2011. Climate change, uncertainty, and natural resource management. The Journal of Wildlife Management 75:6–18.

Nichols, J. D., and B. K. Williams. 2006. Monitoring for conservation. Trends in Ecology & Evolution 21:668–673.

Polasky, S., S. R. Carpenter, C. Folke, and B. Keeler. 2011. Decision-making under great uncertainty: environmental management in an era of global change. Trends in Ecology & Evolution 26:398–404.

Sethi, S. A. 2010. Risk management for fisheries. Fish and Fisheries 11:341–365.

Walters, C. J. 1986. Adaptive management of renewable resources. Macmillan, New York.

Walters, C. J. 2007. Is adaptive management helping to solve fisheries problems? Ambio 36:304–307.

Walters, C. J., and C. S. Holling. 1990. Large-scale management experiments and learning by doing. Ecology 71:2060–2068.

Williams, B. K. 1997. Approaches to the management of waterfowl under uncertainty. Wildlife Society Bulletin 25:714–720.

Catfish in the Courtroom: How Forensic Science Catches Seafood Cheats

Trey Knott
Marine Forensics Program, National Oceanic and Atmospheric Administration,
National Centers for Coastal Ocean Science
219 Fort Johnson Road, Charleston, South Carolina 29412, USA

David D. Stephens*
Michigan State University, School of Criminal Justice
Forensic Wildlife and Environmental Outreach
655 Auditorium Road (450 Baker Hall), East Lansing, Michigan 48824, USA

Key Points

In order for new scientists who pursue careers in emerging forensic areas to produce legally relevant outcomes, they will need to fully understand

- Their specialized roles and responsibilities.
- The level of vigor needed in evidence examination protocols and documentation.
- The importance of being effective communicators.

Introduction

A career in fisheries can obviously take many different professional paths. As a marine biology graduate student with an eye on a career in fisheries management, I (Trey) chose forensic science. My professional career started when I was finishing my master's thesis, which dealt with genetic species identification of marine fish in the hopes of providing a tool to fishery managers. Then an opportunity arose to take a position as a forensic scientist identifying marine species for the National Oceanic and Atmospheric Administration (NOAA). To me, this applied use of genetic techniques learned in graduate school was an appealing combination of conservation and law enforcement. Looking back now, I realize the value of having kept an open mind when I was just starting my career.

The ensuing 15 years as a forensic scientist have presented a number of unusual challenges and opportunities, some anticipated and some definitely not. Forensic science requires a standard of work that will withstand the strictest scrutiny by a defense counsel with regard not only to the rigor of the scientific work, but also the care taken to maintain control and the integrity of the evidence. It requires creative problem-solving skills to discern how forensics can help solve the puzzles presented by any criminal investigation, as well as an ability to remain acutely aware of, and communicate, the limitations of forensics. It also requires an ability to testify in

* Corresponding author: steph340@msu.edu

court—to translate sometimes very technical and complex scientific issues into terms that are understandable to members of a jury who may lack any background in the sciences, without appearing in any way condescending or patronizing or, when seemingly under attack by defense counsel, defensive. Perhaps above all, forensic science provides opportunities to help protect our environment, particularly the fauna within it, and to interact with law enforcement agents and attorneys who are working toward the very same purpose. The unique combination of challenges and opportunities presented by forensic science will appeal to some future fisheries scientists looking to do something just a little different and very rewarding. This chapter elaborates on some of these challenges with a recent representative case that illustrates several of the lessons learned.

Background

My professional career to date has coincided with a marked growth in global demand for seafood. This increased demand has impacted every link in the seafood supply chain from fishermen to processors, importers, wholesalers, and ultimately consumers. Unfortunately, various seafood substitution schemes have emerged as unscrupulous solutions to meeting demand pressures. If there is an inadequate supply of a particular species of fish, other species of fish can be falsely labeled as the unavailable species to meet market demands. If there is a glut in a cheap species of farmed fish, that species may be falsely labeled as a more expensive fish to increase profits. Such activities have negative effects at many points along the supply chain, starting with the resources themselves and ending with the consumers. The resources are harmed in part because this seafood fraud creates the impression that stocks of the species of label are healthier than they are in reality. Fishermen are harmed when law-abiding domestic fishermen are unable to compete with the falsely labeled product. Consumers are harmed when they pay more than they should for product that is unwanted, lower quality, or even unsafe. Driven by greed, those engaged in seafood false labeling ultimately compromise the integrity of the seafood industry itself and consumer confidence therein.

Have you ever had or overheard a dinner conversation like this?

Suzanne: Gee, my sole tastes kind of funny, dear.
Frank: Well, it certainly looks like a white fish filet like sole.
Suzanne: It's actually mushy instead of flaky, and it has a fairly strong fishy taste that I'm not used to with sole.
Frank: Could it be overcooked? We can send it back if you want…or do you want to try some of my grouper first?

Many countries have responded to the increasing demands with a variety of laws, regulations, and tariffs in an attempt to reduce overfishing, prevent seafood product substitutions, and address fair market issues with import tariffs on specific seafood species. For instance, in 2003, the U.S. Government uncovered that farm-raised Vietnamese catfish (*Pangasius* spp.) were being sold on the U.S. market at less than fair market value, a practice commonly known as "dumping." The United States responded by imposing anti-dumping tariffs of up to 64% on some such imports. To avoid the tariff, some foreign exporters and domestic importers falsely labeled shipments of catfish as other fish species. In many of these cases, the false labeling of farmed catfish as "wild-caught grouper" or "sole" allowed importers not only to avoid tariffs, but also to further increase profits by demanding higher prices than the market would have borne for imported catfish. In some instances, the international coordination involved in these schemes is every bit as sophisticated and complex (and profitable) as those seen in international drug trafficking cases. The international illegal seafood trade is now organized crime just like any other.

Case Example—U.S. v. Virginia Star et al.[1]

In late 2004, a special agent with NOAA requested my assistance to sample containers full of frozen fillets for forensic testing as part of an investigation into imported shipments of seafood. As samples of the shipped product, labeled as "conger pike," "common carp," or "sole," were analyzed with DNA sequencing, it became clear that the product was actually from a catfish species (*Pangasius hypophthalmus*) that is commonly farmed in the Mekong Delta in Vietnam. These preliminary results only scratched the surface of a massive seafood fraud scheme that would occupy much of my work for the next nine years.

The ensuing investigation into this fraud expanded and resulted in extensive work for me in the laboratory. When NOAA special agents seized almost 400,000 pounds of frozen fillets as evidence in the case against Virginia Star et al., the investigation required that I perform further sampling and DNA testing of more than 800 samples to statistically define the species composition of these shipments. From the outset, I had to keep in mind that these genetic analyses would very likely end up being scrutinized in a legal setting.

While marine fisheries cases often result in plea agreements, the Virginia Star case proceeded to trial in U.S. District Court in Los Angeles in 2008. The facts at trial ultimately revealed that the conspiracy involved US$15.5 million worth of falsely labeled fillets that were imported and subsequently sold in domestic interstate commerce. This constituted a felony violation of the Lacey Act, the United States' oldest wildlife protection statute (enacted in 1900), which, in part, makes it illegal to falsely identify any fish that has been, or is intended to be, imported, sold, purchased or received from any foreign country or transported in interstate or foreign commerce. The end result was the criminal conviction of 13 individuals and companies who collectively were sentenced to fines of $444,000, prison terms totaling 75 months, 6 months of home confinement, and 174 months of probation. While these sentences were among the longest imposed by a federal judge concerning the false labeling of seafood, numerous other individuals and companies have since either pled guilty or been convicted of similar violations related to seafood substitution.

Punishment in conservation crime convictions

Readers should consider whether they think the punishments in this case example, and in other similar crimes that involve natural resources, are generally adequate. A common complaint among some natural resource professionals and in wildlife enforcement is that our society views conservation crimes as largely victimless and therefore relegates them, in terms of investigation, prosecution and punishment, to a level of importance below other crimes that have the names and faces of human victims.

[1] United States v. Virginia Star Seafood et al (Central District of California 2:07-CR-0049-PSG), 25 ITR 15188 (2008).

What Distinguishes Forensic Science

The case above highlights how fisheries forensics can play an important role at the confluence of fishery sustainability, market forces, and public safety. The involvement of forensic science from the beginning provided me with an opportunity to employ my expertise and training. It also taught me a number of valuable lessons about the distinctions between forensic and other types of science as the whole process unfolded.

In complex cases like Virginia Star, the potential to be called as a witness was real regardless of whether one was involved as a fishery biologist, toxicologist, forensic scientist, or criminal investigator. Unfortunately, too few professionals are familiar with or are formally trained in expert witness testimony. Consequently, such training can come from the school of rough waters, potentially having a negative impact on some cases and generally in how science is perceived. So how does a scientist prepare for the possibility of their work being needed and scrutinized by the legal system? As a potential forensic scientist or expert witness, what are the landmarks that you should expect to encounter so that you can methodically navigate a proper course when being called upon to testify?

While the vast majority of scientists adhere to high technical and ethical standards, there are a number of critical differences between nonforensic research and the work of a forensic scientist whose analyses are intended for and will be scrutinized within the legal system. For the latter, it has been estimated that the actual scientific analysis only constitutes about 20% of casework time while the remaining 80% centers on components that are unique to forensic situations. Many of these components—casework confidentiality, evidence handling, chain of custody, organized and standardized sample testing methodologies, robust quality control systems, proper documentation and recordkeeping, and an objective review process prior to reporting any findings—may not be routine in a research atmosphere, but adherence to them is critical for forensic professionals. Also, linkage of these components to the professional responsibilities of an individual scientist has been further underscored by a number of recent court decisions (see Crawford[2] and Melendez-Diaz[3]) that have emphasized the U.S. Constitution's Sixth Amendment right of defendants to directly confront witnesses; this means that you personally (not a surrogate, supervisor, or appointee) can be called to court to explain and defend your own work.

> For more information regarding general standardized requirements for forensic testing laboratories, see International Standard 17025, which can be found at www.ISO.org. The Scientific Working Group for Wildlife Forensics has also released standards and guidelines specific to the wildlife forensic discipline (www.wildlifeforensicscience.org/swgwild/).

Advice for New Professionals

Whether you, as an emerging science professional and a potential witness, bring truth and clarity to the criminal justice process is almost always determined by a number of factors that you can directly anticipate and influence. These factors, sometimes referred to as Daubert

[2] Crawford v. Washington, 541 U.S. 36 (2004).
[3] Melendez-Diaz v. Massachusetts, 557 U.S. 305 (2009).

criteria, specifically address how a proper foundation for the use of science must be built prior to its employ in investigations, how science is specifically applied to evaluate evidence during an investigation, and how the forensic findings are presented in court and other legal settings. Additionally, during court preparation, the forensic scientist will need to consider the limitations of their findings and whether there is more than one explanation for a given finding. If your work included testing or analysis, make sure to review the sensitivity and specificity of the instruments and the testing protocols. Such testing limitations can be found in instrument manufacturer manuals, in validation studies, and in references and journals when analytical protocols are published or shared. These limitations often become the focus of scientific scrutiny at trial, so it is critical to have considered those possibilities in advance of testimony.

Daubert criteria

Specific rule of evidence requirements related to the validity of expert witness testimony that was established by the U.S. Supreme Court as a result of three legal proceedings in the 1990s.[4,5,6] When considered within the context of a scientific methodology, the court stipulated that the relevancy factors should include consideration of empirical testing, peer review and publication, error rates, test standards and controls, and general acceptance in the relevant scientific community.

Despite significant resource expenditures, not all investigations will result in formal charges being brought, and most formal charges will not result in a trial; however, this reality should never diminish the due diligence and attention to detail that a scientist needs to systematically dedicate to each and every case worked. It is necessary to treat every evidence item or sample as though it may eventually become a court exhibit. The integrity of the evidence is central to forensic casework and makes it essential that secure storage and proper chain of custody procedures are employed. Since your court testimony may not be required for several months or even years after your first involvement, documentation should be detailed to such a level that your notes, observations, data, photographs, scans, graphs, printouts, charts, and other outputs are adequate to enable a competent third party to review and/or repeat your analysis, as well as to refresh your memory if a subpoena does eventually arrive. Since DNA analysis was required for numerous submissions from various agents totaling more than 800 samples in the Virginia Star case, laboratory organization and record keeping were obviously critical. These steps are among the first required of the forensic scientist when handling evidence and may seem elementary, but they will likely be critical in building the legal foundation that provides for the admission of subsequent laboratory results in court.

As investigative time and expense in the Virginia Star case continued from 2004 to 2008, the corresponding work and preparation required of the scientific witnesses and attorneys resulted in the construction of a legal foundation worthy of a trial. Discovery also occurred contemporaneously.

[4] Daubert v. Merrell Dow Pharmaceuticals (92-102), 509 U.S. 579 (1993).
[5] General Electric v. Joiner, 522 U.S. 136 (1997).
[6] Kumho Tire Co. v. Carmichael, 526 U.S. 137 (1999).

Discovery is when the forensic scientist provides case attorneys with everything in the case file, which details all records of work that has been accomplished. Here, "everything" generally means all documentation, such as written notes, observations, data, photographs, scans, graphs, printouts, and charts, but it can also include phone and e-mail logs, instrument maintenance and calibration records, shipping records, quality control records, and anything that is specified in a discovery request (discovery can be voluntary or it can be elicited by subpoena).

This level of scrutiny can sometimes leave the emerging scientist feeling vulnerable, but this will likely pass with more experience and with a better understanding of what is expected. Being aware of the scrutiny beforehand reiterates the importance of performing forensic analysis in an objective and transparent manner from the beginning.

While forensic scientists examine evidence, the scientists themselves are also occasionally examined, in a sense, prior to testifying in the form of background checks. Criminal background checks and searches of personnel records may be required early in the legal process when a scientist has been identified as a potential witness in a criminal trial. These steps are intended to reveal, in advance, any potential conflicts or reasons to question the integrity of the expert witness. Prior to specific testimony about case issues, a potential expert witness is subjected in court to the voir dire process, where at least one of the attorneys will ask you about your specific education, training, skills, experience, and abilities. This process ultimately establishes whether or not you are qualified to offer an expert opinion to the court.

Expert witness

In a general legal sense, an expert witness is an individual that, due to his or her specialized knowledge, training, skills, and/or experience, is allowed to give opinion testimony in a legal proceeding.

As an expert witness, your overarching goal will be to communicate effectively with the jurors as the triers of fact so that they fully understand the importance and relevance of your testimony. Keep in mind that jury pools can vary widely in their ranges of education, age, and backgrounds, so it is wise to address all jurors as if they are largely unfamiliar with the topic of your testimony. As a rule of thumb, it is a good general perspective to consider that the average juror has completed a high school education and would therefore have about the same level of scientific understanding. This basically means that from the start you need to keep it simple, and it is best to adapt your choice of words when testifying to accommodate the locale or juror differences when you recognize them.

One time early in my forensic career, I (David) was subpoenaed to testify in a rural area on a criminal case. The judge asked me point blank if I knew what an expert witness was. I fumbled a bit and basically responded in the negative. "An expert witness," he said with a straight face, "is someone from out of town that wears suspenders and a tie." It was at that point that I realized that being declared an expert witness by a judge is, in part, a matter of perspective.

Connecting well within the legal system also means that you are confident in knowing that you have employed science properly in your analysis. Overall, effective testimony will occur when you remain objective, limit your testimony to the science at hand, prepare properly, and have at least a basic familiarity with the legal process. It is also critical to not be defensive with attorneys who may seemingly or directly question your work and credibility. As science professionals, we all have a certain amount of confidence and pride in our education, our area of expertise, our experience, and our accomplishments, so it may be natural for you to become defensive in response to such a legal tactic. Consider, however, that with our adversarial legal system, it is the implicit responsibility of attorneys to thoroughly question and sometimes attempt to cast doubt on a witness. If you remain calm and simply answer the questions truthfully, your work will likely be well accepted during the questioning process.

If you expect to testify in additional trials in the future, you can certainly develop a plan to become a better witness. Shortly after you testify, critique yourself and consider any shortcomings and strengths—you will likely identify ways to improve your preparation or approach for future cases. You can do this by reconstructing your testimony (if you have a good memory), having a peer that was present give you feedback, or asking one of the attorneys for a copy of your transcript. After the trial has concluded, another route is to ask one of the attorneys (or the judge, if willing) how you did and, specifically, how you can improve as a witness. When used properly, feedback should reposition your next approach to testimony so that you grow more prepared, confident, and comfortable in the expert witness role.

The True Role of the Expert

At the end of the day, your effectiveness and value as a forensic scientist and expert witness should not be defined by the verdict, be it guilty or innocent. Since you were likely not present for the entire trial, did not hear all of the testimony, and did not see all of the evidence that was presented, you cannot and should not form your own personal verdict based upon the limited information that you may have. If you performed sound scientific analyses without bias, and presented your findings in a clear and understandable manner, then you will have conducted your role quite well; you will have greatly assisted the triers of fact in forming their verdict. Always remember that, in the end, science is not science if it is not objective, just as catfish is not sole or grouper.

Summary

As the world becomes smaller but more complex and the pressures on fisheries more pronounced, the incentives for the unscrupulous to commit fisheries crimes from false labeling to illegal fishing will only increase. As responsible resource stewards and consumers, we will likely move toward the expansion of the forensic sciences into new investigative, monitoring, and conservation roles. Such an expansion will create a need for smart, dedicated, creative, and articulate forensic scientists. Future professionals drawn to this niche by its unique challenges and opportunities will need to be technically savvy, there is no doubt, but they will also need to develop and utilize important organizational and presentation skills so that the science is effectively understood in court and in the public arena. Emerging forensic scientists can gain these skills through study and observation, but they can also be gleaned from peers, by initiating communication with legal stakeholders, and through reflection as they evolve as scientific experts.

Acknowledgments

With our thanks, the authors would like to acknowledge National Oceanic and Atmospheric Administration; Law Enforcement Division, Michigan Department of Natural Resources; School of Criminal Justice, College of Social Science, Michigan State University; College of Agriculture and Natural Resources, Michigan State University; U.S. Department of Justice; and our chapter review peer group.

Biographies

Trey Knott is a forensic scientist with the National Oceanic and Atmospheric Administration's NOS/NCCOS Marine Forensics Laboratory in Charleston, South Carolina, where he has worked to identify numerous marine species for law enforcement since 1997. He now focuses primarily on finfish species with an emphasis on seafood fraud and has testified in U.S. District Court on multiple occasions. He currently serves on a number of scientific working groups involved in strengthening the practice of wildlife forensics and expanding its role both domestically and abroad.

David Stephens is an executive in residence at Michigan State University, directing forensic wildlife and environmental outreach within the School of Criminal Justice, College of Social Science, in East Lansing, Michigan. He has served the forensic sciences for more than 25 years as a state police forensic scientist, supervisor, forensic laboratory director, accreditation inspector and assessor, consultant, trainer, and educator. He has also assisted with several international forensic initiatives in the Americas, Europe, the Middle East, and Africa and has testified as an expert witness in several forensic areas in courts on more than 200 occasions.

Disclaimer

In no way should this chapter be construed as representing the positions, opinions, viewpoints, or recommendations, official or informal, of the aforementioned governmental agencies, institutions, organizations, groups, or individuals other than the authors, as this chapter solely reflects the perspectives and opinions of the authors.

Fisheries as Coupled Human and Natural Systems

Abigail J. Lynch* and Jianguo Liu
Center for Systems Integration and Sustainability
Department of Fisheries and Wildlife, Michigan State University
115 Manly Miles Building, East Lansing, Michigan 48823, USA

Key Points

- Fisheries are complex, coupled human and natural systems (CHANS);
- The CHANS approach is useful for fisheries because it explicitly integrates the human and natural systems; and
- Fisheries professionals should consider fisheries as CHANS to effectively study and manage them in a holistic manner.

Introduction

Consider the old saying "two heads are better than one." If those two heads come from two different disciplines, one specializing in human systems and the other specializing in natural systems, they will approach an issue differently, adding depth and breadth to collaborative research and management. However, this collaborative approach is not universally embraced; humans and natural systems are often researched and managed separately.

Fisheries, by definition, couple humans and natural systems. Fisheries are organized efforts that use aquatic resources for human consumption, recreation, and trade. Such human activities influence the natural environment from which these aquatic resources are derived (e.g., bottom trawling impacts benthic habitats, harvesting changes food webs). Changes in the natural environment, in turn, affect humans. Many fish populations have collapsed due to overexploitation and habitat degradation; consequently, many fishermen and related industries have suffered economic, social, and cultural losses. But, fisheries research has often studied the natural systems separately from the human systems.

Although disciplinary fisheries studies have provided many useful insights, for example, into fish biology, ecology, nutrition, and toxicology, we contest that (1) fisheries are complex coupled human and natural systems (CHANS); (2) the CHANS approach is useful for fisheries because it explicitly integrates the human and natural systems into a research framework; and (3) fisheries professionals should consider fisheries as CHANS to effectively study and manage fish and human behavior in a holistic manner. To address these issues, we start with a brief introduction to the CHANS approach, discuss the complexity of fisheries as CHANS using the Lake Whitefish *Coregonus clupeaformis* fishery in the Laurentian Great Lakes as an example, and conclude with how the CHANS approach can support fisheries sustainability and emerging fisheries professionals.

* Corresponding author: ajlync@gmail.com

Coupled Human and Natural Systems Approach

Coupled human and natural systems are integrated systems in which humans interact with natural systems (see Liu et al. 2007a, 2007b for more information). The core of the concept is reciprocal interactions (feedbacks); humans affect natural systems and changes in natural systems in turn affect people. Another key aspect is that CHANS are open systems; they interact with each other across spatial, temporal, and organizational scales. For example, across spatial scales, they interact with other CHANS not only locally (local coupling), but also over distances (telecoupling—socioeconomic and environmental interactions over distances; see Liu et al. 2013 for more information). While such phenomena are not new and have been observed many times around the world, they are rarely systematically and explicitly incorporated into natural resource policy until negative consequences become obvious.

To avoid negative consequences in the first place and create solutions to minimize negative impacts, my collaborators and I (J. Liu) proposed the explicit concepts of CHANS and telecoupling. These were built on my personal experiences. During 1979–1983, I was a college student and assigned to major in plant protection, which emphasized the use of pesticides and herbicides in controlling insect pests and plant diseases to reduce crop damage. Feedbacks later became obvious—natural ecosystems were damaged (e.g., natural enemies and pollinators were killed) and human health was compromised as residuals of pesticides and herbicides were harmful to humans. Such realization led me to choose work with the famous ecologist Shijun Ma at the Chinese Academy of Sciences, who successfully developed a systems approach to control the devastating locusts through managing locust habitat instead of applying pesticides. For my master's thesis, I discovered that the forest belt adjacent to croplands provided an excellent refuge for natural enemies of aphids (a severe insect pest on cotton) in the winter so that no pesticides in the cropland were needed within approximately 100 meters from the forest belt. The natural enemies overwintered in the forest belt and moved to the cotton field in the spring to eat aphids and moved back to the forests after the crops were harvested in the fall. For my Ph.D. study at the University of Georgia, under the guidance of a seven-member committee (including H. Ronald Pulliam and Eugene P. Odum), I developed an ecological-economic model to balance timber production and conservation of endangered species in forested landscapes. For my postdoctoral study at Harvard University, I built a spatially explicit and individual-based model to evaluate long-term dynamics of tropical forests under different management scenarios across heterogeneous landscapes. After joining the Michigan State University faculty in 1995, I have conducted a variety of projects with collaborators and students. These different experiences helped to further develop the CHANS approach and found the Center for Systems Integration and Sustainability, which aims to integrate various natural and social science disciplines, issues, sectors, and scales (spatial, temporal, and organizational) to understand and promote sustainability.

Telecoupling also plays important roles in resource management. Distant interactions have been studied in many fields. For instance, climate scientists have studied teleconnection (climate change in one place can influence climates hundreds or even thousands of miles away) while social scientists have studied globalization among human systems. While these separate studies have provided useful insights, they are largely on either socioeconomic *or* environmental interactions. In contrast, the telecoupling concept focuses on socioeconomic *and* environmental interactions among coupled systems over distances. Similar to the concept of ecosystem services that includes many types of ecosystem benefits to humans, the concept of telecoupling is also an umbrella concept and encompasses many types of telecoupling processes (distant exchanges of information, energy, and matter). They can have profound impacts on coupled systems far away, including those in remote areas. For example, in 2001, my collaborators and I (Liu) found that in protected areas in southwestern China for the world-renowned, endangered giant pandas, activities of local residents (e.g., fuelwood collection and timber harvesting) led to loss and fragmentation of panda habitat. The results were published in *Science* magazine and disseminated through news media worldwide. Information dissemination to distant places and financial support from China's central government in Beijing (payments for ecosystem services) are example telecoupling processes. The financial resources from the government to local residents to protect forests from harvesting constitute feedbacks to help panda habitat recover. Such findings demonstrate that drawing boundaries around a protected area is not enough and telecoupling processes (e.g., external financial support) are critical for panda habitat recovery and for improving the well-being of local residents. So far, concepts of coupled human and natural systems, as well as telecoupling, have been applied to many disciplines to address many important issues, such as trade of food and forest products, species invasion, water transfer, ecosystem services, geography, ecology, and land-change science.

Fisheries research and management can also benefit from the CHANS approach because the key points of coupled systems and telecoupling are *implicit* in traditional fisheries research, but they often are not *explicitly* integrated together as they are in the CHANS approach. As coupled human and natural systems, fisheries exhibit many CHANS characteristics (see Box 1) and fisheries research does consider many of the points of the CHANS approach. Economists and ecologists, for example, have worked together to develop fish harvest models that include fish population models and economic behavior models, stock assessment modelers have considered nonlinear dynamics and surprises in fish production modeling, and sociologists have investigated fisheries as an important source of livelihoods, especially in developing countries. While these separate fisheries studies are useful, they usually are not as integrative as research with the CHANS approach. For example, livelihood studies often focus on fisheries as a source of livelihoods, but the feedbacks from humans to fisheries have received little attention simultaneously. On the other hand, many fish population studies have considered effects of human activities (e.g., harvesting) on fish mortality and age structure but have not evaluated the impacts of fish on livelihoods.

The CHANS approach provides a useful framework for integration of various aspects of fisheries research. It does not seek to replace traditional fisheries methods. Rather, it suggests that linking different traditional fisheries approaches together in a more integrated manner will advance holistic understanding of fisheries. It is a different way of thinking about fisheries as coupled human and natural systems. By explicitly considering reciprocal effects, cross-scale interactions, and telecoupled effects, CHANS research adds new dimensions to the more traditional fisheries methods and provides new perspectives on the interactions within and between fisheries in different places.

CHANS Approach for Fisheries

The ties between Lake Whitefish, their habitat, and the human populations that harvest them nicely illustrate fisheries as coupled systems. Lake Whitefish is an ecologically, culturally, and economically important Great Lakes species. They are benthivores and transfer energy from one benthic food web to other benthic food webs, pelagic communities, and humans. These fish are and have been a staple food source and focus of livelihoods for aboriginal people in

Box 1. Characteristics of a Coupled Human and Natural System

Reciprocal effects and feedbacks: A reciprocal effect (feedback) is when one aspect of a coupled human and natural system (CHANS) impacts another, which results in an impact on itself. Selective fishing pressure for larger, later maturing fish, for instance, can lead to natural selection for smaller, faster maturing fish. This, in turn, may lead to the necessity of larger harvests to maintain profits in the face of dwindling fish populations.

Nonlinearity and thresholds: CHANS relationships are often nonlinear and include threshold effects. The relationship between water temperature and fish productivity, for example, may be nonlinear due to fish metabolic and physiological requirements, but lethal temperature limits are the ultimate threshold for survival.

Surprises: Surprises and unanticipated consequences are common in CHANS interactions. An aquatic invasive species, for instance, may enter a system unexpectedly and seriously modify a fish community and its dependent human systems.

Legacy effects and time lags: Because a CHANS is dependent upon life cycles of organisms, legacy effects and time lags frequently occur. In fisheries, a delayed collapse of a fishery is a common time lag because recruitment to the fishery is not instantaneous. A legacy effect can result if the harvest pressure is reduced or eliminated but the fishery system does not return to its former state.

Resilience: Resilience for a CHANS is the ability to retain similar structures and functions after a perturbation (e.g., fishing, climate change). Fisheries systems that have high variability/capacity for productivity and/or redundant/versatile community connections often have a high tolerance to environmental change.

Heterogeneity: A CHANS exhibits heterogeneity by spanning spatial, temporal, and organizational scales (e.g., management units). Fisheries cross different scales in the form of fish movement, political boundaries, and management regulations. Fisheries affect and are affected by distant systems, such as through telecoupled supply chains.

the region for thousands of years. Today, populations of Lake Whitefish support the most robust and economically valuable commercial fishery in the upper Laurentian Great Lakes (Lakes Huron, Michigan, and Superior), in large part because the fishery and fishery managers now recognize that the fishery demonstrates many CHANS characteristics.

The fishery was not always managed with consideration of the complex interactions between Lake Whitefish and their harvesters. The collapse of the Lake Whitefish populations in the mid-1900s, for example, was a result of overfishing, habitat degradation, and invasive species. But, there was a time lag before the impact on recruitment to the fishery was recognized because Lake Whitefish take at least three years to mature and enter the fishery. With the extirpation of highly productive stocks, the fishery also experienced legacy effects, even once conditions improved.

Similarly, when the Lake Whitefish fishery removed the larger, later maturing individuals, there was a selective advantage for the smaller, quicker maturing individuals. But these smaller fish were not as valuable and produced fewer eggs, resulting in the need to harvest more fish to maintain profits that furthered the trend towards natural selection of smaller fish. Having an awareness of these reciprocal effects and feedbacks between the humans and natural systems in this fishery now helps the fishers, managers, fish processers, distributers, buyers, and consumers prepare for changes and potentially avert collapse of this important fishery due to human-induced feedbacks to the system.

As shown through the Lake Whitefish example, we believe the CHANS approach is particularly useful for fisheries science because of its integrated framework for understanding complexity of natural and human systems. Though not every CHANS will exhibit each complex characteristic of CHANS (see Box 1), the CHANS approach is more integrated than traditional approaches. It can facilitate the linkages among various projects that only address some aspects of CHANS. One major advantage of using the CHANS approach to fisheries sustainability is linking scales and disciplines. CHANS research should not only target complex human–nature interactions within a CHANS in a specific location, but also across different CHANS at distant locations, or telecoupling. CHANS research can likewise integrate factors that traditionally would be external to the system, such as linking ecology, economics, sociology, and other relevant disciplines to provide a more holistic assessment of fisheries.

Studying fisheries using the CHANS approach, however, poses new challenges as it requires systems integration—incorporation of multiple disciplines in natural and social sciences, as well as advanced technologies. CHANS research is more time-consuming than disciplinary research because it requires coordination among researchers from multiple disciplines such as ecology, sociology, economics, demography, and human behavior. However, these challenges are also opportunities. Systems integration can help better understand human–nature interactions in a holistic framework and provide constructive interpretations of the complex relationships. As shown in the Lake Whitefish example, applying the CHANS framework to fisheries research can advance fisheries science by understanding the complex relationships between humans (including fishermen and relevant industries), fish populations, and fish habitat.

CHANS Approach for Fisheries Professionals

Using CHANS methods requires the integration of many natural and social science disciplines, gaining skills to communicate across disciplines, and learning to engage in different disciplinary cultures and contexts. This does not mean being a jack-of-all-trades, but rather having a solid interdisciplinary foundation to know when and where to seek collaboration from other fields. For students in fisheries science, including courses in social sciences, complex systems, systems thinking, and systems modeling in their curriculum will help build and expand this essential knowledge base. Particularly choosing courses that foster teams consisting of both natural and social scientists and conducting thesis or dissertation research from the CHANS perspective will help practice and master interdisciplinary research skills. Courses on complex systems and systems modeling will assist in understanding the complexity of CHANS.

For professional fisheries scientists, it is crucial to reach out and collaborate with social scientists to understand the complexity of human–nature interactions. This will take time and effort to do so. Consider attending meetings and lectures in other fields or seeking out individuals interested in conducting transdisciplinary research. Establishing and maintaining professional connections with colleagues in other disciplines will facilitate collaboration on new research or enhancement of current research.

We encourage more fisheries scientists to use the CHANS approach in their research and become involved in the International Network of Research on Coupled Human and Natural Systems (CHANS-Net.org, which Liu leads). For how integral people are to fisheries, the number of fisheries researchers on CHANS-Net is astonishingly low. CHANS-Net facilitates communication and collaboration among people who are interested in CHANS research and also nurtures a new generation of scholars through the CHANS Fellows Program. Research on CHANS, by nature, should span disciplines and should be conducted in interdisciplinary teams consisting of scientists from different disciplines and stakeholders from relevant sectors. Consequently, communication and networking are integral to introducing potential collaborators. Research and management of fisheries sustainability from the CHANS perspective can expand the aquatic side of CHANS-Net and increase the number and participation of fisheries scientists and other stakeholders in CHANS-Net.

From personal experience through the CHANS Fellows Program (A. Lynch is a former CHANS Fellow) and CHANS research projects, we have had the opportunity to engage with prominent CHANS scientists and develop interdisciplinary professional networks, which will help strengthen our future research. CHANS researchers are trained to see research questions in a holistic manner. Different disciplines address similar problems, like climate change, overfishing, or invasive species. CHANS research takes advantage of that overlap and utilizes collaboration across these disciplinary lines to benefit from the strengths of all the involved disciplines.

When students and professionals enter a workplace (e.g., state and federal management agency), they may encounter barriers in implementing the CHANS approach in research and management, as many agencies and universities are still disciplinary and use silo methods. Convincing senior personnel, especially those in the leadership positions, to take the CHANS approach will reduce the barriers. Also, forming mentor–mentee relationships and other support groups within and across workplaces will be more powerful than individual efforts in overcoming the obstacles. If researchers who embrace CHANS methods are discounted, they can gain legitimacy by securing funding such as National Science Foundation (NSF) funding designated specially for CHANS research (e.g., NSF Program on Dynamics of Coupled Natural and Human Systems). Accordingly, the institutional support for CHANS from NSF and other organizations (e.g., NASA's Land-Use and Land-Cover Change Program) is critical to sustain the CHANS approach.

Concluding Remarks

With the CHANS approach, researchers can examine complexity of fisheries in an integrated framework that is more appropriate—and realistic—than traditional disciplinary work. Although it requires interdisciplinary expertise and is more challenging than disciplinary research, using CHANS methods to study fisheries presents new opportunities to incorporate multi-scale processes; value all aspects, including nontraditional aspects, of fisheries; and expand the aquatic side of CHANS-Net. Fisheries research from the CHANS perspective can improve understanding of the relationships between global change and fisheries and provide crucial information and useful guidance for sustainably managing these important resources in a changing world.

The depth and breadth of CHANS research has changed significantly since I (Liu) began my career. Each passing year brings more integrated projects and continued recognition of the value of CHANS research. The world has gained much in understanding complex human–nature interactions. For emerging professionals, thinking beyond the disciplinary silo is increasingly becoming second nature. We are excited to see the advancement of CHANS research to improve the ability to address complex fisheries problems, now and in the future.

Acknowledgments

We thank Bill Taylor, Nancy Léonard, and three anonymous reviewers for their insightful suggestions to strengthen this vignette; the Center for Systems Integration and Sustainability for their support; and the International Network of Research on Coupled Human and Natural Systems (CHANS-Net.org) for the opportunity to interact with prominent CHANS scientists and build professional networks through the CHANS Fellowship Program.

Biographies

Abigail J. Lynch is a research fisheries biologist at the U.S. Geological Survey's National Climate Change and Wildlife Science Center. She is a recent graduate from Michigan State University with a Ph.D. in fisheries and wildlife; a secondary degree in ecology, evolutionary biology, and behavior; a doctoral specialization in environmental science and policy; and a College of Agriculture and Natural Resources certificate in college teaching. Her dissertation designed a decision-support tool to assist Great Lakes fisheries professionals and fishermen prepare for the impacts of climate change on Lake Whitefish production. Her research interests include marine and freshwater fish conservation with a management focus on fisheries systems.

Jianguo "Jack" Liu is the Rachel Carson Chair in Sustainability, a university distinguished professor of fisheries and wildlife, and director of the Center for Systems Integration and Sustainability at Michigan State University (MSU). Jack came to MSU after completing his postdoctoral work at Harvard University. He also has been a guest professor at the Chinese Academy of Sciences and a visiting scholar at Stanford (2001–2002), Harvard (2008), and Princeton (2009).+ Jack takes a holistic approach to addressing complex human–environmental challenges through systems integration, which means he integrates multiple disciplines such as ecology and social sciences. His work has been published in journals such as *Nature* and *Science* and has been widely covered by the international news media.

References

Liu, J. G., T. Dietz, S. R. Carpenter, M. Alberti, C. Folke, E. Moran, A. N. Pell, P. Deadman, T. Kratz, J. Lubchenco, E. Ostrom, Z. Ouyang, W. Provencher, C. L. Redman, S. H. Schneider, W. W. Taylor. 2007a. Complexity of coupled human and natural systems. Science 317:1513–1516.

Liu, J. G., T. Dietz, S. R. Carpenter, C. Folke, M. Alberti, C. L. Redman, S. H. Schneider, E. Ostrom, A. N. Pell, J. Lubchenco, W. W. Taylor, Z. Ouyang, P. Deadman, T. Kratz, W. Provencher. 2007b. Coupled human and natural systems. Ambio 36:639–649.

Liu, J. G., V. Hull, M. Batistella, R. DeFries, T. Dietz, F. Feng, T. W. Hertel, R. C. Izaurralde, E. F. Lambin, S. Li, L. Martinelli, W. J. McConnell, E. Moran, R. Naylor, Z. Ouyang, K. Polenske, A. Reenberg, G. de Miranda Rocha, C. S. Simmons, P. H. Verburg, P. M. Vitousek, F. Zhang, C. Zhu. 2013. Framing sustainability in a telecoupled world. Ecology and Society 18(2):26.

Aquaculture in the 21st Century: Opportunities for the Emerging Professional

John R. MacMillan*
Clear Springs Foods, Inc.
Post Office Box 712, Buhl, Idaho 83316, USA

Eric A. MacMillan
Earth Resources Technology, NOAA Restoration Center
1315 East West Highway, Silver Spring, Maryland 20910, USA

Key Points

- Aquaculture is here to stay, and production will continue to increase, displacing wild capture fisheries as the world's primary seafood source.
- Aquaculture can be done in an environmentally sustainable way that enhances economic viability.
- The knowledge and skills fisheries professionals have can assist the domestic and international aquaculture industry and governments to further develop and encourage fish farming practices that lead to both environmental sustainability and long-term profitability.

Introduction

You have heard the rumors—fish farming (i.e., aquaculture) is bad for the environment and causes ecological devastation. In some cases, this may be true, but in most cases aquaculture is a form of agriculture that can help feed an ever increasing world population, alleviate malnutrition, improve prospects for wild fish stock sustainability, enhance recreational fisheries, provide sustainable supplies of ornamental species, and have limited negative environmental impacts. Commercial aquaculture can also be profitable, which is essential if commercial fish farming itself is to be viable. Aquaculture is a significant part of the world economy and is here to stay.

We believe aquaculturists (public and commercial) and fishery professionals have much in common and share the same vision, that is, ensuring sustainable fisheries and a healthy aquatic environment. However, past poor aquaculture practices in some parts of the world have led to misperceptions about the current and prospective sustainability of aquaculture. While many of these past practices have now been replaced with more responsible and informed environmentally friendly practices, the negative perceptions persist. In this chapter, we will briefly identify some of the key environmental sustainability challenges confronting aquaculture, why environmentally sustainable aquaculture

* Corresponding author: randy.macmillan@clearsprings.com

is likely to be more profitable than alternative forms of aquaculture, how aquaculture can be part of sustainable fisheries management, and how the emerging fishery professional might consider focusing their career on one of the many scientific, policy, or regulatory facets of aquaculture. We believe emerging fishery professionals can help improve the prospects for aquaculture's long-term environmental sustainability and, in so doing, help feed the world.

Feeding the World

For a variety of reasons, there is a growing demand for animal sourced proteins, including those from aquaculture. It is estimated that the world's current population of seven billion people will increase to as much as nine billion by the year 2050. The Food and Agriculture Organization of the United Nations (FAO) estimates that global food production will need to double in order to meet the ensuing demand. Seafood is currently an important element of the global food supply and helps provide food security. For some people (more than three billion), seafood is their primary source of animal protein. Demand is also rising because of an increasing standard of living and higher economic prosperity in many developing countries. Additionally, more and more people now understand the many health benefits of eating fish, leading to even higher demand for seafood products. It is unlikely that global capture fisheries production will substantially increase to meet the ensuing demand because, despite our profession's best efforts, many wild fish populations are fully harvested, overharvested, or declining. Aquaculture stands out as a critical method to increase seafood supply while, at the same time, alleviating some of the harvest pressure on wild fish stocks. The FAO recently estimated that at some point this year (2013), global consumption of farm-raised fish will surpass that of wild-harvested fish. This trend will continue. By 2030, it is anticipated that fish farms will produce more than 65% of the world's total seafood supply. This increase in production will come from the geographic expansion of aquaculture and increases in the number of species that can be farm-raised, but perhaps more importantly, from the development and application of improved, cost-effective technologies that increase output per farm, reduce losses from disease, and foster environmental sustainability.

Meeting this demand will require the input and collaborative efforts of many different fishery professionals, including those specializing in aquaculture. Our profession itself is quite diverse—one need only examine the variety of scientific and professional subdivisions within the field of fisheries science and management to appreciate that diversity. We believe our professional diversity creates significant opportunity for collaborative efforts among professionals with a shared vision and the ability to marshal their scientific knowledge and policy expertise for a common good. This is one area where emerging fishery professionals can contribute. Arm yourself with the latest scientific understanding and insights through research endeavors of your own, formal educational experiences and by utilizing the knowledge and insight of your peers. The scientific fields of nutrition, health management and biosecurity, reproduction, selective breeding, stress physiology, aquaculture engineering and waste control, and business management and marketing stand out as areas where meaningful contributions are needed. Combine this knowledge with the desire to contribute to aquaculture's success, and the emerging fishery professional has the opportunity to work with a receptive aquaculture community seeking to continually improve its environmental sustainability and economic vitality and help feed the world's growing population.

Environmental Challenges

Most comparative environmental impact analyses (e.g., life cycle assessments) suggest that aquaculture systems outperform terrestrial animal production and large-scale capture fisheries in terms of their environmental sustainability. Compared to terrestrial livestock such as cattle and poultry, fish convert a greater proportion of the food they eat into body mass. Environmental demands per unit mass, or proteins produced, are also lower for aquaculture-based versus terrestrial-based animal production. Additionally, terrestrial animal production places a relatively higher demand on land use for pastures or to grow crops for feed ingredients, often resulting in deforestation, land degradation, and considerable water consumption for irrigation. Intensive terrestrial livestock production is also noteworthy for the relatively high levels of nitrogen, phosphorus, carbon dioxide, and methane produced. Finally, the carbon footprint for aquaculture is significantly less than that of most terrestrial animal agriculture. Concerns with bycatch, habitat destruction, overharvest, and energy use challenge capture fisheries. With that said, aquaculture is not perfect. It does have environmental challenges that must be resolved.

Until recently, much of aquaculture's scientific inquiry was directed at developing and enhancing methods to farm fish and shellfish. Inducing these nondomesticated animals to reproduce in captivity and then grow was a scientific and production challenge. Once this was accomplished for some key species (e.g., shrimp and salmon), the aquaculture industry grew rapidly, oftentimes not stopping to consider the potential environmental impacts of increased production. For example, in the 1980s, frenetic and unregulated black tiger shrimp *Panaeous monodon* farm development in parts of Asia resulted in significant environmental degradation, including the destruction of mangrove habitat and land subsidence from unmitigated groundwater extraction. There was also a general lack of biosecurity, and as a result, nonendemic pathogens were introduced resulting in extensive mortality from disease. The end result, in the early 1990s, the black tiger shrimp industry in Asia collapsed. Similarly, early salmon farms and other marine offshore net pens were sited in locations that caused significant environmental degradation. Had more fisheries and associated professionals collaborated on these issues earlier in the process, perhaps most mistakes could have been prevented, the industry would have prospered, and some of the negative perceptions about aquaculture could have been avoided.

Because of these past experiences, many of the environmental concerns associated with aquaculture are now well understood and some of the areas in need of continued scientific inquiry identified. For example, much of aquaculture depends on capture of a wild, yet limited supply, of small, pelagic, lower trophic level fish species to make fish meal for fish feed. The use of fish meal is particularly important for carnivorous aquaculture species like salmonids and shrimp, but omnivores like carp can use fish meal as well. Feed for the early life stages of carp rely in part on fish meal, and because there is an expansive global carp production industry, fish meal consumption in total can be considerable. Over the past six years, the limited global fish meal supply, as well as an expanding global demand (for aquaculture, swine, and poultry feeds), has caused fish meal costs to increase by more than 300%. In many cases, this can put added stress on already overharvested wild fish populations, but at the same time, it has encouraged fish farmers to seek protein alternatives and better feed use efficiencies. Fish farms can also discharge significant amounts of phosphorus, nitrogen, waste feed, and manure, sometimes resulting in eutrophication and disruption of benthic ecosystems. Other concerns result from escaped farmed fish leading to invasive species issues (e.g., Asian carp in the United States), marine mammal interactions, exotic pathogen introductions, drug discharge, and lack of governmental oversight in developing countries. It is estimated that by 2050, in the absence of significant innovations and improvements in aquaculture management and technology, continued expansion of aquaculture could result in environmental demands (e.g., water use and waste discharge) at least two times greater than current levels.

Technological Innovation

The environmental challenges outlined above are significant but not insurmountable. Fortunately, progress is being made to address all of them, and with continued scientific research and collaborative efforts, we are confident that we can improve the environmental sustainability of the aquaculture industry. For example, overcoming feed ingredient constraints is key to continued global aquaculture development. In most finfish aquaculture, feed accounts for at least 50% of production costs. Over the past 20 years, feed conversion ratios have steadily improved such that today, depending on the farmed fish species, a conversion ratio very close to 1:1 is common. Such success was driven by in-depth investigation of species-specific nutrient requirements, coupled with improvements in feed manufacture and husbandry practices. The farmed salmon industry, for example, has reduced its reliance on fish meal by more than 50% using a combination of soy products, fishery byproducts (e.g., trimmings from fish used for human consumption), and alternative animal proteins. Current research focuses on identifying various additional plant-based proteins that would be suitable for aquaculture feed. This could significantly reduce the industry's reliance on fish meal originating from wild harvested species and help moderate production costs. Developing cost-effective protein alternatives will significantly diminish the need to produce fish meal for fish feed.

In addition to their effect on dietary conversion ratios, feed and husbandry practices also have a significant impact on eutrophication potential. Discharge of phosphorous is problematic because it can stimulate aquatic plant growth. In the Idaho farmed Rainbow Trout *Oncorhynchus mykiss* industry, a 40% reduction in phosphorus discharge has been achieved through conversion to better feeds (i.e., high energy, nutrient dense), better selection of feed ingredients with less total but more biologically available phosphorus (to the trout), and better waste (solids) capture and removal technology. This creates a better rearing environment, which consequently leads to better fish production. Better feed conversion also results in less feed fed for the same level of biomass production. Thus, through time and with technological innovation, we have seen less phosphorous and solid discharge coupled with better production cost control. The end result is reduced environmental impact.

Efforts such as these have increased prospects for both the environmental sustainability and economic viability of the global aquaculture industry. We suggest that sustainability will be a touchstone for all aquaculture innovation in the 21st century. Whether it is to improve biomass conversion ratios, preserve water resources, or to decrease nutrient discharges, new technologies will be in high demand. These challenges present opportunities for emerging fisheries professionals interested in aquaculture. We now know that it is critical to understand the biology of the animals intended for farming. We also now know that as the aquatic animal is domesticated it will lend itself increasingly to successful farming. Domestication will require keen scientific insight into such disciplines as genetics, nutrition, reproductive and stress physiology, and environmental requirements for successful, intensive fish farming. Knowledge of aquaculture engineering and economics will also be helpful. As previously mentioned in regards to helping to feed a growing world population, establishing close ties with other fisheries professionals and creating opportunities for collaboration will remain critical. Such collaborative efforts can occur through university and governmental research programs, business innovation grants such as the USDA Small Business Innovation Research Program, and direct interaction with local fish farmers.

Collaboration with Governments and Industry

While solutions for many of the global environmental degradation issues associated with aquaculture will require technological innovation (e.g., improvements to biomass conversion ratios, water resource preservation, or decreasing nutrient discharge), many of the current concerns about aquaculture result from insufficient regulation and governmental oversight. The differential environmental impacts of aquaculture measured for the same species-production system (e.g., pond, serial reuse raceway, or net pen) operating in different countries strongly suggests that the potential to improve environmental performance with today's technologies already exists. For example, in the United States, many of the negative impacts associated globally with aquaculture have been avoided. Such success is undoubtedly due to enforcement of the Federal Water Pollution Control Act of 1972 (the Clean Water Act) that requires, among other things, the regulation of the amount of pollutants discharged from domestic fish farms. We have discovered that most fish farmers themselves want to be good environmental stewards for several reasons, including their own long-term financial success; they just need to understand what is required and how to utilize existing cost-effective technologies. Fisheries professionals can be a good conduit of information to individual or regional fish farmers when they are equipped with a thorough understanding of environmentally sustainable practices and the associated regulatory requirements of governments.

We believe that fisheries professionals should actively engage with relevant governmental agencies (e.g., in the United States, the National Marine Fisheries Service, the Environmental Protection Agency, the U.S. Department of Agriculture, and state environmental control agencies), the FAO, and the industry to collaboratively develop programs and practices that increase environmental sustainability while preserving individual and industry economic viability. In some cases, such efforts have already been initiated. For example, the FAO has an active program attempting to bolster governmental infrastructure and environmentally conscious decision making in developing countries. Various nongovernmental organization or third-party environmental sustainability certifications also help to globally educate farmers and establish fundamental environmental standards. Similarly, recognizing that not all countries have effective food safety programs, the U.S. Food and Drug Administration (FDA) recently introduced an International Food Safety Capacity Building Program (as part of the Food Safety Modernization Act) to better exchange food safety information between the FDA and other countries and to enhance technical assistance in food safety. Industry in developed nations has also benefited from aquaculture professional outreach and education efforts. For example, in Idaho, the Rainbow Trout industry, in association with the Idaho Department of Environmental Quality and aquaculture scientists, developed "Idaho Waste Management Guidelines for Aquaculture Operations" (Idaho Department of Environmental Quality, no date) to help steer individual producers toward environmentally beneficial and practical waste management practices. Professional associations such as the American Fisheries Society and the U.S. Aquaculture Society, a chapter of the World Aquaculture Society, are proving instrumental in sharing aquaculture technology and encouraging research exchange. We encourage emerging fisheries professionals to join and actively participate in these professional scientific associations. Efforts such as these will help advance the emerging fisheries professional's career, but will also benefit the profession as a whole as the scientific knowledge base grows.

Looking Forward

Aquaculture is here to stay and will continue to expand. The challenge for emerging fisheries professionals is how best to contribute to aquaculture's future success. Scientific, technological, and policy-regulatory needs are significant. How can the strengths of the diverse fishery profession be directed toward aquacultures success? Through good communication and effective collaboration, we believe global aquaculture will become increasingly more sustainable. As outlined above, the primary methods through which this will occur will be a combination of technological innovation and shifts in governmental regulations. As has been shown in the United States, and many other locations around the world, this is possible. Aquaculture production can help meet the nutritional demands of a growing human population. We know that potential solutions, such as increased biomass conversion ratios, must be practical and affordable if they are to be widely adopted. It is also clear that solving the regulatory and enforcement disparity among countries is not something that can be remedied quickly. We encourage emerging fisheries professionals to focus on aquaculture as a means to help feed our growing population. In so doing, the science of aquaculture improves, our fisheries management toolbox expands, and our profession itself becomes enriched. Ultimately, these efforts will increase the long-term economic success of fish farmers and communities around the world while leading to universal application of environmentally sustainable fish farming practices.

Acknowledgments

The authors thank the editors for their assistance. We also thank our anonymous reviewers for their constructive comments.

Biographies

John R. MacMillan is the vice president of research, technical services, and quality assurance at Clear Springs Foods and the immediate past president of the National Aquaculture Association. Clear Springs Foods, a producer of seafood, is the largest producer of Rainbow Trout in North America and is committed to continual environmental improvement.

Eric A. MacMillan works as a government contractor on marine resource issues for the National Oceanic and Atmospheric Administration's Restoration Center in Silver Spring, Maryland.

References

Idaho Department of Environmental Quality. No date. Idaho waste management guidelines for aquaculture operations. Available: www.deq.idaho.gov/media/488801-aquaculture_guidelines.pdf. (October 2013).

Inland Fisheries in Russia: Tales from the Past and Lessons for the Future

DMITRY F. PAVLOV* AND DMITRY D. PAVLOV
I.D. Papanin Institute of the Biology of Inland Water, Russian Academy of Sciences
Borok, Yaroslavl 152742, Russia

Key Points

- Inland fisheries hold economic and social importance for society.
- If the needs of all social and ethnic groups are not recognized, fishing will continue to deplete fish populations.
- Fisheries regulations should manage subsistence fisheries differently than commercial and recreational fisheries.
- Measures should be taken to address the unavoidable conflict of interests between inland fisheries and other water resource users.

Introduction

We, the authors, are father and son. As fishery professionals, we have both witnessed great changes in Russia's inland fisheries industry over the years. In the 1930s, great famine hit the country. For our (grand)father's family, fishing was a means of survival. Considered recreational fishermen, they fished with fishing rods and hooks because traditional commercial fishing with nets was considered poaching, a serious criminal offense. This recreational fishing allowed our family to survive the famine. As a result, fishing was, and still is, something our (grand)father considers sacred.

Our family story was from long ago, but inland fisheries are still important today. We often see a man who sells fish on the street in our town, Borok. We know his story; he is a pensioner and his son is unemployed. Of course, the man and his son receive some money from social services, but this is usually only modest support. They catch fish in extreme conditions, in the summer heat and in winter, when the wind is terrible, the ice is one meter thick, and it could be –30°C. To catch fish, they use cheap monofilament gill nets, and we doubt they have proper licenses for this business. When we buy the fish they catch, we often ponder, "Are we supporting poaching, or are we helping people survive?"

When we participate in meetings about inland fisheries issues, we often hear things like "What is the importance of *your* small-scale negligible fisheries compared to *bigger* issues like water supply, irrigation, energy generation, and flood control?" This viewpoint shows the lack of understanding of the importance of inland fisheries, despite their significant current and historical role in Russian society.

A glance back into our personal histories and professional experiences has allowed us

* Corresponding author: pavlov@ibiw.yaroslavl.ru

to identify problems in Russia's inland fisheries, such as the ones described above. We hope that this chapter will show that valuable lessons can be learned from Russia's past to better prepare for the future of inland fisheries and stress the importance of understanding fishery history. The lessons from the past can help avoid mistakes in the future.

A Brief History of Russian Inland Fisheries

Historically, Russia's inland fisheries were within the European portion of the country, the nation's historical nucleus. With the colonization of Siberia, Russia's fisheries expanded, developing in the large Siberian rivers and lakes. By the end of the 19th century, fish from inland waters became the exclusive source of protein-rich food for millions Russians. However, reliable historical data of inland fish harvests in Russia are scarce. The few known statistics indirectly show that the catches were very significant. For instance, in 1897, the amount of fish harvested in the European part of the Russian Empire was 1,073,600 metric tons, of which, more than 50% were harvested in inland waters.

Russian fisheries were historically considered fairly progressive and well managed. For Russians, the fish and their fishing grounds were indivisible common property. The fishermen obeyed strict regulations in terms of time, place, permitted fishing gear, distribution of harvest, and protection of spawning fish. These organizational systems were based on regional specificity of conditions for fisheries but were not controlled or regulated by the central government; the resident fishermen knew how many fish they needed and how many fish could be sustainably harvested to ensure future yields. An extensive hydrographic network, low population density, presence of numerous remote lakes and rivers serving as refuges for fish stocks, strong recruitment, and a traditional fisheries management system preserved Russia's inland fisheries from depletion. This system proved that fish stocks could sustain certain levels of fishing pressure without additional management efforts (e.g., stocking), as long as fishing pressure remained fairly low and the natural conditions necessary for the successful replenishment of fish populations were maintained.

The socialist revolution hit the Russian Empire in October 1917 and dramatically changed the social and economic conditions for the country. Deformation and disintegration of traditional forms of societal and economic organization in attempts to build a new socialist country also affected inland fisheries. After the revolution, inland fisheries generally were controlled and managed from the top by the central government. In most cases, the management decisions failed to acknowledge the traditional, time-proven systems of fisheries management. The natural conditions and ecological processes of recruitment to fish stocks were also ignored in favor of more ideologically based social and economic needs. To facilitate the intensive industrialization of the new socialist society, dams were built in many of the largest Russian rivers without respecting the natural requirements of inland fish. These dams and their impounded reservoirs modified natural fluvial habitat, promoting strong populations of less valuable, lacustrine fishes, such as Roach *Rutilus rutilus*, European Perch *Perca fluviatilis*, Ruffe *Gymnocephalus cernua* and Blue Bream (also known as Zope) *Abramis ballerus*, while populations of the more valuable anadromous fishes diminished. These dams, over time, have resulted in depletion of inland fisheries and loss of their economic importance.

The second dramatic change occurred in the late 1990s with the collapse of the Soviet socialist model of economy and its transition to capitalism. This transition affected inland fisheries again, as the urge to attain a maximum profit became a major driving force of increased fishing. This transition also led to the collapse of numerous industrial and agricultural enterprises, which resulted in mass unemployment and new poverty in some regions of the country. This, in turn, triggered a rise in poaching, which seriously threatened remaining fish stocks in these regions. Again, as a result, the most valuable fish stocks declined while the less valuable fish species thrived. The central government tried to limit negative consequences of such voracious fishing pressure, but these attempts were not successful.

These two examples of dramatic change show that inland fisheries do not exist in a vacuum; in fact, they largely depend on the social and economic conditions of a country. Ignoring the country's interests in favor of the most important ideological, economical, or social needs can result in the depletion of fish stocks and threaten future sustainability of inland fisheries.

Current Status of Russian Inland Fisheries

Today, Russia remains a significant fish producer, with inland fisheries comprising 22.5 million hectares of natural lakes and 8.9 million hectares of man-made lakes, and rivers totaling 523,400 km in length. These water bodies are inhabited by approximately 360 species of fish, 90 of which are of commercial importance. However, Russian inland fisheries contribute less than 10% of the total fish harvest, with 90% originating from marine commercial fishing and, to a much lesser extent, aquaculture (mostly freshwater).

Why has such a dramatic decrease in Russian inland harvests occurred? Revolutionary changes in Russia's social and economic models resulted in total disintegration of traditional forms of inland fisheries management. These forms were replaced either by centrally controlled and state-owned systems or by collective enterprises (i.e., *kolkhozes*). The latter were independent only in theory—their activities were largely regulated by the state.

The state fisheries' regulatory structure suffered organizational disorder; the fisheries authorities in the ex-Soviet Union and in the Russian Federation were transformed several times. In other words, the rules of the game were changed during the game, which confused the players! Each reorganization meant more disorder in the fishery management and regulatory system. For example, managers entering new positions were often ignorant of local conditions and practices and they needed some time to adapt to new rules in order to make the right decisions. New rules regulated both commercial and recreational inland fisheries nationwide, making the implementation of these rules confusing and not specific to each region.

Now, the role of inland fisheries in the country's gross economy is almost negligible. However, for some people, fishing is still a matter of survival because it remains their main source of food and income. For these certain social and ethnic groups, recreational fishing is, in fact, subsistence fishing, and the amount of fish harvested by these groups is often underestimated. It is also challenging for these groups to abide by the same rules as wealthier social groups, for which fishing is just a hobby. As a result, subsistence fishers are sometimes compelled to become poachers and engage in criminal behavior.

Management and regulation of inland fisheries should be specific to either recreational or commercial fisheries and should acknowledge the subsistence needs for certain social and ethnic groups. Subsistence fishermen, who engage in criminal behavior (forced poachers) according to the current regulations, should be legally allowed to meet their survival needs. This idea has been successfully implemented for managing fish and marine mammal harvest in the past. For example, in Chukotka, the northeastern-most part of Russia separated from Alaska by the Bering Strait, indigenous people are provided with fishing and hunting quotas to meet their food demands. And they do so without depleting the fish and marine mammal populations. In our opinion, inland fisheries require more organizational and institutional stability for proper management and regulation. This stability can be reached by implementing traditional, time-proven fishery management practices that are regionally specific to local ethnic and social groups and traditional practices.

Competition between Inland Fisheries and Other Water Resource Users

Beyond politics, industrialization has also greatly affected inland fisheries. Factories, for instance, often lack proper wastewater treatment facilities, which results in toxic pollution and subsequent eutrophication within many inland waters. Hydroengineering, another example, serves many purposes: hydroengineering connects deepwater waterways; generates electric energy; provides a reserve of freshwater for irrigation, industry, and households; and controls flooding. The growing need for water and energy is widespread and has caused industrial development to increase in many locations. In most cases, this development requires the consideration of certain habitat conditions. In order to keep productive freshwater fish populations, dams must be designed as not to prevent fish migrations. However, the consideration of these conditions often was not of primary importance. As a result, dams prevent successful passage of spawning anadromous fish, causing some of the most valuable fish stocks to become depleted.

Deterioration of environmental conditions for inland fish, depletion of their stocks, underestimation of their potential contribution to food supply, and long-lasting organizational turmoil drove Russian inland fisheries to the brink of extirpation. Unfortunately, this situation is not unique to Russia and is observed in many developing countries. Various uses of water resources are competitive and cause conflicts among stakeholders. The interests of inland fisheries are not always a priority in management decisions.

We think that solving this problem will only be possible if the needs of these various stakeholders are adequately met and protected legislatively. All efforts should be made to coordinate the interests of multiple water resource users in such a way that inland fisheries will not suffer. This may be achieved through better organization of special inter-institutional panels or commissions responsible for resolving conflicts. For example, the water level in the reservoirs could be maintained in the spring season, allowing optimal floodplain conditions for fish spawning and larval development. Sound management decisions such as this could be possible if industry personnel consider seasonal conditions for fish populations and then receive approval by a panel or commission for future development.

We know good examples of such effective coordination in Russia. For instance, in the upper Volga basin, the Rybinsk Authority was established as a special commission to coordinate the activities of various resource users with respect to the interests of local fisheries. This commission includes the representatives of the State Committee for Hydrometeorology and Environmental Monitoring and the Ministries of Energy, Transport, and Agriculture; fisheries managers; and local authorities.

Concurrently, funding for inland fisheries should adequately provide for sustainable harvests, now and in the future. Through proper legislation, this funding may be obtained as investments from wealthier users of inland water resources. Theoretically, this system exists in Russia. However, industrial enterprises are obliged to invest in the fisheries in the form of penalties in order to cover the harm to fish stocks that results from their activity. That is, the harm is planned, but there are no provisional measures to prevent it! The future activity of an industrial or agricultural enterprise must be planned to prevent harm to fish stocks rather than compensate the inevitable damage. For example, a thermoelectric power station built in the early 1950s uses waters from a cooling pond (impounded specifically for power station's operation) as coolant. The pumping stations possess no means of fish protection. However, up until 1990s, it did not bother anyone much since (1) the water reservoir belonged to the power station, and (2) a fish farm built at the warmwater discharge provided valuable fish species (various species of carps and Channel Catfish *Ictalurus punctatus*) to stock for recreational fisheries. These valuable fish were not harmed by the pumping stations because they did not spawn in the reservoir and were large enough to avoid harm from the pump operation. But, in the early 1990s, the water reservoir became federal property. Reservoirs were assigned to penalty categories based on the presence of valuable fish species with the "first fisheries category" paying the highest penalty. The fish assessment programs conducted on this example reservoir in 1989, 1990, and 2013 place it in a very low penalty category (only thousands to tens of thousands of rubles) because the pumping stations only harm species of the least economic value. Therefore, it is much cheaper for the power station to pay these penalties rather than design and build fish protection devices (millions to tens of millions of rubles). In addition to that, the power station must undergo significant renovations that include building cooling towers that would lower water consumption dramatically. These changes would greatly reduce fish kills but would also undermine the fish farm's operation due to the extremely low volume of warmwater discharge.

In our view, the main systemic problems of inland fisheries relate to inefficient regulation and infrastructure-related constraints. There are additional problems, too, for example, the limited effectiveness of management institutions, inadequate legislation, competition with cheap imported products (e.g., farmed fish), insufficient support of inland fisheries by state, inadequate monitoring of the status of fish stocks, overfishing and poaching, poor water quality related to pollution, and no fish passage facilities on major rivers.

Future of Inland Fisheries

The Federal Fisheries Agency of the Russian Federation adopted a fishery management strategy for the Russian Federation that will be in effect until 2020. The objective of this management strategy is the long-term sustainability of fish stocks. This management strategy covers the following measures: (1) recovery and rehabilitation of fish stocks, including the development of artificial propagation and aquaculture; (2) development of more modern infrastructure and technology; and (3) more scientific research contributions. For inland fisheries, the strategy includes a focus on pasturing fish: acquiring juvenile fish, maintaining them in hatchery-like conditions, stocking the larger fish in inland waters, monitoring the stocked populations in their naturalized habitats, and, finally, harvesting the fish. We think these proposed measures are feasible not only for Russia, but for other countries as well.

In our opinion, inland fisheries may play an important role in satisfying the growing demand for food in many countries worldwide. Not only should recreational or trophy fishing be funded and supported in wealthier countries, the interests of social and ethnic groups that rely on inland fisheries for subsistence should also be considered. Solving the modern crisis within inland fisheries is only possible through an understanding of its importance and by considering the needs and wants of commercial, recreational, and subsistence inland fisheries users.

Our Lessons

What have we learned from our stories and experiences as fisheries professionals? It is never too late to learn—even if you are no longer a student! Pay attention to the history of fisheries, for it provides context for why things are the way they are now. Understand the social and economic conditions of a region and the local people to best manage its fisheries. What gear is used in the fishery? What are the fishery's traditions? Why is this fishery socially or culturally important? Determining answers to these questions will surely help avoid mistakes in future management and regulation of inland fisheries.

Because inland fisheries users are often in conflict with other water resource users, you must understand the needs and wants of the other users. Truthfully, we, the authors, wish we were students again. We wish we had spent less time learning fish anatomy and physiology and more time learning about social, economic, and legislative issues related to modern inland fisheries. The problems of hydroengineering, energy generation, agriculture, water supply, waste treatment, and recreation are also the problems of inland fisheries. While you may view these industries as your opponents, seek out opportunities to engage them. Convince them of the importance of fish and the benefits of sustainable fisheries. It is never too late for them to learn either!

Acknowledgments

We are grateful to William Taylor, Abigail Lynch, and Molly Good of Michigan State University for their invaluable help in preparing this paper. Thank you, dear friends!

Biographies

Dmitry F. Pavlov graduated from Rostov State University (Rostov-on-Don, Russia) and became an assistant lecturer there. After moving to I.D. Papanin Institute of the Biology of Inland Waters, Russian Academy of Sciences, he held an array of positions from senior technician to deputy director. He has served as the vice chair of the Upper Volga Fishery Commission among other panels and commissions. Now, Pavlov is a researcher, leading many research projects related to studies on inland fisheries and water quality assessment.

Dmitry D. Pavlov graduated from Yaroslavl State University (Yaroslavl, Russia). Over the years, Pavlov has led and participated in numerous field studies on fish biology in the Russian inland waters. Now, he works as a fisheries research biologist at I.D. Papanin Institute of the Biology of Inland Waters, Russian Academy of Sciences.

The Role of Scientists in Public Policy Development and Advocacy: Advice for Scientists and Policy Makers

Mark Rey* and Adawnice Lucas
Department of Fisheries and Wildlife, School of Agriculture and Natural Resources
Michigan State University, 115 Manly Miles, East Lansing, Michigan 48823, USA

Key Points

- Scientists have clear career choices in regard to their participation in the development of public policy and advocacy within the natural resources arena.
- Policymakers not trained as scientists can most effectively engage scientists and scientific data in decision-making processes by following a few, key principles.

Introduction

Many scientists find themselves in a quandary when trying to evaluate and choose their particular role in natural resources decision making. In our view, this is an unnecessary source of concern because scientists always have multiple career choices. This chapter will explore those choices and discuss how policymakers can best engage scientists.

To assist in planning a career, we describe three options for scientists' participation in natural resources decision making. Individuals should decide which role best fits their personal style, priorities, and career goals. Each role is important and serves a valuable purpose within both the scientific and public policy arenas.

Three Options for Scientists Involved in Natural Resources Decision Making:

1. Scientists can focus exclusively on their research and let the data that they generate speak for themselves. In an information-based society, most data will eventually find theirway into policy-making discussions and processes. Researchers do not necessarily bear the responsibility of doing more than competent research. However, many organizations that sponsor scientific research require more, and many scientists feel some urgency about communicating their results, which brings us to the second alternative.
2. Scientists are often encouraged or pressed into advising policymakers and advocates or engaging the broader public. This imposes more obligations (beyond competent research) and can require some degree of skill in public communications. A few straightforward guidelines can aid scientists in performing these additional functions.
   - First, understand that if you are engaged as a scientist by a policy-making or policy-influencing agency or organization, your employers or sponsors will expect you to

* Corresponding author: markrey8@aol.com

help inform their policy decisions. If you cannot fulfill that expectation, decline the engagement.

- Second, if you accept the engagement, be clear with your sponsors about what your data say, as well as what they does not say. Spell out the known facts and the known unknowns, and stipulate clearly whether there may be unknown unknowns. Your sponsors deserve clear advice in each of these areas.
- Third, although you have not surrendered your first amendment right to speak both freely and (if you choose) publicly as an individual, there is an appropriate manner in which to do so. Be sure to let your sponsor know what you know before you disclose it externally. Thereafter, if you speak out publicly, make it clear that you are speaking as an individual. Also, give credence to the fact—and acknowledge—that there are other data, factors, and considerations that had to be weighed in your sponsor's decision. Your suggestions will likely be appreciated and considered, but another route may still be chosen. Legal and resource constraints often come into play. Thus, be prepared to understand these decisions.

3. Finally, a scientist can decide that he or she knows more about a particular policy issue at stake than either the policymakers or the advocates. Scientists can decide that they are tired of watching both policymakers and advocates flail around. Scientists may conclude, therefore, that they can be—and want to be—more involved as an advocate for a particular policy outcome.

In the interest of formulating more-informed public policy, we believe option three is extremely helpful. We need more advocates with greater scientific training. However, it is important for scientists who cross this threshold to recognize that they are no longer functioning as scientists, but as advocates—of which there is already a surfeit. In addition, it is important for scientists to remember that most policy decisions are rendered in the absence of scientific certainty. Therefore, if an advocate attempts to use his or her status as a scientist to lift an argument, they can expect criticism. Beware of scientists working in the advocacy arena who emphasize their credentials with more force than their arguments.

Many scientists who have entered the advocacy arena believe that their credibility has either been questioned or has suffered. While that has been the effect in some cases, most scientists have improperly diagnosed the cause. Their participation in advocacy did not make them less competent scientists. Rather, we suggest that it likely exposed them as inept advocates. Such an outcome should not be entirely surprising because the skill sets and the rewards associated with the two professions are very different.

Science is the act of discovery. It is inherently empowering. Scientists on the cusp of new discoveries—of new advances in human knowledge—are more knowledgeable than others about the matters in which they are involved. This is both empowering and ego-gratifying.

Advocacy is the art of persuasion. Advocates are forced to share information with people who they are trying to influence, but ultimately who they cannot control. Advocates are not, after all, the ultimate decision makers insofar as policy outcomes are concerned. Effective advocacy is therefore, by necessity, an exercise in humility.

Based upon the very different natures of these endeavors, the best scientists are extremely confident about their intellectual abilities. By contrast, we have found that the best advocates are marked by their sense of humility. Put differently, effective advocates do not spend a great deal of time on their curriculum vitae.

To this point, we have talked about the choices available to scientists for becoming involved in natural resources decision making. For the benefit of scientists becoming involved in policy making, we believe that it is equally useful to explore the question of how policymakers can best engage scientists in making sound public policy decisions. From our experience in this area, we have developed seven principles.

Seven Principles for Policymakers Engaging Scientists

1. If you are a policymaker engaging scientists in difficult decisions, give them clear background and the broader policy context into which their work (and, if requested, their advice) will fit. They deserve a clear understanding of how exactly their work and advice may be used.
2. Give them a complete understanding of their specific role in the particular policy decision. For example, will they be asked to simply do research, analyze, and provide the results, or will they be asked for advice on the scientific credibility of different policy options? Perhaps they will be asked to explain to other audiences the scientific information that provided the analytical platform for the decision. They may also be asked to explain to these other audiences how the scientific information informed the chosen policy outcome. Finally, will they be asked to publicly endorse the chosen policy outcome and thereafter be expected to defend it? Where the role of scientists falls along this spectrum should be both understood and agreed upon before that role is undertaken. During the past decade, the U.S. Forest Service utilized its scientists well to both advise and later explain its decisions involving the management of the national forests of the Sierra Nevada range.
3. As a policymaker, learn to listen actively and intently. Learn to listen for both what your science advisors are telling you, as well as what they are not saying. Do not fill gaps between the two with your own assumptions.
4. Learn to question actively. Do not be afraid to request additional analysis or information. Where the science is uncertain, incomplete, or unclear, there is no such thing as a stupid question. There are only stupid outcomes that come from a lack of questions. Asserting that the science on an issue is settled too often means merely that the decision maker's curiosity has been sated.
5. As a policymaker, you should review your final decision with your science advisors before it is publicly announced. Show them what role their work played in your decision making. Describe the other factors that you were required to consider. Explain why the outcome ended up the way it did, especially if it does not appear to conform to their results.
6. Encourage your science advisors to privately challenge your final decision. Ask them to explain how they would rationalize a different outcome based on their understanding of both the science and the degree of uncertainty involved in this policy area, and the factors that you as a decision maker considered. This could result in a decision that you are better able to defend to the outside world.
7. Finally, do not use your science advisors or their work to prop up a difficult political decision. As a policymaker, in rendering decisions on topics where the science is uncertain or incomplete, (or even where it is clear and you have to ignore it for other reasons), it is your responsibility to own your decisions and not to simply suggest that there were no other options. Many of the scientists engaged in the early stages of the "spotted owl wars" of the 1990s suspected that their political overseers were pushing them to find a scientific sweet spot to embrace in a very difficult political environment. Decision mak-

ers were hopeful that the science would lead to a decision that would at least partially shield them from adverse political fallout.

Invariably, it is best to integrate science into the policy-making process if you can explain why the science presents the best and fairest result. For this outcome to occur, scientists and policymakers must (1) work together in trust, (2) recognize their respective roles and play them, and (3) be willing to stand together and explain the outcome. When this occurs successfully, science is being used properly in decision making, the public interest is being served, and public acceptance of difficult decisions becomes more likely.

Acknowledgment

We are not scientists. Rather we approached these questions as an advocate who retains scientists and uses science in advocacy work, as a public official who oversaw and funded scientific research and engaged (by extreme necessity) scientific advisors to render difficult decisions, and from the viewpoint of a student considering a career in natural resource sciences. We gratefully acknowledge the contributions of the many scientists we have depended upon as advisors, teachers, and mentors.

Biographies

Mark Rey teaches advocacy in the natural resources arena at Michigan State University, which reviews the types of advocacy groups operating in the natural resources arena, tools and techniques used in environmental advocacy, tactics commonly used in advocacy campaigns, and ethical questions that often arise during advocacy work. He also oversees the Demmer Scholars internship program run in conjunction with a natural resource policy class. The internships are with federal natural resources agencies or nongovernmental organizations (both nonprofit and for profit) operating in the natural resources policy arena in Washington, D.C. Rey is the former under secretary for natural resources and environment. In that position, he oversaw the U.S. Department of Agriculture's Forest Service and Natural Resources Conservation Service. Rey served as a staff member with the U.S. Senate Committee on Energy and Natural Resources.

Adawnice Lucas is a Michigan native. Growing up in a rural area, she watched as it was developed into subdivisions and shopping malls. As she witnessed the destruction of habitats and came to the realization that countless plants and animals were being displaced or lost, she decided that she would do what she could to stand up for the flora and fauna that could not speak for themselves. She now studies conservation biology at Michigan State University and aspires to aid in the creation of a world full of people who live sustainably and value and protect our natural resources.

Achieving Funding Needs for Fishery Resources Management and Angler Access

Gordon C. Robertson*, D. Michael Leonard, and Elizabeth A. Yranski
American Sportfishing Association
1001 North Fairfax Street, Suite 501, Alexandria, Virginia 22314, USA

Key Points

- Educate the public and policymakers about the virtues of the existing "user-pay, public-benefit" system to help safeguard it.
- Explore additional funding sources, particularly from non-endemic user groups, that benefit from your work.
- Market your product. "If we build it, they will come" no longer applies, particularly given the nation's changing demographics.

Despite their generous investments, anglers cannot keep up with the costs of fisheries management. Throughout the recent decades, issues and ideas tend to be cyclic and each new generation meets them with enthusiasm. One issue, however, prevails from year to year, and that is maintaining stable funding for fishery and other natural resource programs—funding that gives agency leaders the ability to engage in planning reasonably long-term programs that yield meaningful benefits to the resource and to the public that enjoys it.

The conservation model that the United States has adopted to execute its fish and wildlife management system is unique. It declares that fish and wildlife are public resources, that the "several states" have the primary responsibility for those resources, that they are for the enjoyment of the citizens of the several states, and that the management costs of the system are to be borne, for the most part, by the citizens who enjoy those resources (i.e., fish and hunt them). While fish and wildlife viewing is certainly an enjoyment, those that engage solely in viewing invest very little in any user-pay funds toward management of the resources they enjoy. This means that a large segment of the population has yet to support the natural resources they enjoy. In most states, there are no payment mechanisms for the viewing-only user group, and where there are voluntary systems, revenue is disproportionally low.

The dependency on paying participants makes it essential for the state fish and wildlife agency to engage in an area in which they currently are generally poorly staffed–marketing their product. Regardless of this shortcoming, the model is still the envy of the world because it has produced the best funded and most stable revenues for fish and wildlife conservation. Between a federal manufacturers excise tax on fishing equipment (Box 1) and state fishing license fees, recreational anglers invest about US$1.1 billion annually in fishery resource management and since 1950 invested some $26 billion in the resource.

* Corresponding author: grobertson@asafishing.org

Box 1. How the Sport Fish Restoration and Boating Fund Works

The state fishery agency's receipt of its share, or apportionment, of the excise tax dollars from the Sport Fish Restoration and Boating Trust Fund requires that the state meet certain criteria. The most important criterion is that the state has enacted law that doesn't allow these funds to be diverted for other purposes and that they have a professionally trained staff to conduct the fishery management needs of the state. These tenets are frequently challenged.

Like most issues, funding too has cycles and tends to follow the economic trends of the states and nation. When economic times are good, programs tend to expand, and when economies take downturns, programs contract. Even states such as Missouri that have large and supposedly sustaining funding (in addition to their license fees and sport fish restoration dollars) such as lottery or sales tax proceeds, there is always the inevitable expansion and contraction with the economy, but generally to a lesser extent. Even though these contractions occur, states with substantial additional funding weather economic downturns better than those states that are solely dependent on license fees and excise taxes. Additional, stable revenue streams also make those agencies dependent on general funds more resilient to possible budget cuts that may be proposed by state legislatures seeking to reduce state funding during economic downturns. These cycles have brought creative thinking to funding programs, sometimes in the form of finding new and dedicated funding sources and other times in determining new ways to make existing dollars go further.

Missouri and Arkansas are frequently cited for their innovation as they have successfully used state sales tax dollars to support their natural resource programs. The additional funds have enabled solid public education programs about natural resources and funded research for nongame species. But, even these dedicated funds are not immune from administrative and legislative pressures when funds are tight or when state leaders believe a resource agency has not acted in good faith. In the end, it is probably most fair to have all citizens of a state financially contribute to the management of its public resources. Most states that have created other stable and significant funding sources have invested a sizable portion of those funds in educating their citizens about fish and wildlife resources and their needs. Additional public education and understanding helps garner support for the broader funding needs.

Some states, such as West Virginia, have implemented additional fees to support critical programs. The state's conservation stamp, required by all licensees, is dedicated to capital improvements such as land acquisition and access facilities for anglers and hunters. Acquiring and maintaining a large public land base not only provides public access to public resources, but also allows a state to better manage resources on these public lands. Other states have found various fee-based programs that are readily identified by anglers and hunters as programs that are in the best interest of the resource and those who enjoy it. This demonstration of need and benefits builds constituency support for maintaining the funding model.

Funding fishery programs is subject to many whims and has parallels with the nation's consumer confidence. In many ways, funding depends on the hopes and confidence of the fishing constituency and the public to not only support program funding, but to expand it. The late Ralph Abel, Executive Director of the then Pennsylvania Fish Commission from 1972 into 1987, could instill such confidence. Mr. Abel was an unabashed supporter of his agency and always enjoyed relating this story:

> A man from another state stopped at a gas station in Pennsylvania and asked the man pumping his gas how fishing was in the Keystone State. The man pumping gas proceeded to laud the fishing opportunities in the state and the area and finished by saying the Pennsylvania Fish Commission was the best damn fish agency in the country. Astounded by the long and loyal statement about fishing and the commission, the out-of-state angler asked where the gas station attendant fished. His reply was, "Oh, I don't fish."

Ralph Abel told this story often and used it as an example of how important a positive public perception is and how that translates into funding support and the ability to occasionally raise the price of license fees and expand programs for the fishing public. Sustaining this positive public perception is a balancing act between those that participate and those that do not. As part of the new generation of citizens and fishery professionals, having a positive public perception of your work is even more important in today's culture. Fishery managers and administrators need to understand how to build that positive public perception through today's media and to understand the coming changes and the challenges the future will bring.

While the virtues of the "user pay, public benefit" American model of fish and wildlife conservation are understood and valued by fisheries professionals, very few outside of the field are aware of how management of our public resources is funded. For a variety of reasons, fisheries managers and others in the recreational fishing community have not invested sufficient time and resources over the years in educating the public and policymakers of the vital role anglers play in fisheries resource conservation.

Our increasingly urban society has resulted in a greater portion of the public with little firsthand experience with nature. This has created a large pool of citizens who, while they may enjoy watching nature shows on television or visiting zoos, lack an appreciation for the intrinsic values of spending time interacting with the outdoors. If these people do not understand fishing, much less the economic contributions that anglers make towards fisheries conservation, they are less likely to support policies that may affect fishing access and participation. It is therefore vitally important for the recreational fishing community—angler organizations, industry, and agencies—to redouble its efforts to educate the public of the more than $1 billion that anglers spend annually for fisheries conservation efforts. An individual who has never fished before will become more likely to support pro-fishing policies and programs if they understand the conservation benefits that accompany recreational fishing participation.

In addition to better public education, fisheries managers must become creative with their marketing scheme, putting a heavy emphasis on the recruitment of new anglers. Looking forward, a smaller portion of the public will be born into families and communities where fishing is ingrained in their culture. Our society is not only becoming increasingly urban, but increasingly diverse as well, requiring a different response from fisheries agencies in ensuring a solid base of anglers to fund fisheries programs.

In the coming years, the demographic makeup of the general population will look increasingly less like that of the current angling population, which is 86% Caucasian. By 2050, non-Caucasians will make up more than half of the U.S. population. Hispanics will account for the majority of the growth of the non-Caucasian population, and historically this is an ethnicity that fishes much less than Caucasians. Currently, about 1 in 7 Caucasians fished within the past year, compared only about 1 in 30 Hispanics.

For resource management agencies to remain viable in the future, they must develop strategies to recruit anglers outside of the historical base. Underlying these strategies must be an understanding of what motivates anglers of different ethnicities to fish. Agencies must develop innovative ways to introduce the sport to people who did not grow up in a fishing culture.

Many agencies and recreational fishing organizations have already begun such efforts, and the returns are beginning to pay off, as evidenced by the 11% increase in fishing participation since 2006. To ensure a strong base of anglers, and therefore fisheries management funding, managers must understand and respond to the drastic changes in demographics that our country will experience in the coming years.

Managers must also gain the support of other segments to pay an excise tax. An example is the rapid growth of kayak users. While they already pay an excise tax on their primary fishing equipment and purchase a fishing license, they do not pay an excise tax on their kayak, whereas anglers fishing from boats pay a fuel tax for their conveyance that is ultimately allocated to the state for better fishery management and angler access.

Achieving fishery funding needs will always be a challenge—that is a given. To meet the challenge of funding fishery programs, there are, in our opinion, several key ingredients:

- Maintain your base – Make sure all fund sources are legislatively protected from diversion. In these economic times, every penny counts, and diversions can often occur as governors and legislators look to fill general fund gaps with monies dedicated to fish and wildlife programs. Be ever vigilant, and alert fish and wildlife conservation groups in your state to potential diversions.
- Remember marketing – Today's culture demands drawing attention to and selling your product: fishing opportunity. Use sources like the Recreational Boating and Fishing Foundation to bolster the agency's marketing ability.
- Know the demographics of current and potential anglers and use the knowledge to create more anglers and more funding opportunities – The U.S. Fish and Wildlife Service releases surveys every five years, which can be a guide to the current demographics of recreational anglers. As the demographics of our nation change, so will the demographics of potential anglers. Marketing will need to be done to engage these future anglers and get them excited about the sport. In addition, many independent studies are available and ongoing that document today's anglers.
- Make anglers advocates for fish conservation and access to fishery resources – Anglers have historically been the original conservationists. With new individuals coming into the sport, educational efforts need to be mounted so new anglers understand the importance of conservation and access, and the role anglers play in funding both.
- Seek innovative ways to fund beneficial programs that the angling public and the general public recognize as a public benefit – Do not hesitate to look for success in other states and model those efforts. Learn from what works and what does not.
- Remind legislators that recreational fishing means sustaining jobs and a healthy economy, especially in rural areas – Recreational fishing nationally supports almost 900,000 jobs and provides an economic benefit of $115 billion annually to the nation's economy. Use the American Sportfishing Association's (2013) Sportfishing in America to demonstrate the economic attributes and participation levels of anglers across the nation.

Biographies

Gordon Robertson has spent 42 years in the natural resource management profession. This has spanned forestry, wildlife, and fisheries programs, everything from fieldwork to agency and nongovernmental organization leadership positions. This tenure has given him much opportunity to observe, learn, be mentored, and mentor. He reminds young professionals that coming in to this field has always been for the dedicated, for it is not one that generally brings independent wealth but does convey a rich understanding of the world around us and how resilient, yet fragile, wild things really are.

Mike Leonard is the ocean resource policy director at the American Sportfishing Association, based in Alexandria, Virginia, where he is responsible for the association's activities in a variety of marine resource issues at the national, regional, and state level. He holds a master's degree in fisheries management from Auburn University and a bachelor's degree in fisheries science from Virginia Polytechnic Institute and State University.

Elizabeth Yranski is the policy fellow at the American Sportfishing Association. She has previous research experience at Woods Hole Oceanographic Institution, as well as other academic and biotechnology institutes. She has a master's degree from Johns Hopkins University in environmental science and policy and a bachelor's degree from Northeastern University in biology with a concentration in marine biology.

References

American Sportfishing Association. 2013. Sportfishing in America: an economic force for conservation. American Sportfishing Association, Alexandria, Virginia.

The Changing Landscape in Atlantic Menhaden Assessments: Best Available Science, Uncertainty, and the Tension between Science and Management

DOUGLAS S. VAUGHAN*
214 Shell Landing Road, Beaufort, North Carolina 28516, USA

AMY M. SCHUELLER
National Marine Fisheries Service
Southeast Fisheries Science Center, NOAA Beaufort Laboratory
101 Pivers Island Road, Beaufort, North Carolina 28516, USA

Key Points

- The concept of "best available science" has changed and improved over the years, representing a continuum from expert opinion to analyses based on fishery-independent data derived from an appropriate sampling design.
- Uncertainty exists in all science. Embrace it! But do not allow yourself to weigh the risks to stakeholders, as that is the job of management.
- The interface between science and management is still an area of tension; the precautionary approach is one example that attempts to guide management groups toward less risk-prone solutions.

Introduction

The Beaufort Laboratory of the National Marine Fisheries Service has been at the forefront of Atlantic Menhaden *Brevoortia tyrannus* data collection and analysis since the inception of a detailed biological sampling program for the species in 1955. Although analyses and management questions have changed considerably over the decades, our role of providing sound scientific advice for management based on best available science has not. Beaufort scientists have traditionally taken the lead in conducting stock assessments that inform fisheries management decisions for Atlantic Menhaden. Stock assessment analyses are based on data and models with considerable uncertainty, and that uncertainty may extend to conclusions about stock status. We endeavor to use the best data available and most-appropriate models to address relevant management questions. However, inherent uncertainty exists in the data and also may arise from unmet assumptions of the models. When outcomes from a stock assessment result in negative impacts on user groups, the first criticism is that the data are flawed (sometimes true, but often still the best available data) or that the models are inappropriate. Having a thick skin helps! However, it is important to keep in mind that stock-

* Corresponding author: dvaughan1@ec.rr.com

assessment scientists do not manage fish stocks; they do, however, provide the fundamental analyses for stock management.

Best Available Science

Gradual improvements have been made for the management of fish stocks in the U.S. exclusive economic zone (EEZ), including more-explicit management requirements through Congressional updates to the Magnuson-Stevens Act, such that all EEZ-managed stocks must now have regulations in place to prevent overfishing. However, Atlantic Menhaden is a state-managed fishery in which most of the harvest occurs in state territorial seas 0–3 miles from shore. As such, the Atlantic Menhaden fishery is managed through the Atlantic States Marine Fisheries Commission (ASMFC). The legal requirements are much less strict for intrajurisdictional fisheries, such as those managed by ASMFC; nonetheless, ASMFC has adopted many of the requirements from the EEZ-managed stocks under the federal system, including striving to make management decisions based on best available science.

When I (Vaughan) came to Beaufort in the early 1980s, there was no legal requirement for "best available science," but that was certainly our goal even though it was not formally defined. As model complexity increased over the course of my career, data-input requirements increased concurrently with a more formal construct of best available data and science. One example of increased rigor for the Atlantic Menhaden assessment was the investigation of multiple state fishery-independent seine surveys to create a juvenile abundance index for incorporation into a more complex assessment model structure, with only Maryland data being considered initially.

When I (Schueller) started in 2009, I wish that I had a more concrete thought formed around what "best available science" meant. From what I can tell, "best available science" is a continuum and sometimes means expert opinion or, in other cases, less than ideal statistical analyses with data (which are often less than ideal). At the other end of the spectrum, the goal of an assessment scientist is always to have data collected with a large-enough sample size and the appropriate statistical design. However, as you venture into fisheries management and science, you realize that those properties are rare and you must make do with best available data. There is an overwhelming need for fishery-independent data. For example, for menhaden, a coastwide adult abundance survey would provide an index for use in the stock assessment. Data help scientists to formulate weights of evidence for alternate hypotheses and move "best available data" from expert opinion into a realm that is more objective and scientifically defensible.

> [T]here are known knowns; there are things we know we know. We also know there are known unknowns; that is to say, we know there are some things we do not know. But there are also unknown unknowns—the ones we don't know we don't know.
> —U.S. Secretary of Defense Donald Rumsfeld

As the former U.S. Secretary of Defense put it, the unknown unknowns are the ones that catch us by surprise and can frustrate our best efforts. With the collection of fishery-independent data, we can move more into the realm of known unknowns. For example, the Atlantic Menhaden stock has exhibited apparent periodicity in abundance over time. The typical suspects responsible for these swings are fishing pressure, environmental factors, predator changes, and variable recruitment. However, the weight of evidence available does not point toward any one cause; thus, the best science or data are known, but the best possible science, which could point to the cause(s), is unknown.

As a young scientist entering the field, keep in mind that best available data and science is a shifting baseline over time. Nothing is black and white or straightforward, and the data and science often improve over time. Realize that tasks from management may have an agenda attached and stand firm related to what is and is not supported by best available data and science. Try not to allow yourself to have an agenda as a scientist. You are representing the data, the science, and the resource (although, not advocating for the resource). And finally, always clearly acknowledge the uncertainty inherent to that science.

Uncertainty through the Ages

While "best available data" can be nebulous at times, the issue of uncertainty is a real problem; solutions for dealing with uncertainty have gradually been introduced over time. Early on, sensitivity analysis was the primary tool used to address uncertainty for a few important model inputs. Natural mortality (M) was always a point of conjecture and contention, with a value of 0.2 seemingly applied to all stocks at one time. However, longer-lived stocks were believed to have lower natural mortality (<0.2), and shorter-lived species such as Atlantic Menhaden and other clupeid fishes were believed to have higher natural mortality (>0.2). The early stock assessments for Atlantic Menhaden used 0.45 (see Ahrenholz et al. 1987). Typically, a favored value of M and a lower and upper range were analyzed as part of a sensitivity analysis.

Monte Carlo techniques were developed during the 1980s and 1990s to represent uncertainty in stock assessment results, later incorporated under the term "bootstrapping." These techniques are very computer intensive. More recently, there is concern over whether sufficient sources of uncertainty are incorporated into these analyses or whether these were too optimistic in their evaluation of uncertainty.

Assessment methodologies have also changed considerably, with forward-projection age-structured models being the preferred approach. All recent Atlantic Menhaden and Gulf Menhaden *B. patronus* stock assessments use this method and incorporate bootstrapping techniques within the model structure to address many aspects of uncertainty.

As to whether these characterizations of uncertainty were actually incorporated into the management decisions for these stocks is debatable. Uncertainty has been used as a rationale to delay management action. When a range was provided for stock estimates or management parameters, managers tended to select the end of the range that caused the least initial pain to the fishers. To try to reduce risks of overfishing and to reduce the more risk-prone actions of management, the 1997 reauthorization of the Magnuson-Stevens Act (MSA) included wording based on the precautionary approach. Fundamentally, the precautionary approach states that management should choose more risk-adverse options in order to improve the chances of ensuring sustainability. The precautionary approach is often argued for around the world and is generally implemented in some form in most management schemes, although to different extents from country to country. Although the MSA does not apply directly to Atlantic menhaden, the ASMFC in their management practices sometimes follows the best practices as defined by the MSA.

Uncertainty is a fact in science. Embrace uncertainty! Even the best stock assessments and science have high degrees of uncertainty. Do not take comments about your science personally. Management decisions will always include some uncertainty and associated risk, so as a scientist you need to provide managers with the information that they need to weigh those risks. The outcomes of management can never be known entirely in advance, but managers and scientists can work together to try to decrease uncertainty by gathering appropriate data.

Interface and Interactions between Management and Science

The interface between management and science is always somewhat blurred. This has been the case for management of Atlantic Menhaden. In the past, there were minimal requirements for management of Atlantic Menhaden, and because commercial facilities were closing, scientists and managers alike were not worried about overfishing of Atlantic Menhaden. No management constructs were put in place for Atlantic Menhaden until 2013, except for a bay-wide cap in the Chesapeake Bay. Recently, ASMFC has defined more formal constructs with science questions being addressed by the Atlantic Menhaden Technical Committee (TC) and management decisions being dealt with by the Atlantic Menhaden Management Board (the Board). This formal construct has been set up whereby scientists provide objective, scientific advice and management agents are allowed to incorporate subjective, value-laden considerations because they need to weigh risks to the various stakeholders. Often, managers ask questions that have a highly uncertain answer based on the available data. Scientists provide information on the uncertainty of the numbers, and the managers must decide how to move forward in the face of that uncertainty and the associated risks for the various stakeholders.

For example, the Board requested that the TC use output from the current assessment to project landings levels for the purpose of quota setting. However, the recent assessment was not deemed useful for management purposes by the TC. Nevertheless, managers insisted that a quota be provided. The TC's fallback position was to set quotas based on an ad hoc average of landings for the previous three years. To base a quota on a flawed assessment and subsequent projections would be a contradiction to the original decision made by the TC.

As a scientist, it may be difficult to stand up and provide information based on the limited and uncertain data available without providing your opinion. Our suggestion is not to compromise scientific principles. Even when you know what managers are looking for and why, provide only the information/advice that the data support and its associated uncertainty. Then, allow the managers to weigh the risks of their decisions. Never allow yourself to place value on one outcome over another. The simple point of the science is to provide information to the managers and allow them to make the difficult but informed decisions.

Conclusion: Precautionary Management and Reference Points

The requirements for best available science and characterizing uncertainty in assessment results are related. Until recently, management often viewed assessment results with flexibility, as noted earlier, by selecting from a range of assessment results. This was of great concern and was addressed by technical guidelines:

> Because the NSGs [National Standards Guidelines] were written for a nontechnical audience, they do not provide detailed guidance for the stock assessment scientists who will ultimately be requested to develop many of the conservation and management measures called for, particularly in the Section relating to National Standard 1, and particularly in light of the widely perceived need to adopt a precautionary approach to the management of marine fisheries. The main purpose of this paper is therefore to provide technical guidance on the use of precautionary approaches to implementing National Standard 1 of the MSFCMA in accordance with the NSGs. (Restrepo et al. 1998)

The precautionary approach attempts to shift the burden of proof in favor of sustainability of the resource. Later reauthorizations of MSA tended to strengthen this approach.

This ongoing affirmation of Congress's desire to end overfishing with each reauthorization has to some degree resulted in a much more contentious and laborious process of conducting stock assessments. The commissions are not immune to this. Although the Atlantic States Marine Fisheries Commission is not legally bound by the Magnuson-Stevens Act and its various reauthorizations, it has tried to follow the intent of these reauthorizations in practice. The interface between science and management has also therefore become more difficult. My (Vaughan) view is that Dr. Schueller will have a tougher role to play than I did, at least during my first two decades as a practicing assessment scientist. Since the 2003 stock assessment for Atlantic Menhaden, the bar has been raised because best available science is the standard, uncertainty needs to be acknowledged, and increasingly, management actions are being strengthened. However, because the current assessment process is far more collaborative and open, all of the responsibility and decision making is not solely reliant on the lead stock assessment scientist. If there is any take home message from this narrative, it is the importance of strong quantitative training and the ability to work well with fellow scientist on the federal and state levels, as well as the fishing public.

Biographies

Douglas Vaughan completed his doctoral degree in 1977 at the University of Rhode Island's Graduate School of Oceanography. He spent five years at the Oak Ridge National Laboratory working on power plant effects on fish populations and then almost 30 years as a stock assessment scientist for the National Marine Fisheries Service and was the lead analyst for the Atlantic Menhaden and Gulf Menhaden assessments for most of that time. He retired in 2011.

Amy Schueller completed her doctoral degree in 2009 at Michigan State University. She is currently a stock assessment scientist for the National Marine Fisheries Service and is the lead analyst for the Atlantic Menhaden and Gulf Menhaden assessments.

References

Restrepo, V. R., G. G. Thompson, P. M. Mace, W. L. Gabriel, L. L. Low, A. D. MacCall, R. D. Methot, J. E. Powers, B. L. Taylor, P. R. Wade, and J. F. Witziget. 1998. Technical guidance on the use of precautionary approaches to implementing National Standard 1 of the Magnuson-Stevens Fishery Conservation and Management Act. National Oceanic and Atmospheric Administration, National Marine Fisheries Service, NOAA Technical Memorandum NMFS-F/SPO-31, Silver Spring, Maryland.

Recommended Readings

Ahrenholz, D. W., W. R. Nelson, and S. P. Epperly. 1987. Population and fishery characteristics of Atlantic menhaden, *Brevoortia tyrannus.* Fishery Bulletin 85:569–600.

Cadrin, S. X., and D. S. Vaughan. 1997. Retrospective analysis of virtual population estimates for Atlantic menhaden stock assessments. Fishery Bulletin 95:445–455.

Linder, E., G. P. Patil, and D. S. Vaughan. 1987. Application of event tree risk analysis to fisheries management. Ecological Modelling 36:15–28.

Vaughan, D. S. 1993. A comparison of event tree risk analysis to Ricker spawner-recruit simulation: an example with Atlantic menhaden. Canadian Special Publication in Fisheries and Aquatic Sciences 120:231–241.

Vaughan, D. S., M. H. Prager, and J. W. Smith. 2002. Consideration of uncertainty in stock assessment of Atlantic menhaden. Pages 83–112 *in* J. M. Berkson, L. L. Kline, and D. J. Orth, editors. Incorporating uncertainty into fishery models. American Fisheries Society, Symposium 27, Bethesda, Maryland.

Vaughan, D. S., and J. W. Smith. 1988. Stock assessment of the Atlantic menhaden, *Brevoortia tyrannus*. NOAA (National Oceanic and Atmospheric Administration) Technical Report NMFS (National Marine Fisheries Service) 63.

Future of Inland Fisheries

Robin L. Welcomme*
Long Barn, Stoke by Clare, Suffolk CO10 8HJ, UK

Key Points

- Inland fisheries are important providers of food in many areas of the world.
- Future trends may well depend on the development of social, political, and economic institutions as world demand for food increases, as well as on future climate.
- Emerging professionals will have to adapt their science and knowledge to better understand such challenges as they arise. They will have to adapt to rapidly changing sociopolitical scenarios and to the nature of employment itself.
- Scientific, administrative, and diplomatic work will be required at all levels.

The Importance of Inland Fisheries

Fish from inland waters are an important source of food rich in protein, micronutrients, omega-3 fatty acids, and minerals, with a yield, reported to the Food and Agriculture Organization of the United Nations (FAO) by member countries, of around 11 million metric tons in 2011. Although this only contributes about 2.5% to global animal protein supply, it is extremely important in certain countries, principally in Asia and Africa where many communities are dependent on inland waters for livelihoods and food. About 57 million people are directly employed by the inland food fisheries sector worldwide. In addition, inland water fish support extensive and valuable recreational fisheries and a range of ecosystem services. Inland aquaculture contributed a further 39 million metric tons in 2011 and, together with capture fisheries, contributes nearly 12% of the animal protein consumed worldwide. These figures are possibly unreliable because the geographically diffuse, multispecies, multigear nature of inland fisheries makes the collection of statistics difficult in many member countries, whereas in others the administrative infrastructure for collection and processing of such data is lacking. Based on visits to more than 80 countries in my work for FAO, I feel the reported catches to be low because many smaller fisheries are often omitted from the reports, a view echoed by others who feel that global catches from inland waters are more likely to be about 14 million metric tons per year. The increasing pressures on inland waters make the various lake, river, and wetland ecosystems extremely vulnerable, and the International Union for Conservation of Nature tells us that freshwater aquatic species are among the most in danger of extinction. The social, ecological, and dietary significance of inland fisheries makes it extremely important that our profession try to determine the future direction of the sector to better shape policies for management and conservation.

* Corresponding author: welcomme@btinternet.com

I have been singularly fortunate to have been able to devote my life to inland fisheries at a time when knowledge of the sector was expanding rapidly. I started my career working as an assistant in a water pollution laboratory in 1956 and, upon graduation, worked in the field for 10 years in East and West Africa. For the next 25 years, I worked in the headquarters of the FAO, rising eventually to direct the inland fisheries and aquaculture programs. A combination of focused research and the type of global contacts that the various international fishery bodies, working groups, and projects that the United Nations provided gave my colleagues and me the tools for examining the state and possible direction of inland fisheries worldwide. The years that I worked on inland water fisheries have seen a transition from the early interest in taxonomy and biology of individual species conducted by relatively few scientists to more extensive institutions concerned with social, political, and economic aspects of the sector. It is on the basis of my own studies and those of many collaborators that I make the following statements:

- Inland fisheries are important providers of food in many areas of the world.
- Inland fisheries are under threat because of the demand for water and environmental degradation. This is likely to get worse with climate change.
- Excessive exploitation and environmental changes are leading to the loss of larger species and simplification of fish assemblage structures.
- There appears to be a trend to intensify inland fisheries by stocking to compensate for declining production from natural fisheries and to increase control over harvests.
- There is a trend to decentralize fishery management through comanagement schemes and to privatize many open-access fisheries.
- Excessive exploitation is leading to the loss of larger species and simplification of fish assemblage structures.
- In affluent societies, inland fisheries progress from food fisheries to recreation and conservation.

Future directions of inland fisheries will most likely respond to the increasing human population, the greater trend to urbanization, the increasing demand for quality aquatic food products to satisfy urban demand, a growing international trade for fish and fish products, and the changes in policy and governance that will result. All this has important implications for emerging professionals. They will have to develop strategies to deal with the rapidly changing social and economic background surrounding the fisheries sector as economies develop away from traditional rural economies towards more highly urbanized systems. They will have to provide the type of research and administrative and academic infrastructure that will arise in response to these challenges and the mechanisms to channel the results of such research into policy. This will require an educational background with a mix of social, economic, and biological modules at the graduate level and continued reading and updating of knowledge while working.

Degradation of the Aquatic Environment

The general degradation of inland water environments is a growing problem in much of the world. When I started my working life in the 1950s, the main concern in Europe was water pollution, both for aquatic life and for human health. Many of the rivers were fishless, and fish kills were common. This early work led to a concerted effort to clean up the waters and eventually resulted in the Water Framework Directive of the European Union, under which European countries are obliged to maintain certain standards of water quality and environmental structure. Pollution and eutrophication of waters remain serious issues elsewhere in the world. The increasing discharge of untreated sewage and polluted industrial and agricultural waste waters into rivers, whose discharges are themselves decreasing, has caused widespread contamination. While pollution and eutrophication have largely been addressed in some industrialized societies, declines in water quality are widespread and liable to continue into the near future, at least. The extent to which this will be rectified in the coming years is open to debate. Certainly, countries such as China and India have realized that the current poor state of their inland waters cannot continue, yet others are too poor for any reasonable water treatment infrastructure to be created in the near future. This situation appears to me to be very intractable, and there are even signs that some developed countries are willing to revert to their former practices if the cost of replacing aging infrastructure becomes an issue.

The demand for water and the services it offers is viewed as one of the major drivers of development. Much of the water flowing through rivers is now abstracted and diverted for agriculture, often to the extent that several major rivers have ceased to have any outflow, and some major lakes, such as the Aral Sea and Lake Chad, have dried up to varying extents during my working lifetime. Such water as continues to flow in rivers is often used for power generation and irrigation, altering flow patterns in the channels downstream. By the late 1970s, sufficient information was available from studies, projects, analyses by other authors, and collaborative meetings to allow me to carry out a meta-study on the ecology of river fisheries that identified many of the factors influencing the abundance of fish in rivers. This was supplemented by the river continuum concept and the flood pulse theory into a general framework of river function. One meeting was perhaps the seminal event in the acceptance of the framework—The Large River Symposium held in Ontario in 1986. These theories and evidence gathered through various national and international mechanisms, such as the Commission for Dams, show that changes to water quality and quantity damage inland fish stocks and, consequently, the fisheries that exploit them. Such theories are not sacrosanct, however, and emerging professionals will doubtless revise them as more data become available.

Electricity is seen as the best way to create social and economic development, and hydroelectricity is considered the greenest way to generate power, despite evidence to the contrary. Protests against large dams by social, fisheries, and wildlife interests have met with little success in the face of the overriding need for societies to advance economically. I foresee that efforts to protect such resources will continue to have little impact so long as food prices remain low for the majority of mankind. Lip service to conservation is usually an element in internationally funded dam projects, but the environmental surveys are usually given little weight, especially as they advocate ineffective solutions, such as fish passes, which do not work in many rivers.

Degradation also arises from the conversion of fluvial and lacustrine wetlands to agriculture through drainage and irrigation systems. Such conversions frequently involve deforestation of previously woody environments, destabilizing soils, and simplifying wetland habitats. One of the most common destinations for such drained land is rice culture, which does present some possibilities for cohabitation with fisheries and aquaculture. Rivers and lakes have already been widely degraded through such uses, but it is doubtful whether the trends to use water perceived as available for irrigation, power generation, human water supply, and industry will be reversed in the foreseeable future. Inevitably, the demand for cereal crops will continue to rise, and as groundwater pumped irrigation become less viable because of the emptying of the aquifers, or the rising cost of fuel for pumping, increases in the intensity of cultivation of riparian marginal lands will probably rise. Personal attempts to raise the water quality and quantity need of inland fisheries through national and international forums have usually failed.

I feel it to be extremely unlikely that the current priorities given to grain crops, irrigation, power generation, domestic water supplies, industrial uses, and waste disposal will change, so emerging fisheries professionals are probably going to have to develop mitigating strategies to utilize what aquatic and spatial resources are left from other uses. The rehabilitation of river form and function, management of environmental flows, and pollution monitoring and control will be fertile fields for research, and alliances will have to be sought between fisheries and other governmental and nongovernmental conservation interests to effectively apply the results of such efforts.

Changes in Fish Assemblage Structure

The trends in environmental pressures described above affect the structure and abundance of fish communities. In addition, fishing itself has intensified in many areas of the world, particularly in developing countries, with large increases in the numbers of people participating in fisheries for a living. Fishing itself, when intensively practiced, effectively eliminates the larger species from the fish community or fishery, producing a downward drift in the size of the species present in the fish assemblages of rivers and lakes. This effect was first described from the North American Great Lakes fishery in the 1960s and personal research on the Ouémé River in West Africa. Results from other major fisheries in Africa, Latin America, and Asia indicated that similar effects were experienced wherever fishing pressures rose. This finding is extremely important for the management of multispecies fisheries everywhere as it means that the former classical approaches to management by stock assessment cannot be applied to such fisheries. It is not clear to me to what extent this process is reversible should management regimes change. It seems likely that long-term shifts in the nature of fish communities will result in a progressive simplification of the most affected aquatic systems and a reduction in ecosystem function and services.

This trend is exacerbated by morphological changes to the environment and alterations to the amount and quality of water in the system, all of which tend to eliminate specialized migrant species in favor of more generalized and static forms, thereby decreasing biodiversity and eliminating species that are often preferred for human consumption. This process is accompanied by the introduction and spread of species such as carps and tilapias, often originally made for aquaculture, but escapes and releases have extended the problem to many wild waters.

Research into the responses of complex fish populations to fishing pressure is a particularly fertile area for emerging professionals, as changes in fish population structure only emerge over long periods and require constant monitoring of subject fisheries. However, without such analyses, it will be difficult to decide on management strategies for individual multispecies fisheries as they come under increasing environmental and fisheries pressures.

Intensifying Management of Inland Fisheries

If I am right and fish stocks in inland waters decline as a result of external pressures, there will be a need to increase production from those waters that remain. There is also a need for more stable and predictable products for urban markets, which has favored the expansion of intensive aquaculture. The aquaculture sector has grown rapidly in recent years, particularly in Asia, which accounted for 95% of inland aquaculture in 2011. There has also been a widespread intensification of management of wild fish stocks. Most commonly, this is through stocking of natural ponds, reservoirs, riverine cut-off lakes, and rice paddies with fish seed produced by hatcheries or captured from the wild. The stocked fish then enable such habitats to exceed production levels that would be obtained naturally and give greater control over harvesting and marketing. Early studies we carried out in the FAO showed this practice to be especially widespread in Southeast Asia, South Asia, and China, although it is also common in parts of Africa, the Caspian Sea, and Latin America. Further spread of stocked fisheries is to be anticipated with the increasing control over water and the growing number of small water bodies.

The development of enhanced fisheries is strongly linked to the political and economic system in individual countries. In centrally planned economies and some countries, such as Thailand, stocking was carried out by the state as a subsidy to fishermen. More recently, however, the costs of stocking have been deflected to the private sector and have resulted in the conversion of open-access fisheries on natural fish stocks to culture-based, stocked fisheries. This is primarily because of the financial investment in the purchase of fish seed, which means that the fishery must be reserved for a select group of participant fishers. By implication, in areas where this approach to the fishery is common and expanding, access by the nonprivileged group will be limited and the fishery effectively converted from a public fishery to private ownership. Often, this is through the formation of fisher cooperatives, as in Bangladesh, although wealthy individuals may also intervene, as in the control of the lot fisheries in Southeast and South Asia, whereby control of certain water bodies or sectors of river are auctioned to individuals or groups of fishers for defined periods. Limitations of access of this type are increasingly common in both culture-based and in comanaged systems and have been common for many years in recreational fisheries and would appear to be an inevitable part of the progression from open-access wild fisheries to more heavily controlled production systems. However, the extent to which this occurs is very much a function of the political and financial attitudes to natural resources management generally.

Enhanced fisheries and aquaculture with increased private control over an increasing proportion of the resource may well be the future direction for inland fisheries if they are to continue to supply valuable nutrition elements. Emerging professionals throughout the world will be needed to play an increasing role in the research, development, and management of such fisheries and will need to participate in day-to-day management and conflict resolution as fisheries become intensified and rely increasingly on scientific advice. In this future, developments in inland fisheries will probably parallel those of agriculture over the past few centuries.

Fish for Recreation, Tourism, and Conservation

A number of ecosystem services other than food are provided by fish. For instance, recreation has long been a major use for fish resources. It provided an outlet for the populations of the industrialized towns of Europe and a means of rural escape for many in North America. Its popularity grew rapidly in these areas to the point where it supplanted the use of inland fish for food. Indeed, in some regions, capture and permanent removal of fish from inland systems is illegal. For some time, it was thought that this was uniquely a developed country phenomenon, but more recently, a number of developing countries are converting what were once valuable food fisheries into uniquely recreational-use fisheries for local populations, urban societies, and tourism.

Interest in sport fishing has been paralleled by an interest in conserving inland water habitats to maximize the sporting experience. With the growth of conservation movements in many parts of the world, preservation has become an objective of its own, even supplanting recreational fishing as a prime motive for management. This has given rise to the perception that as wealth increases in societies, there is a trend to move from food fisheries to recreational ones and, finally, to conservation of selected species or communities. I feel that this trend is only possible where alternative sources of food make the value of the fish stocks more valuable as a recreational resource than as a source of food. How long this trend will continue in the face of rising demands for food as populations increase is debatable. Nevertheless, emerging professionals should be aware of such trends. Certainly in northern temperate countries, the majority of professionals currently support conservation and recreation rather than food fisheries, and there is a constant need to assess trends in recreational use in both temperate and tropical areas.

Trends in Governance

During my working life, the whole question of governance of inland fisheries has changed considerably. Originally, most fisheries were regulated by central governments on a one-size-fits-all basis—if, indeed, they were regulated at all. Over the years, it has become apparent that this style of governance is ineffective as it fails to take into account local needs of individual fisheries and also alienates the fisher communities to a point where noncollaboration with the authorities became the hallmark of the fisher. This became apparent in the discussions of the Committee of Inland Fisheries and Aquaculture for Africa where a government such as Zambia, for example, would have to legislate for four major lakes of very different character, one major reservoir, and the diverse floodplains and rapids of the Zambezi River system. More recently, governance has tended towards various types of comanagement whereby fishers are incorporated into the planning, regulation, and enforcement of fisheries at the local level. It is still early days to see whether such initiatives will be generally adopted, as many governments still view such mechanisms as an extension of central government policy.

Whatever the success of devolved management of the fishery, it is clear that the role of central government has not disappeared completely. This mainly arises from the difficulty of local fishers to see themselves in the context of the fishery as a whole. It also arises because the main thrust of management for the sustainability of fisheries no longer lies with the regulation of the fishery per se, but with the condition of the environment. This in turn calls for fishery managers to interact with other sectors in order to negotiate the environmental conditions, be they flow, water quality, or habitat structure, that are needed to sustain fish communities and other stakeholders. Under current regimes, this type of negotiation is rarely effective from the point of view of the fishery because of the diffuse nature and individual weakness of the fishery economic interests and the poor valuation of the contribution the fishery makes to society. This argument is even further extended to international rivers and lakes, where despite the existence of many lake or river basin commissions, few actually address fishery issues. Collaboration among the various interests regulating rivers has rarely been successful in the past, despite our having organized many interdisciplinary forums. Most disciplines are single minded in trying to maximize the yields or financial contribution made by their sectors. As a result, holistic planning to maximize benefits from all sources has yet to be achieved in most river or lake basins. Multidisciplinary approaches will form one of the major challenges for emerging professionals in the future, and this will increasingly require them to work as members of teams to solve particular problems and geographical situations. When I first started work on fisheries, efforts were mainly individual with little collaboration between researchers, mainly because we were relatively few. The more complex needs of modern approaches to environmental sciences, increased analytical capacity provided by modern computers, and improved possibilities for instantaneous communication across the world now make such communal and cross-disciplinary efforts both possible and necessary.

To participate effectively in such multidisciplinary negotiations, efforts to properly value fisheries should increase in the future so that appropriate allocations of water and environmental structure can be made. The rapid modification and loss of habitats to the rapidly changing economic global situation are the major immediate pressures on inland fisheries and will result in the loss of ecosystem and biodiversity over a very short time scale. Countering these trends will involve a considerable amount of research by established and emerging professionals at the national and international level and will require national and international river and lake basin authorities to increase their capacity to include fishery issues.

One of the major areas requiring this international and multidisciplinary approach is that of climate change, although this will probably act over a longer time scale. This will add to the problems of water supply and thus further impact aquatic environments. I foresee that alterations to rainfall patterns will certainly change the current hydrological regimes in rivers, and increased aridity may contribute to the drying out of many seasonal bodies of water. Many of the world's rivers are fed by glacial melt, which stabilizes flow throughout the dry season so the diminution or disappearance of the glaciers will remove this stabilizing influence. Such changes in flow, added to those already produced by water abstractions and damming, may well result in the disappearance of migratory species that depend on specific flow conditions for migration, breeding, and larval downstream drift. Changes in water temperature may also influence the distribution of the many species of fish that have specific temperature requirements for breeding or are particularly sensitive to changes in dissolved oxygen concentrations. These issues will become a major center of concentration for younger professionals entering fisheries, as monitoring and mitigation of such changes, both at national level and at that of individual water bodies, may well become central to fisheries management in many climatic zones.

Conclusion

Prediction of the future of any sector usually involves projecting forward existing trends. This is risky as it does not take into account future changes in public perception and government policy. For instance, will the current interest in sport fishing in industrialized countries continue or will inland fish continue to be appreciated as a food source? Throughout my 60 years in the profession, there has been evidence that both are subject to change locally: inland fish consumption has declined significantly in Western Europe and North America, whereas recreational fishing has expanded in many areas, both as a local sport and tourist activity. Furthermore, changes in political system can bring about far-reaching changes in the way inland fisheries are viewed and managed. For example, it quickly became apparent, through the work of the European Inland Fisheries and Aquaculture Advisory Commission, that the change from centrally planned to free market economies in the 1990s was accompanied by stopping the subsidies whereby inland fisheries were maintained by government-funded stocking. This was replaced by a more market-driven system where management was carried out by the fishermen themselves, leading to a substantial decline in reported fish catch.

In conclusion, our profession is faced with a period of accelerating change that is placing increasing pressures on aquatic environments and the fishery resources they support. In the past 60 years, we have traveled from simple natural history-style explorations of the biology of inland fish, primarily by biologists working on a limited number of species or habitats, to multidisciplinary studies of the resource and its use. This has drawn in other players such as politicians, economists, and sociologists that now contribute a far wider appreciation of the complexities of fisheries and their role in society. As new players have entered fisheries, new roles and disciplines have emerged and will continue emerging as further changes in society and its demands on inland waters and food supply continue to change attitudes toward fish and fisheries. From where I stood in the 1950s, there was no way of anticipating these trends as new world political configurations have arisen and new technologies applied. In this context, emerging professionals will have to adapt their science and knowledge to better understand such challenges as they arise. This will require greater flexibility than in the past, as workers at all levels will have to adapt to rapidly changing sociopolitical scenarios. They will also have to cope with changes in the nature of employment itself, with a move away from long-term employment to shorter-term contracts. Scientific, administrative, and diplomatic work will be required at all levels, from the simple studies of basic biology upon which our understanding of ecological function eventually depends, to studies of the human dimensions of fisheries through economics and sociology, to the national and international politics of natural resource management. Above all though, find joy in your work at whatever level, as I have, and collaborate with others to produce those outcomes that will ensure the survival of humanity in a sustainable ecosystem.

Biography

Robin Welcomme began his career in 1956 working as a laboratory assistant in the Salmon and Freshwater Fisheries laboratory of the UK. On graduating, he went to Africa to work on the fisheries of lakes and rivers of East Africa and, while there, completed a Ph.D. on migration of fishes into a small tributary of Lake Victoria. Welcomme was then transferred to the Republic of Benin where he was commissioned to inventory the inland fisheries of that country. From there, Welcomme moved to Rome where over 26 years he rose to become chief of FAO's Inland Fisheries and Aquaculture Service. During his time with FAO, Welcomme concentrated on advising governments on inland fisheries issues and carrying out technical meta-studies on river fisheries and other aspects of inland fisheries management. After retirement, he joined Imperial College as a senior research fellow, continuing work in these fields. Welcomme still maintains connections with the college as a visiting scientist and is currently associate editor of *Fisheries Management and Ecology*.

Epilogue

Betsy Riley, William W. Taylor, and So-Jung Youn
Center for Systems Integration and Sustainability
Department of Fisheries and Wildlife, Michigan State University
115 Manly Miles Building, East Lansing, Michigan 48823, USA

Mentee versus Chinook

Joining the world of fisheries is sort of like being invited out for your first-ever trip salmon fishing on a charter boat on Lake Michigan. (We'll call the charter boat service Old Grin.) At face value, it sounds like the best way to spend a morning that you can imagine. You get to go out on a boat on Lake Michigan. It's the chance to hang out with people who you think are probably the coolest people in creation, but you would never actually tell them this. And, perhaps most important, it is your chance to catch a really big fish with only your hands and a fishing rod.

You are excited. You are a little nervous. So you pack your sunscreen and granola bars and bottled water. And, just in case, some Dramamine. And you head out at some ridiculous hour of the morning to prove that you know something about fishing.

When you arrive, everything looks exactly like it should, but then you are promptly greeted by multiple people who give you their names that your tired brain (it is 5:00 in the morning, after all) cannot seem to hold onto. You don't really know anyone yet, but everyone seems to know each other, and as the new kid, they all remember you. You have never done this before and have only the faintest idea what you are doing. And there is that little voice inside you that, despite your brave face and bravado, does not actually believe that you can pull 17 pounds of fighting salmon over the side of the boat using only a long piece of wood with some twine and a reel attached.

This is life's mandatory first step. Whether the event is a fishing trip or a conference. Whether your tool is a fishing rod or a grant proposal. Whether you're with people who support you or trying to go it alone. No matter the journey, we all start here. Fresh-faced with our backpack full of random stuff that we hope will prepare us for what lies ahead. Our hearts filled with hope and enthusiasm, but also trepidation and self-doubt and fear because the future is unknown and the problems that we are facing are so much bigger than a tiny human with hope and a backpack.

It is our choice to take on these challenges. Many choose not to. Many would rather not risk making an idiot of themselves, revealing to others, just at that moment, that they might not have the knowledge and skills to succeed. They would rather not be around a lot of people that they have never met. They would rather not be sore in the morning. And so many never catch that first fish. As amazing as pulling that fish on board is, it is not so much the fish, but the *decision* to catch the fish, the *determination* to get up at a ridiculous hour of the morning and board that boat, and the *persistence* needed to keep reeling, long after you were ready to give up, that separates those who have resilience and those who do not.

Resilience is more than just determination and persistence; it is the ability to withstand stress and catastrophe. Just the definition itself is intimidating. And yet, resilience may just be the single most important difference between those who succeed and those who do not.

On my first trip, I (Betsy) almost let go of the rod. I had never done anything like it and my muscles were completely unprepared. Afterwards, I almost threw up. Your first fish is not pretty. You may not even catch it on the first try. Your technique is sloppy. Your arm gets tired quickly. You don't know when to reel and when to let it spin, and the whole time you're thinking secretly that if you just let go of the rod, everyone would yell in dismay, you'd probably have to compensate them for the rod, and you would never ever live it down. But as the fight—the Chinook versus me—dragged on, don't think I didn't consider it.

What stopped me? I wish that I could say that it was my amazing resilience. That I was awesome and that I never gave up and that I pulled the 17-pound beast out of the water using sheer willpower alone. That despite my screaming muscles, I overcame my near defeat as the embodiment of perfect resilience!

The reality? Just when I thought I could not hold onto the rod for another second, the first mate came over on the pretense of checking whether we could see the salmon at the surface yet and ever so subtly helped me hold onto the rod.

Nothing was said. He did not make a big deal about it. And it's not like suddenly the pressure was completely gone and catching the fish was easy. It wasn't. It continued to be mind-blowingly difficult. My muscles were still screaming. I gave up trying to see the salmon. I only knew that I had to keep reeling it in. And, like I said, I almost threw up afterwards. But the task was no longer impossible. Now, knowing that I was not about to lose the fishing rod on top of everything else, I was able to bring the salmon in. Seventeen pounds of pure fish. It was amazing.

And you know what? It was no less amazing because I had help. This is the story of life, for us all. Working together empowers us and allows us the freedom and courage to do things we would not normally do.

Resilience is not something that you magically have that lets you conquer all of life's difficulties just because you want something a lot and try really hard. It is something that you develop by taking risks, failing, and getting back up determined to find ways by which to succeed. It is through this process that we develop the knowledge and skills that we need to keep making progress in our chosen goals. This is life. Sometimes you fail and sometimes you succeed. And often, you succeed because you have people around you that are willing to step in and hold the rod when they see that you are slipping.

We simply cannot, no matter how hard we try, stress too much the value of mentors. It is all well and good to strike out on your own, to test your limits on your own terms. In fact, this can be an invaluable experience, and we highly recommend it. But when you are ready to go after bigger fish, you are going to need someone there to hold the rod when you get tired or discouraged or consider giving in. This is not failure or weakness. It is understanding your limitations and being able to recognize when the situation has moved beyond your control and that victory can only be achieved with the help of others. The relationships that are built when victory is shared serve as the foundation for greater victories, and you gradually begin to wonder how you ever survived on your own, and as you become the mentor and have your own set of earnest mentees, you begin to wonder why you ever wanted to.

The Importance of Resilience

In addition to encouraging and believing in you, mentors are important because they provide mentees with the ways and means to be resilient, and resilience is the key to ultimately overcoming your difficulties and making the most of every opportunity.

Mentors enable you to heal when things do not work out as well as hoped. They provide a path to cultivate your wisdom, creativity, and passion, leading to your ability to have courage as you face the unknown and the tools to becoming a future mentor and leader yourself. Perhaps the most important trait to have is passion for what you are doing. The essential ingredients for passion are enthusiasm, persistence, and love. As is often said, "A great leader's courage to fulfill his or her vision comes from passion not position" (John Maxwell, American author). But courage without knowledge is often wasted, so possessing the courage to act is important in your ability to be resilient. It has been said that "vision isn't forecasting the future, it is creating the future by taking action in the present." However, this is easy to say and very hard to do as Machiavelli pointed out in 1532 in *The Prince*: "There is nothing more difficult to take in hand, more uncertain in its success, than to take the lead in the introduction of a new order of things," but this is exactly what mentors enable! Mentoring gives you the keys for gaining wisdom by providing a safe environment that allows for experimentation without judgment. Such an environment provides for the acquisition of knowledge and experience that will provide you with the tools you need for future success. Without knowledge and experience, little is possible as you will often times waste your energy and passion ineffectively and become disillusioned with yourself, the profession, and society.

Having a great mentor allows you to excel. According to Winston Churchill, "A pessimist sees the difficulty in every opportunity; an optimist sees the opportunity in every difficulty." Mentoring helps you do both and allows for you to strategize about how to mediate the difficulties and optimize the opportunities! The way to do this is to light your creativity on fire; creativity combined with hard work can solve almost any problem! The key to creativity is to have an innate interest in people and how things work. Be curious and nonjudgmental. Hear diverse opinions from all people and all walks of life and be flexible and adaptable; only by doing so will you grow and expand your abilities to be a better mentor and professional.

Ultimately, we have found that a resilient person can only exist in concert with a reliable and supportive network and building that network (both personal and professional) is essential for you to achieve greatness. Leaders thrive on networks, and it is from these networks that leaders obtain the guiding visions, knowledge, and daring that provide them with the mantle of leadership and ability to be good mentors.

Mentoring Is Empowering

Mentoring embraces the full spectrum of all that life has to offer, sheer excitement and pleasure to abject sorrow and pain. To be a mentor means you give of yourself fully and without reservation so that others may prosper and find their pathway to success. Mentoring is not selfish and is not an exercise in self-aggrandizement. Instead, it is sharing of wisdom gained through life experiences that allows one to be part of a larger community and an integral part of a profession and of society in general.

There are a number of traits that make a great mentor, but three of the most important are listening, compassion, and forgiveness. No one goes through life unscathed. Problems will occur, and learning to forgive mistakes, both your own and others, is as important as learning to overcome your problems. Having a space where you can suffer as well celebrate is extremely important to personal and professional growth and self-actualization. Having a true mentor is a wonderful gift, one that no one forgets. So seek out others, share your thoughts, your worries, your failures, your successes, and triumphs. Mentoring is not just scheduling a time to meet. Rather, it is a time to truly interact in meaningful ways that you will remember and appreciate throughout your life, so make sure you celebrate your relationships with your mentors, and remember, they are gaining from their interactions with you as much, if not more, than you are gaining from them.

> It is not the critic who counts; not the man who points out how the strong man stumbles, or where the doer of deeds could have done them better. The credit belongs to the man who is actually in the arena, whose face is marred by dust and sweat and blood; who strives valiantly; who errs, who comes short again and again, because there is no effort without error and shortcoming; but who does actually strive to do the deeds; who knows great enthusiasms, the great devotions; who spends himself in a worthy cause; who at the best knows in the end the triumph of high achievement, and who at the worst, if he fails, at least fails while daring greatly, so that his place shall never be with those cold and timid souls who neither know victory nor defeat.
> —Theodore Roosevelt, except from "Citizenship in a Republic," April 23, 1910

Biographies

Betsy Riley is a university distinguished fellow completing her Ph.D. in global fisheries policy under Dr. William Taylor in the Department of Fisheries and the Center for Systems Integration and Sustainability at Michigan State University. She currently studies fisheries management and governance, looking at the effects of globalization on common pool resource theory and effective governance systems. She received a master's in public policy at the Gerald R. Ford School of Public Policy and a master's of science at the School of Natural Resources and the Environment, both at the University of Michigan, and she received her B.A. in social psychology, with a minor in environmental studies, at Wellesley College.

William W. Taylor is a university distinguished professor in global fisheries systems in the Department of Fisheries and Wildlife and the Center for Systems Integration and Sustainability at Michigan State University. He has been active in the American Fisheries Society throughout his career, serving as president of the society in 1997–1998. He currently holds a U.S. Presidential appointment as a U.S. commissioner (alternate) for the Great Lakes Fishery Commission.

So-Jung Youn is an M.S. and Ph.D. student under Dr. Taylor in the Department of Fisheries and Wildlife and Center for Systems Integration and Sustainability at Michigan State University. She is interested in studying the role of inland fisheries in local food and economic security, particularly in Asia. So-Jung received her B.S. in biology, with a minor in management and organizational design, from the College of William and Mary.